# The Earth System

*An Introduction to E*

# The Earth System

## An Introduction to Earth Science

### David Laing

*University of Maine, Fort Kent*

 **Wm. C. Brown Publishers**

**Book Team**

Editor *Jeffrey L. Hahn*
Developmental Editor *Lynne M. Meyers*
Production Editor *Michelle M. Campbell*
Designer *David C. Lansdon*
Art Editor *Janice M. Roerig*
Photo Editor *Michelle Oberhoffer*
Permissions Editor *Karen L. Storlie*
Visuals Processor *Joyce E. Watters*

 **Wm. C. Brown Publishers**

President *G. Franklin Lewis*
Vice President, Publisher *George Wm. Bergquist*
Vice President, Publisher *Thomas E. Doran*
Vice President, Operations and Production *Beverly Kolz*
National Sales Manager *Virginia S. Moffat*
Advertising Manager *Ann M. Knepper*
Marketing Manager *John W. Calhoun*
Editor in Chief *Edward G. Jaffe*
Managing Editor, Production *Colleen A. Yonda*
Production Editorial Manager *Julie A. Kennedy*
Production Editorial Manager *Ann Fuerste*
Publishing Services Manager *Karen J. Slaght*
Manager of Visuals and Design *Faye M. Schilling*

Cover photo © Wolfgang Kaehler Photography

The credits section for this book begins on page 569, and is considered an extension of the copyright page.

Library of Congress Catalog Card Number: 90–80172

ISBN 0–697–07952–X

Printed in the United States of America by Wm. C. Brown Publishers, 2460 Kerper Boulevard, Dubuque, IA 52001

10   9   8   7   6   5   4   3   2   1

*. . . Earth is so magical with mighty deeds*
*Of little seeming but of great account.*
*Glory is close. There is no need to mount*
*High steeds for long crusades.*
*Here in these glades*
*There are more marvels than the mind can count. . .*

*Dilys Bennett Laing*
**For Lovers of Earth, 1929**

# Table of Contents

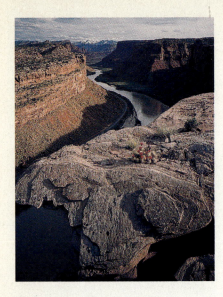

*Part Three*

# Earth's Oceanic Systems  279

## 11

### Water, Salt, and Basalt: The Structure of the World Ocean  280

## 12

### Waves, Currents, and Battered Shores: Processes of the World Ocean   308

**Part Four**

## Earth's Atmospheric Systems   337

## 13

### The Dynamic Sky: Energy and the Atmosphere   338

## 14

### The Sphere of Aeolus: Winds and Deserts   364

*Part Five*

# Beyond Earth   419

*Part Six*

# Earth and Humanity   495

## 19

**Care and Maintenance**   496

How to Use a Living Planet   497
Safety   497

# Preface

*E* arth science has acquired an undeserved reputation for being a dry and dusty subject. As is true of any subject, Earth science can be made to appear dry and dusty if it is presented in the appropriate form, such as a lengthy, encyclopedic listing of our planet's various parts and processes. Textbooks of this sort exist, and they have all the clutter and none of the charm of a department store catalog. At the other extreme are textbooks that simplify, condense, and water down the subject so much that boredom is inevitable simply because it is difficult to discuss the essential concepts of the subject at so simplistic a level.

In *The Earth System,* I attempt to bridge the gap between the encyclopedic and the simplistic approaches to Earth science. My primary objective has been to present just enough information and to develop concepts in just enough depth to allow meaningful discussion while ensuring that the text is "user-friendly" to the student who has no background in science. Beyond this, I have also tried to present the subject in a way that would make Earth come alive for the student and thus kindle a sense of wonder and reverence for the marvels of what many of us dismiss as commonplace. Finally, I have taken pains to ensure that *The Earth System* would be equal to other leading Earth science texts with respect to structure, content, and accuracy.

The text presents the concepts of Earth science in an integrated way that stresses the wholeness and harmony of the Earth system. It focusses on interfaces and processes and on the functional interrelationships among the many parts of the whole. Within that basic framework, I have woven what to some will be a nontraditional theme, but I hope it will also be a welcome one. In view of the growing body of research that underscores the influence of biological processes on Earth systems, I have included discussions of many life processes that have been shown to have such influences. In addition, I have included another theme, not unique to this text, but perhaps with a fresh outlook: an assessment of human activities as they affect and are affected by Earth systems.

The book is arranged into six major parts. Part one, "Impressions of a Living Planet," presents an overview of those special features of Earth that make it a dynamic planet among dead worlds. Here, too, is a look at the overwhelming perspective of geologic time and at the systematic way in which natural systems, both living and nonliving, muster random processes into recognizable, classifiable, and peacefully coexisting phenomena.

The core of the text consists of parts two, three, and four, which present Earth and its environments as an integrated system with solid, fluid, and gaseous components. Part two, "Earth's Continental Systems," views solid matter in terms of three main concepts: structure, material cycling, and energy flow. It also has one major, unifying theme: the recently-crystallized theory of plate tectonics. Chapter 7, "Stories in Stone," for example, places the major rock types and rock-forming processes within their various plate tectonic settings.

Parts three and four, "Earth's Oceanic Systems," and "Earth's Atmospheric Systems," treat water and air as powerful agents of change that interact not only with one another, but also with the solid systems of the continents. Chapter 11, "Water, Salt, and Basalt," pays tribute to water: that most extraordinary chemical without which most of geology and all of biology would be impossible.

Part five, "Beyond Earth," deals with our astronomical setting and Earth's relationship to it. Special emphasis is given to the similarities and differences in form and process between Earth and the other planets and satellites of the Solar System. In part six, "Humanity in the Universe," chapter 19, "Care and Maintenance," focusses the student's attention on the various ways in which Earth can injure us and the various ways in which we can injure Earth. It also presents some thoughts about the historical roots of our counterproductive, human-chauvinistic attitude toward Nature.

The book's seven appendices are intended to serve a dual purpose: learning aid and practical reference. They are designed for use in classroom, lab, and field. Appendices II and III, on minerals and rocks, are accessed by dichotomous keys, as in biology.

Each chapter ends with a "Food for Thought" section. In these, I present topical or controversial concepts in the interest of conveying a sense of the dynamic growth and societal relevance of Earth science in today's increasingly complex world. To balance this forward look with a historical perspective, I also include within each chapter a short biographical sketch ("In the Spotlight") of a major contributor to the subject material of that chapter. Key terms are printed in boldface throughout the text for emphasis and review, and additional terms for honors work are printed in italics. Pronunciation guides and meanings are given for most terms of foreign derivation. Boldface terms are concisely defined in the comprehensive glossary following the appendices. Because of the unusually large vocabulary of Earth science—the inevitable consequence of

its broad scope—the glossary is an especially useful learning aid that should be referred to regularly.

Special acknowledgement and thanks are due to Margaret Laing, without whose patience, support, and help *The Earth System* would have remained a worthy but humanly impossible idea. An inadequate listing of others who have in some way contributed inspiration, direction, criticism, or support includes cousin Richard Goldthwaite; teachers Elso S. Barghoorn, William B. Bull, Stephen Jay Gould, Cornelius S. Hurlbut, Jr., John B. Lyons, Paul S. Martin, Robert C. Reynolds, and Richard E. Stoiber; editor Jeffrey L. Hahn; and associates Anne E. Heller, Nicholas Lampiris, and Steven H. Lumbert. Many reviewers read and critiqued the manuscript in its various stages. The final round of reviewers included:

Dr. Wayne F. Canis
University of North Alabama

Dr. Sue Ellen Hirschfeld
California State University–Hayward

Prof. William F. Kohland
Middle Tennessee State University

Prof. Nicholas H. Tibbs
Southeast Missouri State University

Prof. Don E. Owen
Indiana State University

Prof. Laura L. Sanders
Northeastern Illinois University

Prof. James D. Stewart
Vincennes University

Prof. John Ernissee
Clarion University of Pennsylvania

Prof. Doug Sherman
College of Lake County

Their advice was invaluable, and is sincerely appreciated, but in the end, I accept sole responsibility for any errors of commission or omission. I welcome suggestions for future improvements to the text.

David Laing
Hanover, N.H.

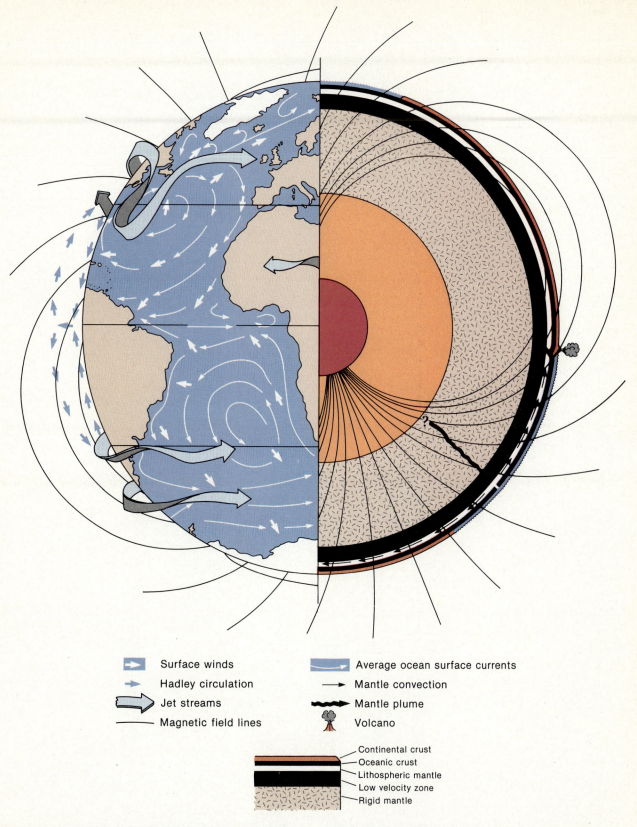

| | | | |
|---|---|---|---|
| ⇨ Surface winds | | ▭ Average ocean surface currents | |
| → Hadley circulation | | → Mantle convection | |
| ⇨ Jet streams | | ∿ Mantle plume | |
| — Magnetic field lines | | 🌋 Volcano | |

Continental crust
Oceanic crust
Lithospheric mantle
Low velocity zone
Rigid mantle

Major themes in the dynamic Earth system. Solar radiation drives the global wind and water cycles of the planet's surface (*left* side), while its internal heat, derived mainly from radioactivity, drives the plate tectonic system (*right* side), and the currents in Earth's outer core that generate the magnetic field. In this cutaway view, the northern edge of the Australian plate is sinking beneath the southeastern margin of the massive Eurasian plate (*right*). Molten rock rises from the sinking plate to form the volcanic island of Java at the surface.

# Part One

# Impressions of a Living Planet

Photo: A glacially sculpted landscape in the Elk Range of central Colorado.

# 1

# No Place Like Home: The Uniqueness of the Third Planet

*S*how thy concern to Earth, and it will instruct thee.

*Job 12:8*

Photo: Vidae Falls, Crater Lake National Park, Oregon.

*Figure 1.1*   A study in planetary contrasts. Closest to the Sun and the solar wind, Mercury (*far left*) has no atmosphere, and a Moonlike, cratered surface. Venus, our closest neighbor, has a dense atmosphere of carbon dioxide. Earth (*center*) is unique in having abundant water on its surface and abundant oxygen in its atmosphere. Mars, with its rusty color and canyoned surface, might at one time have had an Earthlike atmosphere, but its present atmosphere is similar to that of Venus, although much thinner. The giant planet Jupiter (*far right*) is, like the Sun, mostly hydrogen and helium. Except for Earth and Venus, the planets are shown here at different scales.

**Photos by NASA**

## The Living Planet

Throughout history, we have viewed the planet we live on from many different perspectives. To certain tribes of ancient India, Earth was a flat tray resting on the backs of three sacred elephants borne, in turn, on the shell of a giant tortoise. To the Babylonians, Earth was a hollow mountain afloat in a great ocean and surrounded by a crystal dome—the firmament—on which the Sun, Moon, and stars moved in their eternal cycles. In Native American tradition, our planet is the Earth Mother: a living, personal being, who gives life, and commands deep respect from all her children, human and otherwise.

As our outlook has broadened in time and space, our understanding of the planet we live on has evolved and deepened. Through the remarkable decade of the 1960s, Soviet and American manned space flight programs afforded humans their first, breathtaking view of Earth from a cosmic, rather than a terrestrial, perspective. In the ensuing two decades, unmanned space missions extended our extraterrestrial explorations to the farthest reaches of our Solar System and radioed back a rather stark and sobering message to their Earthly home base: *we are it*—the one and only sanctuary of life in the Sun's planetary family and possibly (though not probably) in the entire Universe.

As we have studied the images of planetary and lunar surfaces that these probes have beamed to Earth, we have come to appreciate the extent to which the surface of our own planet has been modified by the presence of life. A simple, visual comparison of Earth with its four closest planetary neighbors, Mercury, Venus, Mars, and Jupiter, underscores this contrast (fig. 1.1). With its brilliant, dynamic, white cloud swirls set above an azure global ocean and its wandering continents with their brown deserts, snow-white mountain chains, and dark green forests, Earth stands in vivid contrast to the cloud-veiled infernos of Venus and Jupiter and the bare, rocky surfaces of Mercury and Mars, which still bear the imprints of colossal asteroid impacts formed in a turbulent age that closely followed the Solar System's birth.

In this first chapter, which generally follows the plan of the book, I give an overview of the major features of the Earth system. The discussion begins with Earth's solid materials and the structures within them, proceeds to the liquid and gaseous portions of the Earth system (the oceans and atmosphere), and concludes with a consideration of some extraterrestrial phenomena. Notice, as you read, that five pervasive cycles underlie most Earthly phenomena. The first of these, the *plate tectonic cycle,* regulates the shifting of continents and ocean basins and the building of mountains. The second, the *rock cycle,* governs the creation and destruction of Earth's great diversity of rock types. The third and fourth, the *hydrologic* and *wind cycles,* embody the transformations among ice, water, and water vapor and the ceaseless movement of these substances over the surface of the globe. The fifth, the *organic carbon cycle,* traces the pathways of carbon through the Earth system as it passes from organic to inorganic reservoirs, each of which exerts powerful controlling influences on planetary processes.

## Continents and Ocean Basins

Figure 1.2 emphasizes the most striking aspect of our planet's surface: that it is divided into two major divisions—high-standing **continents** and low-lying **ocean basins** filled with seawater. As terrestrial beings, we have been preoccupied with the continents. Most who have not made their living by the sea have regarded the oceans mainly as obstacles to efficient travel, transportation, and communication. As a result of this terrestrial

**Figure 1.2** The two primary topographic features of Earth's crust are high-standing continents and low-lying ocean basins. The oceanic ridge is a continuous, broad, submarine mountain range nearly 80,000 km long and offset by numerous transverse fractures. The oceans'

greatest depths are found in oceanic trenches, such as the one off South America's west coast.

*World Ocean Floor,* Bruce C. Heezen and Marie Tharp, 1977, copyright by Marie Tharp 1977. Reproduced by Marie Tharp, 1 Washington Ave., South Nyack, NY 10960.

bias, our knowledge of the nature of the ocean basins—and especially of their hard rock floors—has lagged far behind our knowledge of the continents. Nevertheless, the continents only account for 29% of our planet's surface, whereas the ocean basins account for 71%. Only in the past few decades have we gathered enough information about the hidden landscape of the seafloor to be able to produce maps like figure 1.2. Fortunately, the geology of the seafloor is relatively simple compared to that of the continents, so perhaps another decade or so will suffice to put our knowledge of seafloor geology on an equal footing with our knowledge of continental geology.

## The Structure of Earth's Crust

It should come as no surprise that Earth consists mainly of rock. Most people have encountered enough of this common substance to have an intuitive sense that rock is a hard, heavy material; but for the purposes of Earth science, a more precise definition is desirable. I discuss rocks in depth in chapter 4, but for now, it should be sufficient to point out that a rock is a mixture of minerals. A *mineral,* in turn, is a pure chemical substance

with constant properties, such as composition, melting point, and density. Each rock type is composed of specific minerals, usually one to three dominant ones, in characteristic proportions. Sandstone, for example, is a rock that consists mostly or entirely of small grains of the mineral *quartz,* the crystalline form of the chemical silicon dioxide ($SiO_2$).

We have known for a long time that the kind of rock that underlies the continents is fundamentally different from that which underlies the ocean basins. The continents are composed mainly of **felsic** rock, so-called because it is rich in the mineral **fel**dspar and in **si**lica, or silicon dioxide. Felsic rock is relatively light both in color and in **density,** or mass per unit volume. The density of felsic rock averages about 2.8 g/cm³ (grams per cubic centimeter; see appendix 1 for a discussion of metric units of measurement). The oceans, on the other hand, are composed mainly of **mafic** (*may*-fik) rock, so-called because it is rich in **ma**gnesium and **f**errum (latin for iron). Mafic rock is both darker and denser than felsic rock. The density of mafic rock averages about 3.0 g/cm³, making it noticeably heavier than felsic rock. We have also known for some time that

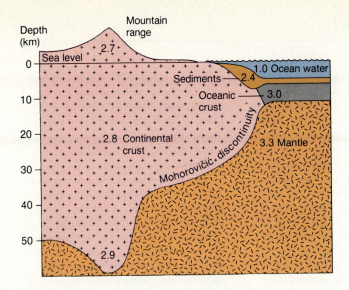

**Figure 1.3** Earth's rocky shell is divided into thick continental crust composed of felsic rocks that are light both in color and density, and thin oceanic crust composed of darker, denser mafic rock. Both of these types of crust are commonly overlain by a variable thickness of sediments and sedimentary rocks of still lower density. Below the Mohorovičić discontinuity, the crust is underlain by mantle rock of high density. Numbers indicate densities.

both these rock types of Earth's *crust,* or surface layer, are underlain by the still denser rock of Earth's upper **mantle,** which has an average density of about 3.3 g/cm³. Figure 1.3 illustrates the general relationships among the rock layers of Earth's crust and upper mantle.

The thickness of the crust above the mantle varies. Mafic **oceanic crust** ranges from 5 to 15 km thick, and averages 7 km. Felsic **continental crust** ranges from 30 to 80 km and averages 40 km. A rather fuzzy boundary, known as the **Mohorovičić (M) discontinuity,** or simply the **Moho,** separates the crust from the mantle (fig. 1.3). It was named for the Yugoslavian seismologist Andrija Mohorovičić (Muh-*hor*-oh-*veech*-ik), who discovered it while studying the behavior of earthquake shock waves as they passed through Earth's interior (see chapter 6).

## Isostasy: The Reason for Highlands and Lowlands

Why do continents stand at higher elevations than do the ocean basins? The principle involved has been called **isostasy** (eye-*soss*-ta-see; Greek for "standing equally"). It was first investigated about 2200 years ago by the Greek philosopher Archimedes, who found that when an object is placed in a liquid, it is buoyed up by a force that is equal to the weight of the volume of liquid that the object displaces. If the object is of lower density than

the liquid, as in the case of ice cubes in a glass of water, it will float. Both ice cubes and icebergs float with 92% of their volume submerged because ice is 92% as dense as water.

Similarly, both continental and oceanic crust float on the denser upper mantle, even though the latter is not a liquid in the ordinary sense. If the mantle were exposed at Earth's surface, "rockbergs" of continental and oceanic crust would float with 85% and 91% of their volumes submerged, respectively, because continental rock is 85% as dense, and oceanic rock is 91% as dense, as mantle rock. This explains the presence of the deep root beneath the mountain range in figure 1.3.

## The Plate Tectonic Cycle: Seafloors that Spread and Sink

One of the more surprising discoveries of the past few decades of investigation of the seafloor is the **oceanic ridge,** a more or less continuous submarine mountain chain 1500–2500 km wide, 1–3 km high, and over 84,000 km long (this length includes two branches in the eastern Pacific Ocean and one in the Indian Ocean; see figs. 1.2 and 1.4). A troughlike depression called a **median rift** runs along the midline of the ridge for most of its length, and the entire ridge is offset perpendicularly to its axis by several hundred prominent fractures in the seafloor. At first, the origin of this remarkable feature was not apparent; but in the early 1960s, some equally remarkable discoveries about the magnetic patterns in the rocks adjacent to the ridge led to the inescapable conclusion that the seafloor is spreading perpendicularly away from the ridge in opposite directions (fig. 1.4)! This odd fact had been anticipated as early as 1929 by the Scottish geologist Arthur Holmes. In that year, Holmes proposed the concept of **seafloor spreading** as a possible explanation to the then unpopular concept—now abundantly confirmed—that the continents are slowly drifting across the surface of the globe. His idea was generally ignored, however, until oceanographic exploration found hard evidence to support it. That evidence led to the modern theory of **plate tectonics** (see chapter 5).

## Global Conveyor Belts: The Mobile Lithosphere

Figure 1.4 shows the essential features of plate tectonics (literally, "plate structures"). In this cutaway view, we see that Earth's rigid, outer shell is made up of **lithospheric plates,** each of which consists of a thinner, upper layer of crustal rock and a thicker, underlying layer of upper mantle rock. Of the four plates illustrated here, three are capped with oceanic crust, whereas the one on the right is capped by continental crust. Most lithospheric plates appear to be roughly 100 km thick except

*I*t was doubtless an unforgettable day for the citizens of Syracuse, Sicily, when no less august a personage than the philosopher Archimedes came tearing stark naked from the public bath screaming *"Eureka! Eureka!"* (Greek for "I have found it!"). What brought on this display of unbridled philosophical exuberance was Archimedes' sudden grasp, as he stepped into the bath, of a method whereby he could determine whether the crown of his friend, King Hieron II, was of solid gold, as advertised by the local artist who had made it, or alloyed with silver. The method was as follows: weigh the crown, and then produce a lump of pure gold and another of pure silver, each having exactly the same weight as the crown. Because of its greater density, the gold lump would be considerably smaller than the silver lump; hence, when immersed in a chalice of water, it would displace less of the water. If the crown, on being immersed, displaced no more water than the gold lump of equal weight, then it must be made of pure gold. If it displaced more water, then it must be alloyed with silver. To the annoyance of King Hieron and the great misfortune of the artist, the crown displaced considerably more water than the pure gold.

Archimedes mustered many other ingenious solutions for the problems that vexed his monarch, among them an array of fearsome engines of war, which held off the Roman seige of Syracuse for three years. Aside from his famous principle and his war engines, Archimedes enriched civilization with many other practical discoveries, including a helical—or screwlike—device for raising water, still in common use today in Egypt and the compound pulley, of which the block and tackle is a modern version. He was by no means a modest genius and on one occasion boasted, "Give me a place to stand, and I will move the world!" As no one rose to the challenge, the boast remained untested.

The intellectual feats of Archimedes were no less impressive. In his exhaustive, lifelong studies of geometry, he not only solved many difficult problems, but in the process developed theorems, or propositions, that differed little from the modern theories of logarithms and integral calculus.

Unfortunately, Archimedes was not as skilled at defending himself as he was at defending his city. As Syracuse finally fell to the army of Marcellus in 212 B.C., he was unceremoniously run through by a coarse Roman footsoldier as he sat in the market place, contemplating a geometrical figure he had drawn in the sand. The irony was all the greater because Marcellus himself had given strict orders that Archimedes was to be spared at all costs.

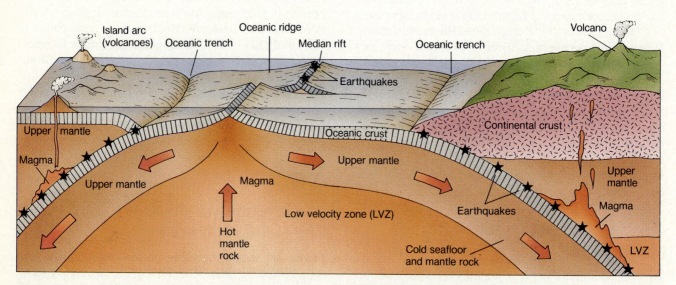

**Figure 1.4** The mechanisms of seafloor spreading and subduction are complementary processes of the plate tectonic system. The lithospheric plates (including crust and upper mantle) slide away from the oceanic ridge on the underlying, soft, low velocity zone of the mantle. New seafloor is created at the ridge as molten magma rises and solidifies on the separating plate margins. Simultaneously, the leading edges of the growing plates are destroyed by subduction beneath oceanic trenches. Partial melting of subducted lithosphere generates magma to feed volcanoes. Convection in the mantle (*arrows*) could be either the cause or the result of seafloor spreading.

near the oceanic ridge, where they become thinner, and beneath the continents where they become thicker. This has been deduced from measurements of the time it takes for earthquake shock waves to be reflected back to the surface from the base of the lithosphere (see chapter 6). The plates overlie a soft, probably partially molten layer in the mantle called the **low velocity zone** (**LVZ**), so called because earthquake waves slow down as they pass through it. The LVZ functions as a lubricant, allowing the overlying lithospheric plates to slide freely (but slowly) over the surface of the planet. Like the lithosphere, this shell of soft rock is also roughly 100 km thick.

### *The Birth and Death of Lithospheric Plates*

In figure 1.4, we find two lithospheric plates spreading apart from each other along the oceanic ridge. As the plates slide slowly away from the ridge crest, the median rift along the crest of the ridge widens, allowing mafic lava to erupt and solidify on the trailing edges of the plates. The identity of the force that causes the sliding is still unclear, but the most likely candidate appears to be gravity causing the plates to move gradually—at about 0.5–10 cm per year—down the flanks of the ridge. The ridge itself appears to be a great, linear blister formed by the rise of hot mantle rock from the depths of the planet.

How can lithospheric plates grow on a spherical planetary surface of constant area? That is clearly a mission impossible unless old plate material is somehow and somewhere destroyed at the same rate at which new plate material is being created. This process is shown happening on both the left and the right sides of figure 1.4. On the left, the western edge of one plate is being forced under the advancing eastern edge of another plate, forming a deep, linear **oceanic trench** along the zone of contact. The underthrust plate sinks beneath the trench in a process called **subduction** ("drawing under"). On the right, another oceanic plate similarly is being subducted eastward under the western edge of a continental plate. Above each of these two **subduction zones,** chains of volcanic islands or mountains have formed. The volcanoes are fed by **magma,** or molten rock, that rises through the overlying lithosphere from a depth of about 80 to 100 km within the subduction zone. The magma is created as the top of the descending slab of lithosphere begins to melt within this depth range.

## The Age of the Seafloor

Another recent and surprising discovery about the seafloor is that it is quite young—a maximum of 200 million years as opposed to a maximum of nearly 4000 million for the oldest continental rocks (fig. 1.5). The continents, therefore, seem to be about 20 times older, on average, than the ocean basins. The reason for this apparent paradox is that continental crust, being of somewhat lower density than oceanic rocks, cannot be readily subducted beneath the denser oceanic lithosphere. Consequently, the continental crust is simply floated over the surface of the globe, whereas the oceanic crust, being so readily subducted, never remains for long, geologically speaking, at the surface.

## Island Arcs and Mountain Belts

People who live in the Appalachian Mountains of eastern North America are familiar with great bodies of granite that, tens to hundreds of square kilometers in extent, lie scattered like meatballs within a spaghetti of twisted and broken layers of ancient mudstone, sandstone, limestone, and lava (fig. 1.6). The layered rocks of this spaghetti have been so intensely heated and compressed that their original composition is often hard to determine. Their once horizontal layers, or *strata,* have been rumpled into folds that range in length, crest-to-crest, from a few micrometers to many kilometers. In some places, they have been piled up on one another like the folds of a cast-off bathrobe (see fig. 6.22). In other places, they have been broken and thrust over one another for distances of up to 100 km (fig. 1.7).

In modern Japan and in the Andes Mountains of western South America, volcanic eruptions and severe earthquakes are familiar and frequent events (fig. 1.8). These events are actually closely related to the type of rock structures found in the Appalachians, but how?

## How Mountains Are Made

Earthquakes and volcanism are surface expressions of deep-seated forces that warp and melt underlying rocks in the process of mountain building. Those forces are directly related to the mechanism of plate tectonics. The earthquake-ridden volcanic islands of Japan have been built gradually by the ongoing subduction of the Pacific plate beneath the eastern edge of the Eurasian plate (see fig. 5.25 and the left side of fig. 1.4). During subduction, both the top of the descending plate and the overlying mantle melt partially, producing magma that rises through the overriding plate margin. Much of this magma erupts at the surface as lava, which builds the chain of volcanoes. Earthquakes result from the fracturing of the descending slab of lithosphere. The same process of subduction produces equivalent effects in the Andes Mountains of South America (see right side of fig. 1.4). The difference between Japan and the Andes is that in the latter case, the overlying plate carries continental crust; therefore, the volcanoes form a chain on land rather than at sea.

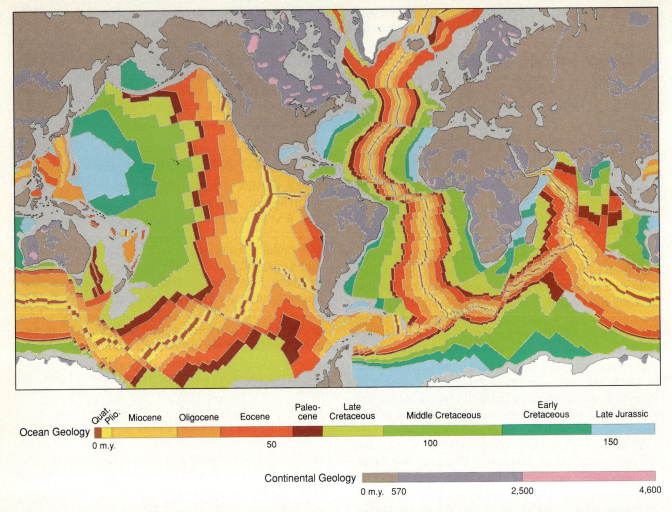

Ocean Geology

| Quat.·Plio. | Miocene | Oligocene | Eocene | Paleo-cene | Late Cretaceous | Middle Cretaceous | Early Cretaceous | Late Jurassic |

0 m.y.        50        100        150

Continental Geology

0 m.y.   570        2,500        4,600

*Figure 1.5*    In this map showing the ages of the rocks of Earth's crust, it is evident that the seafloor becomes progressively older with distance from the oceanic ridge, but no part of it is older than about 170 million years (*blue*). Large areas of continental rocks, on the other hand, are older than 570 million years (*purple*), and smaller areas range from 2500 million to 3960 million years old (*red*).

The end result of these plate tectonic processes can be seen best in regions where subduction has been long inactive. In such regions, erosion has had time to strip away much of the upper crust to a depth of several kilometers, revealing the complex internal structure beneath. Thus, in the Appalachians, we are able to see preserved in the surface rocks the kinds of changes that are now taking place at depths of several kilometers beneath the islands of Japan! At such depths, temperatures and pressures are high enough to allow rocks that would shatter at the surface to be warped and molded as readily as putty. Under extreme conditions, crustal rocks actually melt. In fact, it was the melting of deeply buried mudstone that formed much of the granitic magma that later solidified to form the "Appalachian meatballs."

## An Earth Exclusive

Does the mechanism of plate tectonics operate on other planets? So far, we have found no evidence that it does, but why? Preliminary studies suggest that some other planets might have low velocity layers, but they are too deep, and the planets' lithospheres are too thick for plate tectonics to work effectively. Furthermore, the essential ingredient of actively rising magma seems to be missing on all other bodies of the Solar System except Io, one of the moons of Jupiter (see fig. 17.24), although Mars shows clear evidence of such activity in the distant past (see fig. 17.19). Finally, as I explain in chapter 5, there is good reason to suspect that plate tectonics does not operate in the absence of water. Of all the rocky, inner planets in the Solar System (Mercury, Venus, the Earth-Moon system, and Mars), only Earth retains enough of that remarkable fluid to support plate tectonic processes.

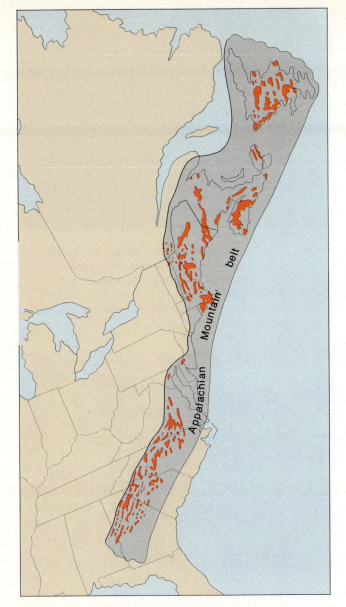

**Figure 1.6** Intense compression during mountain building folded once flat-lying sedimentary rocks of the Appalachian mountain belt (*grey*). Folding forced some rocks to such depths that they melted in Earth's hot interior, creating great masses of felsic igneous rock (*red*).

## Wear and Tear

Plate tectonics tends to build up the land surface with volcanoes and mountain chains. If there were no terrestrial force opposing this tendency, Earth's surface would be considerably more rugged than it actually is. The existence of such a force is attested to by ancient mountain systems, such as the Appalachians. Worn down over time from towering peaks to humble hills, the contorted bedrock of the Appalachians betrays an origin beneath many kilometers' thickness of overlying rocks that have since been stripped away. Because of our ex-

**Figure 1.7** At Lone Rock Point on the eastern shore of Vermont's Lake Champlain, the massive rocks of the cliff have been thrust westward for as much as 100 km over softer mudstones at the cliff base. The mudstone is about 100 million years *younger* than the overthrust rocks.

**Figure 1.8** This infrared image of Kyushu, the southwesternmost of the Japanese islands, was taken by NASA's ERTS–1 satellite in 1973. The volcanoes of the Japanese island chain are being built by magma rising through the overriding plate from an underlying subduction zone that plunges toward the northwest. The reddish color in the image results from heat given off by vegetation, blue-gray patches are cities, and white areas are clouds.

Photo by NASA

tremely short life spans, geologically speaking, it is difficult for us to comprehend the effectiveness of this stripping force. Seen in terms of geologic time, however, and in comparison with the surfaces of other planets and moons, Earth's land surfaces are suffering rapid and frequent changes. Consider, for example, the

**Figure 1.9**  Neil Armstrong's footprint on lunar soil, made on 20 July 1969, could remain as one of the few relics of the human species for millions of years after it has become extinct. Photo by NASA

**Figure 1.10**  Agents of chemical weathering—carbon dioxide, oxygen, water, and organic acids—constantly penetrate rock outcrops along exposed surfaces and cracks. They decompose many of the rock's constituent minerals and produce a rotten rind surrounding a core of fresh, unweathered rock.

**Figure 1.11**  When water enters cracks in rock and freezes, it exerts an overwhelming force that rends the rock apart in the mechanical weathering process of frost wedging.

life expectancy of a footprint in Earthly mud and that of one in lunar soil (fig. 1.9). Under exceptional conditions, a footprint on Earth might last a month or two. On our Moon, the footprints left by the Apollo missions should last for millions of years because they will never be affected by atmospheric gases, water, ice, or wind, the principal agents of change that have modified Earth's surface.

## Weathering: The Rock Breaker

The sculpting of landscapes begins with the slow disintegration of rock (fig. 1.10). In Earth's peculiarly destructive surface environment, rocks are broken down by the process of **weathering.** This weathering process takes two forms: *mechanical* and *chemical.* In regions that are subject to frost, water that seeps into cracks in rocks can gradually shatter those rocks mechanically because water expands about 9% on freezing (fig. 1.11). Water also reacts chemically with some minerals in rocks, causing them to crumble and decay. Atmospheric carbon dioxide, dissolved in all rain water, greatly intensifies this effect as it reacts with the water to form carbonic acid, which is mildly corrosive to most minerals. Oxygen, too, corrodes some minerals, especially those containing iron. Oxidized iron minerals produce the red, yellow, and brown colors seen in many rocks that have been exposed to the atmosphere by erosion (see fig. 3.17*b* and *c*).

Life-forms, especially plants and bacteria, are also powerful weathering agents. Lichens—cooperative complexes of fungi and algae that encrust rocks—produce organic acids that attack many rock minerals, thereby releasing chemical nutrients to the plants (fig. 1.12). The root systems of other plants—especially flowering plants, including hardwood trees—produce carbon dioxide, which reacts with soil water to form carbonic acid. Mechanically, plants are also effective rock breakers because their roots work their way into cracks and pry them open as they grow (see fig. 8.5).

Some of the surface material that has been decomposed and loosened by weathering mixes with decaying organic matter, and is temporarily retained by plant roots as *soil* (see chapter 8). The rest is swept away to the sea as **sediment** by flowing water, flowing ice, or wind.

**Figure 1.12** Gray-green lichens encrust frost-shattered rock on the continental divide in central Colorado. Acids released by lichens slowly attack some of the minerals of which rocks are composed, thereby contributing to the process of chemical weathering.

## Erosion: The Earth Mover

Figure 1.13 shows a landscape whose surface has been etched into an intricately branching network of valleys and ridges. With the exception of a few, minor human modifications, this landscape is mainly the result of two processes of **erosion**—*stream abrasion* and *mass wasting*—that work together to remove weathered bedrock from the land surface, typically as quickly as it forms. Stream abrasion is the lowering of a valley floor by the scouring action of sediment-choked flood waters rushing downstream after rapid snowmelt or heavy rain. **Mass wasting** is the slow to rapid downslope movement of soil and rock under the influence of gravity. The stream drainage pattern in figure 1.13 might appear to be haphazard and random, but careful study reveals that the drainage lines are relatively evenly spaced, allowing them to drain water and sediment from the entire area as efficiently as possible. This spacing reflects the relatively even distribution of rainfall over the landscape.

In addition to terrain sculpted by stream abrasion and mass wasting, landscapes in high latitudes and at high altitudes often show the effects of *glacial erosion*

**Figure 1.13** Stream erosion has carved this landscape in western Colorado into a branching network of stream channels, which remove evenly distributed rainfall and weathered sediment most efficiently from the land. Precipitation and sediment are directed toward the stream channels by the slopes of the ridges.

Photo by Colorado Geological Survey

**Figure 1.14** In this view from the summit of New Hampshire's Mount Kearsarge, smoothed and grooved rock outcrops (*foreground*) and rolling countryside (*background*) reflect the landscaping action of a thick continental ice sheet that advanced southeastward from Canada several thousand years ago. The ice moved in the direction of view.

(fig. 1.14). Armed with rock debris on their undersides, glaciers are slow but effective sculptors of the underlying landscape. *Wind erosion* is considerably less effective as a carver of landscapes than water or ice, but it can be important in desert regions (see fig. 14.17). Another locally important sculpting agent, active at shorelines, is *wave erosion* (see fig. 12.21). In soil and sediment, animals can also be effective agents of erosion.

a.

b.

*Figure 1.15* Two views of mafic Moon rocks. The gray gabbro boulder (*a*), 18 cm long, came from the Moon's rugged, cratered highlands. The photograph in (*b*), taken through a microscope, shows individual mineral grains in a thin slice cut from basalt lava that erupted on the Moon about 3800 million years ago. The minerals appear colored because they are viewed through polarizing prisms. Geologists routinely study such thin sections to identify rocks by their mineral contents.

Photos by NASA

## Genesis in Stone: The Rock Cycle

In comparison to any of our planetary neighbors, Earth is blessed with a much greater variety of rock types to fascinate (and sometimes confuse) the student of Earth science! Only two major rock types are found on our Moon (fig. 1.15), and these are the same two rock types that make up most of Earth's oceanic crust: dark, fine-grained, mafic *basalt* and its coarse-grained equivalent, *gabbro.* Lunar *breccia* is a derivative rock type formed from basalt or gabbro by meteorite impacts. It consists of fine- to coarse-grained fragments of preexisting rocks. Lunar *glass* is formed by the complete melting and sudden chilling of moon rock during meteorite impacts. So far, our limited investigations of the geology of other planets indicate that they have few, if any, rock types other than the basic two we have found on the Moon. Earth, on the other hand, boasts hundreds of different types, of which the most significant are described and illustrated in chapter 4.

Why this staggering diversity of terrestrial rocks? The answer lies mainly in the equally staggering variety of natural processes at work on Earth's surface. Plate tectonics constantly stirs and renews Earth's lithosphere. Water provides an ideal medium for complex chemical reactions. Living organisms further complicate those reactions, as do the gases of the atmosphere. Working together, these four factors provide the great diversity of processes needed to produce such a variety of rocks out of a few, fundamental, primitive types.

## Pathways of Change in Rocks

Rocks fall into three broad classes: igneous, sedimentary, and metamorphic. **Igneous rocks** (from Latin, *ignis,* "fire") are formed by the solidification, or freezing, of magma; **sedimentary rocks** are layered deposits formed at Earth's solid surface by the consolidation of sediment; and **metamorphic rocks** (from Greek, *metamorphosis,* "transformation") have undergone physical or chemical changes, or both, under the influence of temperatures and pressures higher than those that normally prevail in Earth's surface environment. Figure 1.16 illustrates a few of the many chains of events that can affect Earth rocks and change them from one class and type into a variety of others. Here, by following the arrows, we can trace the various transformations undergone by basalt as it passes through the plate tectonic cycle and the hydrologic cycle.

On the right, the partial melting of mantle rock yields mafic magma, which wells up within the median rift of the oceanic ridge. The magma cools and solidifies on the edges of the spreading lithospheric plate, forming basalt and gabbro. As the seafloor spreads, hot seawater circulates freely through fractures in the plate, leaching out some elements, such as calcium, and adding others, such as sodium, and forming a low-grade metamorphic rock called *spilite.* Subduction of the lithosphere subjects the rock to higher temperatures and pressures, which further alter it to the higher-grade metamorphic rock types *amphibolite* and *eclogite* (see figs. 4.39*b* and *c*). At a depth of about 100 km in the mantle, the lithosphere melts partially, regenerating mafic magma. This magma is subsequently erupted in volcanoes, producing lava, which cools to form basalt. Here, then, we

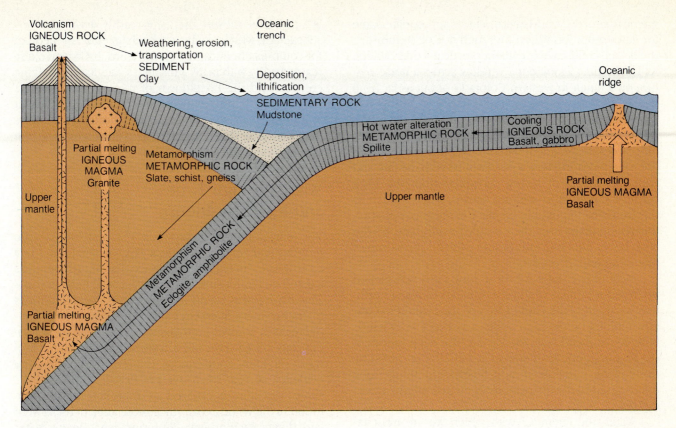

**Figure 1.16** Pathways of change in rocks. Rock classes are shown in bold letters; processes of change and representative rock types appear above and below the rock classes, respectively.

have come full circle, from magma through igneous and metamorphic rock back to magma and igneous rock. This cyclic regeneration of rock classes has been called the **rock cycle.**

Unlike the basalt of the submarine oceanic ridge, however, this newly erupted basalt is exposed to the atmosphere. Consequently, it is subjected to weathering, which reduces it to *clay,* a sediment. The clay is then eroded and transported to the ocean by streams, where it is deposited. As more layers of sediment accumulate, the lower layers are compacted and often cemented by chemicals dissolved in seawater. In addition, the clay crystals tend to dissolve where the pressure on them is greatest and to grow where it is least, a process called *recrystallization.* These three processes—compaction, cementation, and recrystallization—work together to convert soft sediment into hard rock. Collectively, they are included under the general term of **lithification** (Latin for "rock making"). The product of lithification is sedimentary rock, in this case, a *mudstone* (see fig. 4.18*g*).

Some of this mudstone can become buried so deeply that it is subjected to metamorphism. Because its chemistry is different from that of basalt, the metamorphic rocks produced from mudstone are corre-

spondingly different. With increasing intensity of metamorphism, the mudstone is accordingly converted to *slate, schist,* and *gneiss* (see figs. 4.34*a, c,* and *e*). If these rocks are carried to depths at which the temperature is high enough, they can melt partially and be converted to felsic magma, which eventually cools to form the igneous rock type, *granite* (see fig. 4.5*a*).

The chain of events illustrated in figure 1.16 is an oversimplification of a complex process. Nevertheless, it does suggest how, through the operation of the plate tectonic and hydrologic cycles, felsic continental crust can be formed over long intervals of geologic time from the mafic materials of oceanic crust.

## Water: The Humble Provider

Because of its great abundance on Earth's surface, we tend to take water for granted. If we were suddenly deprived of it, however, we would quickly appreciate how dependent most Earth processes are on that common fluid. It has even been suggested that because 71% of our planet's surface is covered with the 1.4 quintillion metric tons of the world ocean, a more appropriate name for Earth might be Water! Although it is often indirect, the ocean's influence is evident practically everywhere. As I write this, I look out a window across a horse pasture on a ranch in western Colorado, about as far from the sea as one can get in North America. On this late June afternoon, nevertheless, the ground is wet with fresh rain, the grass is green, and cottonwood and box-elder trees, resplendent with foliage, line the banks of the nearby Roaring Fork River, swollen with the rain and with the waters of spring snowmelt from the high country. On a world without oceans, a scene such as this would be impossible. In place of the grassy pasture, the trees, and the river, there would be only bare, barren rock, fewer hills and valleys, and a liberal sprinkling of meteorite impact craters.

Most of the water for these June rains was picked up from the surface of the sea by winds blowing northwestward across the Gulf of Mexico. Last winter's snow, now melting on the high peaks, was swept in by winds blowing eastward from the Pacific. The deep valleys of the Roaring Fork River system are still being carved by the flowing water itself as it flushes crumbling rock debris and soil downstream in its eternal pilgrimage to the sea.

Water, the chemical with which we are most familiar and take most for granted, is one of Nature's great wonders. Let us consider a few of the ways in which this extraordinary chemical makes planet Earth unique in the Solar System.

## Water as a Chemical

By the end of the eighteenth century, Henry Cavendish and Antoine Lavoisier (La-*vwaz*-yay) had shown by experiment that water consists of two parts of hydrogen (H) to one part of oxygen (O). Water is, in other words, the liquid form of the chemical *hydrogen oxide,* which we have come to know by the familiar formula $H_2O$ (this may also be written as $OH_2$). Somewhat later, it was found that hydrogen oxide exists as discrete atomic groupings, or **molecules** (from Latin, *molecula,* "little mass"). Each of these molecules has two hydrogen atoms attached, unsymmetrically, to a single oxygen atom (see fig. 11.18). In chapter 11, I discuss the physical and chemical properties of hydrogen oxide in detail. Here, we need only focus on four properties that together give $H_2O$ a vitally important role in the functioning of Earth's living and nonliving systems.

First, hydrogen oxide is the only common substance that exists in all three possible states of matter: solid (ice), liquid (water), and gas (water vapor), under the range of temperature that prevails on Earth's surface (see figs. 11.18 and 11.19). This fact alone accounts for most of Earth's extraordinary diversity as compared with other planets of the Solar System. Second, liquid water

is a powerful *solvent,* that is, it readily dissolves many substances, which is why it is such an efficient medium for complex chemical reactions. Third, hydrogen oxide reacts readily with many substances, facilitating both the creation and the breakdown of minerals and organic matter. Fourth, unlike most substances, $H_2O$ is denser as a liquid than it is as a solid, and therefore, ice floats in water. If it did not, icebergs would sink, and the oceans would soon freeze solid!

## Water through Time and Its Implications

The temperature range between 0° C and 100° C—the melting and boiling points of $H_2O$ at sea level—is a narrow one, yet we have evidence that liquid water has existed on Earth's surface for at least 3800 million years. That evidence is found in the form of water-laid rocks of that age in southern Greenland and 3500-million-year-old rocks in northwestern Australia. Neither of these rock deposits could have been formed anywhere but in a water medium. Furthermore, the record of *fossils*—the evidence of past life-forms preserved in the rocks—shows that life has enjoyed an unbroken span of evolution in the world ocean since those ancient rocks were laid down. Finally, few life-forms can tolerate water temperatures above 40° C or below 0° C for long. The remarkable implication of these facts is that in spite of many disturbing influences, our planet's average surface temperature must have remained between 0° C and 40° C for at least 3800 million years! In a following section, I consider a possible reason for this fortunate circumstance.

## Water on the Wind: The Hydrologic Cycle

As the example of snow and rain in Colorado suggests, water is by no means confined to the ocean basins. In figure 1.17, which portrays the **hydrologic cycle,** the general trends of water movement on Earth's surface are shown. A steady influx of solar energy induces continual changes of state (solid, liquid, and gas) in hydrogen oxide, which keep it moving through the various pathways of this cycle.

The cycle begins as incoming solar energy heats the surface layers of the ocean, imparting enough energy to some of the water molecules that they break loose and fly away from the water surface to form water vapor, which dissolves completely in the overlying air. There is a limit to the amount of water vapor that can dissolve in a given volume of air. That limit, called **saturation,** increases with temperature, but it is rarely met or exceeded at the ocean surface. On the average, about 455,000 km³ of water is evaporated from the ocean each year.

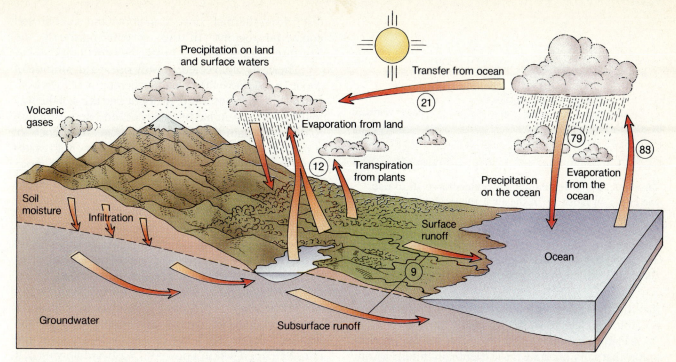

**Figure 1.17** Powered by solar energy, the water cycle is an endless interaction of ice, water, and water vapor with the circulation system of the atmosphere and with gravity. Arrows show major pathways. Circled numbers indicate percentages of the total amount of water evaporated or transferred annually. Note that water losses from the land (12 + 9 = 21) are compensated by transfers from the ocean (21), and that runoff from land plus precipitation on the ocean (9 + 79 = 88) balance evaporation from the ocean (88).

Solar energy also heats this moistened air, causing it to expand and become less dense and more buoyant than the surrounding air, which is cooler because of shading by clouds or cooling by wind. Because of this density difference, the expanded air rises spontaneously by a process called *thermal* (heat) *convection* (from Latin, *convehere,* "carry together"; see chapter 13) and is replaced by cooler, denser air at the ocean surface (fig. 1.18).

As the warm, moist air rises, it expands and cools because in the higher, thinner regions of the atmosphere there are fewer gas molecules in a given volume of surrounding air; therefore the pressure is less. If the air rises high enough, it can cool to a temperature at which it becomes saturated. The altitude at which this occurs depends on how nearly saturated the air was as it began its ascent. If the air rises still farther and cools much below the saturation temperature, it becomes *supersaturated,* and water begins to separate from the air in the form of cloud droplets, a process called **condensation.**

Most of the 455,000 km³ of water that is raised annually from the ocean surface falls back on the oceans as precipitation. A much smaller proportion is carried horizontally by wind over land surfaces. This forces the moist air to rise, expand, and cool (fig. 1.19). As with thermal convection, this too can bring about supersaturation, cloud formation, and precipitation. Some mountainous parts of the continents are high enough

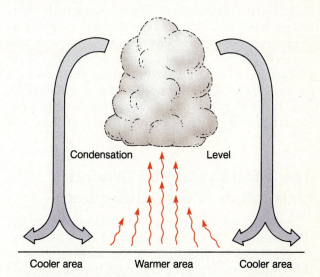

**Figure 1.18** Sunlight heats land and water surfaces unevenly. Warmer areas in turn heat overlying air more than cooler areas do. The warmer air expands and becomes less dense than adjacent air, whereupon it rises and cools. The base of the cloud marks the elevation at which the rising air has cooled to a temperature at which it has become supersaturated. Above this elevation, the air is too cold to hold the moisture it has carried aloft, so water vapor condenses, forming the cloud. On all sides, the rising warm air displaces cooler air, which descends over cooler areas of the land or sea.

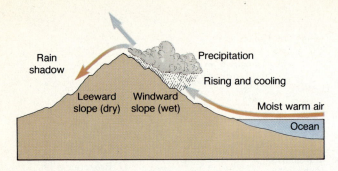

**Figure 1.19** When mountain barriers force air to rise, the effect on moisture in the air is the same as when the rise results from convection. On the windward side of the barrier, moisture condenses, clouds form, and precipitation falls. On the leeward side, the descending air, wrung dry, becomes warmer and absorbs water from the land, creating a rain shadow with dry conditions.

that the air temperature falls well below 0° C, and the precipitation falls as snow.

Clearly, every drop and every flake of the precipitation that falls on land must eventually return to the ocean. Otherwise, in about 40,000 years, all the ocean's water would be transferred to the land, and the ocean would dry up! This return flow is accomplished in three ways (see fig. 1.17). First, approximately 32,400 km³ of water per year returns rapidly to the ocean in the world's rivers. Second, about 13,600 km³ per year returns by slow, underground flow. Third, some 62,000 km³ per year evaporates from the land surface or from the pores and surfaces of leaves. Most of this evaporated water is re-precipitated on the land as rain or snow, and the rest is returned by winds to the sea. From this, it should be clear that whether you are on land or sea, the precipitation that falls on you is derived from *both* land *and* sea!

## The Wind Cycle: The General Circulation of the Atmosphere

If precipitation is to fall on the land, there must be some mechanism to carry it there from the sea. This task is accomplished by **atmospheric circulation.** Early in this century, Norwegian climatologist Vilhelm Bjerknes (*Byark*-nace) and his son Jakob developed a conceptual model of the general circulation of the atmosphere (see chapter 15). A full appreciation of the patterns of atmospheric circulation, however, had to await the Gemini and Apollo space missions of the 1960s. The Apollo 17 picture of Earth in figure 1.20 shows the continents of Africa and Antarctica under mostly fair skies and parts of the Indian and Pacific Oceans under variable cloud cover. A band of patchy clouds stretches over central Africa and the equatorial Pacific. These clouds

result from thermal convection above the solar-heated equatorial regions. In the extreme northeast, the whirling cloud spiral of a hurricane is evident.

Thermally convected air only rises to an altitude of about 18 km above the equator, whereupon it flows both northward and southward. At about one third of the way to the poles, the air has become cool and dense enough to descend again toward Earth's surface. As it does so, it expands and warms, and its capacity to hold water vapor increases, allowing it to absorb whatever moisture is present on the land beneath. This creates parallel belts of deserts on either side of the equatorial cloud belt. Notice the lighter areas of northern and southern Africa flanking the darker, wetter, more densely vegetated central portion of the continent. From these desert belts, *trade winds* return the dry air to the equator. This cycling of air between the equator and the desert belts has been called the *Hadley circulation* (see chapter 14).

Between about 35° and 60° south of the equator, a series of magnificent cloud swirls obscures much of the southern ocean. These swirls are cyclonic storm systems entrained within a powerful, eastward-flowing belt of high-speed, high-altitude winds known as the southern polar jet stream (see chapter 15). Like the Hadley circulation, cyclonic storms also carry equatorial heat, along with a lot of water vapor, toward the poles. Hurricanes also serve this same function. Without these three heat-transferring mechanisms—hurricanes, Hadley cells, and cyclonic storms—the tropics would be unbearably hot and the polar regions unbearably cold, leaving only the midlatitude temperate regions habitable between flanking zones of "fire and ice."

At the bottom of the globe, the continent of Antarctica gleams white beneath a cloudless sky. Strange as it might seem, both the north and the south polar regions are deserts! Despite the vast expanse of the Antarctic ice sheet, which is more than 4 km thick in places, little precipitation occurs on it because of the extremely low water content of the cold, polar air.

## Earth's Strange, Reactive Atmosphere

The presence of liquid water is by no means the only unique and remarkable thing about our planet's surface. Among the nine planets of the Solar System, ours is the only one that has nitrogen, ($N_2$, 78%) and oxygen, ($O_2$, 21%) as major components of its atmosphere, and negligible amounts of carbon dioxide, ($CO_2$, 0.03%). In contrast, the atmospheres of both Venus and Mars are almost entirely carbon dioxide, with only about 3% nitrogen (see fig. 13.4). Their similarity ends there, however, as the atmosphere of Mars is 140 times less dense at the surface than Earth's atmosphere, and that of Venus is 90 times more dense! The atmospheres of the four

**Figure 1.20** A sapphire world with its dynamic, fluid systems of air and water, Earth is without parallel in the Solar System. This magnificent, full-disk view from the Apollo 17 mission shows the main features of Earth's atmospheric circulation system. An equatorial belt of patchy clouds (*above center*) marks the zone where strong solar heating causes moist air to rise. Flanking this are relatively clear zones with dry, descending air and little precipitation. Poleward of these desert belts are turbulent zones where cyclonic storm systems follow the polar jet streams eastward around the planet, bringing equatorial heat toward the poles, where cold, dry air and clear skies prevail.

Photo by NASA

**great planets**—Jupiter, Saturn, Uranus, and Neptune—consist mainly of hydrogen ($H_2$), with smaller amounts of helium (He), methane ($CH_4$), and ammonia ($NH_3$), and they are much more dense, even, than the atmosphere of Venus.

## Light Gases, Heavy Gases, and Reactive Gases: Some Implications

The significance of these differences becomes clear when we consider the natures of the various gases involved. Hydrogen and helium, the two lightest gases in Nature, are also the two principal gases in the Sun, and in the Universe. They are present in about the same proportion in the atmospheres of the great planets as they are in the Sun. This strongly suggests that the great planets originated within the same cosmic cloud of dust and gas from which the Sun condensed long ago (see chapter 17).

Carbon dioxide and nitrogen are much heavier gases. They are rare in the Universe but fairly common in the rocky interiors of the **inner planets,** Mercury, Venus, Earth, and Mars. Oxygen is a component of both water and carbon dioxide; but because it is a highly reactive gas, it seldom occurs free in Nature. Its presence in major amounts in Earth's atmosphere is an indication that our atmosphere is extremely unstable chemically. We know, in fact, that the average length of time that an individual oxygen molecule spends free in the atmosphere (i.e., its *residence time*) is only about 3000 years. This means that if oxygen were not being actively replenished by the regenerative process of *photosynthesis,* it would disappear quickly from our atmosphere.

**Figure 1.21** Volcanic eruptions, such as this one from Washington's Mount St. Helens in 1980, deliver massive amounts of mineral ash, water vapor, and carbon dioxide to the atmosphere along with lesser amounts of nitrogen, sulfur oxides, hydrogen sulfide, and hydrochloric acid.

Photo by R. Hoblitt U.S. Geological Survey

## The Making of Earth's Atmosphere

How did our peculiar, reactive, oxygen-rich atmosphere originate, and why is it so different from those of other planets? Attempting to find reasonable answers to questions about things that happened so long ago may seem presumptuous. Nonetheless, we do have some fairly good clues for guidelines. It is likely that Earth and the other inner planets once had atmospheres like those of the great planets, although they were probably much thinner. It is also likely that these original atmospheres were blown away by an early, hyperactive phase of the Sun. Subsequently, the present atmospheres developed in their place, but how?

The most likely explanation suggests itself when we note that small amounts of most of the gases in the atmospheres of the inner planets, including $H_2O$, are present in most terrestrial (Earthbound) rock types and in many meteorites and comets. Furthermore, we find great volumes of these same gases being expelled from volcanoes. Therefore, it seems likely that the atmospheres of the inner planets have simply leaked from their hot interiors during volcanic eruptions, a process that has been called **volcanic outgassing.** The small planet Mars apparently lacked enough internal heat to produce a dense atmosphere by this process. The rocks of Earth's Moon are bone dry and virtually gas-free, which probably explains why the Moon has no atmosphere. It is likely that the same explanation holds for Mercury, the closest planet to the Sun, which has only a faint trace of an atmosphere, composed of the elements sodium and potassium. Volcanic outgassing still occurs on Earth (fig. 1.21), producing chiefly water vapor (about 70%), carbon dioxide (about 15%), and nitrogen (about 5%). It seems, then, that at the outset, Earth's atmosphere consisted of a mixture of these three gases together with some hydrogen, carbon monoxide (CO), hydrogen chloride (HCl), and hydrogen sulfide ($H_2S$).

## Preparing the Nest: Neutralizing Acid and Cooling the Greenhouse

If our atmosphere had remained as it was when it was first outgassed, life on Earth would not have persisted, as a brief look at Venus will reveal. The temperature at the surface of Venus is about 460° C, hot enough to melt lead and zinc, and whereas Earthly clouds are composed of water droplets, the clouds of Venus are composed of sulfuric acid! The reason Venus is such a hot, acidic desert is that its dense carbon dioxide atmosphere is transparent to solar radiation but opaque to heat rays from the planet itself. In other words, solar radiation penetrates to the planet's surface unchecked, raising its temperature, but the heat reradiated by the Sun-warmed surface is unable to escape through the carbon dioxide atmosphere. Consequently, it heats the atmosphere. The heat therefore escapes into space from the *top* of the atmosphere instead of from the surface of Venus. This has been called the **greenhouse effect** because of its superficial similarity to the way the Sun's rays heat a greenhouse.

The greenhouse effect also operates on Earth, but to a far smaller extent, because Earth's atmosphere contains only about 0.03% carbon dioxide; therefore its temperature has remained well below the boiling point of water. Earth's oceans, furthermore, are not acidic but are in fact slightly alkaline (the opposite of acidic). Why these radical differences between two neighboring planets that are otherwise nearly identical in size, mass, and composition?

The answer is **photosynthesis.** With this humble but remarkable life process, green plants remove carbon dioxide from the atmosphere and combine it with water to make sugar and oxygen, which they return to the atmosphere to replace the extracted carbon dioxide. In chemical symbols, that process (greatly simplified) is

$$CO_2 + H_2O \rightarrow CH_2O + O_2$$

carbon + water yield sugar + oxygen
dioxide

Not only does photosynthesis cool the planet by removing carbon dioxide from the atmosphere, but it also neutralizes acid on a massive scale. Lifeless lakes in the craters of some Indonesian volcanoes are as sour as battery acid; but in lakes where algae are thriving, the water can be so alkaline as to be almost caustic. In the following section, I discuss the **carbon cycle,** a remarkable system that uses photosynthesis to help maintain Earth's surface environment in a condition suitable for life.

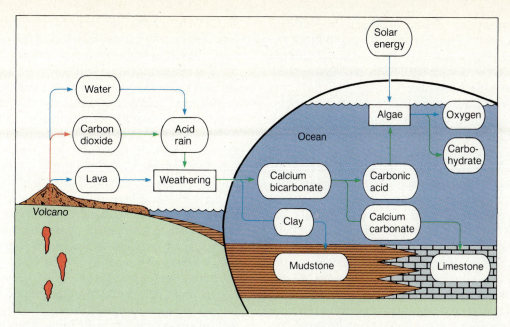

**Figure 1.22** The carbon cycle removes carbon dioxide from the atmosphere and ocean (*solid green arrows*) as fast as it is injected into them by volcanoes (*dashed red arrow*). The cycle converts volcanic products into materials, such as carbohydrate, that are useful to living systems, and waste products, such as mudstone, that are not. Calcium carbonate and oxygen, originally waste products, have found functions in more advanced organisms as skeletal material and a source of metabolic energy, respectively.

## The Carbon Cycle: Life's Control System

Because the discipline of Earth science has traditionally fallen under the physical sciences, the key role of photosynthesis in regulating Earth's surface environment has not been fully appreciated until recently. That role is still disputed, however, by some Earth scientists who prefer the more traditional view that Earth's surface environment would have developed as it has even in the absence of life. Nevertheless, there can be little doubt that the carbon cycle would have been adequate to the task of creating and maintaining our chemically reactive, oxygen-rich atmosphere over the past 4000 million years. Central to this remarkable chain of events is one of the most improbable transformations imaginable: the slow conversion of the products of volcanism—lava, water, and carbon dioxide—into mudstone, limestone, living organisms, and oxygen by the processes of weathering, photosynthesis, and shell-making.

The portion of the carbon cycle that governs this transformation is illustrated in figure 1.22. Because the carbon cycle is involved with the plate tectonic, rock, and hydrologic cycles, I postpone a walk-through of the complete cycle until after these other cycles have been more thoroughly discussed (see fig. 11.31). As we examine the portion of the cycle shown in figure 1.22, note that the solid, green arrows indicate the removal of carbon dioxide from the atmosphere and ocean, a trend that decreases the greenhouse effect and the acidity of the ocean. The dashed, red arrow indicates the reverse effect. Dotted, yellow arrows indicate pathways that do not involve carbon. We begin with the erupting volcano on the left in figure 1.22.

As magma erupts at Earth's surface, it separates into liquid and gaseous components. The gases, mainly water vapor and carbon dioxide, escape to the atmosphere, and the liquid either flows out as lava or solidifies and is blown out in pulverized form as *volcanic ash.* These various components of magma are able to coexist under the high temperatures and pressures of Earth's mantle, but at the surface, they react with one another. Water and carbon dioxide combine in the atmosphere to form *carbonic acid.* This weak acid mixes with water in cloud droplets to form a mild, natural, acid rain, the principal agent of chemical weathering. As carbonic acid reacts with lava and ash in the process of weathering, it creates two products: *calcium bicarbonate*—a chemical similar to baking soda—and *clay.* Both of these products are washed into the sea by rivers. Clay is insoluble, so it settles slowly to the seafloor. There it accumulates in thick deposits, which eventually harden into *mudstone,* the most abundant of all sedimentary rock types. Calcium bicarbonate is soluble and remains dissolved in seawater, but not for long!

In the process of photosynthesis, marine plants—mainly algae—extract carbonic acid from dissolved calcium bicarbonate. With the help of solar energy, they rearrange the atoms of this acid into oxygen and **carbohydrate,** a class of chemicals (such as sugar or starch) composed of carbon, hydrogen, and oxygen atoms. Some of this carbohydrate is used to build the bodies of the photosynthetic organisms (or of animals that eat them). The oxygen is released into seawater and, ultimately, into the atmosphere.

When organisms extract carbonic acid from calcium bicarbonate during photosynthesis, they leave an insoluble residue of *calcium carbonate* (center of oval in fig. 1.22), which either precipitates in milky clouds from the seawater surrounding the organisms or is actively *secreted* by the organisms as tiny, colorless crystals within their bodies, or as protective shells around them. Various marine animals (e.g., clams, snails, sea urchins, and one-celled foraminifera) have also developed the ability to secrete calcium carbonate as shell material. As these organisms die and decompose, the calcium carbonate crystals and shell fragments are dispersed, and they settle to the seafloor, where they harden into layers of the common rock type *limestone.* If it were not for photosynthesis, there would be no limestone on Earth because photosynthesis is the only known process capable of reducing the oceans' acidity to the point at which calcium carbonate can precipitate.

This model of even this small portion of the carbon cycle is, of course, a gross oversimplification. Many other minerals are involved in the cycle, and photosynthesis occurs on land as well as in the sea. Nonetheless, it does indicate the extent to which weathering and life processes have converted dead mineral matter into this extraordinary beautiful living planet on which we are so privileged to exist.

## The Ozone Layer: Earth's Radiation Shield

One of the many fortunate consequences of having an oxygen-rich atmosphere is that a rarefied layer of **ozone** gas forms and persists in the upper atmosphere at altitudes between about 15 and 55 km. Ozone is a variety of oxygen whose molecules consist of three atoms each ($O_3$) instead of the normal two ($O_2$; see chapter 3 for a discussion of atoms and molecules). Ozone is created from normal oxygen by the action of solar ultraviolet rays (see chapter 13), a process that absorbs almost 100% of this life-threatening radiation. If Earth's protective ozone shield were to vanish, life on the continents would suffer severe damage from cancers and genetic mutations. In addition, it is likely that a significant portion of our planet's water would eventually be lost, as water molecules are readily broken apart or photodissociated, by ultraviolet radiation.

Because ozone is derived from oxygen, it certainly could not have existed in significant amounts before photosynthesizing organisms produced an oxygen-rich atmosphere. The question might be asked, how could life have originated in the absence of a protective ozone

shield? A reasonable answer is that the earliest life-forms probably developed in a sufficient depth of ocean water to be protected from severe irradiation.

The creation and destruction of ozone by solar ultraviolet radiation generates heat. The shell of warm air produced by this heat is less dense than the cooler air beneath it, so it tends to remain above the cooler air, forming a stable layer in Earth's atmosphere known as the *stratosphere.* Because of its great stability, the stratosphere acts as a "lid" on the turbulent, convective circulation of the underlying *troposphere,* the lower layer of Earth's atmosphere in which virtually all weather phenomena occur. Without this protective lid on convective circulation, Earth's water might eventually be broken apart, or *photodissociated,* by ultraviolet radiation leaving our planet as hot, dry, and lifeless as its neighbor, Venus. It is possible that Venus might have lost its water because it never developed living organisms to produce an atmosphere rich enough in oxygen that an ozone shield could form. Even if the loss of Earth's ozone layer did not lead to the loss of its water, the general circulation of the atmosphere would certainly be altered in unpredictable ways. Clearly, these disturbing possibilities deserve careful study before we engage, or continue, in any activity that could harm the ozone layer.

Unfortunately, we have already been engaging in such activities for some time. Recently, an ominous hole has begun to develop in the ozone layer over Antarctica in springtime (October in the Southern Hemisphere), showing an alarming 50% drop in ozone concentration within the decade of the 1980s. It now seems clear that human-made **chlorofluorocarbon compounds** (**CFCs**) used as chemical propellants in spray cans, as refrigerants ("freon") in refrigerators and air conditioners, and as inflatants in styrofoam, are drifting slowly upward into the stratosphere. There they are releasing monatomic chlorine, which destroys ozone *catalytically,* in other words, a single atom of chlorine can destroy as many as 100,000 ozone molecules. In view of the vital, protective functions of the ozone layer, this could be the most serious of the many human-caused environmental problems facing our planet.

## A Magnetic Field and a Cosmic Ray Shield

All four of the great planets of the Solar System—Jupiter, Saturn, Uranus, and Neptune—have powerful magnetic fields. Weak magnetic fields also surround Mercury, Venus, and Mars, but Earth is distinctive among the four inner planets in having an especially powerful field, probably generated in dynamo fashion by the flow of electrons within the liquid outer portion of our planet's iron core (see fig. 5.8). This **geomagnetic field** radiates outward from Earth's center, enclosing the planet within a series of lopsided, bagel-shaped "lines" of

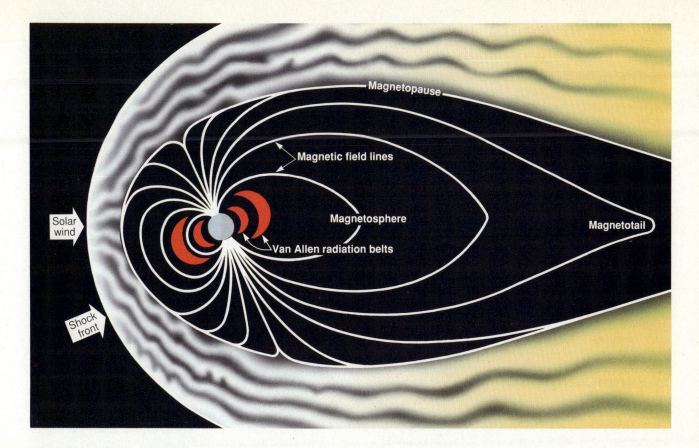

Magnetopause

Magnetic field lines

Solar wind

Magnetosphere

Magnetotail

Van Allen radiation belts

Shock front

**Figure 1.23**   Earth's magnetic field, or magnetosphere, is pushed into a highly unsymmetrical shape by the solar wind, a constant stream of protons and electrons from the Sun. The solar wind, in turn, is distorted by the magnetosphere behind a curved shock front. A sharp boundary, the magnetopause, separates the solar wind from the magnetosphere. The two sets of "parentheses" flanking the planet are the Van Allen radiation belts.

magnetic force called the **magnetosphere** (fig. 1.23). The shape of magnetosphere is lopsided because it has been distorted by the *solar wind,* a powerful flood of electrically charged particles streaming outward in all directions from the Sun.

## The Van Allen Radiation Belts

Within the magnetosphere lie two concentric rings that have a high density of electrically charged particles (fig. 1.23). The outer ring contains protons and electrons from the solar wind that have been trapped by Earth's magnetic lines of force. The inner ring contains mainly protons from natural sources, and in addition, it contains electrons from a nuclear device detonated in the atmosphere in 1962. These **Van Allen radiation belts** were first described in 1959 by the atmospheric physicist J. A. Van Allen. Evidence for their existence came from Geiger counters mounted in research satellites launched the previous year. The density of the inner ring is highest at about 3000 km above Earth's surface, whereas that of the outer ring is highest at about 10,500 km.

These two radiation belts, particularly the inner one, protect Earth's surface from bombardment by **cosmic radiation**—high-energy particles from deep space. These particles (mainly protons) interact with those that are trapped within the Van Allen belts; thus, most of them are prevented from reaching the ground, where they are as damaging to biological systems as ultraviolet radiation.

## Reversals of the Geomagnetic Field

Recently, it has been discovered that the geomagnetic field fluctuates in intensity, and eventually reverses itself on an irregular schedule. If such a **geomagnetic field reversal** were to occur in the near future, compass needles would then point south instead of north. During geomagnetic reversals, of which there have been 24 in the past 4 million years (see fig. 5.9), the geomagnetic field intensity drops to zero. At such times, the Van Allen belts must temporarily decay and vanish, and the cosmic irradiation of Earth's surface must markedly increase.

## Food For Thought

We are a society of specialists. Each of us is concerned primarily with his or her own specialty—be it art, music, finance, engineering, marketing, or Earth science. We let other specialists worry about things that do not seem to fit within the spheres of our own disciplines. From this perspective, it makes sense for someone who plans to specialize in Earth science to study that subject, but what about the rest of us? Why not simply leave all problems of an Earth science nature to the Earth scientists? After all, that is their specialty.

The disquieting answer to that question is that Earth scientists have never even come close to having all the answers, and they are beginning to admit it. Increasingly, they are having to face the embarrassing reality that they might no longer be able to find ways to make Earth fall in line with the desires and demands of an increasing human population. It seems, therefore, that a new role has been added to the professional duties of the Earth scientist: that of educating the public about the real limits of our planet's ability to serve humankind.

In the face of traditional attitudes of long standing, this is not an easy task. For most of human history, we have regarded Earth as a warehouse of miscellaneous, unrelated parts provided by a benevolent deity for us to use as we see fit. This attitude has worked fairly well so far. With it we have been fruitful, we have multiplied, and we have subdued Earth, or at least we would like to think we have.

Now, however, we are beginning to see our planet in a different light. The more we learn about Earth, the more we find that it is anything but miscellaneous, and we are finding less and less support for the view that Earth was specifically designed to serve human needs. We have learned that its parts are related in a precise and efficient way, as if it were an enormous living organism carrying on its life functions on a scale so slow and so vast that to our perceptions, limited in time and space, to that which is small and quick, they are practically incomprehensible. With this new view, we have begun to see ourselves more as what ecologists call a *microsymbiont*: a small organism that lives on, or in, a bigger one, like a mite on an elephant or a cellulose-digesting bacterium in the intestine of a cow. This is not necessarily as demeaning as it sounds. Without its intestinal bacteria, a cow is unable to digest its food, and so it quickly dies, yet it could just as easily die from an overabundance of intestinal bacteria as from a deficiency of them.

This analogy is imperfect because it is by no means clear that human beings benefit Earth in any significant way. Furthermore, if we consider the variety of toxic substances (pollutants) that we excrete into the environment as byproducts of our high-technology lifestyle, our relationship to Earth seems disturbingly closer to that of a pathogenic, or disease-causing, organism. If this is, indeed, our role with respect to the planet on whose continued health and welfare we all depend, can we reverse that role in time to avoid killing our host? If so, how do we proceed?

Subsequent chapters identify various ways in which the presence, abundance, and activities of human beings on Earth are altering the ancient, natural cycles of the planet. The great challenge facing humankind in this and future generations is to understand those natural cycles well enough to be able to draw a reasonable compromise between our desires and Earth's capability to support those desires on an ongoing basis. One sobering reality that we might well learn from the record of species extinctions over geologic time (see chapter 2) is that unless we are especially careful from now on, we might stand a good chance of destroying Earth's ability to support one rather recently evolved species, *Homo sapiens!*

---

The effects of this on terrestrial life are unknown, but it is possible that the rate of genetic mutations would also increase. This could mark geomagnetic reversals as times of relatively rapid evolutionary change.

## Summary

**Mantle** rock is of higher **density** than **mafic oceanic crust,** which is denser and thinner than **felsic continental crust.** The **Mohorovičić discontinuity (Moho)** separates the crust from the mantle. **Isostasy** requires that **continents** stand higher than **ocean basins. Lithospheric plates** are about 100 km thick and consist of crust and upper mantle. They grow away from **oceanic ridges** by **seafloor spreading** at **median rifts,** sliding on a partly molten **low velocity zone** in the mantle and are consumed in **subduction zones** beneath **oceanic trenches.** Because continental crust cannot be **subducted,** the seafloor is about 20 times younger than the continents. This **plate tectonic** process, unique to Earth, produces *mountain belts* on continental margins.

Intense *mechanical* and *chemical* **weathering** occurs on Earth but not on other planets. Agents of **erosion** include gravity (producing *mass wasting*), rain, wind, streams, glaciers, waves, and animals. Earth's unique diversity of **igneous, sedimentary,** and **metamorphic rock** types and the environments in which they form are due to the combined influences of the plate tectonic, hydrologic, wind, and carbon cycles.

These rock types are mutually interconvertible through the **rock cycle. Magma** is molten rock. **Lava** is erupted magma. The **lithification** of sedimentary rock comprises the processes of *compaction, cementation,* and *recrystallization* of **sediment.**

Hydrogen oxide ($H_2O$) is a **molecule.** The persistence of liquid *water* on Earth for 4000 million years implies extreme environmental stability. The **hydrologic cycle** comprises the various pathways of water exchange between land and sea, including *evaporation, thermal convection,* **condensation** and *precipitation,* and stream and ground water flow. *Convective precipitation* occurs in **saturated** air, and dominates the tropics; descending dry air creates deserts to the north and south; *cyclonic storm systems* embedded within *polar jet streams* carry heat poleward. Together, convective and cyclonic systems constitute the **atmospheric circulation.**

The early atmospheres of the **inner planets** were probably created by **volcanic outgassing** and were probably rich in carbon dioxide. The atmospheres of the great planets consist mainly of hydrogen and helium. **Photosynthesis** (the conversion of carbon dioxide into **carbohydrate**), and carbonate secretion remove carbon from the atmosphere and the ocean, thereby counteracting oceanic acidity and the **greenhouse effect.** In the **carbon cycle,** *lava,* water, and carbon dioxide are converted by weathering, photosynthesis, and carbonate secretion to clay, limestone, sugar, and oxygen.

Earth's **ozone layer** creates the *stratosphere,* keeps ultraviolet (UV) radiation out, and keeps water in. Ozone is created and destroyed by UV, and is now threatened by **chlorofluorocarbon compounds, (CFCs).** Interaction of the solar wind with Earth's **geomagnetic field** creates a lopsided **magnetosphere** containing two **Van Allen radiation belts,** which protect the planet against **cosmic radiation. Geomagnetic field reversals** occur irregularly. During reversals, the magnetic field and the Van Allen belts disappear. Earth is a complex, living, *co-evolutionary system* evolved for maximum efficiency. We can kill it if we are not careful.

## Key Terms

continent
ocean basin
felsic
density
mafic

mantle
oceanic crust
continental crust
Mohorovičić (M)
    discontinuity (Moho)

isostasy
oceanic ridge
median rift
seafloor spreading
plate tectonics
lithospheric plate
low velocity zone (LVZ)
oceanic trench
subduction
subduction zone
magma
weathering
sediment
erosion
mass wasting
igneous rock
hydrologic cycle
saturation
condensation
atmospheric circulation
great planets

inner planets
volcanic outgassing
greenhouse effect
photosynthesis
carbon cycle
carbohydrate
ozone layer
chlorofluorocarbon
    compounds (CFCs)
geomagnetic field
magnetosphere
Van Allen radiation
    belts
cosmic radiation
geomagnetic field
    reversal
sedimentary rock
metamorphic rock
rock cycle
lithification
molecule

## Questions for Review

1. Distinguish between continental crust and oceanic crust in terms of position, composition, density, and thickness.
2. Why is the seafloor generally much younger than the continental surfaces?
3. What makes it possible for lithospheric plates to slide over Earth's surface?
4. List six erosional agents and the environments in which they are active.
5. Explain why sedimentary rocks do not occur on the Moon.
6. Describe the differences among igneous, sedimentary, and metamorphic rocks. How can they change from one to another?
7. Identify three atmospheric mechanisms that move heat from the tropics toward the poles.
8. Describe how the carbon cycle transforms lava, volcanic ash, water, and carbon dioxide into clay, limestone, carbohydrate, and oxygen. How does this transformation affect Earth's surface environment?
9. What are the two principal functions of atmospheric ozone?
10. Describe the origin and function of the Van Allen belts.

# 2

# Time, Life, and Systems: The Overwhelming Perspective

*Time is a river of passing events. Strong is its current! No sooner is something brought into view than it is swept away and another takes its place, but in time this, too, will pass.*

Marcus Aurelius

***Meditations***

Photo: A butte of sandstone (*above*) and varicolored mudstone (*below*) in the Circle Cliffs region of southeastern Utah.

# Concepts of Time

If there is one phenomenon that truly symbolizes our civilization, it is the clock face (or its digital equivalent). Even though we live our lives by the clock, however, most of us have a dim concept of just what this odd entity *time* is that clocks measure. Because time is central not only to human civilization but also to Earth's long and varied history, an adequate understanding of the concept of time is essential to the study of Earth science. In chapter 16, I discuss the astronomical phenomena on which our system of measuring time is based. In the following sections, I discuss the impressive accumulation of evidence of the passage of time preserved in the rocks of Earth's crust.

## Humanistic Perspectives of Time

Being human, we make use of recurrent human events in the measurement of time. A typical human life span, for example, is about 70 years. Traditionally, a human generation is reckoned at 25 years. Our human perception of time is strongly influenced by these human recurrence intervals. We think of the roughly 200 year interval since the American Revolution as a long time (about three human life spans or eight generations). Visitors to American Indian archaeological sites, such as Mesa Verde in Colorado, are understandably impressed by tree-ring dates that reveal ages of more than 700 years (ten human life spans) for these ruins. The Parthenon in Greece was completed over 2400 years ago (34 human life spans), and some of the Egyptian pyramids are almost twice as old as that. The origins of human civilization, closely tied to the development of agriculture and irrigation, stretch back about 10,000 years, and the origins of our species, *Homo sapiens,* go back to about 300,000 years ago.

Working with the humanistic perspective of the Judeo-Christian Bible, James Ussher, the Irish Archbishop of Armagh, conducted research that led him to conclude, in 1654, that Earth was created on October 26, 4004 B.C. To most of European society, steeped as it was in biblical tradition, an age of 5658 years for Earth seemed quite reasonable.

## Scientific Perspectives of Time

In the shadow of this prevailing perspective, the perceptions of **James Hutton,** eighteenth century physician, gentleman farmer, and founder of the science of geology, seem all the more remarkable. Hutton was a sensible, logical Scotsman and a particularly keen observer of the geological phenomena he found as he scouted the English countryside. He was looking for an explanation that would better account for those phenomena than the orthodox biblical explanation of the flood of Noah. The key to his search lay in the clear-sighted observations of **Niels Stensen,** a contemporary of Archbishop Ussher's.

More than a century before Hutton, Stensen (also known as Nicolaus Steno), a Danish-born Italian physicist, had established three important natural laws concerning layered, or **stratified,** rocks (fig. 2.1). His **law of superposition** states that in an undisturbed sequence of **strata,** or rock layers, the bottom layer is the oldest, and the top is the youngest. His **law of original horizontality** states that because stratified rocks are originally deposited as sediments within a fluid medium (water or air), they must be laid down with an originally horizontal attitude, perpendicular to the direction in which the force of gravity acts. His **law of original lateral continuity** states that such layers must be laid down as continuous, uninterrupted "sheets" that thin gradually to a feather edge at the margins of the basin in which they are deposited.

Hutton reasoned from these premises that the twisted, broken, and fragmented rock layers he observed in the British Isles and on the continent must have been subjected to profound disturbances. That much was in accord with biblical thinking, but another evident fact was not. As a farmer, Hutton had a strong interest in the process of soil formation. He had observed that the weathering of rocks to soil and sediment was a very slow process. For each rock layer to accumulate as the final result of the erosion, stream transport, and deposition of weathered bedrock, vast amounts of time must be required. Hutton therefore concluded that the deformed, layered rock sequences he observed, in places many hundreds of meters thick, must have accumulated over millions of years. In contrast to Ussher, he saw "no vestige of a beginning; no prospect of an end" to geologic time. To Hutton, then, the age of our planet seemed limitless. Given an unlimited amount of time, all the great chasms, the mountain ranges, and the deformations of rock layers could easily be accomplished by humble forces that can be observed at work on Earth's crust today. Thus, to counter the prevailing theory of **catastrophism,** Hutton advanced a contrasting theory of **actualism,** best expressed in 1802 by his friend and colleague John Playfair:

> Amid all the revolutions of the globe the economy of Nature has been uniform, and her laws are the only thing that have resisted the general movement. The rivers and the rocks, the seas, and the continents have changed in all their parts; but the laws which describe those changes . . . have remained invariably the same.

Some geologists misinterpreted Playfair's use of the word "uniform," and consequently advanced a doctrine of *uniformitarianism,* which argues that the processes

## James Hutton (1726–1797)

Self-motivation and dogged persistence in the face of adversity are traits commonly attributed to Scots. In the case of James Hutton, the attribution appears justified, but with the qualification that he was something of a dilettante. That very trait, however, ultimately stood him and the science he founded in very good stead.

Hutton's father, a merchant and an Edinburgh town officer, died when James was three years old. In spite of this setback, Hutton received a good education and graduated from the University of Edinburgh, where he developed a strong interest in chemistry. After graduation, he apprenticed himself to a lawyer, but distinguished himself chiefly by performing spectacular chemical stunts for the amusement of the legal clerks. At the same time, he began to collaborate with a friend, James Davie, on an industrial process for the manufacture of sal ammoniac (ammonium chloride) from coal soot.

Discharged from his apprenticeship after less than a year, Hutton returned to the University of Edinburgh, where for three years he studied medicine, the closest available discipline to chemistry. After two years' further study in Paris, he obtained an M.D. degree from the University of Leiden in 1749. By then, however, his industrial sal ammoniac process had begun to bear financial fruit, so Hutton chose to retire from medicine before ever practicing it. He settled on a small farm he had inherited in Berwickshire and set about learning the gentleman farmer's trade.

In the course of his farming studies in England and abroad, Hutton developed an interest in rocks and soil and the effects of flowing water on Earth materials. These studies paid off and made Hutton a prosperous farmer. In 1765, he went into partnership with Davie in the manufacture of sal ammoniac. Three years later, he had amassed enough funds to retire from farming and study geology full time at the University of Edinburgh. At the time, geology as a discipline was unknown. Even the name did not come into common use until over 100 years later.

In developing a science of geology, Hutton had to go against not only the prevailing beliefs of most of his contemporaries, both academicians and laypersons, but also the teachings of a dogmatic German contemporary, Abraham Gottlob Werner, who taught at the Freiberg Mining Academy. Although Werner did not make mention of the flood of Noah, his ideas were compatible with that account, as he taught that all the rocks of Earth's surface, including basalt, were chemically precipitated from a universal ocean, which had since drained into underground caverns. Countering this *neptunist* school of thought (named for the Roman god of the sea) were the *vulcanists,* mainly French and Italian scholars who had observed basalt in volcanic settings, and therefore favored a volcanic origin for it.

Hutton entered this controversy with the following observations:

1. Basalt intrusions that had baked coal seams and forced their way into open fractures in sedimentary rocks.
2. Fine-grained, chilled margins in these intrusions where they had been suddenly cooled by the surrounding rocks.
3. Mineralized veins in granite and schist, whose complex assemblages of high-temperature minerals could not possibly have been precipitated from a water solution.

Aided by chemists James Hall and Joseph Black, he analyzed these and countless other relations of rocks in the field and performed experiments, such as the melting and chilling of basalt, to confirm his conclusions.

Attacked for his thoughts by theologians and proponents of neptunism, Hutton responded in 1795 with his two-volume work, *Theory of the Earth,* which is generally acknowledged as the cornerstone of geology. Unfortunately, it was not until Sir Charles Lyell published his immensely popular textbook *Principles of Geology* in 1830 that Hutton's ideas came to be generally accepted. Interestingly enough, at the time of his death Hutton was at work on a book that anticipated Charles Darwin's theory of the origin of species by natural selection, but it was not until 1947 that the manuscript was finally examined.

James Hutton shared with many other great visionaries the ability to bring expertise in a variety of disciplines to bear on a complex and diverse field of study. He also possessed the equally important ability to synthesize a great diversity of observations on natural phenomena into an internally consistent explanation of their origin. The process is known as *inductive reasoning,* and it is central to all valid scientific thought. Werner, on the other hand, arrived at his erroneous conclusions by the process of *deductive reasoning,* in which he allowed his thinking to be controlled by a set of assumptions rather than by actual observations.

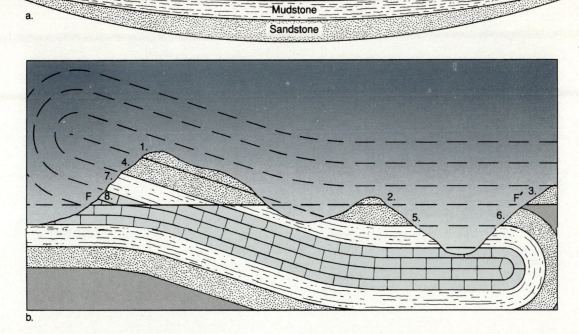

**Figure 2.1** Stensen's laws make possible the interpretation of complex geologic structures. (*a*) Sedimentary strata are laid down horizontally as continuous, unbroken sheets (the lower strata were warped as the basin subsided under the weight of accumulating sediment). The sandstone is the oldest stratum; the overlying mudstone is younger, and the limestone is the youngest. (*b*) Where these same strata have been deformed and eroded, Stensen's laws reveal that layers 1–3 are all the same sandstone, that layers 4–6 are also the same sandstone but inverted (layers 7 and 8 are also inverted), and that the sequence 1–3, 4, 7, and 8 has been transported toward the left along line FF′.

that change Earth's surface operate at *uniform rates.* We now have ample evidence that this is a false assumption. The rates of geological processes do vary with time, but the *natural laws* that govern those processes are constant, which is precisely what Playfair's statement says. The concept of actualism is often expressed in the phrase: *The present is the key to the past.*

Between Ussher's view of a youthful Earth and Hutton's view of an Earth of limitless age, several other attempts have been made to estimate the age of our planet. Georges Louis Leclerc, Comte de Buffon, in his *Histoire Naturelle* (1749), argued for an originally molten Earth with an age of 75,000 years to allow adequate time for cooling. Ninety-seven years later, in 1846, Sir William Thomson, Lord Kelvin, applied a sophisticated thermodynamic analysis to the same argument. This analysis led to an age estimate of between 20 and 30 million years. We now know that the interior of our planet is, in fact, cooling at the very slow rate of about 200° C per 1000 million years (the temperature of the surface has remained essentially constant, however). We also know something that Lord Kelvin, working half a century before the discovery of radioactivity, could not have known: Earth has a source of internal heat in the decay of radioactive elements.

After Kelvin's estimate, others were made, following Hutton's lead, by comparing known rates of sediment deposition with the maximum known thickness of sedimentary rocks and by comparing the *salinity,* or salt content, of the ocean with the rates at which rivers deliver salts to their mouths. Such salts, of which calcium bicarbonate is but one example (see fig. 1.22), are derived chiefly from the weathering of continental bedrock. Because of the enormous uncertainties involved, these estimates ranged from 3 million to 1584 million years. Recent studies indicate, however, that the salinity of the ocean has been practically constant for at least 3000 million years! The ocean, in other words, is a great *open system* from which the salts that are added by rivers and volcanoes are steadily removed by various other processes. Hence, all this figuring was a waste of time!

Our presently accepted estimate of Earth's age is about 4600 million years, more than three times the most liberal of the earlier estimates. Earth scientists arrived at this incredible figure through a variety of analyses, all based on radioactivity, the phenomenon that spelled the undoing of Lord Kelvin's conclusions. Using this technique, geologists have dated Earth's oldest rocks in Canada's Northwest Territories at about 3960 million years (see fig. 1.5). The oldest meteorites, which probably formed at the same time as Earth, date to between

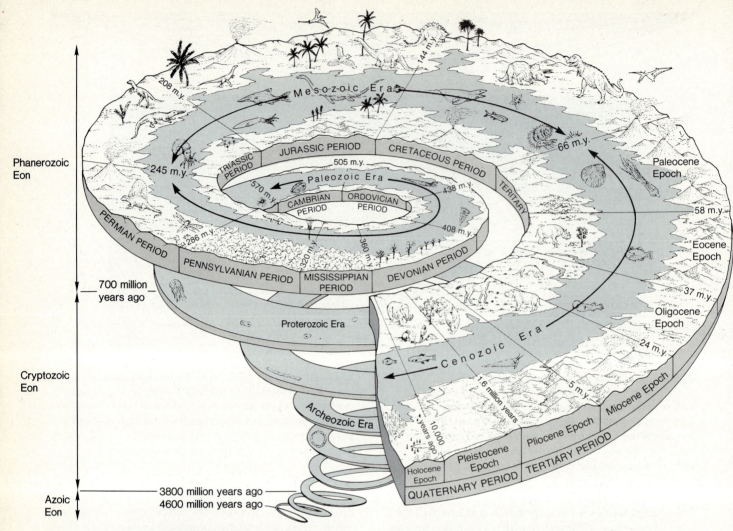

**Phanerozoic Eon**

**Cryptozoic Eon**

**Azoic Eon**

700 million years ago

3800 million years ago
4600 million years ago

*Mesozoic Era*

208 m.y.

144 m.y.

245 m.y.

JURASSIC PERIOD

CRETACEOUS PERIOD

TRIASSIC PERIOD

505 m.y.

*Paleozoic Era*

570 m.y.

CAMBRIAN PERIOD

ORDOVICIAN PERIOD

TERTIARY

66 m.y.

438 m.y.

408 m.y.

286 m.y.

PERMIAN PERIOD

PENNSYLVANIAN PERIOD

320 m.y.

MISSISSIPPIAN PERIOD

360 m.y.

DEVONIAN PERIOD

Paleocene Epoch

58 m.y.

Eocene Epoch

37 m.y.

Oligocene Epoch

24 m.y.

*Proterozoic Era*

*Cenozoic Era*

5 m.y.

Miocene Epoch

1.6 million years

Pliocene Epoch

TERTIARY PERIOD

*Archeozoic Era*

10,000 years ago

Holocene Epoch

Pleistocene Epoch

QUATERNARY PERIOD

**Figure 2.2** The intervals of geologic time are defined on the basis of the kinds of fossil organisms found preserved in the rocks of different ages. Greater detail in the Paleozoic, Mesozoic, and Cenozoic Eras than in the Archeozoic and Proterozoic Eras results from better preservation of more recent rocks and from an increase through time in diversity of life-forms with hard body parts.

Source: "Geologic Time," U.S. Geologic Survey.

4400 and 4600 million years. The rocks recovered from the Moon by the Apollo missions date between 3200 and 4600 million years. These and other lines of evidence point to a common origin for the various bodies of the Solar System around 4600 million years ago.

At some point in the Earth science laboratory, or on an afternoon walk, you could find yourself face-to-face with a 2000-million-year-old rock. Certainly, rocks whose mineral grains have been sitting next to one another in the same relative positions for billions of years deserve a lot of respect! Before you rush out to pay that respect to the local bedrock, however, it might first be helpful to try to visualize the meaning of 4600 million years. Suppose we have a stack of paper in which one sheet is equivalent to one year. If the stack contained one sheet for each year of Earth's age, it would rise to a height of about 360 km, approximately the width of the state of California, or the crow's-flight length of the coast of

Maine! To put this immoderate stack in human perspective, suppose we remove from the top of it one sheet of paper for each year the human species has existed, that is, about 300,000 sheets. This would shorten the stack by 24 meters, still leaving 359.976 km of paper! In other terms, humanity's existence to date amounts to about 1/15,000, or 0.0067%, of Earth's age. Archbishop Ussher's estimate of Earth's age would remove a little less than half a meter of paper from the stack, and the span of American independence would remove only 1.6 cm. Finally, a human lifetime would remove 0.56 cm, being only 1/64 millionth the age of our impressively ancient planet.

## Relative Time: The Geologic Time Scale

Figure 2.2 conveys a vivid impression of the enormous duration of geologic time, with emphasis on the most recent half-billion years. Notice that the upper part of

**Table 2.1**    The Geologic Time Scale (modified from Geological Society of America, 1983)

| Eon | Era | Period | Epoch | Duration ($\times 10^6$ yrs) | Began ($\times 10^6$ yrs ago) |
|---|---|---|---|---|---|
| Phanerozoic | Cenozoic | Quaternary | Pleistocene | 1.6 | 1.6 |
| | | Tertiary | Pliocene | 3.7 | 5.3 |
| | | | Miocene | 18.4 | 23.7 |
| | | | Oligocene | 12.9 | 36.6 |
| | | | Eocene | 21.2 | 57.8 |
| | | | Paleocene | 8.6 | 66.4 |
| | Mesozoic | Cretaceous | | 78 | 144 |
| | | Jurassic | | 64 | 208 |
| | | Triassic | | 37 | 245 |
| | Paleozoic | Permian | | 41 | 286 |
| | | Pennsylvanian | | 34 | 320 |
| | | Mississippian | | 40 | 360 |
| | | Devonian | | 48 | 408 |
| | | Silurian | | 30 | 438 |
| | | Ordovician | | 67 | 505 |
| | | Cambrian | | 65 | 570 |
| | | Ediacarian | | 130+ | 700+ |
| Cryptozoic | Proterozoic | | | 1800 | 2500 |
| | Archaeozoic | | | 1300 | 3800? |
| Azoic | | | | 800? | 4600? |

Source: Geological Society of America, 1983 (modified).

this great time spiral has been divided into segments of various durations. These segments—the divisions of the *geologic time scale*—are also shown in table 2.1. The largest divisions are the **eons** of which there are three: the **Azoic** ("Without Life"), from Earth's beginnings to about 3800 million years ago; the **Cryptozoic** ("Hidden Life"), from about 3800 million to about 700 million years ago; and the **Phanerozoic** ("Visible Life"), from about 700 million years ago to the present.

The two more recent eons have been subdivided into five **eras**: the Cryptozoic comprises the **Archeozoic** ("Ancient Life") and the **Proterozoic** ("Former Life"); whereas the Phanerozoic comprises the **Paleozoic** ("Old Life"), the **Mesozoic** ("Middle Life"), and the **Cenozoic** ("Recent Life"). Notice the emphasis on *life* in all the names of both eons and eras. This reflects the fact that the *fossil record* of extinct life-forms is the basis for our division of geologic time into standard intervals.

The three most recent geologic eras have been further subdivided into a series of *periods*. The names of these periods generally reflect the names of the places where stratified rocks of the respective ages were first described. For example, rocks between 570 and 505 million years old were first identified and described in Wales, an old name for which is Cambria. *Cambrian* rocks, wherever they may occur, are readily identified by their content of fossil species assemblages similar to those of the Cambrian rocks in Wales. The *Ordovician* and *Silurian* Periods are named after ancient Welsh tribes because rocks of these ages were also first described in Wales. *Devonian* rocks are named after Dev-

onshire, England. The *Triassic* Period is named for a three-fold division of its rocks in Germany. The name *Cretaceous* (from Latin *creta,* "chalk") refers to the chalk cliffs on either side of the English Channel (see fig. 4.29). With the exception of *Tertiary* and *Quaternary,* terms left over from an early geologic time scale that had only four divisions, all the other period names in table 2.1 are also place names.

## Faunal Succession: The Key to Geologic Time

The quality that makes the fossil record so useful for the purpose of dating the layered rocks that contain them is expressed in the **law of faunal** (animal) **succession,** which states that characteristic fossil assemblages have succeeded one another over geologic time. From primitive, one-celled forebearers, the great diversity of life has developed as a result of the interplay of changing environments and responding genetics. Throughout geologic time, this interplay has produced an ever-changing, but always internally harmonious life system in which organisms have *co-evolved* together in such a way as to be able to interact as efficiently as possible both with one another and with their physical environments. Thus, at any given time, a particular state of *dynamic perfection* prevailed on Earth that was different from those that preceded and followed it, and each of these states set the stage for what would happen in the next. Traces of these sequential states of dynamic perfection have been preserved as fossils in some of the layered rocks of our planet's surface.

From these traces, we have been able to reconstruct enough of a picture of the way Earth's surface environment has changed through time to give us a fairly detailed calendar, based on those changes. Because this calendar was completed by 1891, before the discovery of radioactivity, it was for many years only a **relative geologic time scale,** based on the laws of superposition and faunal succession. The addition of actual dates, to produce an **absolute geologic time scale,** had to await the development of radiometric dating.

Figure 2.2 illustrates some of the most important fossil species of plants and animals, both marine and terrestrial, whose remains are characteristic of the various time intervals. In addition, geologic environments are indicated, including volcanic activity, mountain building, and the changing relative proportions of land and sea. After the first turn and a half of the spiral (going back in time), however, the detail falls off rapidly. There are three reasons for this:

1. More recent layered rocks have been less extensively eroded than more ancient ones, and hence are more abundant.
2. The diversity of life-forms has increased over geologic time.
3. Before about 570 million years ago, animals had not yet evolved the ability to produce hard skeletons or shells of calcium phosphate and calcium carbonate; hence, they were not as well or as abundantly preserved in sedimentary rocks as more recent species have been.

## Absolute Time: The Radiometric Dating of Rocks

At the end of the nineteenth century, French physicists Pierre and Marja Curie discovered that certain elements, such as uranium, are unstable and tend to break down into other elements, such as radium and helium, a phenomenon they called *radioactivity*. Shortly afterward, other physicists discovered that such breakdowns occur according to a common pattern in Nature called the **exponential decay curve** (fig. 2.3). This curve expresses a simple relationship: for each radioactive substance, there is a certain time interval, called the **half-life,** at the end of which half of the amount of the radioactive element that was present at the beginning of that interval will have decayed. Notice that after one half-life, half of the original amount remains, and after two half-lives, half of that remaining half remains, or one-fourth of the original amount. After three half-lives, again half of the remaining half remains, or one-eighth of the original amount, and so on. This decay pattern contrasts

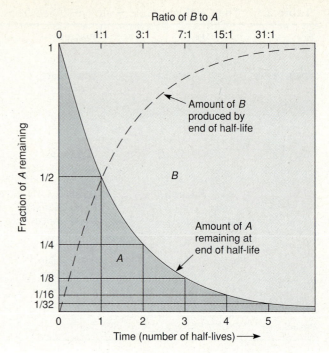

**Figure 2.3** A radioactive element, *A* (*solid curve*), decays exponentially, losing one half of its original substance over a fixed time interval, called the half-life. Over the next half-life it loses half of the remaining half, and so on. Each radioactive element has a different half-life. The dashed curve indicates the amount of decay product, *B* accumulated over time. The ratio of *B* to *A* at the end of any given half-life (*top*) is the same for all radioactive elements.

with that of an hourglass or a candle, both of which spend their substances at uniform rates.

If the half-life of a given radioactive element is known, and if a rock contains both that element and its decay product, the age of the rock can be determined by a simple calculation based on the ratio of decay product to parent element, a procedure known as **radiometric dating.** This assumes that none of the decay product has leaked out of or into the rock, otherwise the calculated age will be too young, or too old, respectively. Table 2.2 lists the various radioactive elements commonly used in the radiometric dating of rocks, together with their half-lives, their decay products, and some of the minerals in which they occur in common rocks. The significance of the numbers attached to the element symbols in the table is explained in chapter 3. Notice that the range in half-lives is extremely broad, from 5730 years for carbon-14 to 47,000 million years for rubidium-87. The length of the half-life determines the age range over which the element is effective as a dating tool.

**Table 2.2**   Radioactive Elements Commonly Used for Radiometric Dating

| Element | Half-life (years) | Effective Range (years BP*) | Decay Product | Minerals | Rock Types |
|---------|-------------------|------------------------------|---------------|----------|------------|
| Carbon-14 ($^{14}$C) | 5,730 | <60,000 | Nitrogen-14 ($^{14}$N) | Organisms | |
| Uranium-235 ($^{235}$U) | $713 \times 10^6$ | $>100 \times 10^6$ | Lead-207 ($^{207}$Pb) | Zircon | Granite, gneiss |
| Potassium-40 ($^{40}$K) | $1300 \times 10^6$ | $>0.1 \times 10^6$ | Argon-40 ($^{40}$Ar) | Biotite, muscovite, K-feldspar | Igneous, metamorphic |
| Uranium-238 ($^{238}$U) | $4510 \times 10^6$ | $>100 \times 10^6$ | Lead-206 ($^{206}$Pb) | Zircon | Granite, gneiss |
| Rubidium-87 ($^{87}$Rb) | $47,000 \times 10^6$ | $>100 \times 10^6$ | Strontium-87 ($^{87}$Sr) | Same as $^{40}$K | |

*Before the present.

## Dating the Rock Record

Using these various radiometric dating techniques, Earth scientists have gradually been able to add quantitative age limits to the divisions of the relative geologic time scale. The dates given in table 2.1 are the latest available, but revisions continue to occur, so these dates should be regarded as still tentative and subject to errors that can range from 3 million to 35 million years.

With the exception of carbon-14, used to date organic material, radiometric dating techniques are virtually limited to igneous and metamorphic rocks. The geologic time scale, however, was established on the basis of layered sedimentary rocks. Applying radiometric dating to such rocks is a problem, and this is a major source of the remaining uncertainty attached to the dates of the geologic time intervals. Nevertheless, this uncertainty has been greatly reduced in recent years.

Figure 2.4 illustrates the necessary conditions for the effective radiometric dating of sedimentary rocks. First, the rocks must be divided into units, or **geologic formations.** Normally, a geologic formation represents a deposit of sediment, such as a sand beach, that was laid down more or less continuously over a relatively short span of geologic time and under a relatively uniform set of environmental conditions. In most cases, the rock unit is designated by the name of the locality where it was first described and by the word "Formation," as in the Wasatch, Mesaverde, Morrison, and Summerville Formations. If the unit is composed of only one rock type, the name of that rock type is used instead, as in the Mancos Shale and the Dakota Sandstone.

The volcanic ash bed in the Morrison Formation (fig. 2.4) has been radiometrically dated at 150 million years old. Therefore, we can say that the Morrison Formation, in which it lies, is also about 150 million years old. The igneous dike (a crack filled with solidified magma) cuts through the Mancos Shale and the Mesaverde Formation, but is itself cut off by the Tertiary Wasatch Formation. The only reasonable explanation for this arrangement is that the magma of the dike shot up through a crack *before* the Wasatch Formation was laid

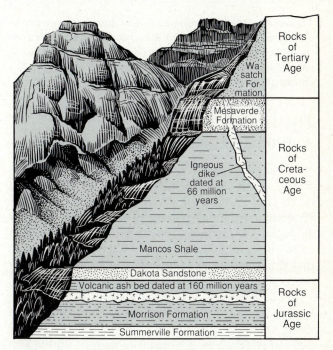

**Figure 2.4**   The only way that sedimentary rocks can be dated radiometrically is to observe their spatial interrelationships with dated igneous rocks. Here, the dated igneous dike must be younger than the Mesaverde Formation, which it cuts, but older than the Wasatch Formation, which truncates it. The volcanic ash bed is clearly contemporaneous with those beds of the Morrison Formation within which it is located.

Source: W. L. Newman, "Geologic Time," U.S. Department of the Interior.

down, otherwise, the crack (and the dike) should extend up into the Wasatch Formation as well. Because the dike is radiometrically dated at 66 million years, the Mesaverde Formation must be *at least* that old, and, by the law of superposition, it must also be *no older than* the underlying, 150-million-year-old Morrison Formation. By similar reasoning, the Wasatch Formation must be younger than 66 million years. By analyzing many such geometric relationships among formations worldwide, the absolute geologic time scale has been slowly and laboriously established and refined.

**Figure 2.5** *Eobacterium isolatum,* found abundantly in a 3200-million-year-old rock formation in South Africa, is the oldest life-form yet discovered. It is quite similar to some modern bacteria. In this view, *Eobacterium* is shown about 80,000 times its natural size.

## The Legacy of Life: A 4000-Million-Year-Old Tradition

Now that we have a time frame within which to view Earth's long history, we can begin to look at the pageant of life and its dynamic evolutionary response to the changing environments of the planet's surface. Sketchy as the fossil record is, it is nevertheless complete enough to give us a picture of the course of biological evolution that is far more complete than what could have been pieced together by looking at living species alone.

Still, it is worth keeping in mind that the circumstances leading to the preservation of evidence of a living organism in sedimentary rocks are exceedingly rare. To avoid decay, the organism must be buried as it dies, or very shortly thereafter. It must also be kept constantly wet and totally isolated from oxygen until its perishable substance has been completely replaced by some chemically stable mineral, such as quartz or calcite, dissolved in circulating groundwater. Least affected by these restrictions, and therefore most commonly found, are *trace fossils,* including tracks, trails, and burrows in sand and mud. Less common are *hard body parts,* such as shells and skeletons, and least common of all are actual organic remains.

### The Oldest Fossils

Among the earliest convincing evidences of life are the fossilized remains of a minute creature found in the 3200-million-year-old Fig Tree cherts in South Africa by Harvard geologist Elso S. Barghoorn, who named it *Eobacterium isolatum* ("Isolated dawn bacterium," fig. 2.5). *Eobacterium* is similar in appearance to some

modern bacteria. Barghoorn also found another organism in the Fig Tree rocks that resembles some modern species of *cyanobacteria* (formerly called blue-green algae), one-celled, photosynthetic organisms that often give a bluish or greenish cast to ponds, streams, soil, bark, and even snowbanks. He named that organism *Archaeosphaeroides barbertonensis* ("Ancient spherelike of Barberton").

Certainly, neither *Eobacterium* nor *Archaeosphaeroides* was Earth's earliest life-form. These are just the earliest ones we have found so far. Probably, life had already been well-established for several hundred million years before these ancient fossils were accidentally preserved in the Fig Tree rocks. One thing we know for certain about the environment in which the earliest life-forms originated is that it must have been virtually free of oxygen. Otherwise, oxygen would have quickly attacked and destroyed the primitive organic molecules of which they were made. Only after the process of photosynthesis evolved could free oxygen begin to accumulate in the atmosphere, and only then could organisms begin to evolve defense mechanisms against the corrosive effects of that deadly gas. Before then, the earliest life-forms probably used nitrates dissolved in seawater in place of oxygen to break down the organic molecules on which they undoubtedly fed.

These primeval organisms that first graced our planet's young ocean were probably not much more than simple, thin, spherical shells of protein. They were among the earliest and the simplest of living systems, consuming high-energy organic molecules in their shallow marine environment and draining from them enough energy to power a very limited range of life functions—mainly the synthesis of protein. Experiments have shown that such molecules can be produced by the action of lightning and UV radiation on the gases that probably made up Earth's primitive atmosphere (mainly carbon dioxide, $CO_2$; nitrogen, $N_2$; and water vapor, $H_2O$). The organisms would likely have consumed these organic molecules faster than they could be produced—a situation analogous to the draining of water from a bathtub at a rate that exceeds the input from the spigot. If this process had continued for long, it would quickly have emptied the larder.

If such a crisis actually occurred, we have no record of it, but we do know that at some point a solution to the dilemma presented itself in the form of cyanobacteria and other photosynthetic organisms. These organisms took over from lightning and UV radiation the task of synthesizing organic molecules, and they did a far more efficient job of it. The main reason they were able to do so was that they were the first representatives of a new, highly efficient system that has characterized all living beings since. That system, the biological **cell,** consists of a protective membrane enclosing all the es-

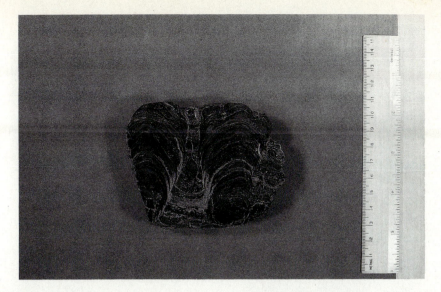

**Figure 2.6** The thinly laminated structure of this limestone stromatolite represents daily growth layers in a cyanobacterial mat that grew along an ancient shoreline. Calcium carbonate, a by-product of photosynthesis, is trapped and bound among the filaments of the cyanobacteria, producing the structure.

Specimen courtesy Department of Earth Sciences, Dartmouth College.

sentials of life: genetic material and mechanisms for food storage, energy production (in plants), energy transfer, protein synthesis, cell division, and a variety of other essential functions.

*Eobacterium, Archaeosphaeroides,* and all of the many other members of this pioneering team of primitive, one-celled organisms, belong in a category of lifeforms that is fundamentally different from all others. Among living organisms, this group, called the *prokaryotes* (pro-*carry*-oats; Greek for "before the nucleus"), is represented only by modern bacteria and cyanobacteria. **Prokaryotic cells** differ from those that compose the rest of the biosphere mainly in the following ways:

1. Their vague to nonexistent cell nuclei are not enclosed by a nuclear membrane.
2. They lack *organelles,* tiny bodies (present in all more advanced cells) that localize such functions as photosynthesis, respiration, food storage, and protein synthesis.
3. Their genetic material is organized into only one hoop-shaped chromosome instead of into several paired, linear chromosomes as in more advanced cells.
4. They reproduce mainly by simple *fission,* or cell division.

Because they do not reproduce sexually, these organisms are relatively insensitive to the kinds of environmental changes that would cause sexual organisms to evolve through natural selection. For this reason, many modern bacteria and cyanobacteria are virtually indistinguishable from their Archaeozoic forebearers. Of the approximately 1.26 million known biological species, only about 3100, or 0.25%, are made up of this kind of cell. Nonetheless, the prokaryotes actually make up well over half of Earth's total biomass!

## The Oxygen Crisis

The rise of photosynthetic ability in organisms is recorded in rocks in several ways. Traces of certain organic chemicals that could only be decomposition products of chlorophyll are present in some *Archaean* rocks (rocks of Archaeozoic age). Other less direct lines of evidence also indicate that photosynthesis was well under way at the beginning of the Archaeozoic Era. Peculiar, laminated structures called **stromatolites** (fig. 2.6) in limestones as old as 3500 million years, are identical with structures being formed today by colonial, photosynthesizing cyanobacteria (fig. 2.7).

Evidence for the first appearance of atmospheric oxygen, one of the products of photosynthesis, has been found in a spectacular, gray and red banded marine rock type known as **banded iron formation** (fig. 2.8). Between 2600 and 1800 million years ago, great thicknesses of this rock type were chemically precipitated from clear, shallow seawater as alternating, thin, gray bands of jellylike *silica,* or silicon dioxide ($SiO_2$) and dark red bands of a variety of iron minerals, mostly iron oxides.

**Figure 2.7** Few places remain on Earth today where cyanobacteria can escape grazing by voracious snails long enough to produce stromatolites. One of these places is Shark Bay, Western Australia, where the seawater is too salty for snails to survive.

**Figure 2.8** Thick deposits of banded iron formation, composed of alternating layers of chert (silicon dioxide) and iron oxides, were formed in shallow marine waters from 2600 to 1800 million years ago. They apparently formed as photosynthetic marine microorganisms disposed of their poisonous oxygen wastes by oxidizing soluble iron in the seawater to insoluble iron oxides.

Carla Montgomery.

For iron oxides to precipitate from seawater, two requirements must be met. First, there must be iron compounds dissolved in the water, and second, those compounds must come in contact with free oxygen. This raises a major problem, however. The only reasonable sources for the vast quantities of iron present in the iron formations would have been either submarine volcanoes or the weathering of vast areas of continental rocks. Both of those sources would have been located far from the shallow marine waters in which the iron formations were deposited, and the iron would have had to be transported over great distances to reach those waters. The problem is that if there had been more than the slightest trace of oxygen in the atmosphere at that time, it would have combined with dissolved iron and converted it to insoluble iron oxide as fast as it was released by weathering or volcanism. Therefore, that iron would never have been able to remain dissolved in the water long enough to reach the sites where the iron formation was deposited. This implies that at the time, Earth's atmosphere must have been *virtually free of oxygen!*

What, then, caused the iron to be deposited as iron formation on the seafloor? Some *local* source of oxygen must have been available. The culprits most probably were microscopic fossils that have been found scattered abundantly throughout the silica bands in the iron formations. If these primitive organisms were photosynthetic, they would have released oxygen into the ocean water as a byproduct of photosynthesis (fig. 2.9*a*). Recall that oxygen is exceedingly reactive and would have been capable of destroying the very organisms that created it. If there had been soluble iron compounds in the ocean water, however, the oxygen would have attacked the iron more readily than it would have attacked the organisms, thereby generating the red iron oxides that make up the iron formations. This suggests that the presence of soluble iron compounds in seawater could have been an essential prerequisite for the evolution of photosynthesis because it would have provided a mechanism for the immediate disposal of poisonous oxygen wastes.

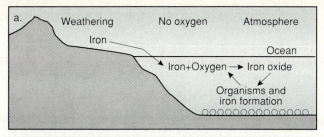

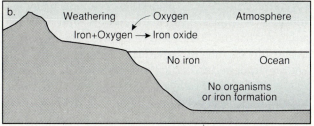

**Figure 2.9** As long as Earth's atmosphere contained no oxygen (*a*), iron compounds weathered from continental rocks could be dissolved in river water and transported to the sea, where they combined with oxygen wastes from photosynthetic organisms, producing the iron oxide for iron formation. After oxygen entered the atmosphere (*b*), iron compounds released by weathering combined with oxygen at the weathering site, and could no longer be transported to the sea in solution.

Since about 1800 million years ago, however, little iron formation has been laid down. Why? The logical answer is that by then, free oxygen had begun to accumulate in the atmosphere and ocean, so soluble iron compounds were oxidized as rapidly as they were released by weathering or volcanic activity (fig. 2.9*b*). Other evidence supports this conclusion. At about this time, we begin to find the same red color of the marine iron formations suddenly showing up in **red beds:** deposits of reddish river sand and mud laid down on the continental surfaces and in shallow marine water (fig. 2.10). In other words, instead of being carried to the sea in soluble form, the iron released by the weathering of terrestrial rocks was, for the first time, combining immediately with atmospheric oxygen at the weathering site.

The pigmenting power of red iron oxide (the mineral *hematite,* see fig. 3.17*b*) is so great that geologists generally regard its presence in red beds as an indication that the rocks were exposed to free oxygen at the time of their formation. Sediments that were laid down in stagnant, oxygen-free environments, on the other hand, can normally be recognized by their gray to green hues.

**Figure 2.10** Red beds (*foreground*) and gypsum (*background*), first deposited about 1800 million years ago, are evidence that by that time enough oxygen had accumulated in the atmosphere to oxidize iron- and sulfur-bearing minerals during the weathering of continental rocks. Because water is unable to leach the insoluble iron oxides from terrestrial sediments, they remain in place, giving the red beds their characteristic color. Gypsum, on the other hand, is soluble, and forms by the evaporation of seawater.

A further indicator of the buildup of free oxygen in the atmosphere is a sedimentary rock type called *gypsum rock* (fig. 2.10). This consists of hydrous (water-bearing) calcium sulfate ($CaSO_4 \cdot 2H_2O$, the mineral *gypsum;* see fig. 3.17*i*). In this chemical compound, the element sulfur (S) is as fully combined with oxygen as it can possibly be (as iron is in hematite). In Earth's interior, by contrast, both iron and sulfur are in a highly *reduced* state (the opposite of oxidized), as is iron. Prior to about 1800 million years ago, no gypsum rock formed because there was no oxygen in the atmosphere to oxidize volcanic sulfur.

The end of iron formation deposition appears, then, to mark a serious crisis in organic evolution. With no more soluble iron in seawater, how could organisms cope with the problem of disposing of deadly oxygen? The answer is that some organisms evolved *enzymes* that render oxygen noncorrosive as it enters and leaves a living cell. Presumably, those organisms that failed to evolve such oxygen-mediating enzymes either did not survive the oxygen crisis, or they retreated to oxygen-free environments, such as swamps, bogs, stagnant marine basins, and some lake bottoms. Prokaryotic organisms that can not survive exposure to free oxygen can still be found abundantly in such environments.

## The First Advanced Cells

In marked contrast to the prokaryotic cell, the **eukaryotic** (you-carry-*ot*-ic; greek for "good nucleus") **cell** has a well-organized **cell nucleus** that contains several pairs of *linear chromosomes* and is enclosed by a *nuclear membrane.* It also contains various organelles.

Exactly when in geologic history the first eukaryotic cells evolved is as hard to pin down as is the emergence of prokaryotic cells, but there are some microorganisms in the 2000-million-year-old Gunflint Chert of Ontario, Canada, that appear to be eukaryotic. Certainly, by 1000 million years ago, the threshold had been passed, inasmuch as true algae, whose cells are eukaryotic, have been found in Australian cherts of that approximate age.

The advantage of the eukaryotic cell is its adaptability. In 3800 million years of evolution, the asexual prokaryotes have produced less than one new species per million years. Such extreme evolutionary conservatism can be an advantage in stable, unchanging environments, but as we have seen, such environments are more characteristic of the Moon than of Earth. Because of its potential for rapid evolutionary change, the eukaryotic cell is much better suited to our highly change-

able Earthly conditions. In the next section, I discuss the reasons why this is so, but first we might consider how the eukaryotic pattern came to be established. A logical first question might be, "How did the organelles get into the cell?"

The fossil record could be too sketchy ever to give us enough evidence to answer that question with confidence, but a compelling hypothesis for eukaryote evolution has recently gained favor among many evolutionary biologists. Lynn Margulis of Boston University suggests that the eukaryotic cell could have developed as a result of small cells invading and living parasitically within larger cells. Gradually, what began as a situation that was of benefit only to the parasite became one of mutual benefit as the host cell came to realize advantages from certain life functions of the parasites. In time, the distinction between parasite and host became blurred. The parasite lost its identity as an independent system, and became a functioning subsystem of the host, losing all but that one special function (e.g., photosynthesis, energy production, protein synthesis) that gave it its particular usefulness to the larger cell. In the final analysis, the ultimate winner in this competition was not the invading parasite, but the invaded host, which managed subtly and gradually to turn an initially threatening situation to its own advantage.

### *Sex and Selection: The Grand Strategy*

If Margulis's hypothesis is correct, the earliest species of eukaryotic cell presumably responded to its useful invaders by wrapping up its own genetic material—the *chromosomes* and their protein-synthesizing instructions, the *genes*—in a nuclear membrane and drawing it into an isolated area where it could not be harmed by the newcomers. This, then, became the cell nucleus, which soon developed the ability to subordinate and control even the genetic material of the organelles.

As with the prokaryotes, eukaryotic cells continued to reproduce themselves by simple fission, resulting in two new cells from a single parent, but now the process had become more involved. It began with, and was dominated by, division of the nucleus, which was essentially complete before the division of the cell body occurred. This process is known as **mitosis** (my-*toe*-sis). Somewhere along the line, a further complication arose: an elaborate partitioning of genetic material known as *reduction division,* or **meiosis** (my-*oh*-sis; fig. 2.11). This results in the production of not two new cells from one parent but four, each with only half the number ($n$) of chromosomes that were present in the parent cell ($2n$). Such cells are called *spores.*

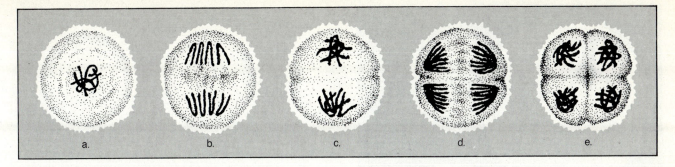

**Figure 2.11** Meiosis involves two sequential cell divisions. In *a,* each chromosome (dark lines in nucleus) has replicated itself. Because there were originally two of each chromosome (*2n*), the result is 4 of each (*4n*). In *b,* two of each chromosome migrate into either half of the cell. The cell then divides (*c*), resulting in two cells, each with *2n* chromosomes. In *d,* one of each chromosome migrates into either half of both cells, which then divide (*e*), resulting in four spore cells, each with *n* chromosomes. When two spore cells recombine, they form an embryo with *2n* chromosomes.

In the cells of the parent organism, chromosomes exist in pairs. Just prior to meiosis, each chromosome replicates itself, resulting in four identical copies of each (*4n*). During meiosis, these are evenly apportioned; hence, all four of the new spore cells are identical, and each contains half the number of chromosomes that its parent had before replication (*n*). These new spore cells then develop either into sex cells, as they do in animals (fig. 2.12*c*), or into adult organisms, which then produce sex cells, as they do in most plants and *protists* (sexually reproducing one-celled organisms; fig. 2.12*b* and *d*). Two such sex cells, preferably from different parents, pair off and fuse together in the process of *fertilization.* The result of this process is a fertilized cell that contains the original number of chromosomes (*2n*), but half of these are from one parent and the other half are from the other parent. This cell then divides repeatedly by mitosis and develops into an *embryo,* which, in turn (if all goes well), grows into an adult organism.

Meiosis, therefore, prevents the number of chromosomes from doubling with each new generation as a result of the process of fertilization. Aside from that essential service, the other principal function of meiosis is to promote *genetic diversity* when sex cells combine to form embryos. Genes determine an organism's physical characteristics. The genes in the chromosomes of one individual are usually slightly different from those of another individual. The continual recombination of different genes in the sexual cycle maintains a variety of genetic types in the overall population of each sexually reproducing species. This can be viewed as a "strategy" used by the species to ensure its survival in the event of environmental change. The chances are fairly good that whatever change might occur, there will be *some* individuals in the population whose genetics are ideally suited to the new conditions. The ultimate irony in this unplanned "strategy" is that because the price of survival is flexibility, the best deal a species can get is that it will escape extinction only to engender a new and different species!

The process of **natural selection** is shown by a simple example in figure 2.13. For each curve (known as a *normal distribution curve* or bell curve), the height of a particular segment above the baseline corresponds to the percentage of individuals in a population of a certain plant species. Annual precipitation increases to the right along the baseline. Each of the letters *A* through *L* refers to a particular genetic type of individual, or *genotype,* in the population.

In the uppermost bell curve, the precipitation is 44 cm. Because the genotype *D* in the plant population grows best under an annual precipitation of 44 cm, it dominates the population. Genotypes *B* and *F,* which grow best under 36 cm and 52 cm precipitation, respectively, are unable to compete successfully with genotype *D.* As a result, there are very few of these two genotypes in the population. Now, if the annual precipitation of this area should suddenly increase to 54 cm, as it has in the second curve from the top, genotypes *F* and *G* would be best adapted to the new conditions and would become dominant at the expense of genotype *D.* At the same time, genetic mutations would occur that would produce genotypes *H, I,* and *J.* These might have been produced before but were unable to survive under the lower precipitation regime.

If yet another increase in annual precipitation should occur, as in the third curve from the top, these new genotypes would in turn become dominant, whereas the original dominant, *D,* would by then have been eliminated from the population. Finally, if an environmental shift should occur rapidly to well beyond either tail of the bell curve, there might be no genotypes in the population that could tolerate the new regime. In this case, the population would become extinct, as on the bottom line of the diagram, in which the precipitation has suddenly dropped to 32 cm.

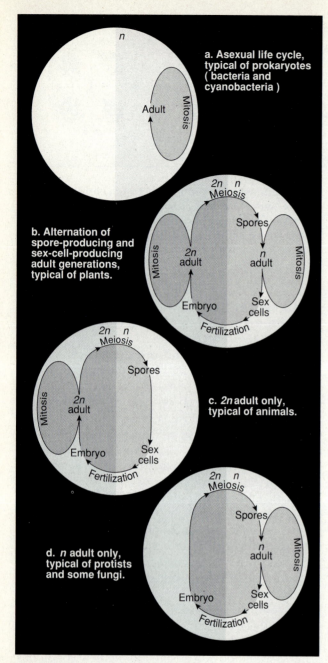

**Figure 2.12** Life cycles of organisms. Right sides of diagrams represent organisms with *n* chromosomes per cell; left sides represent organisms with 2*n* chromosomes per cell. Large central oval represents sexual reproduction (meiosis); smaller, side ovals represent growth by cell division (mitosis). In prokaryotes (*a*), sexual reproduction is absent; therefore, only the *n* generation is represented. In most plants (*b*), the 2*n* adult plants (herbs, shrubs, trees) are large, whereas the *n* adult plants are microscopic, but in some (e.g., mosses), this is reversed. In animals (*c*) there is no *n* adult phase, and in some protists and fungi (*d*), there is no 2*n* adult phase.

Within the life-cycle figure:

a. Asexual life cycle, typical of prokaryotes ( bacteria and cyanobacteria )

b. Alternation of spore-producing and sex-cell-producing adult generations, typical of plants.

c. 2*n* adult only, typical of animals.

d. *n* adult only, typical of protists and some fungi.

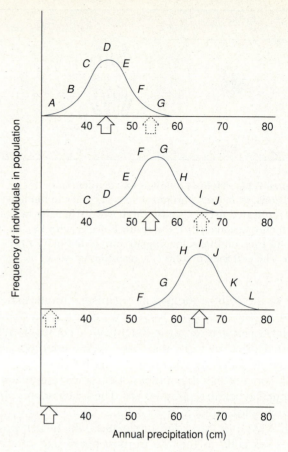

**Figure 2.13** In natural selection, a particular set of prevailing environmental conditions (*solid arrows*) favors certain genotypes (*A–L*) in a species population over others; hence, these dominate the population, as shown by the peaks of the bell curves. Shifts in environment (*dashed arrows*) result in corresponding shifts in genotype or in extinction of the population if environment shifts too far, too fast (*bottom*).

If this same large shift had occurred gradually, however, genotypes *A* and *B,* present in the original population but eliminated in the second and third, might have reappeared. For each species, many such genetic responses to environmental factors are at work in the same way, and each could be represented by its own bell curve. Figure 2.13, in other words, is simply a graphic illustration of a single component of Nature's ever-changing states of dynamic perfection.

# E Pluribus Unum:
# Many-Celled Organisms

Even in the most ancient sedimentary rocks, over 3500 million years old, we have found evidence of the tendency for one-celled organisms to cluster together and form colonies. This tendency has found varying degrees of expression. Cyanobacteria can exist as solitary cells or as long filaments of cells sharing adjacent cell

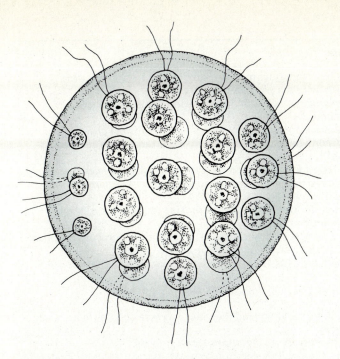

**Figure 2.14** *Pleodorina* is a primitive, many-celled organism in which the component cells have become sufficiently interdependent and functionally specialized that neither the colony nor the individual cells can survive if the cells are separated.

walls. Even in such colonial arrangements, however, the individual cells are functionally independent. In other cases, such as that of the green alga *Pleodorina* (fig. 2.14), the 32 to 128 cells of the colony are definitely interdependent and are unable to survive on their own if removed from the colony. Furthermore, the various cells are somewhat *functionally differentiated,* or altered to serve specific functions within the organism. The ones in the front of the colony are capable only of simple mitotic division, whereas those in the rear are much larger and are capable of both mitosis and meiosis. Human beings represent the extreme of this evolutionary trend in which no individual cell has an independent identity, and all cells have exchanged their ability to survive on their own for the security of being integral parts of a larger organism.

This trend toward greater organic complexity has manifested itself in five essential ways:

1. An increase in the number of component cells in the organism
2. An increase in the functional differentiation of cells
3. A consequent increase in mutual interdependency between the larger organism and its component cells
4. An increase in the subordination of individual cell function to the needs of the larger organism
5. A consequent increase in coordination of function among the cells

In the fossil record, we find that the trend toward greater complexity began slowly and made little progress until about 700 million years ago, when a sudden explosion of highly coordinated, many-celled life-forms began (see fig. 2.2). According to some, that explosion was made possible by a rapid increase in atmospheric oxygen at about that time. It is also possible that it was facilitated when one of the many-celled organisms—probably a snail—proved to be a highly efficient grazer of algae and cyanobacteria, which had until then been the dominant life-forms on Earth.

## The Strategies of Survival

As life advanced from one-celled to many-celled organisms, it also began to move into new environments. At first, the simple prokaryotes inhabited only the near-surface ocean waters and the shallow seafloors near shore. In time, organisms began to invade other environments: the deeper ocean waters; the deeper seafloors; and eventually the land. There were certain requirements that had to be met before each of these invasions could occur. Before the deeper waters and the seafloors beyond the shallow continental shelves could be inhabited, they would have to acquire enough dissolved oxygen to support the new life-forms. Before life could invade the land, there would have to be enough oxygen in the atmosphere to create a protective ozone shield and to sustain the respiratory needs of complex, many-celled animals.

Taken together, these requirements accord well with the evidence, already discussed, that oxygen began to accumulate slowly in the atmosphere about 2000 million to 1800 million years ago and that it continued to increase slowly for about another 1000 million years. By about 800 million years ago, there would have been enough oxygen in the atmosphere to create an ozone shield strong enough to allow a rapid increase in marine plants and a correspondingly rapid increase in oxygen production. That, in turn, would have set the stage for the sudden burst in evolution that appears in the fossil record at this time. This is a good example of a *threshold effect,* the principle whereby no water spills from a glass until the water level rises to the brim. Similarly, the great burst in evolution on Earth had to await the generation of enough oxygen to produce enough of an ozone shield to make the sea surface waters and the land habitable.

A number of new evolutionary strategies accompanied this evolutionary burst. Table 2.3 lists some of these and some are illustrated in figures 2.15 and 2.19. In each case, the new development permitted the irreversible expansion of life into previously unoccupied regions of Earth's surface. *Shells,* for example, provided several benefits: protection from predators and UV radiation; structural support for larger bodies and the development of muscles; and weight to keep the organisms on the seafloor. The earliest many-celled organisms,

**Table 2.3** Evolutionary Strategies through Time

| Strategy | Era or Period | Functions |
|---|---|---|
| Prokaryotic cell | Archaeozoic | Provides basis for all living organisms. Encloses self-replicating protoplasm within a protective membrane. |
| Eukaryotic cell | Proterozoic | Increased range of cell functions. Provides basis for sexual reproduction. |
| Sexual reproduction | Proterozoic | Provides means for species to adapt to environmental change. |
| Multicellularity | Ediacarian? | Low surface/volume ratio (conserves heat). Large size (discourages predation). |
| Digestive system | Ediacarian | Renders large food particles more easily assimilable by cells. |
| Circulatory system | Ediacarian | Supplies oxygen, nutrients, and hormones to cells. Removes carbon dioxide from cells. |
| Respiratory system | Ediacarian | Oxygen transfer into body. Carbon dioxide transfer out of body. |
| Central nervous system | Ediacarian | Rapid response to stimuli. |
| Shell | Cambrian | Protection against storms, predators, and radiation. Structural support and muscle attachment. Antiflotation device. |
| Articulated external skeleton | Cambrian | Body armor. Structural support and muscular control of appendages. |
| Articulated internal skeleton | Cambrian | Structural support and muscular control of appendages. |
| Root system | Silurian | Allows firm foothold on land. Provides access to soil nutrients, water, and oxygen. |
| Shoot system | Silurian | Exposes photosynthesizing organs (leaves) to Sun. |
| Vascular tissue | Silurian | Structural support. Translocation of water, nutrients, and hormones. |
| Seed | Devonian | Protection of embryo against predation, dessication, and fire. Dispersal and propagation. |
| Internal fertilization | Pennsylvanian | Permits sexual reproduction out of water. |
| Amniote egg | Pennsylvanian | Protects embryo against dessication. |
| Flowers | Cretaceous | Exploitation of insects for pollination. |
| Fruit | Cretaceous | Dispersal and propagation of seeds. |
| Grass | Cretaceous | Heavy ground cover reduces erosion. Growth at leaf base permits grazing. |
| Intelligence | Quaternary | Permits manipulation of natural systems to advantage on the basis of understanding how they work. |

living about 700 million years ago, lacked shells, but by 250 million years later, shells were in vogue and were sported by members of most of the ten major subdivisions of the animal kingdom.

Because shells are often preserved in sedimentary rocks, the fossil record suddenly becomes much more complete in the Cambrian, the second period of the Paleozoic Era. Late in the Cambrian, primitive fish of the vertebrate phylum evolved *jointed internal skeletons*, which provided both greater body flexibility and the means for precise muscular control of appendages. Before this, in the early Cambrian, a class of bottom-dwelling *arthropods*, the *trilobites* (*try*-luh-bites, Greek for "three lobed," fig. 2.16), evolved jointed *external skeletons*, which serve the same functions as the internal vertebrate skeleton and are found today in such modern arthropods as insects, spiders, and crabs.

Another important feature of both internal and external jointed skeletons is that they provide a means of locomotion on land, but the first spiderlike arthropod to invade the land did so more than 150 million years *after* the trilobites first provided the necessary structure. This kind of early-appearing structure that later comes to serve a different function in a new environment is known as a **preadaptation.** In the Late Devonian, some 50 million years after the appearance of the spiderlike arthropod, the first vertebrate amphibians used their preadapted internal skeletons to help them crawl out of the sea and onto the land (fig. 2.17).

The *central nervous system* is a strategy that allows multicellular animals to respond almost instantly, and with appropriate muscular actions, to environmental stimuli, such as the presence of danger or food. *Circulatory systems,* using fluids (e.g., sap and blood) as transport media, were evolved by both plants and animals to deliver oxygen, nutrients, and hormones to the various parts of the body, and to remove carbon dioxide and other waste products. In one-celled organisms, in which such deliveries and removals can take place through the cell membrane, circulatory systems are unnecessary. *Respiratory systems* convey oxygen into the interiors of many-celled animals and expel carbon dioxide. Aquatic animals usually accomplish this with semiexternal *gills,* whereas terrestrial animals accomplish it with *lungs* (in vertebrates) or with much-branched *tracheae* (*tray*-kih-ee, Greek for "rough"; in insects and some other arthropods). Tracheal systems

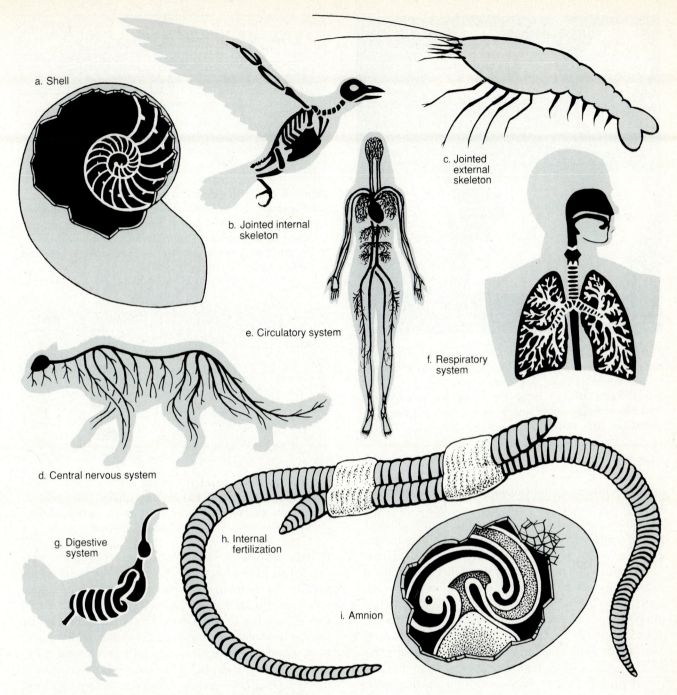

a. Shell

b. Jointed internal skeleton

c. Jointed external skeleton

d. Central nervous system

e. Circulatory system

f. Respiratory system

g. Digestive system

h. Internal fertilization

i. Amnion

***Figure 2.15*** **Some important evolutionary strategies in the animal kingdom.**

deliver oxygen directly to the body cells, whereas lungs deliver it to the circulatory system which then delivers it to the cells. Both lungs and tracheae were necessary preadaptations to terrestrial life because gills would have quickly dried out if exposed to the air. Those land animals (e.g., crabs and isopod crustaceans or "sow bugs") that do breathe through gills must remain in moist places, and they have evolved protective structures for their gills that prevent their drying out.

*Digestive systems* enable animals to capture, ingest, and digest food particles much larger than the cells of their bodies. Some primitive digestive systems have only one opening that serves as both mouth and anus, whereas other, more advanced systems have two. Most have some form of storage unit, such as a *crop* (birds, earthworms) or *stomach* (vertebrates, insects, etc.) and devices, such as *teeth* (most vertebrates) or *gizzards*

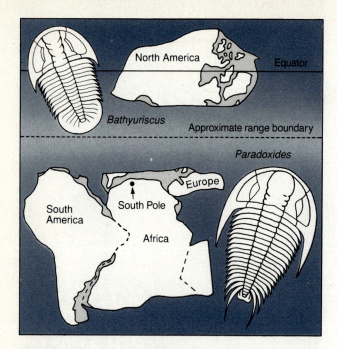

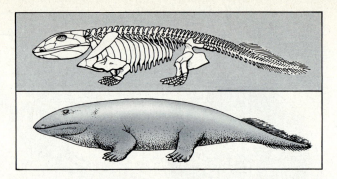

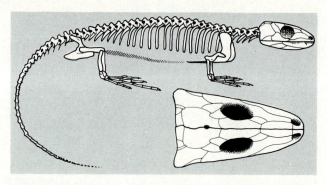

**Figure 2.17** The oldest known amphibian, *Ichthyostega,* whose remains have been found in Upper Devonian sandstones of East Greenland, retains many fishlike characteristics, including a finned tail, weak limbs, and fishlike skull and teeth. *Ichthyostega* measured about one meter in length.

**Figure 2.16** Two Middle Cambrian trilobites. Tiny, half-inch *Bathyuriscus* (*above*), lived in warm, equatorial waters, while its giant, foot-long contemporary, *Paradoxides* (*below*), lived in cold, subantarctic waters. This difference is one of several indications that the continents have shifted over geologic time.

**Figure 2.18** Representative of the earliest reptiles, half-meter-long *Hylonomus* was an unspecialized form with a primitive reptilian skull lacking the openings behind the eyes that characterize more advanced reptiles.

(birds, earthworms), for breaking up food. Others rely on piercing and the sucking of fluids (e.g., most spiders, true bugs, and some flies, including mosquitoes).

*Internal fertilization* and the *amniote egg* are two strategies, as are lungs and tracheae, without which life on land would be practically impossible. Marine plants and animals typically release vast quantities of sex cells directly into the ocean water, where some of them join and become fertile. A few of the resulting embryos survive to grow into adults. Land plants and animals, however, lack a watery medium into which they can disperse their sex cells, so they require strategies to prevent them from drying up. Direct insertion of sperm into the female body was the solution developed by the earliest reptiles in the Lower Pennsylvanian Period (fig. 2.18). The other critical strategy that made it possible for the reptiles to be the first vertebrates to become entirely independent of the sea was a membrane within the egg, the *amnion,* that encloses the embryo in a watery medium during its development. A somewhat later strategy was the retention of the embryo within the female's body until its development is complete, as in most mammals.

Plant strategies for living on land are equally ingenious (fig. 2.19). The development of a branching *root system* allowed many-celled plants to gain a firm hold

in rock and soil. Above ground, the *shoot system* provided a means for exposing their photosynthesizing tissues to the Sun. In most modern land plants, those tissues are located in specialized organs called *leaves* (fig. 2.19*c*), but in the earliest land plants, such as the leafless Devonian *Rhynia* (fig. 2.20) and in modern cacti, photosynthesis occurs in the plant's green stem. Land plants evolved *vascular tissue,* bundles of elongated, tubelike cells that strengthen the plant against wind and gravity and conduct fluids. There are two kinds of vascular tissue (see fig. 2.19*b*). The inner *xylem* raises water and nutrients from the soil to the leaves, whereas the outer *phloem* distributes food and hormones from the leaves to other parts of the plant. Cells on leaf surfaces evolved waxy coatings and other defenses against drying out (see fig. 2.19*c*). Some plants, such as desert ocotillos and sugar maples, developed the strategy of losing their leaves during extended periods of very dry or very cold weather.

Coniferous trees and flowering plants evolved a unique reproductive strategy. By meiosis, the $2n$ adult plant produces spores, which grow into microscopic,

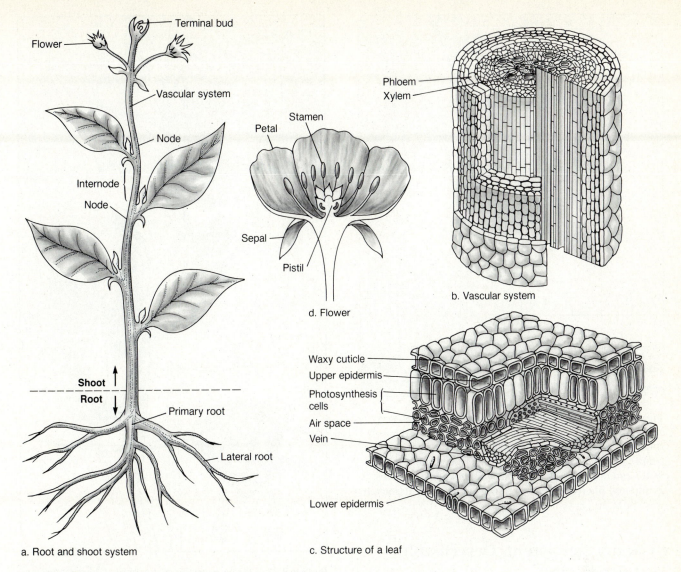

**Flower**

**Terminal bud**

**Vascular system**

**Node**

**Internode**

**Node**

**Shoot**

**Root**

**Primary root**

**Lateral root**

a. Root and shoot system

**Petal**

**Stamen**

**Sepal**

**Pistil**

d. Flower

**Phloem**

**Xylem**

b. Vascular system

**Waxy cuticle**

**Upper epidermis**

**Photosynthesis cells**

**Air space**

**Vein**

**Lower epidermis**

c. Structure of a leaf

***Figure 2.19*** Some important evolutionary strategies in the plant kingdom.

few-celled, *n* adult "sex plants" specialized for sexual reproduction. The female sex plants remain attached to their parent plants, either beneath the scales of cones or at the bases of the central pistils of flowers. The male sex plants grow into four-celled, tough-skinned *pollen grains* that are released from the parent plant and carried by winds or insects to female sex plants of the same species. On contacting the female sex plant, the pollen grains grow long tubes that penetrate and fertilize egg cells within them, thus forming embryos, which develop into *seeds*. In many species, the seeds become surrounded by a fleshy, usually edible, expansion of the ovary called a *fruit*. The consumption of fruit by animals and the subsequent voiding of the indigestible seeds is an effective strategy for plant propagation. Other plants rely on the wind to disperse their seeds. Certain flower parts attached to such seeds develop into wings (as in maples, ashes, spruce, etc.) or parachutes (as in dandelions, salsifys, etc.). If buried in the ground, seeds grow into asexual $2n$ adult plants which then produce sexual spores by meiosis (see fig. 2.12*b*). As in land animals, specialized, protective structures prevent the sex cells of land plants from drying out.

In some cases, the adaptive strategies of plants verge on the bizarre. There are a few species of orchid, for example, each of which so closely resembles the female of a certain species of bee that the male bee attempts to mate with the flower, and thereby pollinates it (fig. 2.21)!

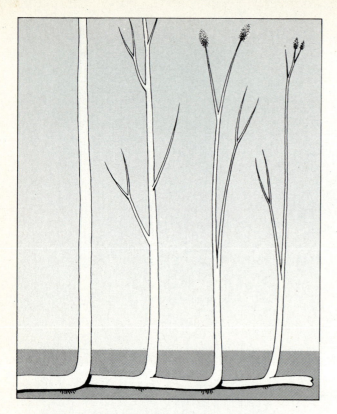

*Figure 2.20* *Rhynia,* a Middle Devonian species, was among the earliest land plants. An underground stem, or rhizome, served in place of roots for support, and a simple vascular system permitted upward transport of water and nutrients. Because *Rhynia* lacked true leaves, it conducted photosynthesis within its green stem tissues. It was about 20 cm tall.

## Systems: Persistent, Describable, and Predictable

As the repeated use of the term throughout this chapter should suggest, Earth science deals exclusively with *systems.* Because many of the phenomena described in the pages that follow are discussed in systemic terms, we should briefly consider the characteristics of natural systems. There are many definitions of the term, but for the purposes of this text, I define a **system** as *any persistent, describable, and predictable arrangement of matter, energy, or both.* It is easy to see that there is very little in Nature to which this definition does not apply. Although some would disagree, even a phenomenon as fickle as Earth's weather (see chapters 13–15) qualifies, as it, too, is at least moderately persistent, describable, and predictable. Indeed, we refer to *weather systems* on the strength of their displaying these qualities. In fact, if challenged to identify something that is *not* persistent, describable, and predictable, we would be hard put to come up with anything other than human imagination!

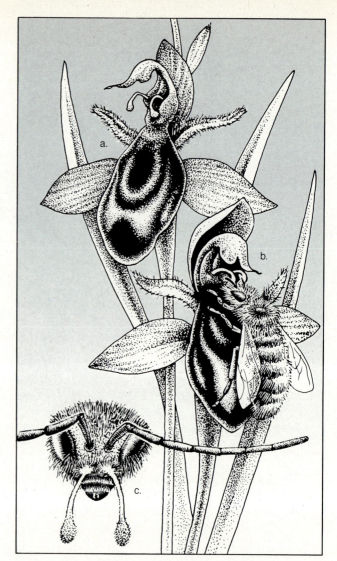

*Figure 2.21* Flowers of the bee orchid, *Ophrys apifera,* (*a*) mimic the female of one species of bee so effectively that male bees, emerging earlier in the season than the females, repeatedly attempt to mate with them (*b*). In the process, pollen sacs become glued to the bees' heads (*c*), and these are then transferred to other flowers during subsequent mating attempts. In this symbiotic relationship, only one species benefits, a condition called commensalism.

Furthermore, it is abundantly clear that not only is "no man an island," as John Donne observed, but no *system,* whether man, woman, bear, tree, brook, or boulder, exists in isolation. All of Nature's systems *co-evolved* over the 4600 million years of geologic time, and therefore, they are all *interactive.* Furthermore, they are also *hierarchical,* or nested one within another, as atoms nest within molecules (see chapter 3), and molecules nest within plant and animal cells, cells within tissues, tissues within organs, organs within organisms, and organisms within ecosystems. Because of this hierarchical, interactive network, Earth itself is one great system in which every part is an interactive subsystem that must behave in an appropriately persistent, describable, and predictable fashion if the greater Earth system is to retain its own persistence, describability, and predictability.

This vital requirement can be stated in the form of a generalization that we might call the **consistency principle:** *if it is to persist, any new system that arises on Earth must be consistent with what has gone before.* This is certainly comforting, because it automatically denies existence to such inconsistent tribulations as pink grass, flying rhinoceroses, and eight-legged rocks. Nature is an orderly affair, and we can be thankful for that! Nonetheless, like Alice, the human species seems to be doing its best to create a Wonderland out of Nature's orderly system by creating many new, *inconsistent* subsystems and finding ways to allow them to persist in spite of Nature's best efforts to eliminate them. I make note of a number of these throughout this book.

## Matter and Energy: The Stuff of Systems

Early in this century, Albert Einstein convinced us, with his well-known equation, $E = mc^2$, that matter and energy are both manifestations of the same thing ($E$ is energy, $m$ is a property of matter called *mass,* and $c$ is the speed of light). In spite of this equivalency of matter and energy, each has certain characteristics not shared by the other. For our purposes, **matter** can be defined as anything that is capable of assuming definite physical form, whereas **energy** is a field, or "presence" capable of changing the motion of matter.

Energy occurs in three fundamental forms: potential, kinetic, and electromagnetic. **Potential energy** is bound to material objects, and it emanates from them in all directions as fields that diminish in strength as the square of the distance from the object. Earth's gravity field is one example of a potential energy field. The electrical field surrounding an electrically charged particle is another, and the magnetic field surrounding a magnet is a third. **Kinetic energy,** also bound to matter, is energy of motion, such as that of a boulder rolling down a mountainside. **Electromagnetic** (or **radiant**) **energy,** in contrast to the other two types, is not bound to matter but is released from it as wavelike pulses whenever electrically charged particles within such matter experience an acceleration, or a change in motion. Electromagnetic energy takes many forms (see fig. 3.1), including radio waves, light waves, and X-rays.

Potential, kinetic, and electromagnetic energy are all completely interconvertible from one to another. In the case of the boulder, before it rolled down the mountainside, it possessed a certain amount of *gravitational potential energy* (GPE) by virtue of its distance from Earth's center, which is also the center of Earth's gravity field. When the boulder finally came to rest in the valley, it possessed somewhat less GPE than before because it had moved somewhat closer to Earth's center. The GPE lost by the boulder in its downhill trip was converted to kinetic energy. That kinetic energy, in turn, was converted to a common variety of electromagnetic energy called *heat* as the boulder repeatedly struck the ground, inducing vigorous acceleration in charged particles within both boulder and ground.

## Isolated, Closed, and Open Systems

There are many varieties of system, of which the simplest is the **isolated system.** A simple example of an isolated system is a bathtub full of water at a constant temperature that has been isolated from its environment by insulating walls (fig. 2.22a). The term *isolated* signifies that neither matter nor electromagnetic energy is either entering or leaving this system.

If we were to remove the insulating walls, the bathtub would become a **closed system,** so called because, like the isolated system, it is closed to matter flowing in or out, but unlike the isolated system, it allows **inputs** and **outputs** of electromagnetic energy (see fig. 2.22b). In any nonisolated system, the presence of such energy inputs and outputs is indicated if the temperature of the system's matter is higher than *absolute zero* ($-273.16°$ C), the coldest temperature possible. The reason for this is that when the atoms of a substance (including air and water) absorb electromagnetic energy, they are forced to vibrate, or oscillate. The phenomenon we call *temperature* is simply a measure of the intensity of such atomic oscillations. These oscillations release electromagnetic energy of their own, which leaves the system as output. A playground swing (fig. 2.23) is another familiar example of a closed system, in which oscillations can only be maintained if they are forced to do so by continual inputs of energy (in this case, kinetic energy).

As long as the rate of input of electromagnetic energy into a nonisolated system remains the same as its rate of output from the system, the system's temperature remains constant, a condition known as **dynamic equilibrium** (fig. 2.24a). *Equilibrium* is a state of no change. *Dynamic* equilibrium is a condition in which the amount of energy in the system remains constant even though energy continues to flow through the system. If the input rate exceeds the output rate, however, equilibrium is disturbed, and the temperature rises (fig. 2.24b). Conversely, if the output rate exceeds the input rate, the temperature falls (fig. 2.24c). Such imbalances between input and output tend to correct themselves, however. For example, if the rate of input of electromagnetic energy into our bathtub system

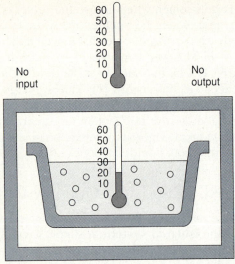

No input

No output

a. Isolated system

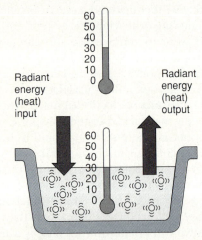

Radiant energy (heat) input

Radiant energy (heat) output

b. Closed system

***Figure 2.22*** The temperature of an isolated system (*a*) can remain different from that of its environment indefinitely because of an isolating barrier preventing exchanges of matter and energy. Truly isolated systems are virtually unknown in Nature. In closed systems (*b*), energy exchange with the environment quickly abolishes any temperature differences between system and environment. The source of the heat output is thermal agitation of molecules caused, in turn, by heat input.

should increase, the oscillation of the water molecules would increase, and so would the temperature. This would increase the rate of output of electromagnetic energy until it equalled the new, increased input rate (fig. 2.24*d*). Similarly, if the input rate should decrease, the molecular oscillation rate and the temperature would also decrease. In turn, this would decrease the output rate until it equalled the new, reduced input rate (fig. 2.24*e*). The ultimate effect of such corrections, then, is *to maintain an equality between rates of input and output.* Systems that behave in this way are called *dynamic equilibrium systems,* or simply **equilibrium systems.**

***Figure 2.23*** **A playground swing is an example of a closed system. Like any pendulum, it must continually be supplied with pulses of energy in order to maintain its oscillating motion against the damping effect of friction with air and bearings.**

Photo of Franchesca Lecesne courtesy of Darius Lecesne

If we now add a spigot and a drain to our bathtub system, thereby allowing water to enter and leave, it becomes an **open system** (fig. 2.25). Let us assume that the tub is empty to begin with, and that we open the spigot gradually. Up to a certain point, water will flow from the drain as fast as it enters from the spigot. As long as this happens, the system really does not exist, as there is no persistence of water within it. Eventually, however, the input rate of water from the spigot will begin to exceed the output capacity of the drain hole, as in figure 2.25*a*. At this point, the system comes into being as water begins to accumulate and persist within the tub. As the water level rises, the accumulating water exerts a pressure on the drain hole, which forces an increase in the output rate.

If we now hold the input rate from the spigot steady, the water level will continue to rise until the pressure becomes sufficient to force water out the drain hole at a rate that is exactly equal to the input rate (fig. 2.25*b*).

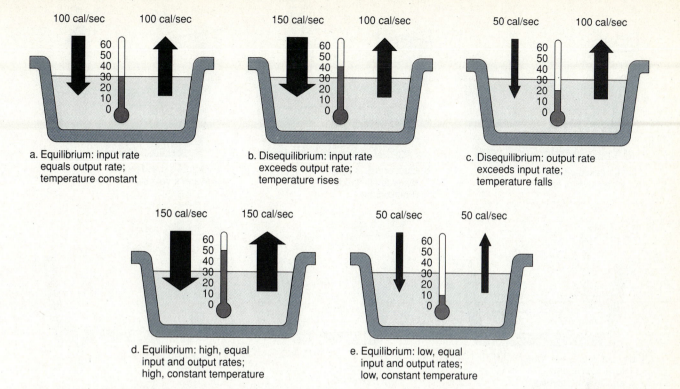

**Figure 2.24** Dynamic equilibrium in a closed system (*a*) is characterized by equal input and output rates and constant temperature. Increases (*b*) and decreases (*c*) in input rate disturb equilibrium, sending temperature higher and lower, respectively. In time, new equilibria arise from such disturbances. Thus, the increased input rate in *b* eventually induces an equally high output rate and a higher equilibrium temperature (*d*), whereas the decreased input rate in *c* induces an equally low output rate and a lower equilibrium temperature (*e*).

At this point, dynamic equilibrium has been achieved, and as long as the input rate remains constant, the water level will maintain a constant elevation.

Note that the flow of energy in this open system is closely tied to the flow of matter. As water flows from the faucet, it possesses kinetic energy, which is converted to gravitational potential energy as the water level rises. The higher the water level, the more GPE is available to force water output from the system, reconverting itself to kinetic energy in the process. Because of this, the water level represents an **equilibrium potential** (EP) or that quantity of potential energy required to force output from the system at the same rate at which input is entering. Note that the output rate from the system is determined *only* by the EP, a fact that is underscored by the narrower tub in figure 2.25*e*, which has the same input rate as the tub in figure 2.25*b*. The amount of water in storage within the narrower tub is only about half that in the wider tub, but the water level and the EP are the same in both systems. If we were suddenly to shut off the tap in either system, the outflow rate would at first remain unchanged, but would soon decrease as outflow drained the water stored within the tub, thus lowering the EP.

Suppose that we disturb the equilibrium in it by dumping a bucketful of water into the tub (fig. 2.25*c*).

This would raise the water level and the EP, thus increasing the output rate. As long as the input rate from the spigot remained constant, however, that increased output rate would quickly lower the water level and the EP to their former position, as in figure 2.25*b*, and output would again equal input; hence, equilibrium would be restored. We could also disturb the equilibrium by reducing the flow from the spigot, thus establishing a lower input rate (fig. 2.25*d*). If we did this, the water would duly lower itself to a new level, and a new EP, that would be just sufficient to produce an output equal to the new, reduced input. Finally, if we shut off the tap altogether, as in figure 2.25*f*, the EP would quickly drain off whatever water remained in storage, and the system would decay.

## Cybernetic Systems: The Strategy of Life

From the foregoing, it should be clear that the fundamental characteristic of dynamic equilibrium systems, whether they are closed or open, is that they maintain equal rates of input and output, and that in order to achieve this equality, they vary their equilibrium potentials. In the case of the closed bathtub system, the equilibrium potential was in the form of temperature; and

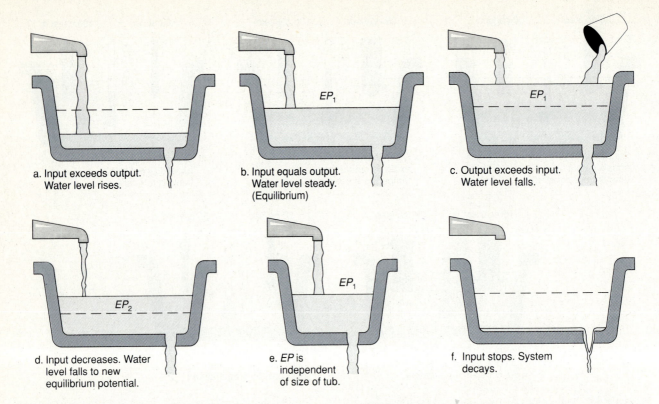

a. Input exceeds output. Water level rises.

b. Input equals output. Water level steady. (Equilibrium)

c. Output exceeds input. Water level falls.

d. Input decreases. Water level falls to new equilibrium potential.

e. *EP* is independent of size of tub.

f. Input stops. System decays.

*Figure 2.25*  In this simple open system, input from the faucet causes the water level to rise (*a*), thereby converting kinetic energy to gravitational potential energy. The potential energy at equilibrium ($EP_1$ in *b*) is just sufficient to drive water from the drain at the same rate at which it is flowing in. Note that the EP depends only on water level and not on volume (compare *e*). Artificially raising the water level (*c*) increases output, which soon restores equilibrium. Reducing input (*d*) results in a lower EP. Shutdown of input (*f*) eliminates the EP altogether.

in the case of the open bathtub system, it was in the form of a water level. Suppose, however, that we wished to maintain a *constant temperature* in the closed system, or a *constant water level* in the open system regardless of fluctuations in the rate of input of electromagnetic energy or water. In order to do this, we could install a thermostat in the closed tub (fig. 2.26*a*), and a water level gage/regulator in the open tub (fig. 2.26*b*). Then, if the temperature of the water in the closed tub should fall below a certain minimum level, the thermostat would sense this and would activate an auxiliary source of electromagnetic energy, such as an electric heater. The heater would then add sufficient heat input to raise the temperature into the acceptable range. If the temperature should rise too far, the thermostat would sense that, too, and activate a fan, which would blow heat away from the system until the temperature again returned to the acceptable range.

Similarly, if the water level in the open system fell below a certain minimum, the level gage would sense this and would activate a diaphragm in the drain hole, which would close enough to restrict output until the water level rose to within the acceptable range, whereupon the gage would deactivate the diaphragm. If the water level rose too high, the gage would open the diaphragm wider than its normal setting until the increased output lowered the water level into the acceptable range.

Such systems, in which energy potentials are maintained within predetermined, optimum ranges instead of being allowed to fluctuate freely in order to equalize input and output rates, are known as **cybernetic systems** (from Greek *kybernetes,* "helmsperson"). The sensory regulating mechanisms that maintain the energy potentials of cybernetic systems in such **steady states** are called **negative feedback loops** because they *reduce* deviations from the steady state. *Positive* feedback loops on the other hand, *enhance* such deviations and thus tend to destroy systems. Two familiar examples of positive feedback are runaway population growth and the growth of capital invested with compound interest, both of which tend to fuel their own fires.

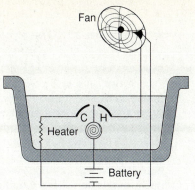

a. Cybernetic temperature control

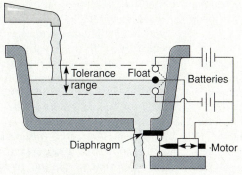

b. Cybernetic water level control

**Figure 2.26** Equilibrium systems can be converted to cybernetic systems by adjusting inputs or outputs in order to regulate energy in storage (EP). In *a,* temperature is regulated by a thermostat. Falling temperature moves the needle to the left (*C*), activating the heater circuit; whereas rising temperature moves it to the right (*H*), activating the fan. In *b,* water level is held within a tolerance range by a float, which activates a motor to close a diaphragm in the drain when the water level rises too high and to open it when the level falls too low.

Why should the phenomena of negative feedback and steady states exist? It seems clear that the only systems that benefit from these modifications to equilibrium systems are living organisms, which require steady states in order to survive. It also seems clear that only living organisms are capable of producing them. Accordingly, it has been proposed that the presence of cybernetic systems should be regarded as the most fundamental sign of life, and that any system is alive to the degree to which its subsystems are cybernetic.

To discuss the dynamics of all Earth's systems in detail would result in a book of unmanageable size. In the chapters that follow, I highlight some significant aspects of many of Earth's systems. From these examples, the interested student should be able to identify equivalent aspects in other systems encountered in this book, and in the "real world."

## Summary

*Time* is a measure of the duration of change in systems. Archbishop Ussher's biblical research in 1654 gave Earth an age of 5658 years. **James Hutton** regarded Earth's age as immeasurable. He countered the prevailing theory of **catastrophism** with that of **actualism** (Nature's laws are constant in time; the present is the key to the past). *Niels* **Stensen's** three laws concerning **stratified** rocks and their strata are those of **superposition, original horizontality,** and **original lateral continuity.**

Earth's currently accepted age, 4600 million years, is based on the decay of radioactive elements. Earth's oldest known rocks are about 3960 million years old. Our planet is about 15,000 times older than the human species. The **relative geologic time scale** is based on the distribution of characteristic fossils and fossil groups in layered rocks and on the laws of superposition and **faunal succession.** The **absolute geologic time scale** is based on the **radiometric dating** of igneous and metamorphic rocks, and is related to the relative scale by noting where sedimentary rocks enclose, or are cut by, igneous and metamorphic rocks. Each radioactive element has a specific **half-life** during which one half of an original quantity of the element decays, a process described by an **exponential decay curve.** A **geologic formation** is a rock unit deposited fairly continuously over a relatively short time under a rather uniform set of environmental conditions.

The earliest life-forms were probably spherical shells of protein. **Cells** enclose genetic material and mechanisms for cell division, energy production and transfer, protein synthesis, food storage, etc. within a *cell membrane.* **Prokaryotic cells** lack *nuclear membranes* and *organelles,* have only one hoop-shaped *chromosome,* and normally reproduce asexually by fission. They have changed little over geologic time. *Cyanobacterial* **stromatolites** give evidence of photosynthesis from 3500 million years ago.

**Banded iron formation** was formed between 2600 and 1800 million years ago by photosynthetic marine organisms whose waste oxygen oxidized soluble iron to insoluble iron oxides. The first appearance of free oxygen in the atmosphere is marked by the appearance about 1800 million years ago of gypsum rock and continental **red beds** containing iron oxide. Organisms that failed to evolve oxygen-mediating enzymes either went extinct or survived in oxygen-free environments.

**Eukaryotic cells,** having *organelles* and a well-developed **cell nucleus,** could have evolved as early as 2000 million years ago, probably by the invasion of larger

We have seen how the complexities of Earth's living, cybernetic systems have tended to increase through geologic time. Formed from the simple gases of the atmosphere, those systems began with space-enclosing, organic membranes and then evolved progressively into prokaryotic cells, eukaryotic cells, and finally into many-celled organisms.

In some instances these many-celled organisms themselves have entered into a higher order of complexity. Among the social insects, the significant unit of the species is not the individual ant, wasp, termite, or bee, but the *colony*. All the members of a particular beehive are brothers and sisters, and they are all *genetically differentiated* into one or another of three different castes: fertile queens, sterile female workers, and fertile males developed from unfertilized eggs. Each of the castes is genetically programmed to perform certain functions in and for the hive, and these functions change for each insect as it grows older.

Parallels have often been drawn between insect and human societies. Such parallels have been criticized because of obvious differences in biology and intelligence between humans and insects. On a purely systemic level, however, the comparison might have some validity. In our society of specialists, each of us has a special function within a human social order whose needs and directions often appear to transcend the needs and directions of its members.

Under this regime, each citizen is, and most often chooses to be, restricted in many ways by the structure of the society. Before the present century, the automobile was a luxurious fantasy available only to the rich and powerful. Today, however, society has been so thoroughly restructured by the potentialities of the internal combustion engine that life without an automobile is almost as unthinkable as life without a house, and for both, we willingly indenture ourselves for most of our working lives. Hence, the cheerful bumper sticker: "I owe, I owe, so off to work I go!" We get our cars and houses, and in return, society gets our brow sweat and elbow grease to further its development.

From a systems viewpoint, perhaps the most interesting question to be asked about this familiar situation is "Who, or what, is in control here?" Certainly it is not the individual, who must have a house to evade the elements and a car to get to work and to market. It is also apparently not the bankers, who depend on the individuals to indenture themselves, and it is also not the administrators, who depend for their positions on the individuals who borrow money (and all too often on the bankers!). This leaves only the system of human society itself, which seems to be choreographing its own complex evolution independently of the desires and sensibilities of individuals, bankers, and administrators. All of these functionaries are being swept along for the ride and used to the best advantage of the system. Furthermore, it seems clear that the *city* is the fundamental evolutionary unit, comparable in some degree to the species population in Nature.

If this is indeed what is happening, then we, as individuals, bankers, or administrators, are in much the same position with respect to our technoindustrial, urban "superorganism" as the cells in our bodies are in with respect to us. The superorganism, in short, might actually be assuming control as it becomes a more complex and functionally coordinated system, and we, as individual human components of it, might simply be obliged to follow its destiny.

There is nothing intrinsically wrong with such a situation, any more than there is anything intrinsically wrong with the subordination of organelles to eukaryotic cells or the subordination of cells to human beings. The joker in the deck, however, is revealed by the large number of mass-extinctions evident in the fossil record. Nature is surprisingly tolerant of new experiments in evolution as they arise within the abundant flow of solar energy. Unless such experiments can hold their own within the whole Earth system under the consistency principle, however, there is no guarantee that they will not wind up in the evolutionary trash bin.

Human beings have been called Earth's first agents of self-awareness. If our human superorganism should choose to alter its evolutionary course to one that is more favorable to peaceful coexistence with the planet on which it depends, then the Earth system will have evolved a completely new kind of driving force in evolution, one that might be called **cognitive selection.** Indeed, it seems clear that this kind of selection is humankind's *only* hope of survival, considering that we have effectively exempted ourselves from Nature's only other "pruning device," *natural* selection. In chapter 19, I examine the prospects for the long-term viability of the urban superorganism.

---

cells by smaller ones that began as parasites and wound up as organelles. Simple cell division, or **mitosis,** duplicates the original cell. Reduction division, or **meiosis,** produces four cells from one, each with half the chromosomes of the original. These can either function directly as sex cells (in animals) or grow into one-to-many-celled adults that produce sex cells (in most plants).

*Fertilization* (the union of two sex cells) produces an *embryo,* which grows into a one-to-many-celled adult, which produces more sex cells by meiosis. The recombination of genes in this cycle of sexual reproduction provides the *genetic diversity* on which **natural selection** acts to drive evolution. In any population of a given species, those *genotypes* best adapted to prevailing environmental conditions are the most numerous.

The trend toward multicellularity in organisms has five main components:

1. More cells in the organism
2. More functional differentiation of cells
3. More interdependence between organism and cells
4. Subordination of individual cell function to the organism
5. Better coordination of cellular functions

The appearance of oxygen in ocean and atmosphere, and of algae-eating snails, allowed many-celled organisms to evolve and expand explosively into virtually all marine and terrestrial environments. Survival strategies for such organisms include *shells, jointed external or internal skeletons, central nervous systems, respiratory systems, circulatory systems,* and *digestive systems.* Survival strategies for many-celled plants include *root and shoot systems, vascular tissue,* and *sex plants.* **Preadaptations** that allowed organisms to invade the land include *jointed skeletons, lungs, tracheae, internal fertilization,* and the *amniote egg.*

**Systems** are persistent, describable, predictable assemblages of **matter** and **energy.** They are *hierarchical,* and interact according to the **consistency principle. Potential** and **kinetic energy** are bound to matter; **electromagnetic energy** is released when charged particles accelerate. **Isolated systems** have no **inputs** or **outputs. Closed systems** have inputs and outputs of energy. **Open systems** have inputs and outputs of matter and energy. At **dynamic equilibrium,** input rate equals output rate. If input fluctuates in an **equilibrium system,** the **equilibrium potential** adjusts to maintain equilibrium. In **cybernetic systems, negative feedback loops** adjust input or output rates to stabilize energy potentials and maintain **steady states.** *Positive feedback loops* tend to destroy systems. The city could be the fundamental evolutionary unit in human social evolution. **Cognitive selection** might be the only workable means of keeping human society in harmony with the rest of the planet.

## Key Terms

James Hutton
Neils Stensen
law of superposition
strata
stratification
law of original horizontality
law of original lateral continuity
catastrophism
actualism
Azoic Eon
Cryptozoic Eon
Phanerozoic Eon

Archaeozoic Era
Proterozoic Era
Paleozoic Era
Mesozoic Era
Cenozoic Era
law of faunal succession
relative geologic time scale
absolute geologic time scale
exponential decay curve
half-life
radiometric dating
geologic formation

cell
prokaryotic cell
stromatolite
banded iron formation
red bed
eukaryotic cell
cell nucleus
mitosis
meiosis
fertilization
natural selection
preadaptation
system
consistency principle
matter
energy
potential energy

cognitive selection
kinetic energy
electromagnetic energy
isolated system
closed system
input
output
dynamic equilibrium
equilibrium system
open system
equilibrium potential
cybernetic system
steady state
negative feedback loop

## Questions for Review

1. What percentage of geologic time passed before the Ediacarian "explosion" of many-celled life forms? Before the first reptile? Before the first human being?
2. Give two examples from Nature that illustrate the theory of actualism.
3. Describe a situation for each of Stensen's laws in which that particular law is needed to explain the appearance of layered rocks.
4. If a rock contains 3 times as much decay product as parent element, how old is it? (See fig. 2.3; assume half-life equals $1300 \times 10^6$ yr.)
5. A sandstone formation overlies a body of 319-million-year-old granite, and is cut by a 291-million-year-old basalt dike. In which period, era, and eon was the sandstone deposited (see table 2.1)?
6. In what ways do eukaryotic cells differ from prokaryotic cells?
7. Would you expect to find banded iron formation and red beds of the same geologic age? Why, or why not?
8. How does meiosis differ from mitosis? What is the function of each?
9. What advantage does sexual reproduction have over asexual reproduction?
10. Describe the specific function of each of the survival strategies of multicellular animals and plants.
11. Explain the consistency principle.
12. Distinguish among isolated, closed, and open systems in terms of inputs, outputs, and energy potentials, and give an example of each.
13. How do equilibrium and cybernetic systems differ?
14. Give an example of a negative feedback loop, and describe how it works.
15. What is the likely reason for the appearance on Earth of cybernetic systems?

***Figure 3.1***
wavelength
(even divisi
radio waves

## What th

Most scient
as much as
easier it is t
goal have
entire Univ
ticles, *phot
trinos*.

   **Photon
netic energ
energy bun
like particle
ticle, howev
matter to re
by all subst
lute zero (−
light (about
the tempera
emitted ph
which they
**magnetic s**
about $10^{11}$
energy end
than $10^{18}$ cy
Our eyes ca
trum called
The reason?
sunlight, th
useful form
figure 3.1 in
radiation de

## Electron Clouds

As protons are added to an atomic nucleus, electrons must be added, *one-for-one,* in surrounding orbits, to balance the positive charges of the protons and maintain electrical neutrality. Accordingly, hydrogen, with one nuclear proton, has only one orbital electron. Helium, with two protons, has two electrons. Lithium, with three protons, has three electrons, and so on.

The region surrounding the atomic nucleus is highly complex, but can be visualized as a series of **electron shells** representing different energy levels. For simplicity, these shells are shown in figure 3.3 as concentric circles; but in reality, they are far more elaborate than this. Each shell consists of a number of *orbitals,* spherical or dumbbell shaped regions within which the electrons move. When filled, each orbital holds two electrons spinning in opposite senses. The number of orbitals in a given shell is equal to the square of the shell number. For example, the first shell consists of a single orbital ($1^2$), the second of four ($2^2$), the third of nine ($3^2$), and the fourth of 16 ($4^2$). The fifth, sixth, and seventh shells are never filled to capacity because atomic nuclei large enough to fill those shells with electrons do not exist in Nature; hence, the numbers of orbitals in these shells are 16, 9, and 1, respectively.

You might think that with this nicely balanced arrangement of positive and negative particles, atoms would be quite stable and "content." Actually, most atoms tend either to lose electrons from the outermost shell, which leaves them with an excess of positive charge or to gain electrons within that shell, which confers on them an excess of negative charge. They express these paradoxical tendencies in a variety of ways, but why do they do so?

## Atomic "Discontent" and the Molecular Compromise

There are six elements—all gases—that have just enough electrons to fill their outer shells to capacity. These are the *noble gases,* helium, neon, argon, krypton, xenon, and radon (see fig. 3.3 for helium and neon). If you compare helium with hydrogen, you will see that hydrogen's single electron shell is only half-filled. If we momentarily allow ourselves the unscientific luxury of looking at electrons in human terms, we might conclude that the single electron in this half-filled shell is "lonely" and would be much "happier" if it could find a companion electron to share its orbital. Hydrogen actually accomplishes this objective in a remarkable way. Two hydrogen atoms pair off and share their two electrons so that each spends half its time in the orbital of one atom and half its time in the orbital of the other (fig.

3.5*a*). This arrangement allows hydrogen to satisfy the requirements of both electrical neutrality and a full shell at the same time.

Atoms that behave in this manner are called *chemically active.* Most other gases, including nitrogen, oxygen, fluorine, and chlorine, are chemically active and also share electron pairs, although more than one pair can be shared (fig. 3.5*b*). The six noble gases, however, do not pair off either with themselves or with other elements because they do not have the "problem" of unfilled outer shells. Accordingly, they occur in Nature only as single, independent atoms. All other elements are chemically active and must share at least one electron pair per atom.

When two hydrogen atoms combine with one another the result is a hydrogen molecule: $H_2$ (fig. 3.5*a*). As I note in chapter 1, molecules are small groups of atoms in fixed proportion to one another. If a molecule contains more than one kind of atom, it is called a **compound.** Three simple compounds are shown in figure 3.5*c–e.* There are also other forms besides molecules in which atoms can combine to form compounds.

## Bonding Mechanisms: Electron Sharing, Electron Transfer, and Electron Soup

The combining of atoms to form compounds is called *bonding.* The type of bonding just described, known as **covalent** (Latin for "equal strength") **bonding,** is typical of gases (except the noble gases), $H_2O$, all organic molecules, and many minerals, especially the *silicates* which make up over 90% of the rocks of Earth's crust.

Another common form of bonding involves the *physical transfer* of outer-shell electrons from one atom (or group of atoms) to another instead of the sharing of electron pairs. The result of this process is two electrically charged atoms called **ions** (Greek for "that which has gone"). In the case of sodium chloride, or table salt (NaCl), sodium (Na) has one "lonely" electron in its outer shell and chlorine (Cl) lacks one electron in its outer shell (fig. 3.6*a*). Both atoms are "satisfied" by the transfer of the single electron from the sodium to the chlorine. The transferred electron, instead of orbiting both the sodium and the chlorine, now becomes the *sole property* of the chlorine. This, however, creates an electrical charge imbalance in both atoms. By losing an electron, the sodium atom acquires a net charge of $+1$, and becomes a *sodium ion,* $Na^{+1}$. By gaining an electron, the chlorine atom acquires a net charge of $-1$ and becomes a *chloride ion,* $Cl^{-1}$. The symbol for an ion is always written with a superscript indicating *the number of protons in the ion minus the number of electrons.*

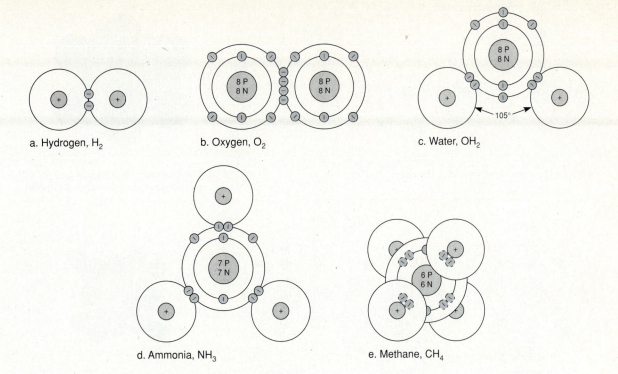

a. Hydrogen, $H_2$

b. Oxygen, $O_2$

c. Water, $OH_2$

d. Ammonia, $NH_3$

e. Methane, $CH_4$

**Figure 3.5** Covalent bonding. Molecules are formed when atoms combine in fixed proportions, sharing the electrons in their outer shells and thus filling those shells to capacity. Compounds *a* and *b* are molecules containing atoms of only one element each. Molecules *c–e* contain more than one element each.

The problem of the charge imbalance is solved by the mutual attraction between the oppositely charged sodium and chloride ions. This attraction creates a powerful **ionic bond** that causes the ions to stick together. In so doing, they arrange themselves not into discrete molecules of NaCl, but into an "endless," three-dimensional, rectangular network in which each sodium ion is surrounded symmetrically by six chloride ions. Each chloride ion is, in turn, surrounded symmetrically by six sodium ions (fig. 3.6*b*). This 6-for-6 arrangement (not obvious in fig. 3.6 because only a small, cubic fragment of the network is shown) maintains equal numbers of positive sodium and negative chloride ions in the network, as is required for electrical neutrality. Such three-dimensional ionic networks are called **crystal structures.** In this case, the structure is that of the mineral *halite* (see fig. 3.17*j*). In the case of calcium fluoride (fig. 3.6*c* and *d*), the process of electron transfer is identical; but because each calcium atom has *two* outer-shell electrons to lose and each fluorine atom can accept only *one,* the ratio of fluoride ions to calcium ions must be 2 to 1. This fact is reflected in the chemical formula for calcium fluoride, $CaF_2$. The corresponding mineral is *fluorite* (see fig. 3.19*f*).

In addition to bonding by electron sharing and transfer, there is a third possibility in which the bonding electrons are shared not by specific ions but by *all* the ions within the mineral. The ions, in other words, are bathed in a sort of "electron soup" that flows readily from one place to another within the mineral. This is the **metallic bond.** As its name suggests, it is typical of such metals as gold, silver, and copper. Again, the shock you get from an iron fence wire is nothing more than bonding electrons flowing through the metallic crystal structure of the wire (and you) from an electron source and into the ground. Such a flow of electrons, called *electricity,* is possible because it makes no difference to the metal which electrons are present as long as the *right number* of them is present at any given moment to maintain electrical neutrality in the metal. A metal wire with an electric current flowing through it is a particularly good example of an open system in which the throughput is measured by the *amperage,* or rate of flow of electricity, and the equilibrium potential (EP) is the *voltage* that drives the output of electricity.

A fourth type of bond, important because it is found in ice and clay minerals, is the **residual bond.** Unlike the other three types, this bond does not involve electrons but depends instead on the asymmetrical distribution of electrical charge in many molecules and

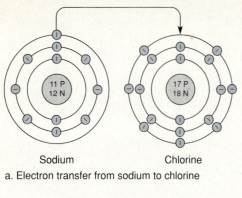

Sodium          Chlorine

a. Electron transfer from sodium to chlorine

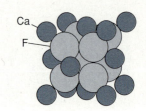

Cl

Na

b. The arrangement of sodium and
chloride ions in sodium chloride (halite)

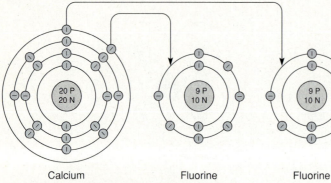

Calcium          Fluorine          Fluorine

c. Electron transfer from calcium to fluorine

Ca

F

d. The arrangement of calcium and fluoride
ions in calcium fluoride (fluorite)

*Figure 3.6*  Ionic bonding involves the transfer of outer shell electrons from one atom to another, forming electrically charged ions that bind themselves into crystal structures due to their mutual electrostatic attraction. In sodium chloride (*a, b*), each sodium atom requires only one chlorine to accept its one outer shell electron. In calcium fluoride, however (*c, d*), each calcium atom requires two fluorines to accept its two outer shell electrons. The crystal structures of halite and fluorite reflect these differences.

crystal structures. This distribution tends to give these molecules and structured compounds a "positive end" and a "negative end," each of which attracts the other. Figures 11.18 and 11.21 illustrate residual bonding in ice.

## Anions: The Building Blocks of Crystals

When an atom gains one or more electrons, the resulting ion is larger than the original atom (fig. 3.7), and it is called an **anion** (*an*-eye-on; Greek for "that which has gone up"). When an atom loses electrons, on the other hand, the resulting ion is smaller than the original atom, and it is called a **cation** (*cat*-eye-on; "that which has gone down"). In most minerals, therefore, the anions are larger—often much larger—than the cations. As in sodium chloride and calcium fluoride, the large anions pack together like oranges in a box, and the smaller cations fit into the leftover spaces. Just as the pattern in a brick wall depends on the sizes and shapes of the bricks of which it is made, so the pattern in the crystal structure of a mineral depends on the sizes and

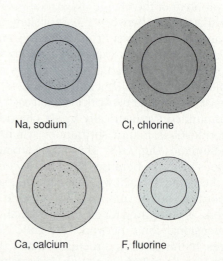

Na, sodium          Cl, chlorine

Ca, calcium          F, fluorine

*Figure 3.7*  Relative sizes of atoms and their corresponding ions. Positive cations (speckled, *left*), formed by the removal of one or more outer-shell electrons, are smaller than the original atoms (unpatterned, *left*). Negative anions (speckled, *right*), formed by the addition of one or more outer-shell electrons, are larger than the original atoms (unpatterned, *right*).

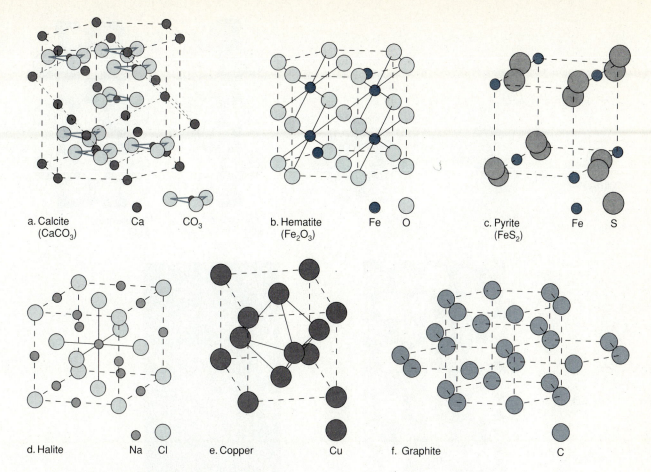

a. Calcite
(CaCO$_3$)          Ca        CO$_3$

b. Hematite
(Fe$_2$O$_3$)        Fe        O

c. Pyrite
(FeS$_2$)           Fe        S

d. Halite            Na        Cl

e. Copper                      Cu

f. Graphite                    C

**Figure 3.8**  Six important nonsilicate crystal structures. These are "exploded" diagrams. In reality, the ions are packed tightly together.

Source: After W. H. Dennen and B. R. Moore, *Geology and Engineering.* Copyright © 1986 Wm. C. Brown Publishers, Dubuque, IA.

shapes of the ions of which the crystal is made, and most particularly on the anions. Figure 3.8 illustrates some important nonsilicate crystal structures.

The **silicate minerals** make up over 90% of Earth's crust and are, oddly enough, the most complex in terms of crystal structure. The anion that is fundamental to all silicate structures is the **silica tetrahedron** (fig. 3.9), in which a single silicon atom is symmetrically enclosed by four oxygen atoms. This unit can exist as a simple anion, $SiO_4^{-4}$, balanced by such cations as $Fe^{+2}$ and $Mg^{+2}$, as in the *isolated silicates.* It can also combine with itself by sharing from one to four of its oxygens with adjacent tetrahedra by means of covalent bonds (fig. 3.10). This "chain-linking" of silica tetrahedra is called **polymerization,** and it is the same process whereby carbon atoms are linked into complex organic molecules. In the silicate minerals, it combines silica tetrahedra into *pairs, rings, single and double chains, sheets,* and *frameworks.* The most highly polymerized of all silicate minerals is also one of the most abundant and one of the simplest chemically, namely *quartz* ($SiO_2$), a framework silicate (see figs. 3.11, 3.12, and 3.16*g*).

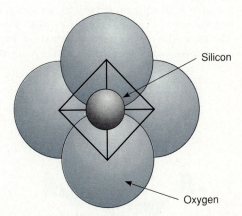

Silicon

Oxygen

**Figure 3.9**  The silica tetrahedron is so named because the centers of its four oxygen atoms, symmetrically surrounding the central silicon atom, are arranged as if they were at the corners of a regular tetrahedron (a four-sided figure).

Systems in Miniature: The Nature of Minerals    **63**

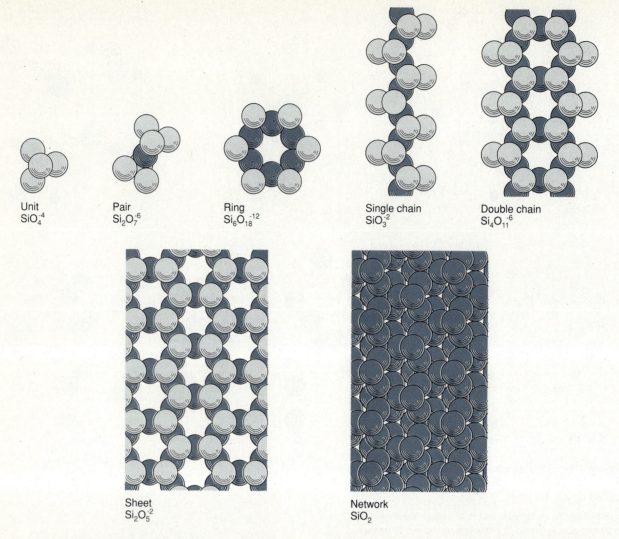

Unit
$SiO_4^{-4}$

Pair
$Si_2O_7^{-6}$

Ring
$Si_6O_{18}^{-12}$

Single chain
$SiO_3^{-2}$

Double chain
$Si_4O_{11}^{-6}$

Sheet
$Si_2O_5^{-2}$

Network
$SiO_2$

**Figure 3.10** Silica tetrahedra readily share one or more of their oxygens with adjacent tetrahedra, a process called polymerization (shared oxygens are shown with a darker shade). The lower the temperature of formation, the greater is the degree of polymerization. Except for the framework silicates, these structures all have excess negative charges (indicated by superscripts on the formulas), which must be balanced by the addition of cations.

**Figure 3.11** The hexagonal arrangement (note dashed lines) of fully linked silica tetrahedra in quartz gives crystals of this framework silicate a hexagonal form. Here, oxygen atoms have been omitted for clarity, and their centers are indicated by the points of the tetrahedra. (See also figs. 3.12 and 3.16g.)

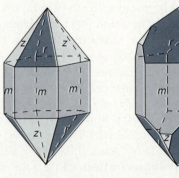

**Figure 3.12** Accidents of growth often favor one set of faces over another set, but the angle between adjacent faces is always the same. On the quartz crystal on the right, the faces labelled *r* have been favored over those labelled *z*.

Source: G. J. Brush and S. L. Penfield, *Determinative Mineralogy,* 17th ed. Copyright © 1898 John Wiley & Sons, Inc., New York, NY.

## Acceptable and Nonacceptable Substitutes

Just as an oversized brick cannot be fitted into a wall, so an oversized ion cannot be forced into a crystal structure. In crystals, however, the restrictions are even more rigid than this. Not only do crystal structures reject ions that are too big, but they also reject ions that are too small or of the wrong electric charge. This is why minerals are such well-behaved systems, and it accounts for the great purity of most natural mineral crystals.

Some substitutions are permissible, however. If a "foreign" ion is close in size to one of the ions in a crystal structure, it may replace that ion if it also has the same electric charge. A case in point is *olivine* (see fig. 3.16*a*), a common isolated silicate mineral that accommodates both magnesium (Mg) and iron (Fe), and therefore, has the formula $(Mg,Fe)_2SiO_4$. Because the $Fe^{+2}$ cation is only slightly larger than the $Mg^{+2}$ cation and because they both carry a charge of $+2$, they freely substitute for one another in all proportions. A given specimen of olivine, therefore, can have any conceivable proportion of magnesium and iron. This kind of continuous ionic substitution is called **solid solution.**

As it happens, the most abundant mineral group of Earth's crust, the *feldspars* (from *Feldspat,* German for "field crystal"; see figs. 3.16*e* and *f*), even manages to defy the rule that a substituting ion must always carry the same charge as the ion it replaces. They accomplish this feat by substituting *two* ions whose combined charges add up to the charge of the replaced ion. To see how this works, you first need to understand that the feldspars are essentially variations on the theme of quartz; and like that mineral, they are also framework silicates. Because the formula for quartz, $SiO_2$, is simply a ratio, we may quadruple it to obtain $Si_4O_8$. In the feldspars, either one or two of the four silicon atoms in this quadrupled formula are replaced by aluminum ions ($Al^{+3}$). This results in a charge imbalance because aluminum has one less positive charge than silicon ($Si^{+4}$). At lower temperatures, where only one silicon out of four is replaced, either potassium ($K^{+1}$) or sodium ($Na^{+1}$) accompanies aluminum to balance the charge, resulting in *K-feldspar* ($KAlSi_3O_8$) or *Na-feldspar* ($NaAlSi_3O_8$). At higher temperatures, where two silicons out of four are replaced, calcium ($Ca^{+2}$) is the accompanying atom, resulting in Ca-feldspar ($CaAl_2Si_2O_8$).

Because sodium and calcium ions only differ in size by about 2%, solid solution in all proportions occurs between Na- and Ca-feldspar; hence, neither mineral ever occurs pure in Nature. Because of this, they are usually referred to under the single name *plagioclase feldspar* (Greek for "obliquely splitting" because they tend to split easily along two directions that are not quite mutually perpendicular). Solid solution between K-feldspar and plagioclase is very limited, however, because the potassium ion is about 30% larger than the sodium and calcium ions.

## Crystals as Systems

Crystals are isolated systems. They can, however, be converted into closed systems by inputs of electromagnetic energy and even into open systems where they come into contact with magma or hot water solutions that add atoms to them and remove atoms from them. Under the right conditions, crystals can form and grow within both magma and hot water and within gases as well. In both the liquid and the gaseous states, atoms vibrate with enough energy to prevent bonds (and hence crystal structures) from forming. If a liquid or a gas cools sufficiently, however, this thermal agitation can be reduced enough that bonds will begin to form. Atoms then combine into small groups called *nucleation centers* that can grow into crystals. In the case of crystals growing from a water solution, such dissolved ions as $Na^{+1}$ and $Cl^{-1}$ must become concentrated enough in the water that their electric charges can attract each other through the thick films of water molecules that surround the ions. When this happens, crystals form and are precipitated.

## Mineral Environments

Like everything else in Nature, minerals obey the consistency principle (see chapter 2). No mineral, however rare, has ever appeared anywhere without a good reason. In other words, the occurrence of a given mineral at a given place and time is a record of the environmental conditions, at that place and time, that gave rise to the mineral.

Among the many factors that influence the occurrence of minerals, the most important are *temperature, pressure, time,* and *the availability and concentration of ions in solution.* Different minerals are stable under different conditions of temperature and pressure. Clearly, a given mineral can form only if the ions it is made of are present in the environment. Ion concentrations in solution must be high enough to permit the formation of the embryonic nucleation centers required for crystal growth. The amount of time available for crystal formation and the amount of water present determine the size of the crystals. Large crystals result from long, slow cooling of magma or from high water content, both of which favor the free migration of ions. Rapid cooling and low water content, on the other hand, suppress ion migration, and result in small or even microscopic crystals. Sudden chilling can even prevent crystal growth altogether, resulting in **glass,** a solid that cooled too quickly for its atoms to move into ordered crystal structures (see fig. 4.3*c*).

# The Nature of Minerals

By now, you should have a clear enough grasp of what constitutes a mineral to be able to appreciate the definition. A **mineral** is a *naturally occurring, inorganic, solid substance,* with a *definite range of chemical composition,* and a *definite, ordered internal structure.* This definition excludes all gases, liquids, human-made and organic substances, chemical mixtures, and glasses. Let us now consider some of the important characteristics whereby minerals can be classified and identified.

## Physical Properties of Minerals

The ways in which a mineral interacts with energy flows, and with other systems, are unique and distinctive enough that they can be used to identify the mineral. Such interactions are called **physical properties.** They depend on the *chemical composition,* the *crystal structure,* and the *type of bonding* present in the mineral.

### Specific Gravity

In chapter 1, I discuss the densities of the major rock types of Earth's crust in connection with the principle of isostasy. Mineral densities are also worth noting but for a different reason. As one of its least variable physical properties, a mineral's density is especially useful in its identification. For convenience, however, mineral densities are usually expressed in terms of **specific gravity,** or the ratio of the mineral's density to the density of water at 4° C. The specific gravity of a substance has the same numerical value as its density because the density of water is 1.000 g/cm³, and dividing by 1.000 leaves the value unchanged. The convenience of specific gravity lies in the fact that in the division process the cumbersome unit "g/cm³" is lost, and the value can (indeed, must) be written as a simple number.

Specific gravity depends on both chemistry and crystal structure. Minerals that contain mainly elements of low atomic mass, such as hydrogen, oxygen, sodium, and chlorine, have low specific gravities, whereas those that contain mainly heavy atoms, such as iron and lead, have high specific gravities. The effect of crystal structure is much less influential, but in general, more expanded crystal structures have lower specific gravities than more compact structures. Specific gravities are given for all the minerals listed in table 3.3.

### Color, Streak, and Luster

Some atoms, including iron, manganese, and copper, impart strong colors to minerals. Because such elements can occur either as essential constituents or as impurities, color is not a reliable criterion in the identification of mineral crystals. In powdered form, however, minerals are less susceptible to the coloring effects of impurities, so the **streak**—the color of the powder

**Table 3.2** The Mohs Scale of Mineral Hardness

| Mineral | Relative Hardness | Other Common Materials |
|---|---|---|
| Diamond | 10 | |
| Corundum | 9 | |
| Topaz | 8 | |
| Quartz | 7 | Steel file (6.5) |
| Feldspar | 6 | Glass (5.5) |
| Apatite | 5 | |
| Fluorite | 4 | |
| Calcite | 3 | New penny (2.5–3) |
| Gypsum | 2 | Fingernail (2+) |
| Talc | 1 | |

formed by rubbing a mineral on unglazed porcelain—is routinely used as an aid in mineral identification. A related property, **luster,** is the appearance of the mineral's surface under reflected light. Luster is related not only to the kinds of atoms present but also to the surface texture of the specimen, which is strongly influenced by the crystal structure of the mineral. Typical lusters are illustrated in the mineral photographs of figures 3.16–3.19.

### Magnetism

A few iron-bearing minerals are *magnetic,* the only important ones being the iron oxide *magnetite,* $Fe_3O_4$ (see fig. 3.19a) and the iron sulfide *pyrrhotite,* $Fe_{1-x}S$.

### Hardness, Melting Point, and Tenacity

These three physical properties are all related to the strength of the bonds that hold the atoms of a mineral together. In ionic bonded minerals, the stronger the bond, the harder is the mineral, and the higher is its melting point. For example, sodium chloride has a melting point of 801° C, but magnesium oxide (the mineral *periclase,* MgO), whose ions have twice as much electric charge as those of sodium chloride, melts at 2800° C.

In 1824, an Austrian mineralogist, F. Mohs, established a scale of mineral **hardness** (table 3.2) that is still in use today. Each mineral in this list can scratch the ones below it and can only be scratched by the ones above it. For most mineral identifications, the common materials listed on the right of the table are adequate.

*Tenacity,* a mineral's resistance to being broken or deformed, reflects not only bond strength but also the type of bond. Most metals deform easily without breaking because their atoms are not locked into rigid crystal structures by ionic or covalent bonds. Ionically and covalently bonded minerals, such as halite and quartz, tend to be brittle because of their rigid structures.

## Crystal Faces, Cleavage, and Fracture

Crystals form in shapes that reflect their internal structure. These forms often consist of smooth, flat growth surfaces parallel to planes of atomic layering in the crystal structure. Such planes are known as **crystal faces.** In the seventeenth century, Niels Stensen, whose laws governing stratified rocks are described in chapter 2, recognized that the various crystal faces of a given mineral species always bear the same angular relationship to one another, regardless of the size or overall shape of the crystal (fig. 3.12). This led Stensen to propose the **law of constancy of interfacial angles.** On the strength of this law, crystallographers have set up a standard classification of minerals based on the angular interrelationships of crystal faces.

Most minerals have planes of weakness along which they can be easily split or *cleaved.* Like crystal faces, these **cleavage planes** (fig. 3.13) are parallel to, and result from, the regular atomic layering in crystal structures, and they are extremely useful in mineral identification. Some minerals, such as quartz and olivine, lack such planes, and they break along irregular surfaces called **fractures.** The kind of fracture shown by quartz is called *conchoidal* because its curved fracture surfaces resemble the inside of a marine conch shell (fig. 3.14).

## In the Spotlight:

### René Just Haüy (1743–1822)

**C**lumsiness is seldom considered an asset. On several occasions, it left its indelible mark on Ludwig van Beethoven's piano, as the maestro had a vexing habit of upsetting his inkwell into the instrument, which stood next to his writing desk. In the case of the French priest René Just Haüy (Ah-you-ee), however, clumsiness turned out to have an unexpected benefit. One day while viewing a friend's mineral collection, Haüy dropped a prized specimen of calcite. To his great dismay, the crystal shattered into a myriad of angular fragments of various sizes. As Haüy apologetically gathered up the wreckage, however, he noticed the fragments were all of the same shape, that of a rhomb, or squashed cube (see fig. 3.17*f*).

Fascinated by this interesting result, Haüy then went on a shattering spree, breaking all the calcite crystals he could lay his hands on. Out of this odd binge came a new understanding of the internal structure of crystals. Every specimen he broke shattered into rhomb-shaped fragments whose interfacial angles were identical to those of the first one that he had broken accidentally. From this observation, Haüy concluded that crystals must be made of exceedingly small units that have the same shape as the larger fragments. As the atomic theory had not yet been proposed, he had no way of telling that the shapes of those units were determined by the ways in which different sized atoms fit together (see fig. 3.8). Aside from this detail, his conclusion was essentially correct. It served as the basis for a classification of minerals, which he published in 1809 at the request of Napoleon Bonaparte.

About a decade earlier, Napoleon had become aware of Haüy's work and considered it important enough to have him appointed professor of mineralogy at the Museum of Natural History in 1802 and to commission a number of works on crystallography. Today, Haüy is regarded as the founder of the science of mineralogy, much as Hutton is regarded as the founder of the more inclusive science of geology.

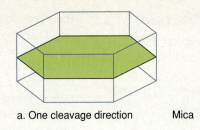

a. One cleavage direction      Mica

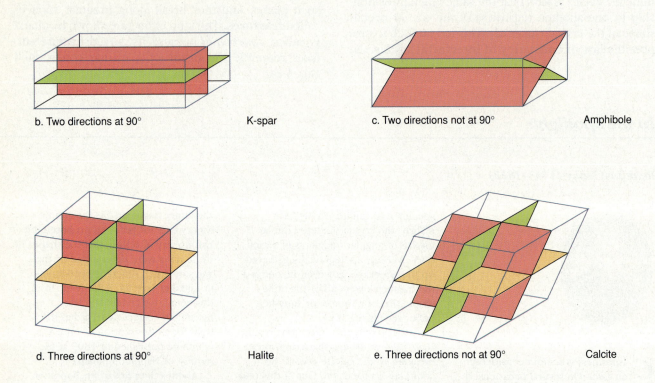

b. Two directions at 90°      K-spar

c. Two directions not at 90°      Amphibole

d. Three directions at 90°      Halite

e. Three directions not at 90°      Calcite

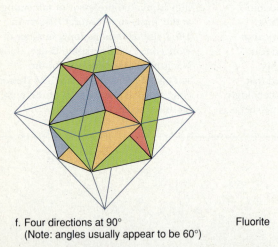

f. Four directions at 90°      Fluorite
(Note: angles usually appear to be 60°)

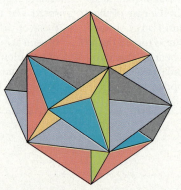

g. Six directions at 90° and 60°      Sphalerite

**Figure 3.13**   Many minerals cleave, or split, readily along one or more sets of parallel planes determined by the arrangement of ions or ionic groups within their crystal structures. The seven possible varieties of cleavage are illustrated here with names of representative minerals.

## The Rock-Forming Minerals

There are about 2800 known mineral species; but of this total, only about 30 are of major importance in Earth's crust. These are listed in table 3.3 according to their dominant anions, and illustrated in figures 3.16–3.19. For purposes of discussion, I divide these common minerals into four categories, igneous, sedimentary, metamorphic, and accessory and ore, reflecting their typical modes of occurrence within the rock cycle, but many of them originate in two or more of these environments, and a few of them (e.g., calcite) originate in all four.

**Figure 3.14** Substances in which ions are not arranged along flat planes do not cleave, but fracture instead. Here, glass exhibits a common variety of fracture called conchoidal. Photo by C. C. Plummer

**Table 3.3** Common Rock-Forming and Accessory Minerals of Earth's Crust

| Minerals[a] | Chemical Formulas | Rock Class[b] | Specific Gravity | Cleavage Planes[c] | Hardness | Other Features |
|---|---|---|---|---|---|---|
| **Silicates** | | | | | | |
| **Isolated** | | | | | | |
| *Olivine* | $(Fe, Mg)_2SiO_4$ | Ig | 3.2–4.4 | None | 6.5 | Green to black, glassy |
| *Garnet* | $X_3Y_2(SiO_4)_3$ | Met, Acc, Ig | 3.6–4.3 | None | 6.5–7.5 | Granular crystals |
| | (X = Ca, Fe, Mg, Mn; Y = Fe, Al, Cr) | | | | | |
| *Al-silicates* | $Al_2OSiO_4$ | Met | 3.2–3.7 | 0–1 | 5–7.5 | Lozenge or blade shape |
| **Paired** | | | | | | |
| *Epidote* | $Ca_2Al_3OSiO_4Si_2O_7OH$ | Met | 3.4–3.5 | 2 @ 65° | 6–7 | Pistachio to deep green |
| **Single Chain** | | | | | | |
| *Pyroxene* | $XY(SiO_3)_2$ | Ig, Met | 3.2–3.6 | 2 @ 87° | 5–7 | Stubby crystals |
| | (X = Mg, Fe, Ca, Na; Y = Mg, Fe, Al) | | | | | |
| **Double Chain** | | | | | | |
| *Amphibole* | $X_2Y_5(Si_4O_{11})_2(OH)_2$ | Ig, Met | 2.9–3.6 | 2 @ 56° | 5–6 | Bladelike crystals |
| | (X = Mg, Fe, Ca, Na; Y = Mg, Fe, Al) | | | | | |
| **Sheet** | | | | | | |
| *Biotite Mica* | $K(Mg,Fe)_3AlSi_3O_{10}(OH)_2$ | Ig, Met | 2.8–3.2 | 1 | 2.5–3 | Thin sheets, brown to black |
| *Muscovite Mica* | $KAl_2 AlSi_3O_{10}(OH)_2$ | Ig, Met | 2.8–2.9 | 1 | 2–2.5 | Thin sheets, colorless |
| *Kaolinite Clay* | $Al_2Si_2O_5(OH)_4$ | Sed | 2.6 | (1) | 2 | Earthy, sticky when wet |
| *Montmorillonite Clay* | $Al_2(Si_2O_5)_2(OH)_2 \cdot nH_2O$ | Sed | 2.5 | (1) | 1–1.5 | Earthy, sticky when wet |
| *Illite Clay* | Similar to muscovite | Sed | 2.8–2.9 | (1) | 2–2.5 | Earthy, sticky when wet |
| *Talc* | $Mg_3(Si_2O_5)_2(OH)_2$ | Met | 2.7–2.8 | (1) | 1 | Slippery feel |
| *Serpentine* | $Mg_3Si_2O_5(OH)_4$ | Met | 2.3–2.6 | (1) | 3–5 | Green, often fibrous |
| *Chlorite* | $(Fe,Mg)_6(Si_2O_5)_2 (OH)_8$ | Met | 2.6–3.3 | (1) | 2–2.5 | Small, dark green flakes |
| **Framework** | | | | | | |
| *K-feldspar* | $KAlSi_3O_8$ | Ig, Met | 2.5–2.6 | 2 @ 90° | 6 | Pink, white, or green |
| *Plagioclase feldspar* | $(Na,Ca)Al_{1-2}Si_{2-3}O_8$ | Ig, Met | 2.6–2.8 | 2 @ 86° | 6 | White to gray, with striations |
| *Quartz* | $SiO_2$ | Ig, Met | 2.65 | None | 7 | Hexagonal crystals |

**Figure 3.16** Minerals of predominantly igneous origin.

*a.* Olivine, $(Mg,Fe)_2SiO_4$. Crystals granular; greenish to black; luster glassy.

*b.* Pyroxene, $XY(SiO_3)_2$. Crystals stubby columns with squarish cross section; usually green to black; luster glassy to pearly.

*c.* Amphibole, $X_2Y_5(Si_4O_{11})_2(OH)_2$. Crystals stubby columnar to needlelike, lozenge-shaped cross section; usually green to black; luster glassy to silky.

*d.* Mica, $K(X)_{2-3}AlSi_3O_{10}(OH)_2$. Crystals sheetlike; muscovite is colorless, biotite black; luster shiny.

*e.* K-feldspar, $KAlSi_3O_8$. Crystals stubby columnar to tabular; white, gray, pink, or green; luster glassy (often dull).

*f.* Plagioclase feldspar, $(Na,Ca)Al_{1-2}Si_{2-3}O_8$. Crystals tabular; white (sodium rich) to dark gray and iridescent (calcium rich); luster glassy to pearly (often dull).

*continued*

*g.* Quartz, $SiO_2$. Crystals hexagonal stubby columns; colorless (crystal), pink (rose quartz), purple (amethyst), dark gray (smoky quartz), yellow (citrine), white (milky quartz), or crystals microscopic and variously colored or banded (chalcedony); luster glassy.

Specimens courtesy of Earth Sciences Dept., Dartmouth College

Early-formed minerals from the upper ends of the reaction series often sink within the magma chamber as they crystallize, and are therefore no longer able to react with the residual magma. Because these minerals are relatively rich in iron, magnesium, and calcium, this depletes the magma progressively of these elements and enriches it in silicon, aluminum, and water, all major ingredients of the low temperature minerals at the base of the series.

The beauty of Bowen's diagram is that it encompasses virtually all the important igneous rock-forming minerals; and as such, it is an admirable example of a *synthetic theory.* In its infancy, a science is invariably preoccupied with finding new things, such as minerals, and classifying them. The science begins to reach maturity when its practitioners begin to draw these seemingly unrelated things together into patterns, such as Bowen's series, that show the various things as interactive parts of a larger system.

## Sedimentary Minerals

Sedimentary minerals differ from igneous minerals in that they are formed at low temperatures and pressures at Earth's surface in contact with oxygen, carbon dioxide, and water. Because such surface conditions are virtually unique to Earth, the same thing is true of this class of minerals. As far as we know (with one or two exceptions), they occur nowhere else in the Solar System.

With the exception of quartz, the igneous minerals are unstable under the environmental conditions that prevail at Earth's surface, and they are rapidly decomposed by those conditions. Plagioclase feldspar, for example, is formed deep in Earth's crust at temperatures of 1200°–1500° C and pressures of some 3000–7000 atmospheres. When Ca-feldspar is exposed to the environment of Earth's surface, however, it is weathered by carbonic acid and water to yield *clay* (an old English term; fig. 3.17*a*) and dissolved calcium bicarbonate. The latter is, in turn, removed from solution by organic processes (see fig. 1.22) and converted into protoplasm, oxygen, and calcium carbonate in the forms of *calcite* (from *calx,* Latin for "lime"; fig. 3.17*f*) and *aragonite* (named for Aragon, Spain; fig. 3.17*g*).

These same two processes of weathering and precipitation from solution produce a variety of other sedimentary minerals. The most significant are *chert,* or *flint,* $SiO_2$ (origins of terms obscure; fig. 3.17*d*); *dolomite,* $CaMg(CO_3)_2$ (after French geologist D. de Dolomieu; fig. 3.17*h*); *gypsum,* $CaSO_4$ (from *gypsos,* Greek for "chalk," a misnomer; fig. 3.17*i*); and *halite,* $NaCl$ (from *hals,* Greek for "salt rock"; fig. 3.17*j*). Because each of these minerals forms an important rock type, their modes of formation are discussed in chapter 4. The formation of *hematite* (from *haimatites,* Greek for "bloodlike"; figs. 3.8*b* and 3.17*b*) by the oxidation of iron-bearing minerals is discussed in chapter 2. With the addition of water, reddish hematite is converted to yellowish *goethite* (*ger*-thite, with a silent r, after German naturalist Johann Wolfgang von Goethe; fig. 3.17*c*). Goethite is responsible for the yellowish or brownish color of many soils and weathered mineral veins. *Gibbsite* (after American mineralogist George Gibbs; fig. 3.17*e*) is a deposit of white aluminum hydroxide, usually stained with red polka dots of hematite. It is formed, often abundantly, by the intense leaching of aluminum-bearing bedrock by heavy rainfall in tropical climates (see chapter 8).

A glance at table 3.3 shows that, with the exception of the oxides, the sedimentary minerals have substantially lower hardnesses and specific gravities than the igneous minerals. When you consider that the sedimentary minerals are formed under much lower pressures and temperatures, this difference makes sense. High bond strength gives the oxides a compact structure, which accounts for their relatively greater hardness and heaviness.

**Figure 3.17** Minerals of predominantly sedimentary origin.

*a.* Clay, $Al_2(Si_2O_5)_{1-2}(OH)_{2-4} \cdot nH_2O$. Crystals submicroscopic; white (*kaolinite*), to off-white (*montmorillonite*); luster earthy.

*b.* Hematite, $Fe_2O_3$. Crystals hexagonal, tabular, often microscopic in small rounded grains or large rounded masses; metallic gray to reddish brown to reddish purple; luster metallic to earthy.

*c.* Goethite, $Fe_2O_3 \cdot H_2O$. Crystals microscopic or finely fibrous; yellow to dark brown or black; luster earthy to brilliant.

*d.* Chert and flint, $SiO_2$. Crystals microscopic and granular, light (chert) to dark (flint), variously colored; luster waxy.

*e.* Gibbsite. Crystals usually microscopic; white, often discolored by hematite, as here; luster earthy.

*f.* Calcite, $CaCO_3$. Crystals hexagonal columns, hexagonal pyramids, or rhoms (squashed cubes); often occurs as microscopic coatings on objects, such as the pine cone illustrated here; usually colorless or white, or variously colored; luster glassy to earthy.

*continued*

*g.* Aragonite, $CaCO_3$. Crystals usually twinned, forming pseudohexagonal columns, often microscopic; colorless, white, or variously colored; luster glassy.

*h.* Dolomite, $CaMg(CO_3)_2$. Crystals rhombic, often microscopic; colorless, white, or tinted; luster glassy.

*i.* Gypsum, $CaSO_4 \cdot 2H_2O$. Crystals elongated tabular or fibrous, often microscopic; colorless, white, or tinted; luster glassy to silky.

*j.* Halite, NaCl. Crystals cubic; colorless or variously tinted; luster greasy.

Specimens courtesy of Earth Sciences Dept., Dartmouth College

## Metamorphic Minerals

Minerals can undergo changes in chemistry or crystal shape as a result of changes in environmental pressures, temperature, or chemical solutions. By these criteria sedimentary minerals could be classed as metamorphic, but we exclude them by drawing arbitrary (and sloppy) lines at about 150° C and about 2000 atmospheres of pressure. Minerals formed above these limits are considered metamorphic, whereas those formed below them are considered sedimentary. Because the pressures, temperatures, and chemical environments under which metamorphic minerals form are similar to those under which igneous minerals form, the physical properties of the metamorphic minerals are generally quite similar to those of the igneous minerals.

The most important metamorphic minerals not already mentioned under the section on igneous minerals are *garnet,* $X_3Y_2(SiO_4)_3$ (from *granatus,* Latin for "grainlike"; fig. 3.18*a*), a red, yellow, brown, or green isolated silicate of variable composition; the *aluminum-silicates,* $Al_2OSiO_4$ (fig. 3.18*b*), a group of white, reddish, or blue isolated silicates; the green paired silicate *epidote,* $Ca_2Al_3OSiO_4Si_2O_7OH$ (from *epididonai,* Greek for "increase"; fig. 3.18*c*); the white sheet silicate *talc,* $Mg_3(Si_2O_5)_2(OH)_2$ (a Persian name; fig. 3.18*d*); the green sheet silicate *serpentine,* $Mg_3Si_2O_5(OH)_4$ (from the snakelike markings on some specimens; fig. 3.18*e*); a second green sheet silicate *chlorite,* $(Fe,Mg)_6(Si_2O_5)_2(OH)_8$, (Greek for "greenstone"; fig. 3.18*f*), and *graphite,* C (Greek for "writing stone"; figs. 3.8*f* and 3.18*g*), a soft, slippery, silver-black mineral. In addition to these, pyroxene, amphibole, biotite, muscovite, all the feldspars, quartz, calcite, agaronite, and dolomite also occur within the metamorphic environment.

**Figure 3.18** Minerals of predominantly metamorphic origin.

*a.* Garnet, $X_3Y_2(SiO_4)_3$. Crystals granular, 12-sided; red, yellow, brown, or green; luster glassy to greasy.

*b.* Al-silicates, $Al_2OSiO_4$. Crystals columnar and of square cross section in andalusite and sillimanite, bladelike in kyanite (shown here); white to reddish or greenish in andalusite and sillimanite, blue in kyanite; luster glassy to pearly.

*c.* Epidote, $Ca_2Al_3OSiO_4Si_2O_7OH$. Crystals flattened columnar, more often microscopic; pistachio to dark green; luster glassy.

*d.* Talc, $Mg_3(Si_2O_5)_2(OH)_2$. Crystals usually microscopic; white, gray, apple green; luster pearly to greasy.

*e.* Serpentine, $Mg_3Si_2O_5(OH)_4$. Crystals submicroscopic, often fibrous; various shades of green; luster greasy, waxy, or silky.

*f.* Chlorite, $(Fe,Mg)_6(Si_2O_5)_2(OH)_8$. Crystals sheetlike, often microscopic; various shades of green; luster glassy to pearly.

*continued*

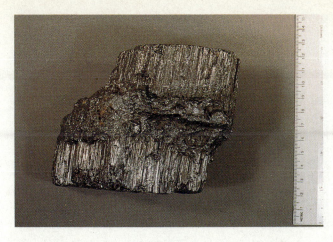

*g.* Graphite, C. Crystals sheetlike, often microscopic; black; luster metallic to earthy.

**Specimens courtesy of Earth Sciences Dept., Dartmouth College**

Extensive studies of mineral changes during metamorphism have led to an understanding of how increasing pressure and temperature influence metamorphic changes. The nature of these changes depends on the type of mineral assemblage in the rock that is undergoing metamorphism. If the rock consists entirely of a single, *nonreactive* mineral (e.g., quartz in sandstone, or calcite in limestone), the only change that normally takes place is a gradual growth of the mineral grains and a welding of their boundaries. Under sufficient pressure, however, calcite is converted to aragonite because of its more compact crystal structure.

If the rock consists of one or more *reactive* minerals, metamorphism can produce chemical changes as well. Under increasing pressure, sodic plagioclase, for example, is converted to the green sodic pyroxene *jadeite,* which is much prized as an ornamental gemstone under the common name jade. Clay is converted to *kyanite* (Greek for "bluestone"; fig. 3.18*b*), the blue, high-pressure variant of Al-silicate. In both mudstone and basalt, high pressure often produces the striking, blue sodic amphibole *glaucophane* (Greek for "bluish appearance") from a variety of other minerals (see fig. 4.34*d*).

The effects of increasing temperature on minerals are more extensive than those of pressure, frequently producing long chains of reactions with as many as seven different mineral species before melting finally occurs (see fig. 7.32). The most important of these temperature-driven metamorphic reaction chains are as follows, with original minerals shown at the left:

1. Organic material → graphite
2. Assorted calcium-rich minerals → epidote → calcic plagioclase
3. Olivine, pyroxene, or amphibole → serpentine → talc → amphibole → pyroxene
4. Clay → chlorite → biotite → garnet → staurolite → Al-silicate (*Staurolite,* $Fe_2Al_9O(SiO_4)_4(O,OH)_2$, is an iron-bearing Al-silicate.)

Because the ranges in temperature and pressures under which each of these minerals is formed are reasonably well known, the assemblage of minerals present within a particular metamorphic rock serves as a good guide to the pressure and temperature conditions under which the rock formed.

## Accessory and Ore Minerals

The igneous, sedimentary, and metamorphic minerals previously mentioned compose the bulk of Earth's crust. In addition, there are a few less abundant minerals that are important either because of their constant association with the more common rock-forming minerals or because of their economic value to our technoindustrial society.

There are certain ingredients typically found in a given rock type. In granite, for example, K-feldspar, Na-feldspar, and quartz are always present, and commonly clear muscovite mica, dark biotite mica, and amphibole also occur. Anything else is "foreign" to the normal granite system. Those minerals that the granite system tolerates enough to allow a light sprinkling of them to occur in the rock are called **accessory minerals.** Examples are *hematite,* $Fe_2O_3$ (fig. 3.17*b*) and *magnetite,* $Fe_3O_4$ (Greek for "magnet stone"; figure 3.19*a*).

There are many minerals that rock systems do not tolerate at all, however, because they contain ions, such as lead and sulfur, which are the wrong size to fit into the crystal structures of the common rock-forming minerals. These so-called **ore minerals** are flushed away in hot water solutions as their parent magma crystallizes. Typically, ore minerals crystallize with quartz or calcite in rock fractures to form thin, flat bodies of mineral material called **veins** (fig. 3.20). It is perhaps ironic that the minerals that rock systems tolerate least are the ones we value most!

Examples of ore minerals include the sulfides of various metals, including *pyrite,* $FeS_2$ (Greek for "firestone"; figs. 3.8*c* and 3.19*b*), a light-brass-yellow sulfide of iron that usually forms cubic crystals; *chalcopyrite,* $CuFeS_2$ (from *chalcos,* Greek for "copper" + *pyrite;* fig. 3.19*c*), a golden, usually massive sulfide of copper and iron; *galena,* PbS (Latin for "lead ore"; fig. 3.19*d*), a silver-gray sulfide of lead that forms cubic crystals; and *sphalerite,* ZnS (from *sphaleros,* Greek for "deceptive stone"; fig. 3.19*e*), a resinous, brown sulfide of zinc that forms cubic crystals with six good cleavages. Another common vein mineral that is not a sulfide is *fluorite,* $CaF_2$ (Latin for "flow stone," in allusion to its low melting point; figs. 3.6*d* and 3.19*f*), a variously colored fluoride of calcium that forms cubic crystals with four good cleavages.

**Figure 3.19** Common accessory and ore minerals.
*a.* Magnetite, $Fe_3O_4$. Crystals granular with triangular faces; iron black; luster metallic. Often magnetic.

*b.* Pyrite, $FeS_2$. Crystals cubic or granular, faces square or five-sided; pale brass yellow; luster shiny metallic.

*c.* Chalcopyrite, $CuFeS_2$. Crystals usually microscopic; deep brass yellow, often tarnished purple; luster metallic.

*d.* Galena, PbS. Crystals cubic; lead gray; luster shiny to dull metallic.

*e.* Sphalerite, ZnS. Crystals granular, often with triangular faces; usually dark yellowish-brown; luster greasy.

*f.* Fluorite, $CaF_2$. Crystals cubic or granular with square or triangular faces; light green, light amber, blue-green, violet; luster glassy to greasy. Cleavage fragments are shown at right.

Specimens courtesy of Earth Sciences Dept., Dartmouth College

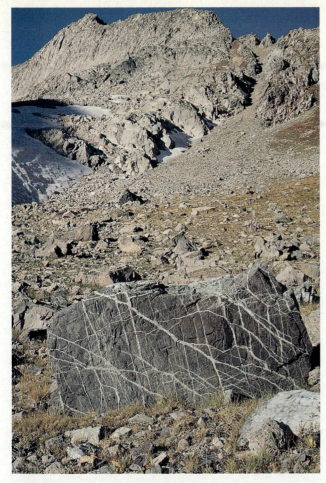

*Figure 3.20* A network of calcite veins cuts through a large limestone boulder on the slopes of K–2 Mountain in the Elk Range of Colorado.

## Summary

The Universe is made of **photons, electrons, protons, neutrons,** and **neutrinos.** Photons are massless energy particles vibrating at a wide range of frequencies that make up the **electromagnetic spectrum.** Electrons are the fundamental units of electricity. Quarks combine in groups of three to produce positive protons and neutral neutrons, each of which has about 1840 times more **mass** (resistance to change of motion) than the electron. Neutrinos are massless, uncharged particles.

**Atoms** are stable systems of oppositely charged protons and electrons mutually attracted by **electrostatic force.** The short-range **strong force** binds protons together at high temperatures, forming **atomic nuclei.** In the lighter nuclei, about half these protons are converted to neutrons, more than half in the heavier. Neutrons pad the nucleus, counteracting proton-proton repulsion. Electrons, one for each proton, surround the nucleus in a series of **electron shells** consisting of spherical or dumbell-shaped *orbitals* containing a maximum of two electrons each. The $n$th shell contains $n^2$ orbitals. All atoms of the same **element** have the same **atomic number** of protons in the nucleus. The number of neutrons in an element can vary, giving rise to **isotopes.** The proton number plus the neutron number is the *mass number.* The weighted average of the mass numbers of all isotopes of a given element is the element's **atomic mass.**

All atoms except the noble gases tend to lose or gain electrons in order to achieve a complete outer shell, thus becoming **ions.** If an atom loses outer-shell electrons, it shrinks and becomes a postively charged **cation.** If an atom gains outer-shell electrons, it expands and becomes a negatively charged **anion.** In **ionic bonding,** cations and anions are mutually attracted into alternating positions within a rigid **crystal structure.** In **covalent bonding,** pairs of outer-shell electrons are shared by adjacent atoms, binding them into *molecules.* Molecules containing more than one element are **compounds.** In **metallic bonding,** electrons are shared among all atoms. Asymmetrical charge distributions produce **residual bonds.** Large anions, like $Cl^{-1}$, and anionic groups, like $CO_3^{-2}$, make up the frameworks of crystal structures into which the smaller cations fit.

The **silica tetrahedron** is the fundamental unit of the **silicate minerals. Polymerization** of silica tetrahedra results in a diversity of silicate crystal structures. Different cations may substitute for one another in crystal structures only if they are about the same size and if the substituting ion(s) have the same electric charge as the replaced ion. **Solid solution** results from ionic substitution in all proportions. **Glass** lacks crystal structure.

A **mineral** is a *naturally occurring, inorganic, solid substance with a definite range of chemical composition, and a definite, ordered, internal structure.* Accretion of crystals on *nucleation centers* results from cooling of a liquid or gas, or when the concentration of ions in a solution rises. The **physical properties** of minerals depend on chemical composition, crystal structure, and bond type. **Specific gravity** is the weight of a mineral relative to the weight of an equal volume of water. A mineral's **streak,** the color of its powder, is a more characteristic feature than the color of a solid specimen. **Luster** is the appearance of a mineral's surface under reflected light. Some iron-bearing minerals are magnetic. Bond strength determines a mineral's **hardness,** *melting point,* and *tenacity,* or resistance to breaking. The *Mohs scale of hardness* is useful in mineral identification. Minerals split along **cleavage planes** of weakness in the crystal structure. **Crystal faces** develop parallel to those same planes, and obey the **law of constancy of interfacial angles.** Some minerals lack cleavage and instead show irregular **fractures.** Metallic bonded minerals conduct electricity because of the mobility of their bonding electrons.

To varying degrees, we rely on different minerals as natural resources. Feldspar is used in ceramics, quartz in a variety of applications from watches to glass. Garnet is an excellent abrasive, and calcite is used to make portland cement. Gypsum rock is mined mainly for plaster of Paris and wallboard. Supplies of these minerals are ample enough that at present there is no cause for concern over their exhaustion. There is, however, one mineral which, though still available today, will probably soon be depleted. Ironically enough, it is the one we can least afford to do without.

Within the past few decades, the "green revolution" has done wonders in making submarginally fertile lands agriculturally productive. A large part of this remarkable progress is due to the application of phosphate fertilizer, derived from phosphate rock. Phosphorus is an essential nutrient; and if its concentration in soils is too low, plants cannot grow. Most of our agricultural soils in the United States today are artificially fertilized with phosphate. Without this treatment, their productivity would decline and, in many cases, cease altogether.

Phosphate occurs naturally as the mineral *apatite,* $Ca_5(PO_4)_3(F,Cl,OH)$ (Greek for "deceptive stone"). Apatite deposits form mainly on the upper continental slope and outer continental shelf where cold, deep ocean water relatively rich in dissolved phosphate ion ($PO_4^{-3}$) rises to replace surface water that has been blown out to sea by offshore winds (see fig. 12.14). Here, phosphate ions tend to replace the carbonate ions in limestone. Apatite also occurs as a minor accessory mineral in most igneous, sedimentary, and metamorphic rock types.

Occasionally, it can become concentrated into large deposits in various igneous rock types, such as the one in a deposit of sodium-rich igneous rocks on the Kola Peninsula in northwestern Russia. Bird guano is another minor source of apatite.

Examination of trends in mining and production of phosphate over the past 40 years and of trends in the amount of proven economic world reserves of phosphate rock suggests the disturbing likelihood that these reserves will be exhausted within a few decades. There might be enough undiscovered and subeconomic reserves in addition to stretch the depletion time into the second half of the twenty-first century, but that simply postpones the inevitable. The fact still remains that in the near future, phosphate rock will be mined out and no longer available. When this happens, it will be necessary to reduce world population to a number that can be supported indefinitely on land that is naturally endowed with an adequate phosphate supply. That number has not yet been determined; but it is significant that in 1970, when the world population was about 4 billion, roughly half of us were being fed from land that required artificial phosphate fertilization. This suggests that 2 billion is a practical maximum for an indefinitely sustainable human population on Earth. Demographers predict a population of about *8* billion by 2010 A.D. Ironically, the longer the delay before the problem of depletion arises, the more severe the problem will be when it finally hits.

Could improved methods of conservation and recycling solve the problem? Unfortunately, the extremely reactive nature of the phosphate ion makes this unlikely. Phosphate is so actively and vigorously bound by cations in the soil that out of every 2.7 kg of phosphate applied to crops only 1 kg is actually captured by plant roots before being immobilized by soil cations. This insoluble, bound phosphate is then flushed into rivers, lakes, and the ocean and dispersed. Once dispersed, it is too expensive and energy-consumptive to reconcentrate. Currently, there are few uses for phosphate other than for fertilizer (principally detergents), so there is little to be gained in rediverting its other uses. Because the assimilation of phosphate by animals is only a moderate percentage of the still more moderate percentage assimilated by the plants they eat, more efficient recycling of animal (including human) wastes would likewise add little to available future reserves.

Are there substitutes for phosphate? No. As a fundamental ingredient in cell membranes, in some proteins, in the vital energy transfer molecule adenosine triphosphate (ATP), and in genetic material (DNA and RNA), phosphorus is an essential element for all life. No other element can function in its place.

What about the distribution of world phosphate reserves? We might be comforted by the knowledge that the United States possesses about one-fourth of them. Another fourth is scattered among various countries. The remaining half is in northwest Africa. This fact could have interesting consequences for the balance of world power in the twenty-first century.

Mineral formation is influenced by temperature, pressure, available ions, concentration of ions in solution, and time. Igneous rock-forming minerals crystallize from a molten state at high temperatures and pressures, according to the **Bowen reaction series.** The **continuous reaction series** governs the plagioclase feldspars and involves increasing substitution of sodium and silicon for calcium and aluminum with decreasing temperature, whereas the **discontinuous reaction series** governs the mafic minerals and involves increasing polymerization with decreasing temperature. Crystal size is a function of time and water content. Igneous and metamorphic minerals tend to be harder and denser than sedimentary minerals. The latter form only at Earth's surface at low temperatures and pressures by the weathering of preexisting minerals and by the precipitation of weathering products from solution. Metamorphic minerals form by the alteration of preexisting minerals under moderate to high temperature and pressure and in the presence of chemical solutions. **Accessory minerals** are normally associated in small amounts with the major rock-forming minerals. **Ore minerals** consist of ions that do not fit in the crystal structures of rock-forming minerals; hence, they are flushed away in hot water solutions to form **veins.**

The exhaustion of our world reserves of phosphate rock will force a drastic reduction in world population early in the twenty-first century because phosphate is an essential fertilizer for most of our agricultural lands.

## Key Terms

| | |
|---|---|
| photon | cation |
| mass | silicate mineral |
| electromagnetic spectrum | silica tetrahedron |
| | polymerization |
| electron | solid solution |
| proton | glass |
| neutron | mineral |
| neutrino | physical property |
| electrostatic force | specific gravity |
| atom | streak |
| atomic nucleus | luster |
| strong force | hardness |
| atomic number | crystal face |
| element | law of constancy of interfacial angles |
| isotope | |
| atomic mass | cleavage plane |
| electron shell | fracture |
| compound | Bowen reaction series |
| covalent bond | continuous reaction series |
| ion | |
| ionic bond | discontinuous reaction series |
| crystal structure | |
| metallic bond | accessory mineral |
| residual bond | ore mineral |
| anion | vein |

## Questions for Review

1. Which of the fundamental particles is (are) positively charged? Negatively charged? Uncharged? List them in order of decreasing mass.
2. What determines an element's identity? What distinguishes one isotope from another?
3. How are cations and anions formed?
4. Describe the functions of the bonding electrons in ionic, covalent, and metallic bonds.
5. What accounts for the great structural complexity of the silicate minerals?
6. Define a mineral. Give five examples of minerals and five examples of substances that are not minerals.
7. For each of the physical properties, state whether chemical composition, crystal structure, or bond type is (are) most influential.
8. How do the continuous and discontinuous branches of Bowen's reaction series differ in terms of crystal structure?
9. Compare and contrast the environmental conditions under which igneous, sedimentary, and metamorphic minerals are formed.
10. What is the principal difference between an accessory mineral and an ore mineral?

# Earth's Brittle Shell:
# Rocks of the Lithosphere

## Outline

*A*nd this our life, exempt from public haunt,

Finds tongues in trees, books in the running brooks,

Sermons in stones, and good in every thing.

*William Shakespeare*

***As You Like It***

Photo: Stream-eroded sandstone, White
Canyon, Utah.

**Figure 4.1** Chips off the old block. A massive outcrop of bedrock from which fragments, or clasts, of a variety of sizes have fallen away.

## Rocks and Their Environments

The term **rock** has a special meaning in Earth science. Most people think of a rock as a chunk of hard, heavy material that you can hold in your hand or perhaps lift with a bucket loader. Earth scientists, on the other hand, use the term in the commonly understood sense of *bedrock*, or ledge (fig. 4.1). Earth scientists refer to fragments of bedrock as **clasts** (from Greek *klastos*, "broken"), in other words, pieces that have become somehow detached from much more extensive bodies of rock.

In terms of composition, a rock is an *aggregate of minerals or minerallike substances*. The aggregate can consist of a single mineral, as in *quartz sandstone* (quartz), or of two or more minerals, as in granite (K-feldspar, sodic plagioclase, and quartz). In addition, some rocks consist partly or wholly of glass; hence, they are, strictly speaking, nonmineralic because glass lacks an ordered crystalline structure.

There is some debate as to whether ice should be regarded as a rock, and if so, whether it is igneous (it crystallizes from a molten state), sedimentary (it is deposited from a fluid medium—air), or metamorphic (in glaciers, it becomes highly deformed and recrystallized). If it is a rock, it differs from all of Earth's other rock types in its impermanence in the geologic record. For this reason, one waggish geologist suggested that it be given the name *kilroyite* because it leaves marks that persist long after the ice itself has vanished (see chapter 10).

There are three broad categories of rocks: *igneous, sedimentary, and metamorphic* (see chapter 1). To review briefly, the igneous rocks are aggregates of minerals that have cooled and solidified from molten magma. The sedimentary rocks consist of the disintegration products of preexisting rocks (or of living organisms in the case of coal) that have been reassembled at Earth's surface. The metamorphic rocks are aggregates of minerals that have been physically and chemically altered by conditions of pressure, temperature, and chemistry that differed markedly from those under which the aggregates were originally formed.

## The Rock Cycle

The igneous, sedimentary, and metamorphic rock types are interrelated through cyclic interchanges that are unique to Earth. Figure 4.2 shows the various pathways of these interchanges in a generalized way. Like the hydrologic cycle, the rock cycle is an open, equilibrium system (see chapter 2) that depends on energy inputs for its maintenance.

A convenient place to enter the cycle is at the lower center, where melting occurs within the crust or lower mantle. Melting can result from any of several factors or from a combination of them:

1. Release of pressure on rock that is near its melting point
2. The rise of hot, semifluid rock into regions of lower pressure
3. The injection of water or other fluids into hot rock
4. Extreme metamorphism

After rocks melt, the resulting magma begins to rise to higher levels in the crust because of its relatively low density with respect to the nonmolten rocks surrounding it. The process is similar to the rise of blobs of oil through partially mixed Italian salad dressing. Earth's lithosphere corresponds to the vinegar, and the magma corresponds to the oil.

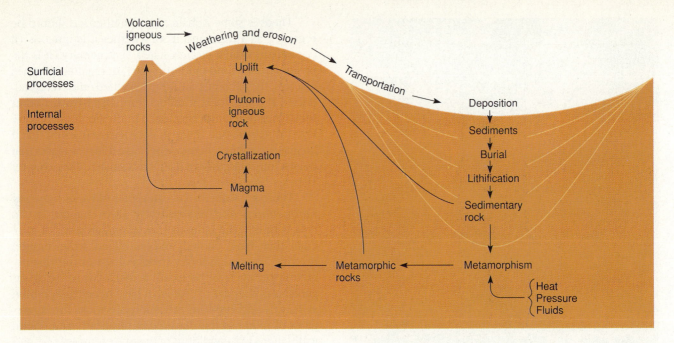

**Figure 4.2** The rock cycle, unique to Earth, illustrates the changes that can affect rock types over time. Notice that sedimentary and metamorphic rocks can go through a short cycle of uplift and erosion, or they can take the longer path through melting and transformation to igneous rock.

Either of two fates can befall rising magma. If it is significantly lower in density than the surrounding lithospheric rock, it is likely to rise to Earth's surface, where it can erupt from volcanoes as **volcanic igneous rock** (fig. 4.2, *left*), such as basalt. If the density difference between the magma and the surrounding rock is small, however, it will not float all the way to the surface but will slowly crystallize below ground to form a mass of **plutonic igneous rock** (after *Pluto,* Roman god of the underworld; see fig. 7.20), such as granite.

Eventually, Earth forces can uplift even deep-seated, plutonic rocks high enough that they become exposed at the ground surface as overlying rocks are worn away. During this process, the *weathering agents*—water, carbon dioxide, and oxygen—attack some of their constituent minerals, causing them to decay and crumble into loose *sediment.* Subsequently, this sediment can be *eroded,* or removed from its source area by water, ice, and wind. These same erosional agents can also *transport* sediment downslope until they become too weak to move it any farther, whereupon the force of gravity takes over and the sediment is *deposited.*

In the case of stream transport, coarse gravel is deposited closest to the source area, sand at a greater distance, and fine mud farthest from the source. The main reason for this is that the *gradients,* or slopes, of stream channels become gentler downstream. Gravity is most effective in moving material on steep gradients; hence, the farther the sediment is transported from the source area, the less effective is the force of gravity in assisting the water to move large clasts. In addition, clasts tend to become broken into finer fragments during transport.

After it is deposited, sediment is buried by the accumulation of subsequently deposited layers. The resulting pressure increase begins the process of *lithification* (see chapter 1) in which unconsolidated sediment is transformed into sedimentary rock. In this complex process, the sediment is subjected to **compaction** by the weight of overlying deposits and to **cementation** by the deposition of chemicals dissolved in seawater or groundwater. Such chemical cements include silica, calcium carbonate, hematite, and clay minerals. As the pressure of overlying sediment increases, mineral grains often also undergo **recrystallization.** In this process, they dissolve at points of greatest pressure and grow in areas of least pressure. This causes the grain boundaries to become sutured together as the spaces among adjacent grains are eliminated.

With still deeper burial, mountain-building activity, or both, these sedimentary rocks can be subjected to sufficient heat, pressure, and mineralizing fluids to change them into metamorphic rocks. If metamorphism is sufficiently intense, it can induce partial melting and the regeneration of magma, thereby closing the rock cycle. As the arrows leading from the words "sedimentary rock" and "metamorphic rock" to the word "uplift" in figure 4.2 indicate, this cycle can be short-circuited at any point by mountain-building activity, which can raise deep-seated rocks toward the surface, where they can be exposed to weathering.

In a broad sense, the rock cycle shows how Earth's diverse rock types are interrelated. From this holistic perspective, let us now turn to the differences that distinguish those rock types from one another.

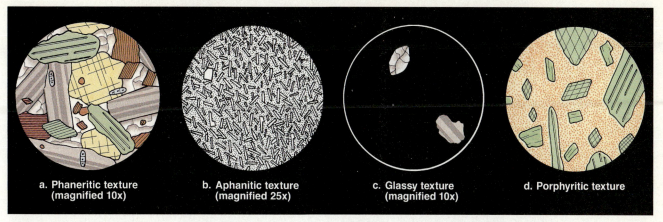

**a. Phaneritic texture** (magnified 10x)

**b. Aphanitic texture** (magnified 25x)

**c. Glassy texture** (magnified 10x)

**d. Porphyritic texture**

***Figure 4.3*** In phaneritic igneous rock texture (*a*), interlocking mineral grains are visible to the unaided eye. Earlier-formed minerals (***blue, dark brown***) often show straight-sided, angular forms; later-formed minerals (***yellow, white***) fill intervening spaces. Aphanitic texture (*b*) is similar, but grains are too small to be seen without magnification. Glassy texture (*c*) shows no grains even with a microscope, but occasional large crystals are often present. Porphyritic texture (*d*) has large crystals scattered within a finer crystalline matrix.

## Igneous Rocks

The classification of igneous rocks is based on texture and composition. The **texture,** or grain size and shape, of a particular igneous rock depends on the rate at which the rock cooled from a molten state. Its *composition* depends, in part, on the composition of the rock from which the parent magma was derived, but more importantly on the temperature at which the magma was generated. Let us consider texture first.

## Igneous Rock Textures

The most characteristic textural feature of igneous rocks is the tendency of mineral grains to interlock perfectly along smooth to uneven sutured boundaries, leaving no space among adjacent grains (fig. 4.3). This interlocking granular texture expresses itself on all scales, from submicroscopic crystal aggregates to groups of giant crystals measuring several meters across. Two factors, time and water content, determine the average size of crystals in igneous rocks. The longer it takes for magma to cool, and the higher its water content, the larger will be its crystals.

If magma remains confined to a chamber deep within Earth's crust, it cools slowly and crystallizes as a plutonic rock with a **phaneritic** (fan-uh-*rit*-ic; from Greek *phaneros,* "visible"), or coarse-grained, texture, like that of *granite* (figs. 4.3*a* and 4.5*a*). In slowly cooling magma, ions have plenty of time to migrate from one place to another. Consequently, the first nucleation centers to form are able to gather most of these migrating ions and grow to a large size before many other competing centers can form. This results in the growth of up to a few thousand crystals per cm³.

In contrast, if magma rises rapidly to the surface, as is the case with lavas erupted in oceanic ridges or volcanic vents, it will cool rapidly to a volcanic rock with a fine-grained or **aphanitic** (af-uh-*nit*-ik; from Greek *aphanes,* "unseen") texture, like that of *basalt* (figs. 4.3*b* and 4.11*b*). Note, however, that the chemical compositions of granite and basalt are different. Consequently, they do not normally crystallize from the same type of magma. Rapid cooling inhibits the migration of ions through the magma; hence, many nucleation centers are able to form before the earliest-formed centers can grow very large. This results in the growth of millions of tiny crystals per cm³. The line that divides phaneritic texture from aphanitic texture lies between rocks whose grains are and those whose grains are not distinguishable with the unaided eye.

The sudden chilling of magma causes extremely rapid and wholesale growth of nucleation centers, and allows virtually no time for ion migrations. This results in a *glassy* texture. The most common example of this texture is *obsidian* (figs. 4.3*c* and 4.5*d*), a glass of granitic composition. At the opposite extreme is the extremely coarse-grained texture of *pegmatite* (see figs. 4.5*b* and 4.7), in which individual crystals can measure as much as 15 m in length! The reason for the coarse grain of pegmatites is the abundance of superheated water in the magmas from which they form. Ions can migrate easily and rapidly to nucleation centers in such magma; hence, the first few centers to form grow so rapidly that they prevent the development of any other competing centers.

*Porphyries* (named after the Greek philosopher Porphyrius) are igneous rocks with a hybrid texture, called **porphyritic** (figs. 4.3*d* and 4.9*b*). Porphyries

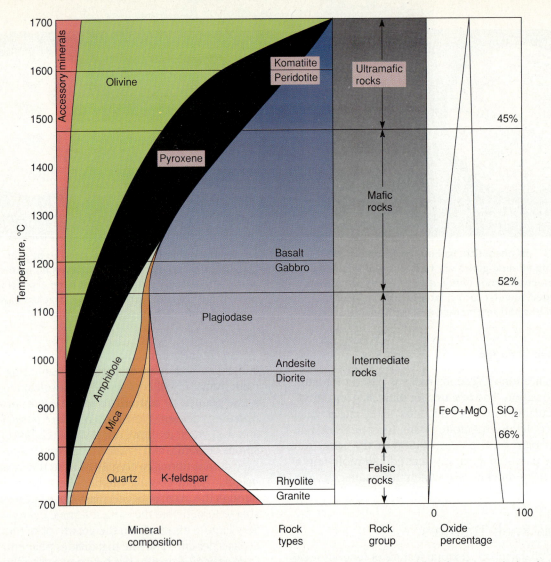

**Figure 4.4**  Mineral composition (*left*), darkness (*right center*), and content of silica, (SiO₂) and iron and magnesium oxides, FeO + MgO, (*far right*) can all be used in classifying igneous rocks. Temperature scale (*left*) and boundaries of mineral fields are based on averages.

Rock types are shown with aphanitic varieties (e.g., rhyolite) above phaneritic varieties (e.g., granite). Compare this diagram with that of the Bowen reaction series in figure 3.15.

have large crystals, called **phenocrysts** (Greek for "showy crystals"), embedded in a finer-grained *matrix,* which can be phaneritic, aphanitic, or glassy. Porphyritic texture results when magma cools slowly at great depth, allowing a few large crystals to form. Subsequently, it rises rapidly to a shallower depth, where the remaining liquid cools more quickly to a finer texture.

## Igneous Rock Composition

The composition of a magma depends on the temperature at which it formed (fig. 4.4). Magmas generated at high temperatures have compositions corresponding to the minerals at the upper end of the Bowen reaction series, i.e., olivine, pyroxene, and calcic plagioclase. Magmas generated at intermediate temperatures have

compositions corresponding to minerals having intermediate positions in the Bowen series, i.e., mainly pyroxene, amphibole, and intermediate plagioclase. Magmas generated at low temperatures consist largely of the relatively highly polymerized minerals located at the lower end of the Bowen series, i.e., amphibole, biotite, muscovite, quartz, K-feldspar, and sodic plagioclase.

From the bottom of figure 4.4 to the top, the minerals of the Bowen series decrease regularly in their content of silica (SiO₂), as shown by the curve labelled SiO₂ on the right side of figure 4.4. Because of this, the silica content of igneous rocks is used as a convenient basis for classifying these rocks into four broad groups. *Felsic* igneous rocks, at the bottom of the series, contain over 66% silica by weight. In most of them, some of that

a.

b.

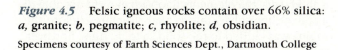

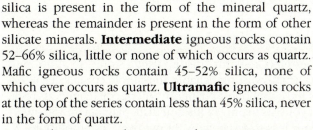
c.

d.

**Figure 4.5** Felsic igneous rocks contain over 66% silica: *a*, granite; *b*, pegmatite; *c*, rhyolite; *d*, obsidian.

Specimens courtesy of Earth Sciences Dept., Dartmouth College

silica is present in the form of the mineral quartz, whereas the remainder is present in the form of other silicate minerals. **Intermediate** igneous rocks contain 52–66% silica, little or none of which occurs as quartz. Mafic igneous rocks contain 45–52% silica, none of which ever occurs as quartz. **Ultramafic** igneous rocks at the top of the series contain less than 45% silica, never in the form of quartz.

As silica content decreases with increasing temperature of igneous rock formation, iron and magnesium content increases. This is shown by the curve labelled FeO + MgO (iron and magnesium oxides) in figure 4.4. Because of this increase and because the mafic minerals (olivine, pyroxene, amphibole, and biotite) tend to be dark colored, rocks that form at higher temperatures are darker than those that form at lower temperatures. Hence, the overall darkness of a rock is often a good index of its position on the scale from ultramafic to felsic. Note, however, that ultramafic rock can be pale green if it is composed of magnesium-rich olivine (see fig. 4.13*a*).

## Felsic Igneous Rocks

*Granite* (Latin for "seed-stone"; fig. 4.5*a*) is the most abundant phaneritic variety of silicic igneous rock. It consists of a roughly equal mix of quartz, K-feldspar, and sodic plagioclase (see fig. 4.4), and is light in color because all three of these minerals are also light. If the K-feldspar is pink, as it often is, that color is imparted to the rock. Small amounts of muscovite or biotite mica (or both) and amphibole are usually present, the latter two sprinkling the rock with dark specks. The specific gravity of granite is low, averaging 2.67. Because granite is a plutonic rock, it is typically found underlying large, irregularly-shaped land areas on the continents. These areas represent the exposed upper portions of solidified magma chambers that have been unroofed by erosion (fig. 4.6).

As the processes of cooling and crystallization within a granitic magma chamber proceed, water and other volatile substances become more concentrated in the residual, uncrystallized magma. This volatile fluid, in which granitic ingredients are dissolved, often runs into cracks and voids in the surrounding rock and solidifies, forming coarse-grained *pegmatite* (Latin for "massive stone"; figs. 4.5*b* and 4.7). Because the volatile fluid also dissolves many ions that do not fit into the crystal structures of granitic minerals, a variety of rare gem minerals is often found in pegmatites, including tourmaline, emerald, garnet, topaz, zircon, ruby, and sapphire. Pegmatite also occurs in situations in which it appears to have replaced solid rock by a process that is still not clearly understood.

**Figure 4.6** This 325-million-year-old granite in southeastern Vermont crystallized far below ground and has been exposed gradually by erosion. Gently curved fracture surfaces, reflecting the lack of any structure or grain in the rock, are typical of granite.

**Figure 4.7** Pegmatite typically fills large fractures in the bedrock surrounding magma chambers (note small offshoot at top). It also occurs in fractures within the magma itself, indicating that it forms after most of the magma has already crystallized.

**Figure 4.8** Rhyolite, the volcanic equivalent of granite, is usually found near granitic magma chambers, either exposed or below ground. The rhyolites shown here forming the steep walls of the canyon of the Yellowstone River were erupted on the Yellowstone Plateau, a vast region in northwestern Wyoming and southwestern Montana.

The aphanitic, volcanic equivalent of granite is *rhyolite* (Latin for "stream-stone," an allusion to its frequently flow-banded appearance; see fig. 4.5c). Rhyolite is a pasty, often explosive lava that typically blankets large tracts of land near felsic magma chambers (fig. 4.8). It usually appears somewhat darker than granite and is quite variable in color, ranging from gray or greenish through yellow to deep red. Frequently, rhyolite contains phenocrysts of quartz or K-feldspar, and it is often banded as a result of flowage. If felsic magma has an extremely low gas content, it can be extruded as a jet black or brown, glassy rock called *obsidian* (after

Obsidius, its Roman discoverer; see fig. 4.5d). As is true of all glasses, the elements in obsidian were not able to organize themselves into crystal structures before the magma was chilled to the solid state. In time, however, obsidian gradually *devitrifies* (i.e., looses its glassy character) mainly due to slow reaction with water, forming a dull gray rock called *perlite*.

## Intermediate Igneous Rocks

*Diorite* (Greek for "distinguished stone" in allusion to its usually distinguishable amphibole crystals; fig. 4.9a) is the commonest phaneritic intermediate igneous rock. It consists mainly of intermediate plagioclase (50–60%) with lesser percentages of pyroxene or amphibole, K-feldspar, biotite, and quartz. Diorite is darker than granite because of its higher content of the dark, mafic minerals pyroxene and amphibole, and because intermediate plagioclase is darker than K-feldspar and sodic plagioclase. Its specific gravity averages 2.84. Its occurrence is similar to that of granite, except that besides being found on the continents, diorite is also often found beneath volcanic islands, such as Japan and Indonesia. *Andesite* ("Andes-stone"; fig. 4.9b) is the aphanitic, volcanic equivalent of diorite. In colors and darkness, it is similar to rhyolite, from which it is most readily distinguished by the large phenocrysts of light plagioclase feldspar or of dark amphibole or pyroxene that are usually present in andesite and by its lack of flow-banding. Andesite occurs typically as lava flows mixed with ash layers in great volcanic peaks, such as those of Chimborazo and Cotopaxi in the Andes Mountains, from which the rock type derives its name (fig. 4.10).

a.

b.

**Figure 4.9** Intermediate igneous rocks contain 52–66% silica: *a,* diorite; *b,* andesite.

Specimens courtesy of Earth Sciences Dept., Dartmouth College

**Figure 4.10** A view eastward from Skylab 4 across the southern part of the South American continent. The Gulf of Corcovada on the coast of Chile is in the lower left foreground, bordered above by the snow-capped peaks of the Andes Mountains. In the upper part of the photograph, the desert plains of Patagonia stretch eastward to the Gulf of San Jorge on the coast of Argentina.

Photo by NASA

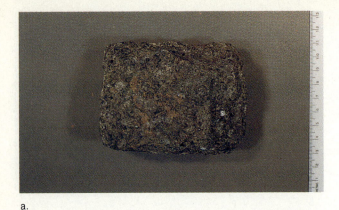

a.

b.

***Figure 4.11*** Mafic igneous rocks contain 45–52% silica: *a,* gabbro; *b,* basalt. Note gas bubbles, some of which have been filled with calcite and other minerals.

Specimens courtesy of Earth Sciences Dept., Dartmouth College

## Mafic Igneous Rocks

*Gabbro* (named after Gabbro, Italy; fig. 4.11*a*) is the principal plutonic, phaneritic variety of mafic rock. It consists mainly of calcic plagioclase, pyroxene, and often olivine. Because of its high content of dark minerals, gabbro is a dark rock, ranging in color from black to greenish-black. Its specific gravity averages 2.98. Gabbro occurs sparingly on the continents, but abundantly as a continuous layer beneath the ocean floor at depths between 1 km and the Moho (see chapter 1), which lies about 7 km beneath the ocean floor. *Basalt* (from Latin, *basaltes,* "dark marble," a misnomer; fig. 4.11*b*), the aphanitic, volcanic equivalent of gabbro, is dull, dark gray to black, sometimes with phenocrysts of glassy green olivine. Often, basalt contains gas bubbles that form near the tops of lava flows. It is the most abundant rock type of Earth's surface, forming the bedrock of the ocean floor as well as voluminous lava flows that cover many areas of the continents and oceanic islands, such as Hawaii (fig. 4.12).

## Ultramafic Igneous Rocks

*Peridotite* (French for "olivine-stone"; fig. 4.13*a*), is composed of olivine, with lesser amounts of pyroxene and calcic plagioclase. It varies in color from light yellow-green to black, depending on the relative percentages of olivine and pyroxene, respectively, and on the ratio of magnesium to iron in the olivine. Its specific gravity is high, averaging 3.23. Rare at Earth's surface, peridotite is nonetheless the most abundant phaneritic ultramafic rock, forming the entire thickness of Earth's upper mantle. When it does occur at Earth's surface, it is usually found along sutures between colliding lithospheric plates where slices of seafloor have been caught within the collision zone and thrust upward. *Komatiite* (koh-*matty*-ite, after Komatipoort, South Africa; fig.

***Figure 4.12*** A basalt flow on the north shore of the Bay of Fundy in Nova Scotia shows irregular fracture surfaces caused by cooling. The reddish coloration is due to hematite stain in thin coatings of silica.

4.13*b*), the volcanic, aphanitic equivalent of peridotite, is a geological curiosity. It is a rare ultramafic lava found only in parts of Africa, Australia, India, and Canada. These lavas were formed during the Archaeozoic Eon, when temperatures in Earth's mantle were still high enough to melt peridotite.

## Pyroclastic Igneous Rocks

As magma erupts at Earth's surface, it is drastically decompressed, allowing dissolved gases—mainly water vapor and carbon dioxide—to be released on a massive scale. This sudden release of dissolved gases accounts for the great violence of many volcanic eruptions. During such eruptions, lava is either atomized (blown into a fine spray) if still fluid, or shattered into fragments if already solidified. Flowing lava also often shatters if it solidifies as it moves. These materials, when subsequently lithified, are known as **pyroclastic** (Greek for "fire-broken") **rocks.** Coarse pyroclastic rocks, whose

a.

**Figure 4.13** Ultramafic igneous rocks contain less than 45% silica: *a,* peridotite; *b,* komatiite.

Specimen in *a* courtesy of Earth Sciences Dept., Dartmouth College
Specimen in *b* courtesy of Geological Survey of Canada

b.

a.

b.

**Figure 4.14** Pyroclastic igneous rocks are produced by volcanic explosions: *a,* volcanic breccia (note angular basalt clasts); *b,* tuff.

Specimen in *b* courtesy of Earth Sciences Dept., Dartmouth College

fragments average over 64 mm in diameter, are deposited near volcanic vents and are called *volcanic breccia* (*bretch-*ah; Italian for "gravel"; figs. 4.14*a* and 4.15). Fine pyroclastic rocks, whose fragments average less than 2 mm in diameter, are laid down in blankets over large expanses of territory and are called *tuff* (rhymes with "woof"; from Italian, *tufa*). Pyroclastic rocks containing fragments of intermediate sizes (2–64 mm) are less common. Many extensive deposits of what was once identified as rhyolite have subsequently been reidentified as *ash-flow tuff* (fig. 4.16), a pyroclastic rock consisting of minute shards of glass blown from a volcanic vent as clouds of frothing, sizzling magma droplets. If the shards become fused together as they settle to the ground, the deposit is known as *welded tuff* (fig. 4.14*b*).

**Figure 4.15** Dark, basaltic volcanic breccia overlies pink sandstone in a sea cliff on Cobequid Bay, Nova Scotia. Note large, angular basalt clasts.

**Figure 4.18** Clastic sedimentary rocks are composed of fragments of preexisting rocks: *a,* conglomerate; *b,* breccia; *c,* quartz sandstone; *d,* arkose; *e,* graywacke; *f,* clastic limestone with fossil fragments; *g,* shale (a variety of mudstone).

Specimens courtesy of Earth Sciences Dept., Dartmouth College

**Figure 4.19** The rounded clasts of this modest outcrop of red, arkosic conglomerate in northern Maine preserve evidence of the uplift of a massive mountain range immediately to the east in New Brunswick, Canada, about 375 million years ago. Following their deposition, the conglomerate strata were tilted gently westward by a second, milder mountain-building event.

gravels accumulate and can eventually become lithified into *conglomerate* (Latin for "balled together"), the coarsest of the clastic rock types (fig. 4.18*a*). Because each clast in a conglomerate is usually an aggregate of mineral grains, it can often be identified as having been weathered and eroded from a specific rock formation located somewhere in the vicinity. This kind of geological sleuthing can be of great value in reconstructing the sequences of events that occurred during the building of ancient mountain systems (fig. 4.19).

Conglomerate occurs in other settings as well. Following heavy cloudbursts, mountain streams often experience *mudflows,* destructive torrents of mud mixed with trees, boulders, and often assorted human artifacts as well. Deposits from such events are frequently lithified into conglomerate. Another potential source of conglomerate is sediment transported by a flowing glacier and consisting of rock fragments spanning a range of sizes from clay to boulders (see fig. 10.17).

These environmental settings, involving high-energy fluid transport, all produce gravel with relatively well-rounded clasts, but several other processes that do not involve fluids produce gravel consisting of angular fragments. Earthquakes shatter bedrock, and the force of gravity produces rockfalls and landslides on steep slopes and causes the collapse of underground caverns.

**Figure 4.20** These strata of the Lower Cretaceous Dakota Sandstone in western Colorado (see also fig. 2.4) were steeply uptilted during the building of the Rocky Mountains about 70 million years ago. The quartz sand that was lithified to form the Dakota Sandstone was originally deposited as a beach bordering a vast, shallow, inland sea.

Sedimentary rocks created by the lithification of such deposits are called *breccias* (fig. 4.18*b*).

## Sandstones

Finer-grained clastic sediments are deposited under the influence of gentler flows of wind and water. Beaches, windy desert regions, river channels, some river deltas (see fig. 7.27), and the deep ocean floor adjacent to continental margins are typical environments for the deposition of sediment with sand-sized clasts (1/16–2 mm). Such deposits can become lithified into *sandstones.*

Mineralogically, sandstones can be fully as diverse as conglomerates, although an individual grain in a sandstone, because of its small size, rarely consists of more than a fragment of a single mineral crystal. Sand that has not been transported far from its source can consist of practically anything, from shell fragments to fragments of basalt. Indeed, the beaches of tropical islands often consist of sands having either or both of those compositions.

Given a fair amount of transport by wind or water, however, minerals that are soft (e.g., talc, gypsum), fragile (e.g., mica, calcite), soluble (e.g., gypsum, calcite), or relatively unstable at Earth's surface (e.g., olivine, pyroxene, amphibole, calcic plagioclase) break down, leaving only the more resistant minerals (e.g., quartz, clay) as a residue. The sand in some pure sandstones has been so extensively transported by streams and waves that it consists of almost 100% silicon dioxide and can be used to make glass and fine ceramics. Sedimentary rocks containing more than 90% quartz are known as *quartz sandstones* (figs. 4.18*c* and 4.20).

Impure sandstones that contain less than 90% quartz are even more abundant than pure sandstones. The kinds of impurities present in these rocks depend on the type of rock present in the source area. If the source is granitic continental crust that has been uplifted into mountain ranges or high cliffs, the resulting rock is rich in feldspar and is known as *feldspathic sandstone*. If feldspathic sediment undergoes very little transport, the feldspar content can exceed 25%, in which case it can become lithified into a common variety of feldspathic sandstone known as *arkose* (a French term of obscure origin; figs. 4.18*d* and 4.21). Arkose sometimes resembles the granite from which it was derived so closely that it can only be distinguished from it by careful examination under a hand lens. Such examination reveals that the grains of the rock are not tightly interlocked along sutured grain boundaries, as they are in igneous rocks, but are more or less rounded by abrasion during transport and cemented together by a matrix of fine-grained material.

**Figure 4.21** The weathering and erosion of high, granitic mountains to the east produced this massive deposit of red arkose in the Elk Range of western Colorado about 300 million years ago.

Where mountain uplifts consist of such erosion-resistant, aphanitic rocks as slate or lava, the weathering products eroded from them are rich in lithic (rocky) fragments; hence, the sedimentary rocks formed from those products are known as *lithic sandstones*. If a lithic sandstone contains a large proportion of mud among the lithic grains, as is usually the case, it is called a *graywacke* (from the German *Grauwacke,* "gray stone"; see fig. 4.18*e*). Typically, lithic sandstones are deposited in deep marine basins, whereas feldspathic sandstones are deposited on land surfaces or shallow continental shelves.

A different, but common type of "sandstone" is **clastic limestone** (see fig. 4.18*f*), a rock consisting wholly of sand-sized fragments of calcium carbonate cemented together either by opaque calcite mud or by clear, crystalline calcite. The fragments can be any of the following:

1. Broken shell material
2. Chips of lime mud
3. Small, probably chemically deposited beads of calcite called *ooliths* (*oo*-uh-liths, Greek for "egg stones")
4. Pellets of limy excrement from marine worms and other bottom-dwelling organisms

Some clastic limestone is being formed today on the Bahama Banks and on the shallow seafloors flanking many tropical islands. Vast quantities of it were produced during the Paleozoic Era when large regions of North America were submerged beneath warm, shallow seas. Occasionally, quartz sand is brought into areas of clastic limestone formation by marine currents. This can result in a hybrid sedimentary rock type known as *calcareous sandstone*.

## Mudstones

Clastic rocks consisting of fragments less than $\frac{1}{16}$ mm in diameter are known as *mudstones*. Mud is a sediment of tranquil, low-energy environments. Because of their small size, particles of silt and clay can only settle and be deposited when the water or wind transporting them becomes calm enough that swirling, turbulent currents can no longer keep the particles suspended.

Included under the general category of mudstones are the varieties *siltstone,* with grains predominantly of silt size ($\frac{1}{16}$–$\frac{1}{256}$ mm); *claystone,* with grains predominantly of clay size ($<\frac{1}{256}$ mm); and an exceedingly abundant, finely laminated mudstone called *shale* (Teutonic for "shell," perhaps because fossil shells are frequently found in shale; figs. 4.18*g* and 4.22). Shale is

**Figure 4.22** Partially mantled by clasts eroded from the overlying Mesaverde Sandstone, a thick formation of gray shale in southern Utah (see the Mancos Shale; fig. 2.4) displays an erosional pattern typical of shale: sharp-crested ridges and steep-walled gullies.

composed of silt or clay or both. Unlike siltstone and claystone, which tend to break into blocky chunks, shale is *fissile,* meaning that it splits into thin flakes parallel with its laminated structure.

A variety of siltstone whose sediment—mainly quartz silt—is transported and deposited by wind, is *loess* (*lurss,* with a silent "r"; from Swiss *losch,* "loose," in allusion to its soft character; fig. 4.23). Loess blankets much of the Mississippi, Ohio, and Missouri River valleys (see fig. 10.36). It appears to have been derived from the extensive deposits of mud carried southward by these rivers during the melting of the last great continental ice sheet. Loess deposits also cover large areas in Europe, China, and southern South America.

The mineralogy of most mudstones is considerably simpler than that of sandstones. Their silt- and clay-size fragments are readily altered during transport because of their small size and the great distances over which they are transported from the source area. Accordingly, most mudstones consist of (1) chips of quartz, which compose most of the silt fraction of the sediment, and (2) clay minerals, which compose most of the clay fraction. Hematite is sometimes present, imparting a reddish color to the rock. If the mud was deposited in an oxygen-free environment, undecomposed organic matter can be present, imparting a blackish color. In tropical environments, calcium carbonate, produced by marine organisms, sometimes mixes with clay or silt to form a calcareous mud called *marl.*

## Chemical and Biochemical Sedimentary Rocks

Whereas clastic sedimentary rocks are made of eroded, transported, and deposited fragments of preexisting rocks, **chemical** and **biochemical sedimentary rocks** are made of various cations and anions dissolved in river, lake, and ocean water. As the names suggest, chemical sediments are deposited without, and biochemical sediments with, the aid of living organisms. Figure 4.24 illustrates the ions most abundantly dissolved in river water, lake water, and seawater. It also illustrates the ionic minerals that commonly form from them. (This figure should make it clear that the dissolved calcium bicarbonate illustrated in figure 1.22 exists in the ocean not as calcium bicarbonate "molecules," but as independent, dissolved calcium cations and bicarbonate anions.) Most of these ions originate in weathering, volcanic activity, and organic processes. They are removed from solution in one of three ways:

1. By *replacement,* in which one species of ion in a mineral crystal is gradually replaced by another in solution
2. By *inorganic precipitation* from solution when the constituent ions of a particular mineral become too concentrated to remain in solution
3. By *organic precipitation,* in which living organisms actively remove ions from solution.

**Figure 4.23** Loess is silt blown by winds from desert areas or the floodplains of rivers that flow from melting glaciers. Vertical joints, or cracks, in loess allow it to stand in steep bluffs. In China, dwellings are often carved into the easily worked material.

Photo by H. E. Simpson, U.S. Geological Survey

**Figure 4.24** Common ions (*blue*) dissolved in river, lake, and seawater and the common ionic minerals formed from them by replacement (*yellow*), by chemical deposition (*brown*), and by biochemical deposition (*green*). Note that calcium enters into four different compounds, bicarbonate and chloride into three, and the remaining ions into one each.

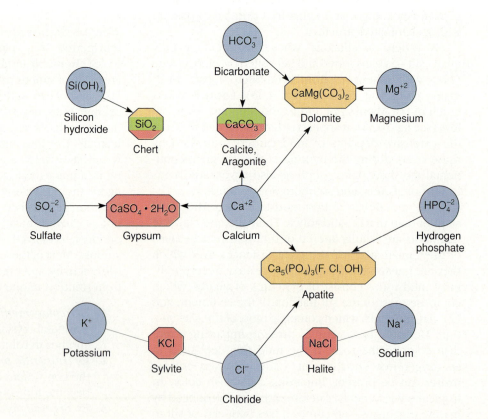

**Figure 4.25**    Logs of the conifer *Araucarioxylon* (Ar-row-car-ry-oh-*zy*-lon), an ancient relative of the Norfolk Island "pine," lie in the desert of southeastern Utah. The resistant logs were eroded from the Triassic Chinle Formation (*background*), within which they were petrified by silica released in great abundance by the weathering of ash from nearby volcanoes. The Triassic forests in which these trees grew were also lush with ferns, among which a variety of primitive reptiles prowled.

## Replacement Deposits

Probably the most familiar example of a **replacement deposit** is *petrified wood* (from *petrifactus*, Latin for "made into stone," fig. 4.25). In this instance, the cellulose in the cell walls of logs that have been buried in wet mudstone for millions of years has been gradually replaced by silica dissolved in the circulating groundwater. Under relatively acid conditions, silica can also replace some of the calcite in limestone to form nodules and layers of chert (see fig. 3.17*d*). Along arid tropical seacoasts, calcite in limestone is often replaced by dolomite to form *dolomite rock* (see fig. 3.17*b*) and by apatite to form *phosphate rock*.

## Evaporites

Rapid evaporation of seawater in isolated areas of the sea with restricted circulation can raise the concentration of dissolved ions to the point at which ionic minerals, including gypsum, halite, and sylvite (potassium chloride), begin to precipitate, forming **evaporite deposits.** *Gypsum* is the earliest of these to form and is by far the most abundant. For this reason *gypsum rock* (fig. 4.26) can be a dominant bedrock type over fairly large regions, as in parts of maritime Canada and the western United States. Ion concentrations seldom rise to the point at which halite is precipitated and even less often to the point of precipitation of sylvite. Nonetheless, vast deposits of *salt rock* have been laid down at various times in the geologic past.

**Figure 4.26**    Wave action carved this jagged arch in an outcrop of gypsum rock on the north shore of Cape Breton Island, Nova Scotia.

Both calcium carbonate and silica are precipitated inorganically on the continents. *Travertine* (after Tivoli, Italy, fig. 4.27), is an encrusting deposit of calcite that forms by the evaporation of calcium carbonate-charged water in caves, streambeds, beaches, pools, waterfalls, and hot springs. In some caves, as at Carlsbad Caverns in New Mexico, these incrustations form spectacular curtains and columns (see fig. 9.34). In arid regions, calcium carbonate is deposited in soils as groundwater evaporates from the soil surface. In places, these deposits of *caliche* (kah-*lee*-chay) can be extensive and thick enough to constitute a significant rock type (see fig. 8.16*e*).

In hot springs and geysers (see fig. 9.37), heated groundwater emerges at Earth's surface. This water frequently contains dissolved calcium carbonate or silica, either of which can be deposited as incrustations surrounding the hot springs or geyser vents. Where the deposit is calcium carbonate, it is travertine; where it is silica, it is called *geyserite.*

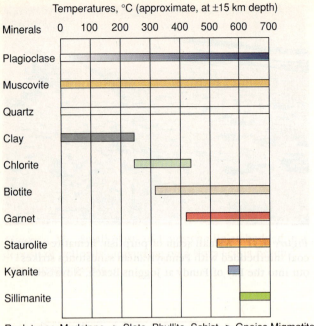

**Temperatures, °C (approximate, at ±15 km depth)**

Minerals | 0 100 200 300 400 500 600 700
Plagioclase
Muscovite
Quartz
Clay
Chlorite
Biotite
Garnet
Staurolite
Kyanite
Sillimanite

Rock types: Mudstone → Slate–Phyllite–Schist → Gneiss Migmatite

*Figure 4.33* Approximate metamorphic changes in mudstone (*bottom*) and its constituent minerals (*above*) under conditions of medium pressure and increasing temperature.

## Conspicuously Foliated Metamorphic Rocks

Because mudrocks contain a high percentage of sheet silicate minerals, they are usually strongly foliated. Figure 4.33 shows typical metamorphic changes with rising temperature in a mudrock under the kind of pressure that prevails at a few kilometers depth within Earth's crust. The minerals usually present in an unaltered mudrock are clay, quartz, plagioclase, and mica (in approximate order of decreasing abundance). At about 250° C, clay is converted to chlorite, and muscovite mica begins to recrystallize with the flat surfaces of its crystals oriented perpendicularly to the direction of any major stress acting on the rock. The rock splits readily along this foliation. Because of this property, *slate* (from Old French, *esclate,* a chip; figs. 4.34*a* and 4.35) is often used as a construction material. In Chester, Vermont, for example, many beautiful old houses were not only roofed with slate, but also built with large slabs of slate for siding (fig. 4.36). In *phyllite* (Greek for "leaf-rock," fig. 4.34*b*), a rock of slightly higher metamorphic grade than slate, the barely phaneritic muscovite flakes impart a silvery sheen to the rock, and often a wavy surface as well.

At about 400° C, most of the chlorite in slate has been converted to biotite mica. The parallel growth of both muscovite and biotite has proceeded to the point at which the mica crystals are phaneritic. At the same time, the growth of phaneritic crystals of quartz and feldspar among the mica flakes has begun to distort the flakes, further enhancing their wavy appearance. A rock with these characteristics is called a *schist* (from Latin, *schistos,* "splittable"; figs. 4.34*c* and 4.37). In subduction zones where pressures are high but temperatures are low, a distinctive variety of schist occurs (fig. 4.34*d*). Its name, *blueschist,* reflects the striking blue color of *glaucophane,* a sodium-rich amphibole that characterizes this rock type.

At slightly higher temperatures, chlorite and biotite are converted to garnet, which can form large, conspicuous crystals, called **metacrysts,** that visibly distort the foliation within the rock (see fig. 3.18*a*). If these are large and pure enough, they can be of gem quality. With increasing metamorphic intensity, metacrysts of *staurolite,* $Fe_2Al_9O_6(SiO_4)_4(O,OH)_2$ (Greek for "cross stone," in allusion to its tendency to form cross-shaped, twin crystals) and the two Al-silicates *kyanite* and *sillimanite* (the latter after American chemist Benjamin Silliman) can also appear within the rock, in that order, if it was originally clay-rich (see fig. 3.18*b*).

At still higher temperatures, the mobility of ions increases to a point at which the dark, platy micas become segregated from the light, granular quartz and feldspars, producing a distinctive, banded appearance. Metamorphic rocks having this feature are known as *gneiss* (*nice;* a German mining term; figs. 4.34*e* and 4.38). With further heating to about 700° C, partial melting occurs, forming granitic magma, which mixes with the unmelted portion of the gneiss to create a strikingly banded rock called *migmatite* (Greek for "mixed rock"; fig. 4.34*f*).

## Poorly Foliated and Nonfoliated Metamorphic Rocks

In some rocks, such as basalt and dolomite, metamorphism does not produce an abundance of linear or platy minerals. Such rocks do not develop conspicuous foliation. The original minerals of basalt—chiefly calcic plagioclase and pyroxene—are altered by low-grade metamorphism to sodic plagioclase, epidote, calcite, chlorite, and green amphibole. During the early stages of this process, the original appearance of the basalt is not greatly changed, and the resulting metamorphic rock is called a *spilite* (Greek for "spotted rock"; see fig. 1.16). In more advanced stages, the outlines of the original minerals become lost, resulting in a pale- to dark-green, practically structureless rock called *greenstone* (fig. 4.39*a*). Deeper in the oceanic crust—especially above subduction zones where magma is rising—more intense metamorphism produces a dark, inconspicuously foliated rock called *amphibolite* (fig. 4.39*b*). Amphibolite is composed of sodic plagioclase and amphibole. Sometimes the two minerals are well mixed;

a.

b.

c.

d.

e.

f.

*Figure 4.34* Foliated metamorphic rocks: *a,* slate; *b,* phyllite; *c,* biotite schist; *d,* blueschist; *e,* gneiss; *f,* migmatite.

Specimens courtesy of Earth Sciences Dept., Dartmouth College

in other instances, they are segregated into layers, giving the rock a more foliated appearance. Under extreme pressure, basalt and its metamorphic equivalents undergo a mineralogical change (but not a chemical one) to *eclogite* (Greek for "select-rock"; fig. 4.39*c*), a dense, spectacular rock composed of ruby-red garnet and jade-green sodic pyroxene.

In most *monomineralic rocks* (rocks composed of a single mineral) metamorphism simply results in recrystallization. The two most common examples of this situation are *marble* (from Greek *marmaros,* "sparkling stone"; figs. 4.39*d* and 4.40) and *metaquartzite* (fig. 4.39*e*), which are metamorphosed limestone and quartz sandstone, respectively. In each case, the orig-

Figure 4.35    An outcrop of Ordovician slate in northern Maine.

Figure 4.37    An outcrop of tightly folded mica schist on the coast of Maine.

Figure 4.36    A slate-sided and slate-roofed house in Chester, Vermont.

inal mineral grains have grown larger and formed a hard, compact mass of interlocking crystals. Marble has a coarser texture than most limestones. It is frequently pure white, whereas most limestones are gray on fresh surfaces because of organic impurities. Such impurities are baked out during metamorphism.

Metaquartzite and marble appear similar enough that they are often confused, but metaquartzite does not effervesce, or "fizz," in hydrochloric acid, as both limestone and marble do; and it scratches glass and steel, which neither limestone nor marble does. Metaquartzite can be distinguished from a well-cemented quartz sandstone by the fact that on fracturing, metaquartzite breaks *through* the sand grains. Sandstone, on the other hand, breaks *around* the grains.

*Metaconglomerate* (fig. 4.39*g*), the metamorphosed equivalent of conglomerate, exhibits the metamorphic properties of whatever rock types are represented among its clasts. In metaconglomerate, the originally rounded clasts have often been elongated or flattened by extreme deformation of the rock.

Figure 4.38    A massive cliff of Proterozoic gneiss rises across a glaciated valley from the continental divide in the Rocky Mountains of Colorado.

Low-grade metamorphism of ultramafic igneous rocks produces a palm- olive-green rock called *serpentinite* (Latin for "snake-rock"; fig. 4.39*f*), in which olivine and pyroxene have been altered to serpentine. For many years, serpentinite was mined for *asbestos,* a fibrous variety of serpentine that was used in construc-

a.

b.

c.

d.

e.

f.

g.

**Figure 4.39** Nonfoliated metamorphic rocks: *a,* greenstone; *b,* amphibolite; *c,* eclogite; *d,* marble; *e,* metaquartzite; *f,* serpentinite; *g,* metaconglomerate.

Specimens courtesy of Earth Sciences Dept., Dartmouth College

**Figure 4.40** A large, glacially transported boulder of marble in western Vermont.

tion as an insulating and fireproofing material. Recently, the inhalation of asbestos fibers has been found to be associated with lung cancer, because of which the mineral is no longer being mined. In many cases, it is being removed from older structures at great expense. Still more recent studies indicate, however, that serpentine asbestos is far less hazardous in this regard than another type of seldom-used asbestos called *crocidolite,* a variety of sodium- and iron-rich amphibole. This finding should provide some comfort to those who have despaired at the high cost of asbestos abatement, but little to those who have already incurred that cost or to those who have gone into the abatement business to take advantage of it.

# Food for Thought

Table 4.2 presents a summary of the relative abundances of the elements, minerals, and rocks in Earth's crust on a volume basis. All of those substances found only in Earth's crust but neither on the Moon nor on the other bodies of the Solar System are given in italics.

This table holds some surprises. First, it is clear that at least in terms of volume, Earth's crust is almost 94% oxygen! By weight percent, oxygen is a more reasonable 46.6%, but still far ahead of the next most abundant element, silicon (27.7% by weight). Oxygen is a large, light ion, as is

potassium, and that is why it occupies such an immodest volume in the crust. Silicon, on the other hand, is a small, dense ion, and that is why it occupies only a little less than 1% of the volume.

A second surprise is that the eight elements listed make up 100% of the

**Table 4.2** Percent by Volume of Elements, Minerals, and Rocks in Earth's Crust

| Elements | Percent | Minerals | Percent | Rocks | Percent |
|---|---|---|---|---|---|
| Oxygen | 93.8 | Plagioclase | 39.0 | Basalt/gabbro[1] | 42.5 |
| Potassium | 1.8 | *K-feldspar* | 12.0 | *Gneiss* | 21.4 |
| Sodium | 1.3 | *Quartz* | 12.0 | *Granodiorite*[2] | 11.2 |
| Calcium | 1.0 | Pyroxene | 11.0 | *Granite* | 10.4 |
| Silicon | 0.9 | *Amphibole* | 5.0 | *Schist* | 5.1 |
| Aluminum | 0.5 | *Mica* | 5.0 | *Mudstone* | 4.2 |
| Iron | 0.4 | *Clay* | 4.6 | *Limestone*[3] | 2.0 |
| Magnesium | 0.3 | Olivine | 3.0 | *Sandstone* | 1.7 |
| | | *Calcite* | 1.5 | *Marble* | 0.9 |
| | | Magnetite | 1.5 | *Syenite*[4] | 0.4 |
| | | *Dolomite* | 0.5 | Peridotite[5] | 0.2 |
| | | Other | 4.9 | | |
| | 100.0 | | 100.0 | | 100.0 |

1. Includes amphibolite and granulite.
2. Intermediate between granite and diorite. Includes quartz diorite.
3. Includes dolomite rock.
4. Like granite, but lacks quartz.
5. In lunar mantle.
Substances not found in lunar rocks are italicized.

Sources: Data from A. B. Ronov and A. A. Yaroshevsky. 1969. "Chemical Composition of the Earth's Crust," American Geophysical Union Monograph No. 13; and B. Mason. 1966. *Principles of Geochemistry,* 3rd ed. New York: John Wiley & Sons, Inc.

## Summary

**Rock** is an aggregate of minerals or minerallike substances. **Clasts** are rock fragments. Clast size decreases with distance from the source area. *Lithification* involves **compaction, cementation,** and **recrystallization.** Low density magma often erupts as **volcanic igneous rock.** Higher density magma usually crystallizes below ground as **plutonic igneous rock. Phaneritic texture** results from slow cooling, **aphanitic** from rapid cooling, **porphyritic** from slow followed by rapid cooling. *Porphyries* contain large **phenocrysts** in a finer *matrix. Felsic* igneous rocks contain over 66% silica, **intermediate** rocks 52–66%, *mafic* rocks 45–52%, and **ultramafic** rocks less than 45%. The darkness and iron and magnesium content of igneous rocks increase with increasing temperature of formation.

*Granite* is a felsic, phaneritic, continental, plutonic rock containing quartz, K-feldspar, and sodic plagioclase and is of light color and density. *Pegmatite* is an extremely coarse-grained granite. *Rhyolite* is the aphanitic, volcanic equivalent, and *obsidian* the glassy equivalent, of granite. *Diorite* is an intermediate, phaneritic, plutonic rock typical of island arcs. It contains intermediate plagioclase and pyroxene or amphibole and is darker and denser than granite. *Andesite* is the aphanitic, volcanic equivalent of diorite. *Gabbro* is a mafic, phaneritic, plutonic rock of ocean floors containing calcic plagioclase, pyroxene, and often olivine and is darker and denser than diorite. *Basalt* is the aphanitic, volcanic equivalent of gabbro. *Peridotite* is an ultramafic, phaneritic, plutonic mantle rock containing olivine, pyroxene, and calcic plagioclase and is denser than gabbro. *Komatiite* is a rare, volcanic, aphanitic equivalent of peridotite.

volume of the crust. What of such elements as carbon, sulfur, chlorine, and phosphorus, which are essential ingredients of limestone, gypsum rock, salt rock, and phosphate rock, respectively? The answer is that if the percentages are only calculated to the nearest tenth, as they are in table 4.2, these elements are all below the level of detection. Only titanium (Ti) comes close to being noticeable among the dominant eight. Even though layers of the previously mentioned sedimentary rock types cover vast areas of the planet, they are only a thin veneer over a more massive foundation of crystalline basement rocks that accounts for the bulk of the crust's total volume.

Third, it appears that although these eight elements are present in the crusts of both Earth and Moon, all but four of the listed minerals, and all but Moon's two of the listed rock types, are absent on the Moon. As far as we know, they are absent on the other bodies of the Solar System as well. Looking at this situation more closely, you might note that the four minerals present in the lunar crust are of igneous origin, as are the two rock types. This is not surprising, as there is no evidence that any sedimentary or metamorphic processes have ever been active on the Moon. Furthermore, all of the lunar minerals and rock types belong to the upper, high-temperature end of the Bowen reaction series. They are all high in iron and magnesium, low in potassium and silicon, and bone dry.

Can any generalizations be similarly drawn for those minerals and rock types that are unique to Earth's crust? At first blush, it seems like a heterogeneous assemblage of unrelated materials. Among the minerals of igneous origin, we have K-feldspar, quartz, amphibole, and mica; among the igneous rocks, we have granite and its close cousins *granodiorite* and *syenite*. The minerals of sedimentary origin include clay, calcite, and dolomite; and the sedimentary rock types include mudstone, limestone, and sandstone. Minerals of metamorphic origin are represented among the clays and the minerals in the igneous list; metamorphic rocks are represented by gneiss, schist, and marble.

A possible clue emerges when, in searching for some common thread, we find that three of the minerals— amphibole, mica, and clay—contain the *hydroxyl group* (OH) in their chemical formulas. This clue leads to the desired answer: with the possible exception of syenite, a quartz-poor relative of granite, *all these minerals and rock types require the presence of water for their formation.* In the case of granite and granodiorite, water is necessary to lower the melting point of the mixture of raw materials (typically mudstone or its metamorphosed equivalents) from which these rock types form. In the case of quartz, water under high temperature and pressure is required to dissolve that mineral out of other silicates. Amphibole and mica both crystallize within water-rich magmas or metamorphic rocks. Clay is a water-rich weathering product of feldspar. Sandstone and mudstone are eroded, transported, and deposited by flowing water. Calcite and limestone are mainly byproducts of vein formation and marine photosynthesis.

This result supports my statement in chapter 1 that water is the essential ingredient that permitted not only the evolution of life on Earth, but also the evolution of a diversity of minerals and rock types.

**Pyroclastic rocks** are formed as dissolved gases explode from decompressed, erupting magma. Coarse *volcanic breccia* is deposited near volcanic vents. Fine *tuff* is more widely deposited. *Ash-flow tuff* consists of glass shards, *welded tuff* of fused shards. Most igneous rocks form by the **partial melting** of preexisting rocks.

Clasts in **clastic sedimentary rocks** can be rock fragments, resistant mineral grains, or chemically altered materials. Minerals high in the Bowen series are unstable at Earth's surface. Sedimentary formations are blanket-shaped, feather-edged, and stratified. *Conglomerate* consists of gravel-sized clasts transported by high-energy water flow or by moving ice. *Breccia* consists of gravel with angular clasts formed by earthquakes, rockfalls, landslides, and cavern collapse. *Sandstone* consists of sand-sized clasts transported by energetic wind and water flows. *Quartz sandstone* contains more than 90% quartz. *Arkose,* a *feldspathic sandstone* derived from the erosion of granite, contains over 25% feldspar. *Graywacke* is a muddy, *lithic sandstone* derived from the erosion of aphanitic rocks. *Clastic limestone* consists of sand-sized calcium carbonate clasts. *Mudstone* consists of clasts (mainly quartz and clay) less than $\frac{1}{16}$ mm in diameter deposited in low-energy environments. *Siltstone* has grains of silt size, and *claystone* grains of clay size. *Shale* is a *fissile,* laminated mudstone. *Loess* is wind-transported quartz siltstone.

**Chemical** and **biochemical sedimentary rocks** consist of ionic minerals, the former deposited without, the latter with, the aid of organisms. **Replacement deposits,** in which one species of ion is replaced by another, include *petrified wood, chert, dolomite rock,* and *phosphate rock.* **Evaporites,** precipitated inorganically from concentrated ionic solutions, include *gypsum rock, salt rock, sylvite,* and some *chert* (all marine), and *travertine, caliche,* and *geyserite* (all continental). *Micrite* *limestone* is a biochemical rock precipitated by algae. Rocks precipitated by floating microorganisms include *calcareous ooze* and *chalk* (both calcium carbonate), and *siliceous ooze* and some *chert* (both siliceous). *Reef limestone* is produced by corals and calcareous algae. The only important **organic sedimentary rock** is *coal,* which forms from *peat.* Plant remains are evident in peat, but seldom evident in coal.

**Foliated metamorphic rocks** develop as linear or platy minerals grow perpendicularly to applied stress in clay-rich rocks. Fissile *slate* has microscopic chlorite and muscovite flakes. Silvery, wavy *phyllite* has larger flakes. *Schist* has phaneritic muscovite, biotite, quartz, and feldspar, and, at higher grades, **metacrysts** of garnet *staurolite, kyanite,* and *sillimanite. Glaucophane*-rich *blueschist* forms in subduction zones. *Gneiss* and partially melted *migmatite* are banded. Nonfoliated metamorphic rocks lack linear and platy minerals. Metamorphosed basalt includes *spilite* (with sodic plagioclase, epidote, calcite, chlorite, and green amphibole), *greenstone* (same, but forms of original minerals lost), *amphibolite* (with sodic plagioclase and amphibole), and *eclogite* (with garnet and sodic pyroxene). *Marble, metaquartzite,* and *metaconglomerate* are metamorphosed limestone, sandstone, and conglomerate, respectively. *Serpentinite* is metamorphosed peridotite.

Earth's crust is mostly oxygen. All lunar minerals and rock types belong to the high-temperature end of the Bowen reaction series. All minerals and rock types that are unique to Earth require the presence of water for their formation.

## Key Terms

| | |
|---|---|
| rock | pyroclastic rock |
| clast | partial melting |
| volcanic igneous rock | clastic sedimentary rock |
| plutonic igneous rock | clastic limestone |
| compaction | chemical sedimentary |
| cementation | rock |
| recrystallization | biochemical |
| texture | sedimentary rock |
| phaneritic | replacement deposit |
| aphanitic | evaporite deposit |
| porphyritic | organic sedimentary |
| phenocryst | rock |
| intermediate | metamorphism |
| ultramafic | foliation |
| | metacryst |

## Questions for Review

1. Why is a stone not a rock? Identify two natural, solid substances that are not rocks, and explain why they are not.
2. Is a low-density magma more likely to crystallize with a phaneritic or an aphanitic texture? Why?
3. Describe the various fates that can befall granite as it moves through the rock cycle. How would basalt differ?
4. Describe two main trends associated with increasing temperature of formation in the igneous rocks. Explain those trends in terms of minerals present.
5. Which is more likely to leave a permanent sedimentary record after being weathered and eroded, granite or basalt? Why?
6. What must happen to the sediment produced by the weathering of a granite cliff if it is to become a. an arkose, b. a quartz sandstone, c. a calcareous sandstone? Could it become a graywacke? Why or why not?
7. Compare and contrast the conditions that result in the formation of a. a conglomerate, b. a breccia, c. a sandstone, and d. a shale.
8. Compare and contrast the conditions that result in a. replacement deposits, b. evaporites, c. biochemical sedimentary rocks.
9. Describe the various forms in which calcium carbonate and silica are deposited on land and in the sea, and the process responsible in each case.
10. Describe the structural and mineralogical changes undergone by a. a mudstone, and b. a basalt during increasing metamorphism.

# The Rock Engine: Plate Tectonics and the Lithosphere

# 5

## Outline

*The common growth of Mother Earth Suffices me . . .*

*Wordsworth*

*Peter Bell*

Photo: A prominent fault in northeast Baja
California marks a portion of the
transform fault boundary between the
Pacific and North American plates.

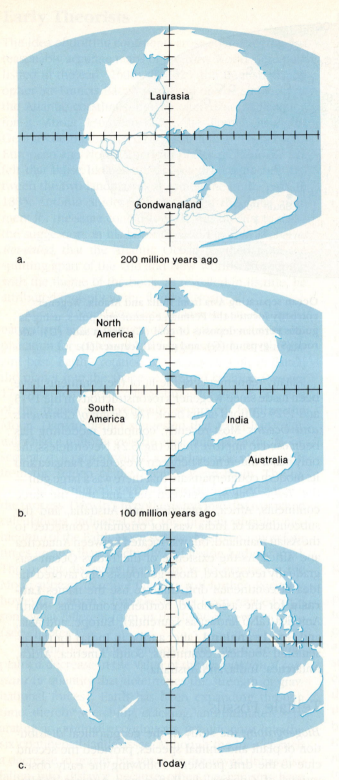

a.      200 million years ago

b.      100 million years ago

c.      Today

***Figure 5.2*** A succession of modern, computer-generated maps shows the progressive breakup of the supercontinent Pangaea over the past 200 million years. Map *a* differs from Wegener's map in figure 5.1 mainly in showing the broad Tethys Ocean that divided the eastern portions of Laurasia and Gondwanaland.

dominated by large *scale trees,* whose modern living relatives are the delicate little ground pines that often grow on dark forest floors of North America and Europe.

The scale trees were definitely *tropical* plants, however, as indicated by their lack of annual growth rings, a feature reflecting the annual recurrence of a winter season in temperate climates. Their presence in most of North America and Asia in the Late Paleozoic suggests that the equator was located in Laurasia at that time (fig. 5.3). In contrast, the flora of Gondwanaland was dominated by large *seed fern* trees, especially of the genus *Glossopteris* (Greek for "tongue-wing"), which had large, straplike leaves (fig. 5.4). The stem wood of these trees shows well-developed annual growth rings. Clearly, therefore, the environment of the Gondwanaland continents, where the seed fern flora grew, must have been temperate.

The present distribution of the fossil seed fern flora is patchy, or **disjunct.** If the continents were located in the Late Paleozoic as they are today, some means of transporting the heavy seeds of the seed ferns from one continent to another would have had to be available. Such a means has yet to be identified. On the other hand, if the Gondwanaland continents had all been united at that time, the difficulty would be avoided.

Two modest-sized reptiles provided some even more convincing evidence (fig. 5.4). The earlier is a primitive, freshwater animal called *Mesosaurus* that lived in the Permian Period (245 to 286 million years ago). Not only was *Mesosaurus* unsuited to the marine environment, but it was also rather flimsily built, and it would probably have fared poorly in a transatlantic crossing. Nevertheless, *Mesosaurus* fossils are found in both Argentina and South Africa.

*Lystrosaurus* was a stout, tanklike beast from the Triassic Period. This reptile was wholly terrestrial, and yet it, too, occurs in disjunct locations separated by thousands of miles of open ocean. One of the great tests of the drift hypothesis was to see whether or not *Lystrosaurus* could be found on Antarctica, as it had already been discovered on all the other Gondwanaland continents. In December 1969, *Lystrosaurus* fossils were indeed found by an American expedition (although the search for them was not deliberate) about 650 km from the South Pole in the Transantarctic Mountains.

## Mountain Ranges that Go Nowhere

Among Alfred Wegener's observations was the fact that many major mountain belts are abruptly terminated at continental margins. When he reassembled the continents that border the Atlantic Ocean basin, he found that two such truncated belts could be joined perfectly across the suture between Newfoundland and the British Isles. A second pair could be similarly joined between

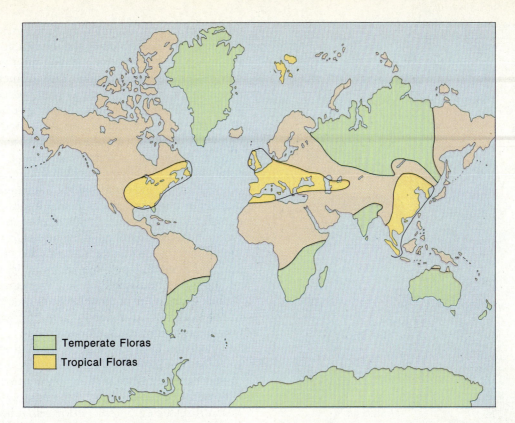

Temperate Floras
Tropical Floras

**Figure 5.3** The distribution of tropical (*yellow*) and temperate (*green*) fossil floras of the Late Paleozoic suggest that the contemporary equator passed not through South America and Africa as it does today, but through Argentina and South Africa. Not only were the directional trends the same, but the rock types and the geologic ages of the belts were the same as well: Early Paleozoic in the case of the northern belt and Late Paleozoic in the case of the southern belt. Spurred by this discovery, a South African geologist, Alexander du Toit, extended Wegener's investigation of the southern mountain belt into Antarctica and Australia and found that this belt could be traced across all these continents when they were reassembled in a particular way (fig. 5.5). More recent studies have fleshed out this compelling, but sketchy picture begun by Wegener and du Toit. Figure 5.5 shows the mountain belts that were formed early and late in the process of assembling Pangaea.

In addition to the rocks of mountain belts, another group of rocks provided further evidence for continental drift. A distinctive sequence of sedimentary and volcanic strata called the *Gondwana rock succession* is found not only in northern India, where it was first described and named, but also in Antarctica, eastern South America, southern and eastern Africa, and Australia (fig. 5.6); hence, the name Gondwanaland was given to the ancient landmass comprising these five continents because of the presence on all of them of the Gondwanaland rock sucession. Typically, the sequence begins

North America and Europe. The disjunct, or fragmented, distribution of floras on either side of the North Atlantic and in the Southern Hemisphere strongly suggests continental drift. (Polar regions are highly distorted.)

with deposits of lithified glacial gravel called *tillite*. It then passes upward into sandstones and mudstones interbedded with coal deposits formed from the undecomposed remains of the *Glossopteris* seed fern flora. At the top of the sequence are thick flows of basalt lava.

It is highly improbable that this unusual sequence of strata, spanning the 200-million-year interval between the Mississippian and Cretaceous Periods could have been deposited simultaneously on the five Gondwanaland continents if they had been in their present, widely disjunct positions at the time. If, however, the continents had been assembled into the supercontinent of Gondwanaland, the strata could easily have been deposited as a single, unified sequence of formations laid down within a clearly defined (and much smaller) region of Earth's surface.

## A Glacier that Came from the Ocean?

One of the most perplexing aspects of the Gondwana succession is that on the glacially scoured and polished bedrock beneath the tillite at the base of the sequence there is evidence that in several places the ice sheets advanced *inland from the sea* (fig. 5.7)! This odd but inescapable conclusion was drawn on the basis of

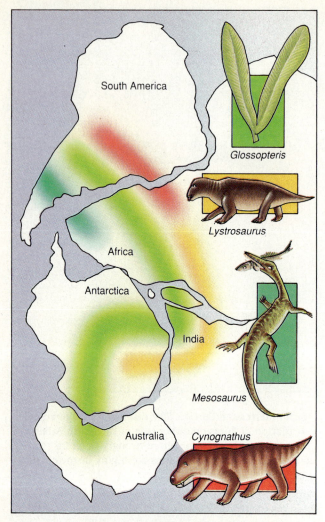

**Figure 5.4** The disjunct distributions of the four Permian and Lower Triassic plant and animal genera illustrated here provide compelling evidence, but not proof, of the breakup of Gondwanaland.

grooves and scratches on the bedrock made by the passage of rocks embedded in the sole of an ice sheet. Inasmuch as modern glaciers invariably form on upland areas of high snowfall, the idea that these Late Paleozoic ice sheets might have formed on the sea surface and then moved inland seemed bizarre. Again, the reassembly of the continents to form Gondwanaland presented a logical solution. Apparently, the ice sheets formed on Antarctica and South Africa and flowed outward from those continents into South America, India, and Australia.

## Glaciers in the Sahara, and Jungles under Ice

A meteorologist by profession, Wegener was alert to the likelihood that ancient climates would provide good evidence for continental drift. If the kinds of fossils and rock types preserved in the rocks indicated that the past

climate of a particular region differed greatly from the climate of the present day, then the case for drift might be strengthened. Wegener studied the rock and fossil records for their implications about contemporary climates and came up with maps of the **paleoclimate** of the Late Paleozoic. Not only did he find evidence of ice in southern Gondwanaland, he also found many other climatic anomalies that make no sense in terms of Earth's present geography. Among these is a belt of coal deposits (see fig. 5.1) that passes through the eastern U.S., central Europe, and central China. Flanking this belt and parallel to it are two others in which salt, gypsum, and desert sand are present, all indicators of an arid climate. From these distributions, Wegener inferred that the Carboniferous equator was located along the coal belt. The arid belts represent the zones of dry, descending air about 30° north and south of the equator. Clearly, the tidy distribution of these parallel paleoclimatic belts would be greatly disrupted if it were laid out on the continents as they are positioned today.

Evidence from other geologic periods also supports the concept of drift. Paleomagnetic evidence (discussed in a following section) shows that during the Ordovician Period the South Pole was located in northwest Africa (see fig. 5.32b). Ordovician tillites show that there were ice sheets where the Sahara desert is today. In contrast, Greenland, now covered with ice, supported luxuriant forests as late as the Upper Cretaceous. In Antarctica, 250-million-year-old coal from Permian jungles now lies beneath two miles of glacier ice.

## Wandering Poles

Of the planets of the Solar System, Jupiter, Saturn, Uranus, Neptune, and Earth have powerful magnetic fields. Spiraling currents within the molten iron of Earth's outer core are thought to be the cause of the geomagnetic field. An electric current in motion generates a magnetic field. An electrical conductor moving through such a magnetic field, in turn, generates an electric current. Complex interactions of this sort probably occur in the outer core. Because such interactions are controlled and maintained by Earth's rotation, the geomagnetic field tends to align itself with the rotational axis. *On average,* therefore, the geomagnetic and geographic poles coincide, but at any given time, they normally do not. At present, for example, they lie about 11.5° apart (fig. 5.8). Slow shifting of the geomagnetic poles relative to the geographic poles gives rise to corresponding shifts in **geomagnetic declination,** the angular difference between the directions of geographic north and geomagnetic north as measured from a given point on Earth's surface. A correction must always be made for geomagnetic declination when a compass is used to determine direction.

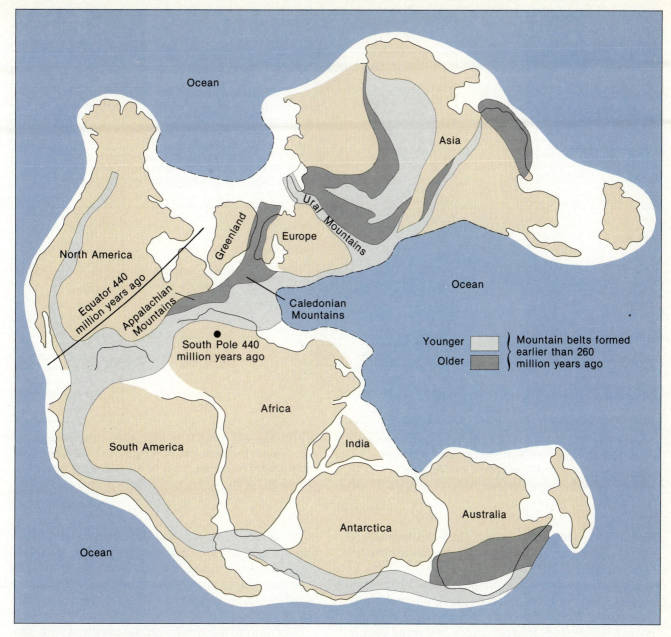

**Figure 5.5** Reconstruction of Pangaea shows ancient, stable continental masses (*brown*) partially welded together by Lower Paleozoic (*dark gray*) and Upper Paleozoic (*light gray*) mountain belts, formed mainly during the assembly of Pangaea by compression of marginal sedimentary rocks. Separation of Early Ordovician equator and South Pole by only about 20° indicates that Pangaea must have been assembled since that time (440 million years ago).

Notice, in figure 5.8, how the geomagnetic lines of force are parallel to the planet's surface near the equator, but that with increasing **latitude** or angular distance north or south of the equator, they dip at progressively steeper angles to the surface. This dip, known as **geomagnetic inclination,** is of little importance to navigation but of major significance to geology because it can be recorded in rocks by crystals of magnetite, and a few other iron-bearing minerals, which behave essentially like small compass needles. When they crystallize within a magma or are deposited in a clastic sediment these crystals align themselves roughly with the pre-vailing geomagnetic field. Sensitive magnetometers can detect such ancient magnetic alignments, and on the basis of this, the new science of **paleomagnetism** has arisen.

The principle behind paleomagnetism is illustrated in figure 5.8. For any given latitude on Earth's surface, magnetic grains deposited at that latitude have a corresponding, characteristic geomagnetic inclination, with their magnetic axes aligned with the lines of force. The inclination of magnetic axes can be determined from oriented rock samples. If the geologic age and the

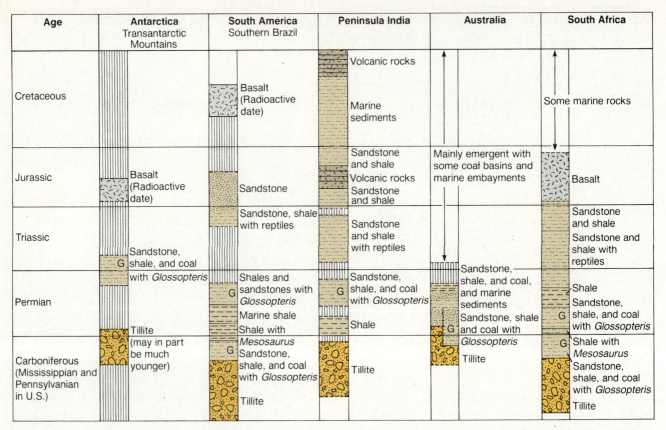

| Age | Antarctica<br>Transantarctic Mountains | South America<br>Southern Brazil | Peninsula India | Australia | South Africa |
|---|---|---|---|---|---|
| Cretaceous | | Basalt (Radioactive date) | Volcanic rocks<br>Marine sediments | | Some marine rocks |
| Jurassic | Basalt (Radioactive date) | Sandstone | Sandstone and shale<br>Volcanic rocks<br>Sandstone and shale | Mainly emergent with some coal basins and marine embayments | Basalt |
| Triassic | Sandstone, shale, and coal with *Glossopteris* (G) | Sandstone, shale with reptiles | Sandstone and shale with reptiles | | Sandstone and shale<br>Sandstone and shale with reptiles |
| Permian | Tillite | Shales and sandstones with *Glossopteris* (G)<br>Marine shale<br>Shale with *Mesosaurus* | Sandstone, shale, and coal with *Glossopteris* (G)<br>Shale | Sandstone, shale, and coal, and marine sediments<br>Sandstone, shale and coal with *Glossopteris* (G) | Shale<br>Sandstone, shale, and coal with *Glossopteris* (G)<br>Shale with *Mesosaurus* |
| Carboniferous (Mississippian and Pennsylvanian in U.S.) | Tillite (may in part be much younger) | Sandstone, shale, and coal with *Glossopteris* (G)<br>Tillite | Tillite | Tillite | Sandstone, shale, and coal with *Glossopteris*<br>Tillite |

**Figure 5.6** The same sequence of rock types occurring at the same time on five widely separated continents is one of the most powerful pieces of evidence for continental drift. Tillite (*bottom*) records a Carboniferous to Early Permian South Polar glaciation; sandstone-shale-coal (*middle*) records estuaries with coastal swamps; and basalt (*top*) records the onset of the rifting that preceded continental drift.

source of the sample are known, the **paleolatitude** of the source area at the time the rock was formed can be determined.

Equally significant to geology and paleomagnetism is the curious fact that occasionally the geomagnetic field disappears. When it reappears after a short interval, its *polarity* is reversed. In other words, following such a **geomagnetic reversal,** the north end of a compass needle would point southward. Geomagnetic reversals were first noted in lavas as early as 1906, but little attention was paid to them until after the end of World War II. Since then, geophysicists have undertaken the often tedious chore of measuring the orientations of magnetite grains from rocks that had been dated by radiometric methods. From this work, a detailed chronology of geomagnetic polarity reversals has emerged (fig. 5.9). In conjunction with radiometric techniques, that chronology has proved invaluable in the dating of volcanic rocks.

Many geophysicists had hoped that paleomagnetic studies would show that the positions of the continents have been fixed through geologic time with respect to Earth's rotational axis, disproving once and for all the "silly" notion of continental drift. Ironically, their results showed just the opposite! Because the geophysicists had been the most outspoken of the adversaries of drift, their abrupt change of attitude was a crucial step in the general acceptance of this "outrageous hypothesis."

For each of the continents, thousands of rock samples of various geologic ages were studied. In each case, the conclusion was inescapable: the positions of the geomagnetic poles have shifted drastically over geologic time. At first, it was supposed that the geomagnetic poles might have "wandered" relative to the rotational axis, so the paths of apparent pole migration came to be known as **polar wandering curves.** If this had been the case, however, and if the continents had always had the same positions on the globe relative to each other as they have today, then the polar wandering curves for the individual continents *should all coincide*. As figure 5.10 shows, however, this is not the case. The curves have similar shapes, but they diverge backward in time. If two such curves are rotated into coincidence, however, it turns out that the curves match quite well

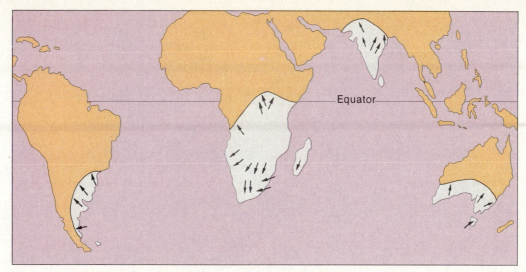

a.

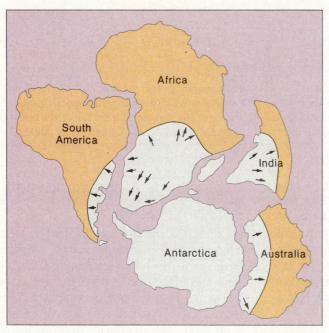

b.

*Figure 5.7*   Striations on glacially polished bedrock in
South America, India, and Australia suggest that glaciers
advanced inland from the ocean (*a*), a highly improbable
event. When the southern continents are reassembled (*b*),
the problem disappears, as these advances are seen to
have resulted from ice flows originating in south-central
Africa and Antarctica.

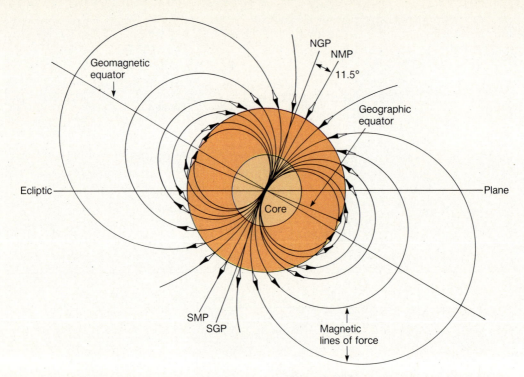

**Figure 5.8** At Earth's geomagnetic equator, a freely suspended compass needle parallels the ground. North of the equator, the needle inclines increasingly downward along geomagnetic lines of force, whereas south of the equator it inclines increasingly upward. At the geomagnetic poles, the needle's inclination is 90°. Note that the north and south geomagnetic poles (*NMP, SMP*) are currently separated from the north and south geographic poles (*NGP, SGP*) by an angle of about 11.5°.

up to about 200 million years ago, prior to which they again appear to diverge backward in time. The only possible explanation for such behavior is that the continents were once joined and that about 200 million years ago they began to split apart and follow separate paths. This conclusion seemed inevitable by 1960, but because a suitable driving mechanism for continental drift was still lacking, most geologists remained reluctant to jump on the "drift wagon."

## The Crucial Testimony

The reluctance of the geological community to accept the concept of continental drift without such a mechanism reflects on the extreme conservatism of science. Distasteful as such conservatism might be to the enthusiastic proponents of a new scientific concept, it ultimately works to the concept's advantage by ensuring that no reasonable doubt as to its validity remains. This screening process has good precedence in Nature itself in the form of the consistency principle (see chapter 2). Let us now look at the evidence that finally pushed continental drift through the geological "blarney filter" and established it as the fundamental, unifying principle of Earth science.

## A Submarine Mountain Chain 84,000 Kilometers Long

Prior to 1920, the standard method for studying the topography of the seafloor was the sounding line. Because each measurement required that the vessel be stopped and the line let down to the bottom, little progress in mapping seafloor topography was made until the German *Meteor* expedition of 1925 to 1927 made use of an *echo sounding* device to probe the configuration of the ocean bottom. The echo sounder projected a sound signal toward the ocean bottom and measured the time required for the signal to return to the ship after being reflected off the seafloor. Using this simple device, a research vessel can make a continuous topographic profile of the ocean floor rapidly and efficiently without having to stop (fig. 5.11). Since World War II, the technology of echo sounding has improved greatly and so has the national and international commitment to oceanographic research. In particular, the efforts of the International Geophysical Year (IGY) in 1957–1958 brought in a wealth of new information on the morphology (form) of the ocean floors.

By far the most remarkable discovery about the seafloor is the oceanic ridge system (figs. 1.2 and 5.12). This great submarine mountain chain can be traced from the Arctic Ocean off northern Siberia, past the North Pole,

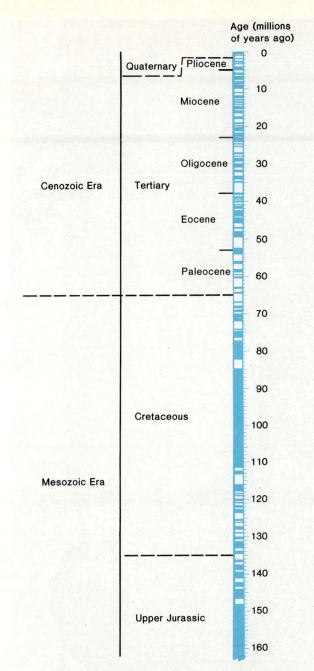

**Age (millions of years ago)**

| | | |
|---|---|---|
| Quaternary | Pliocene | 0 |
| | Miocene | 10 |
| | | 20 |
| | Oligocene | 30 |
| Cenozoic Era | Tertiary | 40 |
| | Eocene | 50 |
| | Paleocene | 60 |
| | | 70 |
| | | 80 |
| | Cretaceous | 90 |
| | | 100 |
| Mesozoic Era | | 110 |
| | | 120 |
| | | 130 |
| | | 140 |
| | Upper Jurassic | 150 |
| | | 160 |

**Figure 5.9** A detailed chronology of geomagnetic polarity reversals extending back into the Jurassic Period has been constructed from analysis of lavas of various ages. Major epochs (*thick bands*) and minor events (*thin lines*) of normal (*blue*) and reversed (*white*) polarity are defined by changes in the orientations of magnetic minerals within such lavas over time. A less reliable chronology extends well back into the Proterozoic Era.

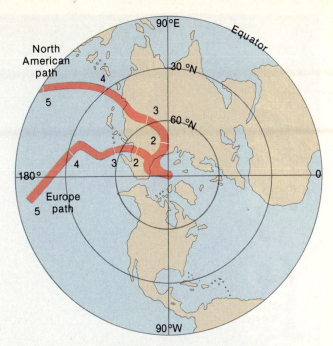

**Figure 5.10** Polar wandering is an apparent effect caused by continental drift. Here, paleopositions of the north geomagnetic pole are plotted for two continents, Europe and North America. Numbers are radiometric ages in hundred million years. Width of pole paths is an expression of uncertainty in the data.

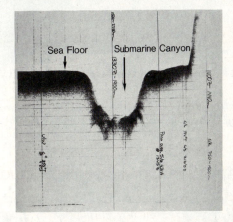

**Figure 5.11** An echo sounding profile of a submarine canyon made by reflected pulses of sound emitted by a research vessel.

Photo courtesy of Geological Data Center, Scripps Institution of Oceanography, University of California, San Diego.

south between Greenland and Scandinavia, down the midline of the Atlantic Ocean, east between South Africa and Antarctica, branching into the Indian Ocean, and on below Australia into the south Pacific, where it turns northeastward into the Gulf of California. A prominent trough, the *median rift,* runs along the axis of the ridge

system in most places (see fig. 1.4). This feature may be as much as a kilometer or more deep, measured from crests of the flanking escarpments. The entire range is offset by a series of several hundred major fractures that run perpendicularly to the main axis.

**Figure 5.14** The 146 m research vessel *JOIDES Resolution,* like its smaller predecessor the *GLOMAR Challenger,* is designed to obtain drill cores from the sediment and underlying basalt of the ocean floor.

Photo courtesy of Ocean Drilling Program, Texas A&M University

from these volcanoes has shown that they become progressively older toward the northwest. The volcanoes of Hawaii are presently active. The ones at the great bend in the chain became extinct about 40 million years ago, and the ones near the Aleutian trench have been extinct for about 75 million years. In the south Atlantic, a symmetrical pair of volcanic seamount chains extends from the oceanic ridge: one northeastward toward Africa, and the other (less well defined) northwestward toward South America (see fig. 5.12).

In 1962 and 1963, Canadian geologist J. Tuzo Wilson argued that such volcanic seamount chains are direct evidence of motion of the seafloor above **mantle plumes** of magma rising within Earth's mantle (see fig. 5.24). According to this idea, the Hawaiian-Emperor seamount chain is a continuous record of 35 million years of northward drift of the Pacific Ocean floor followed by 40 million years of northwestward drift. The volcanic chains in the south Atlantic, moreover, appear to record the progressive lateral drift of the South Atlantic Ocean floor symmetrically away from a mantle plume that is now located beneath the volcanic island of Tristan da Cunha.

## In Quest of an Unknown Earth: The Glomar Challenger

In 1968, a special research vessel, the *Glomar Challenger,* began the seemingly mundane process of drilling hundreds of holes up to 1.7 km deep in the ocean floor by means of an oil derrick mounted amidships and a long string of jointed, flexible drill pipe. Where the water is up to 7 km deep, this can be a for-

midable challenge, which the *Challenger* met with four water jets that kept the vessel precisely located over the drilling site on the seafloor. A cylindrical core of sediment with or without underlying basalt was removed from each drill hole and studied to determine such factors as the age and magentic inclination of the basalt and the thickness, composition, and fossil content of the overlying sediment.

This international *Deep Sea Drilling Project* (DSDP) was funded by the National Science Foundation and managed by the Joint Oceanographic Institutions for Deep Earth Sampling (JOIDES). The DSDP was terminated in 1983, but its work is being carried forward by the Ocean Drilling Program (ODP), employing a much larger and more sophisticated drilling ship, the *JOIDES Resolution* (fig. 5.14).

Results of the DSDP have already greatly expanded our knowledge and understanding of the seafloor. Two of the most significant findings are (1) that the thin blanket of marine sediment that overlies the basalt increases in thickness away from the oceanic ridge and (2) that the age of the basalt also increases with distance from the ridge. Proponents of continental drift took these two facts as powerful evidence for a drift mechanism first proposed by Arthur Holmes in 1928 and in greater detail in 1960 by Harry Hess of Princeton University. The mechanism in question, called *seafloor spreading,* involves the lateral growth of lithospheric plates away from the oceanic ridge as the median rift widens and new basalt, gabbro, and peridotite are added continually to the separating walls of the rift (see fig. 1.4). No other hypothesis seemed adequate to explain the decrease in both age and thickness of sediment cover toward the ridge.

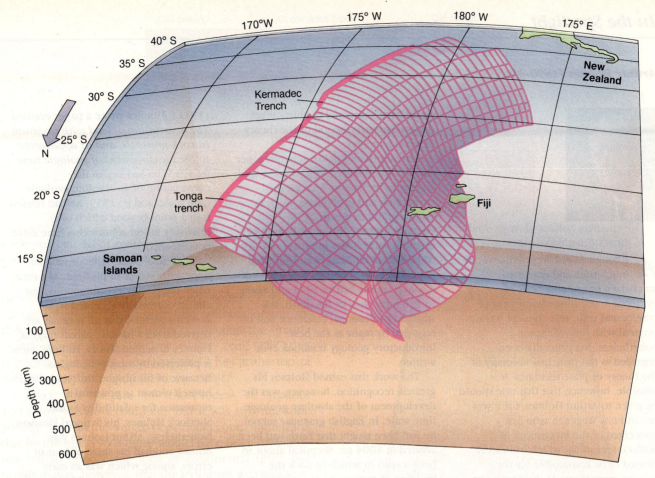

**Figure 5.15** A three-dimensional computer plot of earthquake foci defines the Benioff zone beneath the Kermadec and Tonga trenches in the southwest Pacific Ocean. The earthquakes outline the upper surface of the Pacific plate as it is subducted northwestward beneath the trenches (note north arrow, *left*). The chain of volcanic islands that lies parallel with the trench above the subduction zone is not shown. The Fiji Islands lie far to the west (*right*) of that chain.

From ''The Earth's Mantle'' by D. P. McKenzie. Copyright © 1983 by SCIENTIFIC AMERICAN, Inc. All rights reserved.

There remained one major problem, however. If the ocean floors are indeed spreading at the oceanic ridge, what happens to these growing lithospheric plates at their far edges? Clearly, there is no evidence for lithospheric plates advancing, bulldozerlike, over the surface of the globe or extending, winglike, off into space! Logically, the growth of new seafloor at the spreading median rifts required either an expanding planet to accommodate the increasing area of oceanic lithosphere or some mechanism for destroying old lithosphere at the same rate at which new lithosphere is being created. As so often happens in science, the key to this sticky lock had been in our hands for years, but we had failed to recognize its significance. That key was the earthquake activity beneath island arcs.

## Earthquakes beneath Volcanic Mountain Chains

Early in the 1930s, K. Wadati in Japan and Hugo Benioff in the United States studied the distribution of earthquakes beneath volcanic island chains and discovered an interesting and significant pattern. The *foci* (*foe*-sigh; singular, *focus*) or points of origin, of the earthquakes they studied were distributed along planes that descend at steep angles from oceanic trenches and extend far beneath their associated volcanic island chains (fig. 5.15). As the reality of seafloor spreading became evident, "drifters" suddenly came to appreciate the implication of these curious features, which came to be known among English-speaking geologists as *Benioff zones*.

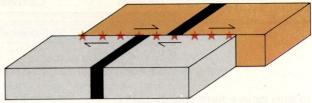

a. Strike-slip fault

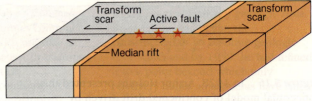

b. Transform fault

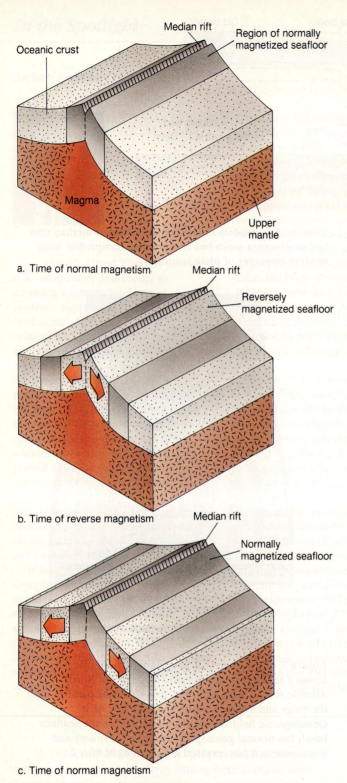

a. Time of normal magnetism

b. Time of reverse magnetism

c. Time of normal magnetism

**Figure 5.18** New additions of basalt to the oceanic lithosphere are imprinted with the polarity of the geomagnetic field in effect at the time they are emplaced.

**Figure 5.19** A strike-slip fault (*a*), separates blocks of rock that have moved laterally in opposite directions, progressively offsetting the severed ends of any feature that crosses the fault (*black band*). Earthquakes (*red stars*) are generated along the full length of the fault plane. The transform fault in *b* superficially resembles a strike-slip fault, but it does not progressively offset the median rift segments. Earthquakes occur only on the active portion of the fault between the offset ridge segments. In both *a* and *b,* the gray block is moving toward the left, the brown block toward the right.

figure 1.5 shows the ages of seafloor rocks and the oldest continental rocks. From this map, you can see that, compared to the continents, whose rocks approach an age of 4 billion years, the ocean basins are much younger, being nowhere older than about 200 million years!

The second piece of critical evidence came from the great **fracture zones** in the seafloor that offset the oceanic ridge (see fig. 5.12). At first, geologists assumed that the fracture zones were vertical breaks in the lithosphere along which the oceanic ridge had been offset to one side or the other by differential movement. Such breaks are commonly found on the continents and are known as *strike-slip faults* (see chapter 6). In such a fault, any linear feature that crosses the fault line is correspondingly displaced in the same direction, and for the same distance, as the movement on the fault (fig. 5.19*a*).

An obvious problem arose with this simple interpretation, however. Many of the ridge offsets are hundreds of kilometers long, yet the ends of the fracture zones die out quietly on either side of the offset without any evidence of the crumpling of the lithospheric plate that would be required to produce such an enormous displacement in the midsection of the fault. Furthermore, there was evidence of earthquake activity along the central portions of the fracture zones between the displaced ridge segments, but no evidence of such activity along the ends of the fracture zones

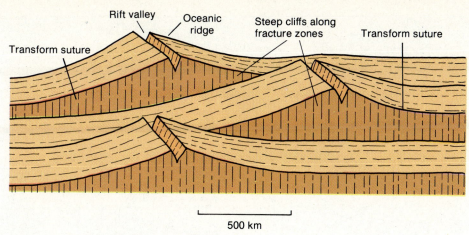

Rift valley
Oceanic ridge
Steep cliffs along fracture zones
Transform suture
Transform suture

500 km

*Figure 5.20*   Because of elevation differences along the oceanic ridge, the transform sutures on the inactive ends of fracture zones are often marked by steep cliffs. Until 1965, it was mistakenly thought that such cliffs were the result of vertical motion along the fracture zones.

beyond the ridge offsets. This led to the seemingly impossible implication that differential movement between the two sides of the fracture was occurring in the central section, but none was occurring on either end!

In 1965, Canadian geologist J. Tuzo Wilson offered an explanation. If the oceanic ridge is the site of seafloor spreading, he suggested, then the motion in the central portion of the fracture zone between the displaced ends of the ridge should be the *reverse* of the motion that would be required to offset the ridge by strike-slip faulting (fig. 5.19*b*). Furthermore, no relative motion would be expected on the ends of a fracture zone because here the plate sections on either side of the fracture should both be moving away from the median rift at the same rate. In other words, the fracture zone ends are really not fractures at all, but *sutures* formed by the joining of two different sections of the same plate that were generated at two adjacent, offset ridge segments (fig. 5.20)! Wilson gave the name **transform fault** to this special variety of ridge-related fault.

## The New Global Synthesis

In 1966, Lynn Sykes of the Lamont-Doherty Geological Observatory at Columbia University took up the challenge of testing Wilson's hypothesis. Analyzing earthquake records from the oceanic ridge, Sykes found that the actual motion on the transform faults was exactly as predicted by Wilson. With this benchmark discovery, the resistance of the geological community to the hypothesis of continental drift finally collapsed. Almost overnight, the diverse lines of evidence that had been accumulating over the years in support of drift were organized and synthesized into one of the most comprehensive, unifying concepts in all of natural science, the theory of *plate tectonics.*

## Earth's Migrating Armor

Because so little was known in his day about the nature of the seafloor, Alfred Wegener could only envision his drifting continents somehow plowing their way slowly through the denser basaltic crust of the ocean basins. The discovery of the oceanic ridge system, the Benioff zones, and the mechanism of transform faulting along the fracture zones rendered this improbable scenario unnecessary. All at once, it became obvious that the fundamental unit of motion was not the *continent,* but the *lithospheric plate.* The appearance of continental drift was simply a part of the more comprehensive reality of plate growth and destruction. The emerging theory of plate tectonics provided a new vision of lithospheric plates growing at the ridge system, being consumed in the Benioff zones, and slipping past one another along transform faults. That vision quickly resolved itself into the world map of seven major, and as many minor, tectonic plates shown in figure 5.21.

In chapter 1, I note that these plates are, on average, about 100 km thick; that they consist mainly of a thick, basal layer of upper mantle peridotite capped by a thinner layer of oceanic or continental crust, and that they overlie the partially molten *low velocity zone,* the low-strength layer within Earth's upper mantle that allows the plates to slide freely (see fig. 1.4). Lithosphere comes in two varieties, each with a different type of crust forming its upper layer. *Oceanic lithosphere* carries an average thickness of 7 km of oceanic crust, comprising, from bottom to top, layers of gabbro, basalt, and deep sea sediment. *Continental lithosphere* carries continental crust, comprising an average thickness of 40 km of igneous rock ranging in composition from granite to diorite, metamorphic rock of similar composition, and clastic sedimentary rocks.

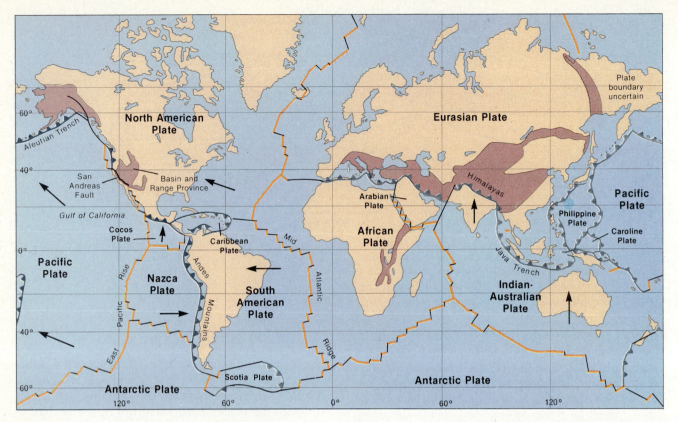

**Figure 5.21** Earth's lithosphere (crust and upper mantle) is broken into seven major and several minor tectonic plates that grow at the oceanic ridge (*orange*), slide past one another along transform faults (*black*), and are consumed by subduction in oceanic trenches (*dark blue, barbed lines*). Barbed lines on land indicate compressional faulting (see chapter 7) in continental collision zones. Arrows indicate directions of plate motion.

Source: U.S. Geological Survey, "Our Changing Continent," 1986 (181–404–155/40006), U.S. Government Printing Office.

Some lithospheric plates consist of a single type of lithosphere, whereas others consist of a combination of the two types. The Pacific, Nazca, and Cocos plates are entirely oceanic (except for a narrow sliver of southwestern California on the northeastern margin of the Pacific plate). There are no plates consisting entirely of continental lithosphere, although the Arabian plate comes close. The South American plate (fig. 5.21) is a typical example of a mixed type. The continent of South America forms only the western portion of the plate. The eastern part, between the edge of the continental shelf and the oceanic ridge, consists of the Atlantic Ocean floor. The entire South American plate therefore, moves *as a unit,* spreading westward from the oceanic ridge. On the western edge of South America, a subduction zone plunges eastward beneath the coast and the Andes Mountains. Here the lithosphere of the Nazca and Antarctic plates is being drawn down into the mantle.

With this brief introduction to Earth's lithospheric plates behind us, let us now take a closer look at the three types of boundaries along which they interact and at the processes that characterize each of them.

## Divergent Boundaries

Boundaries along which the adjacent lithospheric plates are moving away from each other, adding new oceanic lithosphere to their trailing edges as they go, have been called **divergent plate boundaries** (fig. 5.22). They are characterized by *extension,* or stretching, of the crust. Early stages of divergent plate boundaries are evident in the great East African rift system that extends southward from the Red Sea through East Africa. Along this rift zone, sets of *extension faults* (see chapter 6) have developed parallel to the rift (fig. 5.22*a*). Between these faults, elongated slivers of crust called *fault blocks* have subsided, and basaltic lava has erupted intermittently along many of the faults. In time, the African continent east of the rifts will probably split away from the western part and begin a slow drift toward the northeast. As this happens, a narrow ocean basin will develop within the rift zone as oceanic lithosphere grows progressively against the diverging continental margins (fig. 5.22*b*).

This more advanced stage of divergence is represented by the modern Red Sea and the Gulf of Aden separating Africa and Arabia. Both are embryonic ocean basins in the earliest stages of development. They are floored by basaltic crust with a median rift identical with

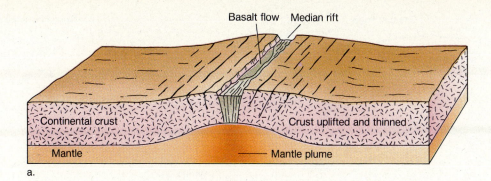

Basalt flow   Median rift

Continental crust          Crust uplifted and thinned

Mantle                     Mantle plume

a.

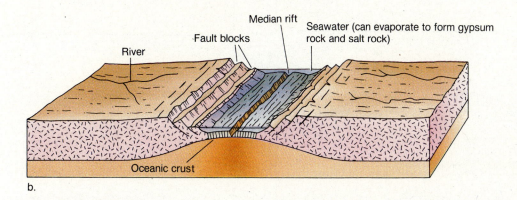

Median rift

Fault blocks      Seawater (can evaporate to form gypsum
                  rock and salt rock)

River

Oceanic crust

b.

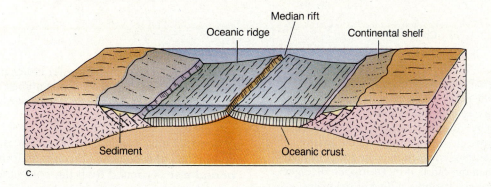

Median rift

Oceanic ridge          Continental shelf

Sediment          Oceanic crust

c.

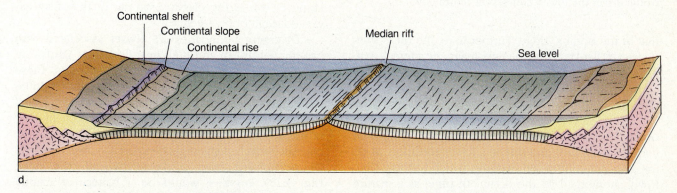

Continental shelf
Continental slope          Median rift
Continental rise                       Sea level

d.

**Figure 5.22**   Evolution of a divergent plate boundary. In *a,* a mantle plume arches, stretches, and fractures the continental crust, permitting the eruption of basalt lava. In *b,* fault blocks subside along extension faults as oceanic crust grows at the median rift and seawater invades the rift basin. In *c,* an oceanic ridge forms as the older seafloor cools and sinks, and clastic sediments begin to build continental shelves on the continental margins. In *d,* prisms of sediment have also begun to accumulate on the continental rises.

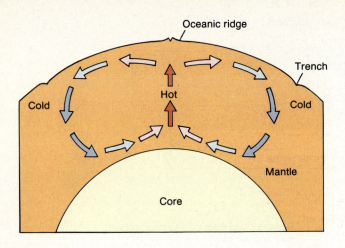

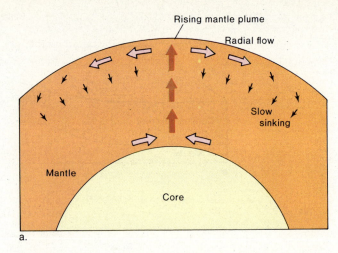

Rising mantle plume

Radial flow

Slow sinking

Mantle

Core

a.

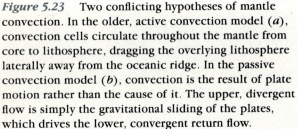

Oceanic ridge

Trench

Mantle

Core

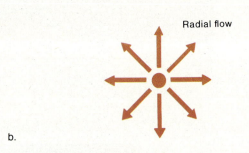

Radial flow

b.

*Figure 5.23* Two conflicting hypotheses of mantle convection. In the older, active convection model (*a*), convection cells circulate throughout the mantle from core to lithosphere, dragging the overlying lithosphere laterally away from the oceanic ridge. In the passive convection model (*b*), convection is the result of plate motion rather than the cause of it. The upper, divergent flow is simply the gravitational sliding of the plates, which drives the lower, convergent return flow.

*Figure 5.24* The driving force for either active or passive convection could be mantle plumes rising from great depths (*a*). In *b*, a plume is viewed from above to show its radial flow pattern.

that of the oceanic ridge. Furthermore, the rift is symmetrically flanked by striped geomagnetic anomaly patterns. To the southeast, it passes into a branch of the oceanic ridge that splits off in the southern Indian Ocean.

As the new ocean basin widens (fig. 5.22*c*), the older, outer margins of the oceanic lithosphere become cooler and denser, and sink to a lower elevation, as do the adjacent continental margins, which become shallowly submerged by the ocean. In time, as a mature ocean basin develops, each of the two submerged continental margins accumulates clastic sediment. Eventually, this sediment constructs a broad, shallow **continental shelf** characterized by sandstones, mudstones, and limestones (fig. 5.22 *c* and *d*). A thick prism

of deep sea mud accumulates above the suture between continental and oceanic lithosphere, forming a **continental rise** separated from the continental shelf by a relatively steep **continental slope.** The modern Atlantic Ocean is the classic example of this most advanced stage of divergence.

Divergent boundaries are always associated with basaltic volcanism and high heat flow from Earth's interior. The basaltic volcanism suggests that these boundaries form over regions where mantle peridotite is melting partially in the asthenosphere. The high heat flow from the crust suggests that some mechanism is bringing heat toward the surface from the lower mantle. One such mechanism might be a pair of convection cells whose upward flows converge beneath the ridge (fig. 5.23). Another would be a battery of mantle plumes beneath the ridge similar to the plume that has produced the Hawaiian-Emperor seamount chain (fig. 5.24). In either case, the source of the heat needed to drive such a mechanism is unclear, as is the depth at which it originates.

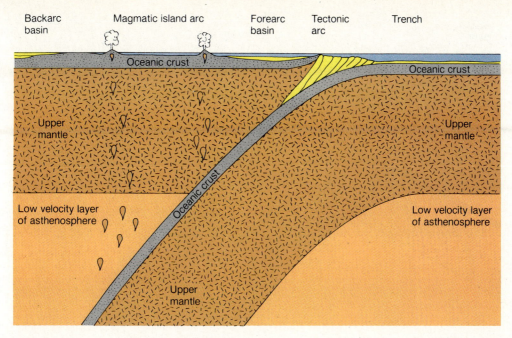

Backarc
basin    Magmatic island arc    Forearc    Tectonic    Trench
                                basin      arc

Oceanic crust                                              Oceanic crust

Upper                                                           Upper
mantle                                                         mantle

Oceanic crust

Low velocity layer                                  Low velocity layer
of asthenosphere                                    of asthenosphere

Upper
mantle

**Figure 5.25**    The essential features of an ocean-ocean
convergence zone.

## Convergent Boundaries

Plate boundaries along which the adjacent lithospheric
plates are moving toward each other are called **convergent plate boundaries.** These include subduction
zones and continental collision zones (see figs. 5.25–
5.27). Convergent boundaries are characterized by
*compression,* which produces *folds* and *compression
faults* in the rocks of such margins (see chapter 6).
As noted previously, it also produces earthquakes and
volcanoes. There are three types of convergent boundaries, each named for the types of crust present on the
opposing plates: ocean-ocean, ocean-continent, and
continent-continent.

Examples of *ocean-ocean convergence,* in which
both the subducted and overriding plates carry only
oceanic crust, are present along most of the northern
and western margins of the Pacific plate (see fig. 5.21).
The most visible feature of this type of boundary
is the chain of volcanic islands that normally develops
on the overriding plate margin parallel to the trench
(fig. 5.25). Such island chains are known as **magmatic
arcs** because they are built by magma rising from the
underlying subduction zone and because they typically
have gently curved, or arcuate, trends. Often, a **tectonic
arc** forms between the trench and the overriding plate
margin, but it is usually more subdued than the magmatic arc and less likely to rise above sea level and produce islands. The tectonic arc consists of chaotically

jumbled slices of ocean floor, deep sea sediment, and
clastic sediment from the magmatic arc that have been
stuffed under the leading edge of the overriding plate
during subduction.

Between the tectonic and magmatic arcs is a *forearc basin* where clastic sediment from the erosion of
both arcs accumulates. Sediments from the magmatic
arc can also accumulate in a *backarc basin,* which normally forms behind the magmatic arc.

Where a subduction zone coincides with a continental margin, an *ocean-continent convergence* results
(fig. 5.26). The best example of this type of boundary
is the western edge of the South American plate where
it overrides the eastward-spreading Nazca plate (see
fig. 5.21). Other examples are the southwest coasts of
Mexico and Central America, the north coast of New
Guinea, and the south coast of Alaska. The most obvious differences between ocean-ocean and ocean-
continent convergences can be seen to the left of the
tectonic arc in figure 5.26. Where the overriding plate
carries continental crust, the magmatic arc and backarc
basin are normally elevated above sea level. This, and
the presence of continental crust, results in generally
coarser-grained and more felsic rock types at ocean-
continent convergences than at ocean-ocean convergences (see chapter 7).

Notice that in figure 5.26 the angle at which the
subducted plate is descending is much shallower than
it is in figure 5.25. *Subduction angles* can vary from

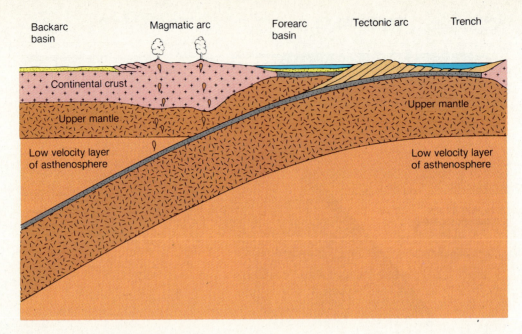

**Figure 5.26** The essential features of an ocean-continent convergence zone. The western boundary of the South American plate (west is to the right) has this structure.

about 30° to 90°, the average being about 45°. Shallower angles result if the subducted lithosphere is (1) moving fast, driven by a high rate of seafloor spreading or (2) relatively young, and hence warm and buoyant. Slower-moving or older, colder, and hence less buoyant lithosphere tends to sink faster, and it, therefore, has a steeper angle of descent. Where the angle is shallow, the distance between the trench and the magmatic arc is much greater than it is where the angle is steep.

On the far right side of figure 5.26, a continental margin can be seen approaching the ocean-continent convergence zone. When this margin encounters the tectonic arc, the buoyancy of its continental crust will prevent it from being subducted. Instead, it will crush the tectonic arc and forearc basin between itself and the opposing continental margin, thereby creating a *continent-continent convergence* (fig. 5.27). Within this colossal continental vise, the marginal sediments and sedimentary rocks of both continents are compressed, metamorphosed, and squeezed upward and outward like putty. Usually slices of oceanic lithosphere are thrust up as well. Before the mechanism of plate tectonics was understood, these slices were a source of considerable mystery and heated debate.

The best example of a continent-continent convergence is the junction between the Eurasian and Indian-Australian plates where the subcontinent of India has collided with the southern margin of Asia (see fig. 5.21). In this case, the northern edge of India has been *slightly* subducted, probably because of an unusually forceful

impact resulting from the Indian-Australian plate's abnormally rapid northward drift rate. The collision formed a double thickness of continental crust expressed dramatically in the Tibetan plateau and the world's loftiest mountain range, the Himalaya (Hih-*mahl*-ya; Sanskrit for "home of snow"). Toward the west, the northeastern boundary of the Arabian plate is colliding with the Eurasian continent. The rocks of the two adjacent continental margins have been compressed and crumpled, forming the Zagros Mountains of southwestern Iran.

### Transform Fault Boundaries

Although divergent and convergent boundaries are the main structures of lithospheric plate margins, **transform fault boundaries** provide the fine tuning. There are three types of transforms: ridge-ridge, ridge-trench, and trench-trench (fig. 5.28). Examples of *ridge-ridge transforms* abound all along the oceanic ridge (see fig. 5.21). Two examples of *ridge-trench transforms* are the eastern end of the ridge bordering the southern side of the Nazca plate west of South America and the northeast edge of the Pacific plate off the coasts of British Columbia and southeastern Alaska. Their function is to permit the spreading plate to move past adjacent plates. The third type, the *trench-trench transform,* is represented by the Alpine fault of New Zealand, which separates a northwestward-dipping subduction zone on the northeast from a southeastward-dipping subduction zone on the southwest. The northern and southern boundaries of the Caribbean plate are probably also of this type.

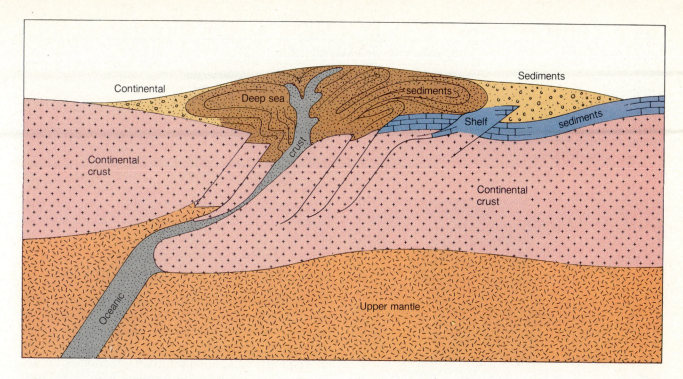

**Figure 5.27** Detail of a continent-continent convergence. Both oceanic crust and tectonic arc sediments have been thrust upward within the collision zone.

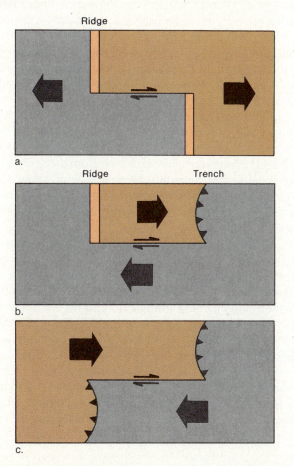

**Figure 5.28** The three types of transform boundaries: *a,* ridge-ridge; *b,* ridge-trench; and *c,* trench-trench.

## Triple Junctions and Failed Arms

A quick scan of the map in figure 5.21 reveals the existence of several **triple junctions,** places where three plate boundaries intersect in a common point. There are nine possible kinds of triple junctions involving divergent, convergent, and transform plate boundaries. In theory, any combination of the three types of plate boundaries can enter into a triple junction except three transform faults.

The most basic form of the triple junction is an intersection of three median rifts. An excellent example of a junction of this type that has been generating new ocean floor for many tens of million years can be seen where the African, Eurasian, and Indian-Australian plates join in the Indian Ocean. A much younger example is evident at the southern end of the Red Sea, where it has just begun the process of splitting the Arabian plate away from the African plate. One arm of the triple junction forms the Red Sea rift, another forms the Gulf of Aden, and the third trends southwest into the East African rift system. If this third arm continues to be active, this triple junction will evolve into a situation similar to that of its neighbor to the southeast, and it will be completely surrounded by spreading ocean floor. If it does not remain active, however, the third branch of the rift system would become what is known as a *failed arm.* Often, a major

**Table 5.1** Some Triple Junctions

| Plates Involved | Boundary Types Involved | | |
|---|---|---|---|
| | *Divergent* | *Transform* | *Convergent* |
| Eurasian, Arabian, Indian | | 2 | 1 |
| Pacific, Eurasian, Phillipine | | | 3 |
| North American, Pacific, Cocos | 1 | 1 | 1 |
| South American, Antarctic, Nazca | | 1 | 2 |
| African, Antarctic, Indian | 3 | | |

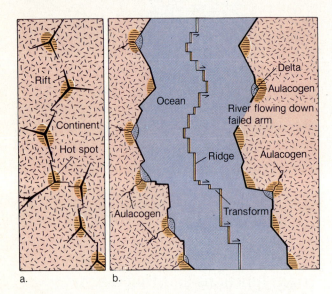

**Figure 5.29** Mantle plumes rising beneath continents are likely causes of rifting. In *a*, triple rift junctions form in the continental crust over a chain of hot spots. Intersections between the arms of adjacent triple junctions create a continuous rift (*b*). The failed arm of each junction directs drainage from the continental masses into the expanding ocean basin.

river occupies the site of a failed arm, and thick deposits of sedimentary rock can fill the linear depression along its trend. Such deposits have been called *aulacogens* (aw-*lack*-oh-gens; Greek for "furrow makers"). The Mississippi Valley is an example of a failed arm.

The geometry of triple rift junctions suggests that they might originate from mantle plumes. The kind of radial fracture pattern found in a triple junction is exactly what should be formed by a column of hot magma rising beneath and uparching the overlying, brittle crust. A series of mantle plumes would produce a corresponding series of triple junctions in the crust. If these plumes and junctions were even roughly aligned, and closely enough spaced, then one arm of each triple junction would intercept an arm of each adjacent triple junction, thereby forming a continuous rift (fig. 5.29).

It is possible that most median rifts form in this way. Examples of other types of triple junctions are listed in table 5.1.

## Spreading Axes and Spreading Poles

The function of ridge-ridge transforms becomes clear when you consider that all the lithospheric plates are fragments of a *spherical* surface. Because these spherical plate fragments move as *pairs,* either diverging from, converging toward, or slipping past one another, the geometry of such motions on a sphere is such that they can only occur as rotations around imaginary axes that pass through Earth's center (fig. 5.30). These are called **spreading axes,** and the points where such axes emerge on Earth's surface are known as **spreading poles.** These poles and axes are completely independent of the rotational poles and axis of the planet. For example, the spreading poles for the plates of the Atlantic Ocean are located near the southern tip of Greenland and south of Australia at about 62° north and south, respectively.

Points on a plate that are farther from the spreading poles (and axis) must move faster than points that are closer to the poles. This means that both spreading and subduction rates must increase away from the spreading poles. Observed spreading rates range from about 1 to about 20 cm/yr for both plates of a pair. (The *half rate* at which a single plate spreads away from a ridge would be half the full value.)

As J. Tuzo Wilson observed, the separation between any two offset segments of the oceanic ridge does not change with time. It therefore seems probable that these separations were formed initially as a result of a non-linear arrangement of the mantle plumes that presumably first caused the rifting (see fig. 5.29). Subsequently, each ridge segment has tended to align itself along a *meridian* to the spreading axis (any of the lines that converge radially on the spreading poles in fig. 5.30) because spreading can only take place perpendicularly to such a meridian. Similarly, transform sutures (fracture zones) must lie along *parallels* to the spreading axis (circles crossing the meridians at right angles).

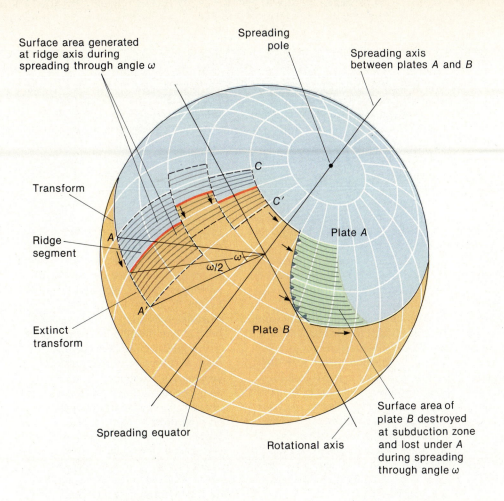

Surface area generated at ridge axis during spreading through angle ω

Spreading pole

Spreading axis between plates *A* and *B*

Transform

Ridge segment

Extinct transform

Spreading equator

Rotational axis

*C*

*C′*

*A*

*A′*

ω

ω/2

Plate *A*

Plate *B*

Surface area of plate *B* destroyed at subduction zone and lost under *A* during spreading through angle ω

## The Driving Forces: Gravity and Convection

Because we are unable to see what is actually going on beneath the lithosphere, much conjecture has centered around the forces responsible for plate tectonic movements. Arthur Holmes's concept of large convection cells rising beneath oceanic ridges, driving the lithosphere laterally away from the ridges and descending again into the mantle with the lithosphere in tow remains appealing. It also remains problematical, however, because analysis of earthquake waves reveals that the mantle has a layered structure (see chapter 6), which would be destroyed by large-scale convection.

A hypothesis that presents fewer problems combines the effects of mantle plumes with the force of gravity (fig. 5.31). Plumes rising beneath the oceanic ridge raise the seafloor, creating a slope away from the ridge axis. Because of the presence of the plastic asthenosphere beneath, the lithosphere actually slides down this gentle slope under the influence of gravity.

At the subduction zones, the descending slab is also pulled down into the mantle by gravitational force. This might seem illogical at first because the upper surface of the slab consists of basalt and deep sea sediments,

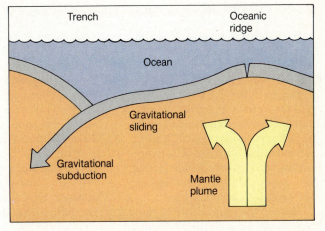

Trench

Oceanic ridge

Ocean

Gravitational sliding

Gravitational subduction

Mantle plume

**Figure 5.31** Current thinking favors the idea that tectonic plates move entirely under the influence of gravity. Elevated by the upwelling of magma beneath the ridge and lubricated by the soft asthenosphere beneath them, the plates are able to slide down the gentle incline from ridge to trench. It is possible that the high-pressure conversion of seafloor basalt to dense eclogite also helps to draw the descending plate down into the mantle.

**O**n the dashboard of his VW camper, my college roommate stuck a piece of red Dymo tape that bore the exhortation, "May the Appalachians rise again!" At the time (1964) the idea of the Appalachians rising again was little more than a jest, but that was before the development of plate tectonic theory. Today, the idea seems quite possible, if not probable in the light of the Appalachian's complex plate tectonic history, recently unravelled by paleomagnetic studies. A brief review of that history follows.

We have evidence that an ancestral "Atlantic" Ocean basin opened as a result of continental rifting and sea-floor spreading in the Late Proterozoic (orange band in fig. 5.32*a*). This ancient ocean began to close again in the Middle to Late Ordovician when a convergent plate boundary formed on the deep ocean floor (barbed line in fig. 5.32*b*) and was driven rapidly northward—probably by a new divergent boundary that formed off the African coast (orange band in fig. 5.32*b*)—subducting the North American plate beneath it. A complex of tectonic and volcanic arcs developed in and around a group of islands of continental lithosphere located above this subduction zone. This arc complex was swept rapidly northward by the seafloor spreading until the North American continent collided with the subduction zone and the arc complex (fig. 5.32*c*). This resulted in the notorious *Taconian orogeny,* or mountain-building event, named for the Taconic Mountains on the Vermont-New York border.

Then, in the Early Devonian, North America was rammed again, this time from the east by northwest Africa (fig. 5.32*d* and *e*), in a major continental collision called the *Acadian orogeny.* The "Atlantic" Ocean then opened for a second time as Africa slid away to the southwest in the Middle Devonian along a transform fault boundary

(fig. 5.32*e* and *f* ), leaving a large piece of itself plastered against the arc complex that had been plastered on in the Ordovician. The ocean then closed again as Africa returned from the south to ram North America for a second time in the Late Pennsylvanian *Appalachian orogeny,* which created the supercontinent Pangaea (fig. 5.32*g*). The present Atlantic Ocean basin began to open when continental rifting began in the Triassic (fig. 5.32*h*).

This restless opening and closing of the Atlantic led to the concept of the **Wilson cycle** (after J. Tuzo Wilson), comprising the formation and destruction of an ocean basin. It would seem likely that the Atlantic Ocean is now in the middle of yet another Wilson cycle, which began with the rifting of Pangaea in the Triassic, and should end with a closing of the ocean basin at some time in the future. That closing should result, in turn, in yet another uplift of the Appalachian mountain belt, or, more precisely, of the thick prism of sediments that has been deposited on the continental margin since the Jurassic.

While a graduate student at the University of Chicago, geologist Christopher Scotese undertook a computerized voyage into the future of continental drift. His computer model shows a progressive widening of the Atlantic Ocean that ends about 150 million years from now when a new subduction zone develops off the eastern seaboard of North America.

What would cause the sudden development of a subduction zone? Having no modern examples of this situation, we can only speculate, but two factors could be especially significant. First, the weight of the sedimentary prism accumulating on a rifted continental margin might cause the underlying lithosphere to rupture. Second, as the growing plate becomes larger and larger, it becomes increasingly difficult for that portion of the plate that still lies on the slope of

the oceanic ridge to push the flat-lying portion and keep it moving. This should encourage the lithosphere to break at its weakest point—off the continental margin where it is loaded with a thick sedimentary prism. Then, a subduction zone would form off the eastern seaboard, paving the way for yet another bashing by the unabashed African continent. The fact that the eastern seaboard is slowly sinking lends strength to this concept.

Scotese also makes a prediction that might spell gloom for geologists, but perhaps rapture for biologists. Eventually, the process of plate tectonics must grind to a halt as Earth's internal heat dwindles to a level below which the asthenosphere becomes rigid. At that point the destructive processes of erosion will begin to dominate the constructive processes of orogeny. Gradually, the continents will be worn down to monotonous lowlands, which might conceivably even become totally submerged beneath an all-enshrouding world ocean. Then we would have even less justification than now to call our planet "Earth!" Terrestrial life-forms obviously would suffer under such a circumstance, but marine life would flourish. The circulation of the world ocean, unencumbered by continental masses, would keep the planetary surface uniformly warm. Under these conditions, *biological diversity* (the number of species per unit area) would increase markedly. The rule is this: the more stable and uniform the climate (as in the world's tropics), the higher is the species diversity. Conversely, the more variable and seasonal the climate (as in the polar regions), the lower is the species diversity. Of course, all this is contingent on whether or not we still have an ocean by then, and that, as I suggest in chapter 1, could be dependent on what we do to our protective ozone shield in the next few years.

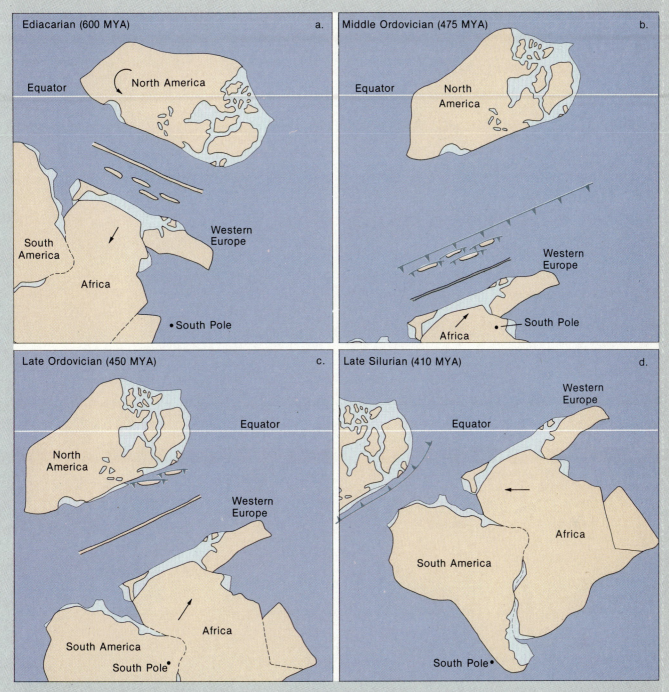

**Figure 5.32** The evolution of the Appalachian mountain belt. Only those continents directly involved are shown. Arrows indicate drift directions and rotations. Double lines are median rifts; barbed lines are subduction belts. Light blue areas are continental shelf. (*a*) Gondwanaland is rifted from North America, and drifts to the south. Small slivers of continental crust form islands south of the rift. (*b*) Gondwanaland begins to drift northeastward. Subduction zones form near the old rift and among the islands, and a new rift forms between the islands and Gondwanaland. (*c*) As Gondwanaland drifts northeastward, the new rift drives the islands and the subduction zone against the coast of North America, resulting in the Taconian orogeny. (*d*) Gondwanaland begins to drift westward, and an ocean-continent convergence zone forms off the "east" coast of North America.

*continued*

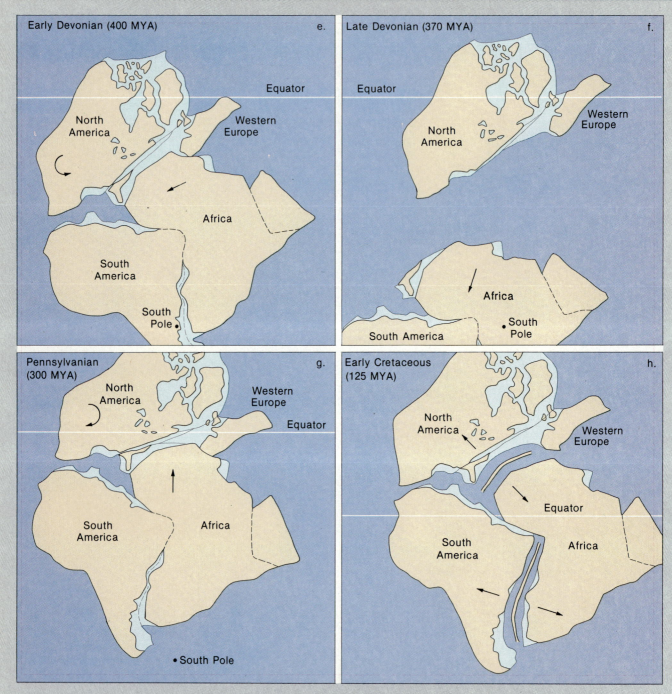

**Figure 5.32 continued**   (*e*) Gondwanaland collides with North America in a continent-continent convergence called the Acadian orogeny. (*f*) Gondwanaland slides away to the southwest, leaving western Europe (including southern England and Ireland) attached to North America.

(*g*) Gondwanaland returns from the south for a second continent-continent convergence called the Appalachian orogeny. (*h*) Rifts form along what becomes Africa's west coast, leaving the Florida peninsula attached to North America and opening the Atlantic Ocean basin.

both of which are of lower density than mantle peridotite! Four facts come to our rescue, however. First, these light rock layers form only a thin veneer over what is essentially a slab of mantle peridotite that forms the bulk of the lithospheric plate. Second, the slab is cold, and therefore relatively dense. Third, the asthenosphere is hot, and therefore relatively less dense. Finally, the high-pressure conversion of basalt to eclogite below a depth of about 75 km further increases the slab's density. The result is that the descending slab is actually denser than the surrounding asthenosphere and is therefore susceptible to being dragged down by gravity. The discovery that the upper portion of the descending slab is often under tension supports this idea. Old slabs that have "bottomed out" on the base of the asthenosphere, however, do not exhibit tension in their upper parts.

Calculations have shown that a lithospheric plate can slide away from a ridge under the influence of gravity alone on a slope of as little as 0.019° (1 in 3000) at a velocity of 4 cm/yr. Because these values are typical of actual slopes and spreading rates for oceanic lithosphere, the gravity sliding mechanism seems to be a reasonable choice for the driving force of plate tectonics. If it is the driving force, convection can still occur in the upper mantle, but it would be an *effect,* and not the cause, of plate motion (see fig. 5.23*b*). The moving plate would exert a viscous drag on the underlying asthenosphere, bringing mantle material toward the subduction zones. This would in turn necessitate a slow, ridgeward return flow of mantle material at lower levels within the asthenosphere. The subsequent rise of this hot, deep peridotite might then be an additional source of uplift and partial melting of the oceanic ridge.

## Summary

Parallelism of the Atlantic coastlines and similarities in modern and fossil flora between Europe and North America spurred early speculations about continental drift. In the early 1900s, Alfred Wegener studied the problem in depth, assembling evidence from geography, geology, biology, and climatology. He reassembled the supercontinent **Pangaea** by joining the continents at the edges of the continental shelves. Geophysical disproof of Wegener's *Polarfluchtkraft* drift model delayed acceptance of continental drift for over 30 years. Key *continental* evidence for drift includes the following:

1. **Disjunct distributions** of the tropical scale tree flora of the northern continents (**Laurasia**); the temperate seed fern flora of the southern continents (**Gondwanaland**); the Permian fresh water reptile *Mesosaurus;* and the Triassic land reptile *Lystrosaurus*
2. Mountain belts truncated at continental margins

3. The disjunct Gondwana rock succession, including evidence of glaciers apparently advancing from the sea onto the land
4. **Paleoclimatic** evidence of drastic climatic changes in a given location
5. **Paleomagnetic** evidence of **polar wandering.**

Key *oceanic* evidence for drift, obtained mostly in the late 1950s and 1960s, includes the following:

1. The oceanic ridge with its median rift and transverse **fracture zones**
2. **Seamount chains** that become older and deeper in a direction away from their volcanically active ends
3. The increase in age of the seafloors and in thickness of deep sea sediment away from the oceanic ridges
4. Benioff zone earthquakes beneath **magmatic arcs**
5. Symmetrical **geomagnetic anomaly** stripes on the seafloor parallelling the oceanic ridge axis and corresponding to **geomagnetic reversals**
6. The motion of **transform faults.**

**Geomagnetic declination** is the angular difference between the directions to the geographic North Pole and the geomagnetic North Pole. **Latitude** is the angular distance north or south of the equator. **Paleolatitude** can be determined from **geomagnetic inclination,** the angle at which a freely-suspended compass needle dips below the horizonal. Earth's surface is composed of seven major and seven minor 100-km-thick, *lithospheric plates* overlying the low velocity zone at the top of the low-strength **asthenosphere.** The plates are composed mainly of thick mantle peridotite, capped by either a ±7 km thickness of oceanic crust (*oceanic lithosphere*), or a ±40 km thickness of continental crust (*continental lithosphere*), or both.

The median rifts of the oceanic ridge are **divergent plate boundaries** where the ridge is elevated by upwelling of basaltic magma due to upward-converging **convection cells,** to **mantle plumes,** or to both. The resulting slope allows the lithospheric plates to slide away from the ridge axis, which is under extensional stress. **Continental shelves, continental slopes,** and **continental rises** form on rifted continental margins as they spread away from median rifts, sink below sea level and accumulate sediments. In **convergent plate boundaries,** lithospheric plates are subducted into the mantle. In an *ocean-ocean convergence,* one plate of oceanic lithosphere is subducted beneath another. Typically, a **tectonic arc** and a *forearc basin* form between the trench and the magmatic arc, and a *backarc basin* forms on the magmatic arc's opposite side. In an *ocean-continent convergence,* oceanic lithosphere is subducted beneath continental lithosphere. Old, cold, slow-moving plates have steeper *subduction angles* than young, hot, fast-moving plates. In a *continent-continent convergence,* two plates of continental lithosphere collide. Because continental crust is too light to be subducted, the two margins act like a vise, compressing their tectonic arcs between them. Seafloor slices are often thrust up onto the continent. Slight subduction of continental lithosphere produces a double thickness of continental crust, as in the Himalayas. **Transform fault boundaries** along fracture zones permit a spreading plate to move past adjacent plates.

Plate boundaries end in **triple junctions,** involving combinations of convergent, divergent, and transform fault boundaries. Triple rift junctions probably form over mantle plumes. All lithospheric plate movements must occur around **spreading axes,** which pass through Earth's center and "emerge" on the surface as **spreading poles.** Rates of spreading and subduction increase away from a spreading pole. Spreading rates range between 1 and 20 cm/yr (for both plates of a pair). At the end of the present **Wilson cycle,** the Atlantic Ocean will probably close again, producing a new east coast mountain range.

## Key Terms

| | |
|---|---|
| convection cell | fracture zone |
| Pangaea | transform fault |
| Laurasia | divergent plate |
| Gondwanaland | boundary |
| disjunct distribution | continental shelf |
| paleoclimate | continental rise |
| geomagnetic declination | continental slope |
| latitude | convergent plate |
| geomagnetic inclination | boundary |
| paleomagnetism | magmatic arc |
| paleolatitude | tectonic arc |
| geomagnetic reversal | transform fault |
| polar wandering curve | boundary |
| seamount chain | triple junction |
| mantle plume | spreading axis |
| asthenosphere | spreading pole |
| geomagnetic anomaly | Wilson cycle |

## Questions for Review

1. Compile as complete a list as you can of arguments that support the concept of continental drift without referring to the seafloor.
2. What do you think Alfred Wegener should or should not have done in order to save his hypothesis from becoming a laughing stock?
3. Name and describe six critical seafloor features supporting continental drift. Of these, which could be interpreted in no other way?
4. You remove a core from an Upper Ordovician basalt lava flow in northern Maine and find that the paleomagnetic field in the core dips very gently toward the east-southeast. On the basis of this observation, describe the location and orientation of the east coast of North America with respect to the equator in the Late Ordovician.
5. Make a table, listing the seven major plates shown on figure 5.21, and for each, tabulate major lithospheric type(s), number and type(s) of major boundaries, number and type(s) of triple junctions, and drift direction relative to the African plate.
6. Suggest two different driving mechanisms for plate tectonics.
7. Describe the differences between magmatic and tectonic arcs, and their modes of origin.
8. What happens to a subduction zone after it experiences a continental collision?
9. If the spreading poles of a pair of diverging plates are located on Earth's rotational equator, in what directions are the two plates moving?
10. Explain why the Appalachians might rise again.

# Stress, Strain, and Structure: Earthquakes and Mountain-Building

## Outline

*E*arth felt the wound; and Nature from her seat,

Sighing through all her works, gave signs of woe

That all was lost.

John Milton

**Paradise Lost**

Photo: Minor faults in Nubian Sandstone, Israel.

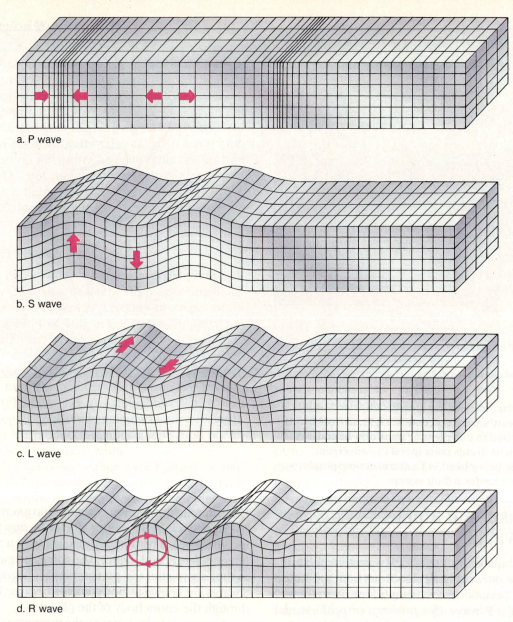

**Figure 6.5** The four types of seismic waves.

From *Inside the Earth*. By Bruce A. Bolt. Copyright © 1982 by
W. H. Freeman and Company. Reprinted with permission.

## Seismographs: Locating Epicenters and Measuring Intensity

A **seismograph** (fig. 6.6) is a device for locating and measuring the relative intensity of an earthquake. In principle, sudden movements of Earth's crust are measured in relation to a heavy suspended weight, which is held motionless by its own inertia, while the base of the instrument moves with the bedrock to which it is attached. This relative motion is recorded on a rotating drum. The resulting record, called a **seismogram**, shows seismic shock waves as deflections from a straight line. The greater the energy of the shock wave, the greater is the amplitude of the deflection. The seismogram in figure 6.7 shows the arrival of the various different seismic waves from an earthquake located in the Japan trench about 9200 surface kilometers due north of the recording station in Hobart, Tasmania (south of Australia). The P waves arrived 12.9 min after the earthquake occurred. At 23.3 min after the earthquake, the S waves arrived followed by the surface waves at 36.5 and 42.0 min, respectively. Notice the generally increasing amplitudes of the four different types of signals.

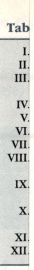

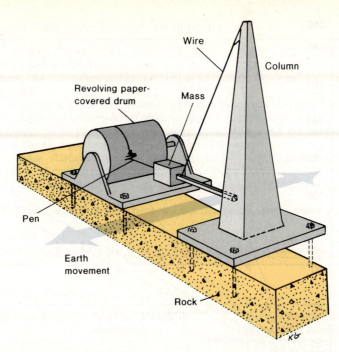

*Figure 6.6* In this schematic diagram of a seismograph, a heavy metal block (mass) mounted on the end of a freely swinging arm is suspended by a wire from the top of a tall column. During an earthquake, the stable weight holds still due to its high inertia, while the instrument shakes around it. The shaking causes deflections in the line being drawn on the rotating drum by a penpoint attached to the weight.

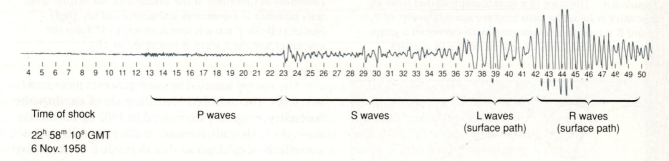

| Time of shock | P waves | S waves | L waves (surface path) | R waves (surface path) |

22ʰ 58ᵐ 10ˢ GMT
6 Nov. 1958

*Figure 6.7* The four kinds of seismic waves are evident in this seismogram of a major earthquake in the Japan trench. The recording station is located in Tasmania, Australia.

The farther a station is from an epicenter, the greater is the separation of the arrival times of P, S, L, and R waves. Figure 6.8 shows three seismograms from the same earthquake as recorded at three stations located at 2500, 5000, and 7500 km from the epicenter. The seismograms have been placed on a graph that relates the travel time of seismic waves to distance from the epicenter. With this graph, it is a simple matter to determine the distance of a particular seismograph station from an earthquake epicenter. The seismogram from that station is placed against the graph and shifted up or down until the beginning points of each wave are aligned with the corresponding curves. The distance is then read from the scale at the left of the graph.

By itself, the distance to the epicenter is not a particularly useful piece of information. For a given station, the epicenter could be located anywhere along a circle, centered on the station, whose radius is equal to that distance. In order to fix the exact location of an earthquake epicenter, at least three stations are required (fig. 6.9). Circles are drawn around each, with radii equal to the corresponding distance values. The intersection point of the three circles is the location of the epicenter. Notice that if only two stations are used, there are two possible epicenter locations at the points where the two circles intersect. The third circle removes the ambiguity.

**Figure 6.13** Inclined strata dip toward the left into a pool along a mountain brook in Maine's Baxter State Park. Intersections of bedding planes with the water surface are lines of strike. The dip is the angle that the plane makes with the water surface, measured perpendicularly to the strike.

## Bending the Crust: The Mechanics of Folding

When horizontal compressive or shear stresses are applied slowly and steadily over a long period of time to thick sequences of stratified rocks, creep tends to predominate over fracturing. This process of slow, internal readjustment results in the gradual *folding* of the rocks. Figure 6.14 shows the effects of slow continental collision on Devonian schist exposed on the upper slopes of Mount Washington in New Hampshire. Figure 6.15 shows the more intense effects of a Cenozoic continental collision on sedimentary strata in the Swiss Alps.

Folds occur in a variety of forms, the more important of which are illustrated in figure 6.16. Basic to describing fold geometry are three features common to all folds: the **hinge,** or line of maximum flexure on a given bedding plane; the **limbs,** or sides of the fold; and the **axial plane** (AP), or the surface that connects the hinges of all the strata of a given fold. *Symmetrical folds* (fig. 6.16*a*) are the simplest. In this type of fold the limbs dip at equal angles on either side, and their axial planes are vertical. In *asymmetrical folds* (fig. 6.16*b*), the axial planes are inclined, and one set of limbs dips more steeply than the other, as they do in the folds on Mount Washington. If the axial planes are inclined to such a degree that one set of limbs is rotated past the vertical (fig. 6.16*c*), the folds are described as *overturned.* If the axial plane is approximately horizontal (fig. 6.16*d*), they are described as *recumbent.*

In general, the axial planes of folds that are close to the source of the stress that created them incline away from the vertical more than do the axial planes of folds that are more distant from the stress. Thus, the folds shown in figure 6.16*d* would have been formed closest to a continental collision zone, and the folds shown in figure 6.16*a* would have been formed farthest away. In the folds in figure 16.6*b, c,* and *d,* the stress would have acted from the left.

Compressional folds have been further classified into archlike folds, or **anticlines** (literally, "dipping apart") and troughlike folds, or **synclines** ("dipping together"). In figure 6.16*a–c,* for example, the crests are anticlines (*A*), the troughs are synclines (*S*), and their limbs blend smoothly into one another. In figure 16.6*d,* however, an obvious problem exists, because in this case there are neither arches nor troughs. In order to resolve this dilemma, the terms have been more precisely defined. An *anticline* is *a fold that encloses older strata within younger strata,* and a *syncline* is *a fold that encloses younger strata within older strata.* These definitions hold true even if the folds are turned upside down, a not too infrequent occurrence.

**Figure 6.14** High on Mount Washington, strata of schist in the Devonian Littleton Formation have been thrown into broad folds by compressional forces. The troughlike fold on the left is a syncline; the archlike fold on the right is an anticline. Both folds are slightly asymmetrical. The rock is gray, but appears greenish here due to encrustations of lichen, a primitive plant.

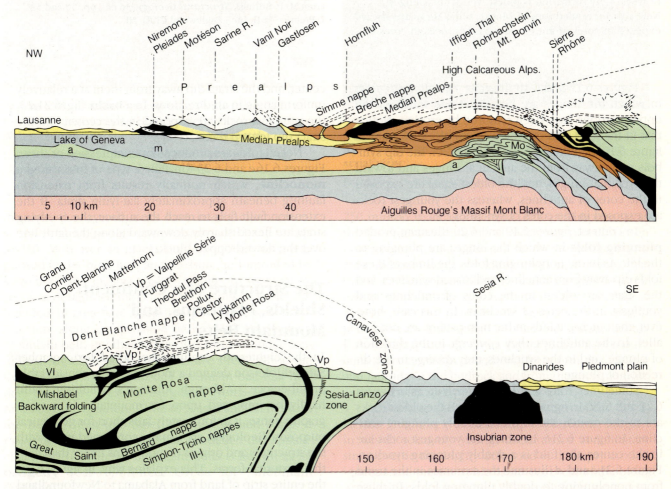

**Figure 6.15** When continents collide, the thick piles of plastic sedimentary strata on their margins are compressed and thrown into complex folds, as illustrated in this cross section through the Swiss Alps. More brittle rocks of the ancient continental nucleus (*salmon*) respond mainly by faulting.

From E. W. Spencer, *Introduction to the Structure of the Earth*, 2d ed. Copyright © McGraw-Hill, Inc., New York, NY.

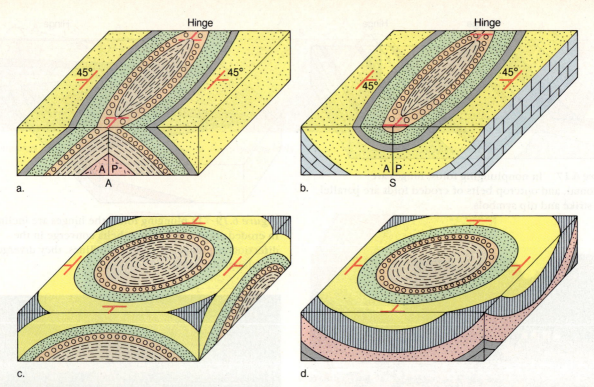

a.

b.

c.

d.

**Figure 6.21** In eroded doubly-plunging folds, outcrop belts converge around the axial plane in both directions. Note that the map patterns of the plunging anticline (*a*) and the plunging syncline (*b*) are identical; hence, either

the dip of the strata or their relative ages must be known before these two different structures can be distinguished. The same condition applies to the dome (*c*) and basin (*d*), whose map patterns are also identical.

**Figure 6.22** A satellite view of the vicinity of Harrisburg, Pennsylvania reveals the folded structure of the Appalachian mountain belt, outlined by ridges of erosion-resistant, lower Paleozoic sandstone. Several doubly-plunging folds are evident in the central portion of the fold belt.

Photo by NASA

**Figure 6.23** The Waterpocket Fold in southern Utah is a well-known example of a monocline. The valley wall on the left consists of the upper surface of a white sandstone formation that is nearly horizontal on the top of the slope, but dips with increasing steepness toward the base, where it disappears beneath younger, overlying formations. At depth, its dip flattens out again, as is evident from the nearly horizontal attitude of the overlying strata at right.

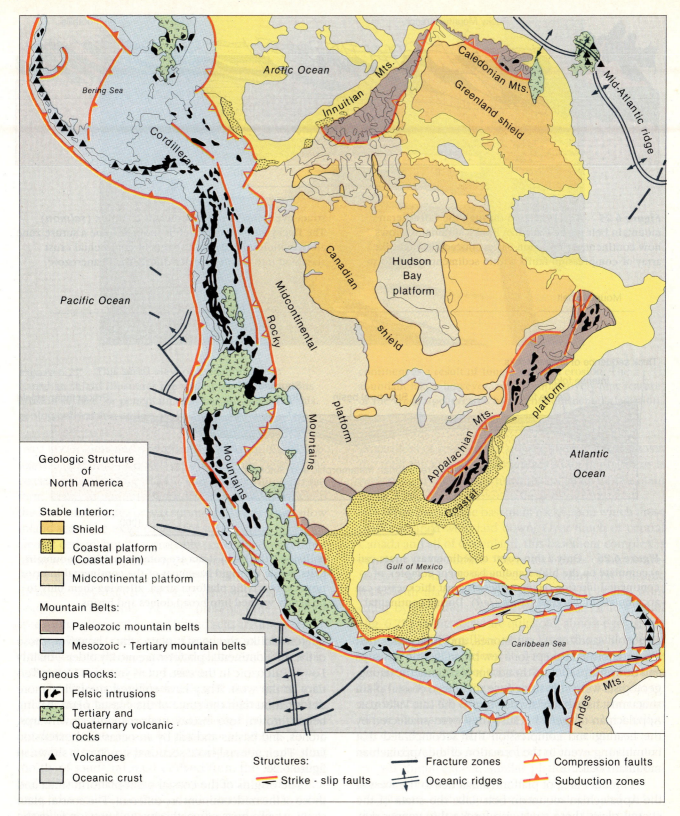

**Figure 6.24** The tectonic structure of a typical continent is well displayed by North America. The Canadian shield consists of the deeply-eroded roots of ancient mountain belts welded together to form a central, stable continental nucleus of low topographic relief. The midcontinental platform is an area where a thin cover of little-deformed sedimentary strata overlies the shield. Marginal mountain belts have formed around the shield-platform complex of the continental interior. Portions of these belts have been overlain by coastal platform sediments.

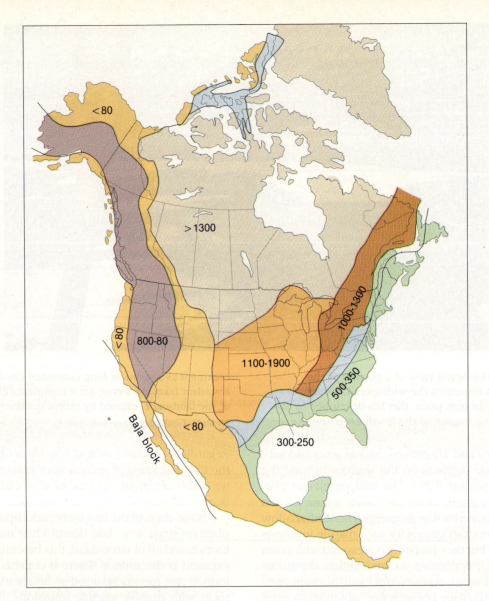

**Figure 6.28** North America is a collage of continental provinces with ages that generally decrease away from the continental nucleus. The ages shown in the diagram represent the oldest granites dated within each of the provinces. Ages are in million years.

## The Growth of Continents

Figure 6.28 is a map showing the distribution of the radiometric ages of the oldest granite intrusions within the Canadian shield and its extensions beneath the platform strata of the North American continent, as well as ages of mountain belts younger than 500 million years. These dates tend to cluster into several distinct *age provinces,* the oldest of which is Archaeozoic and Early Proterozoic (>1800 million years). This province occupies a broad region including the north central United

States, most of central Canada, and Greenland. Most of it consists of rocks between 3800 and 2500 million years old, but scattered through it are mountain belts dating between 2300 and 1800 million years. The older, *Archean* rocks (the term used for rocks of Archaeozoic Age) consist mainly of greenstones and massive bodies of granitic rocks that intruded and fragmented the greenstones. In addition, rocks of this province include the rare ultramafic lava komatiite.

The Lower Proterozoic rocks of this province differ in many respects from the Archaean rocks. Greenstones are in the minority, and sedimentary rock types like those of the Atlantic coastal plain predominate. These sandstones, mudstones, and carbonates, many of which have been metamorphosed to quartzite, schist, and marble, were derived from the long-continued weathering and erosion of the older granitic intrusions. In addition, the younger rocks include feldspathic sandstone, which indicates that rifting and seafloor spreading were under way by about 2000 million years ago. At this same time, the earliest occurrences of red beds and evaporites mark the first appearance of free oxygen in the atmosphere.

Bordering the western edge of the Appalachian mountain belt is a narrow belt of Middle Proterozoic rocks about 1000 to 1300 million years old. Rocks in this belt are mostly remetamorphosed Archean rocks with a small admixture of coastal plain sediments and about 20% of an unusual (and beautiful) variety of gabbro called *anorthosite,* consisting almost entirely of large crystals of dark, often iridescent plagioclase feldspar. The crystals are thought to have floated to the tops of large magma chambers that formed above mantle plumes in the continental interior.

In the south-central United States is a province of Lower to Middle Proterozoic rocks poorly exposed beneath Paleozoic platform strata. These rocks bridge the time gap between the younger province to the northeast and the older province to the northwest.

These age provinces of the Canadian shield show, in a general way, how the North American continent has been built up by the growth of a series of mountain belts *accreted,* or welded progressively to its margins by plate convergence over geologic time. The Appalachian mountain belt was added during the Paleozoic. The Mesozoic to Cenozoic Cordilleran mountain belt bordering the continent on the west represents the latest addition.

In chapter 5, I mention that the Andes Mountains on the west coast of South America are the result of orogeny at a convergent plate margin. The same is true of the North American Cordillera although the west coast of the United States has changed by now from a convergent boundary to a transform boundary, outlined by the "Baja block" in figure 6.28. Similarly, all the older mountain belts that make up the basement of North America were located at convergent plate boundaries at the time they were formed.

## Unconformities: Keys to Mountain-Building

Stensen's principle of original horizontality has provided geologists with a convenient tool for unraveling the often appallingly complex structures produced in bedrock by directed stress during orogenies. James Hutton observed that following an orogeny, new sedimentary strata are often deposited, horizontally, on erosion surfaces that truncate older strata that have been rotated out of their originally horizontal attitude by directed stress. This produces what is called an **unconformity** between the two sets of strata. Figure 6.29 shows the sequential development of a series of unconformities in the Medicine Bow Mountains of southeastern Wyoming during the Lower Tertiary *Laramide orogeny.* In figure 6.29*a,* erosion following initial uplift of the Medicine Bow range has exposed an ancient unconformity between granite gneiss and younger "pre-Medicine Bow" sedimentary formations. A discontinuity such as this, in which sedimentary strata have been deposited on igneous or metamorphic rock (rather than on sedimentary rock), is known as a *nonconformity.* The Medicine Bow Conglomerate at the top of the sedimentary sequence consists of clasts eroded from the uplift.

By Late Paleocene time (fig. 6.29*b*), orogeny had warped the pre-Medicine Bow formations into a broad syncline, which was subsequently truncated by erosion and then overlain by sediments of probable Paleocene age ("Paleocene?"). This has resulted in an *angular unconformity* between the two sets of strata, in which overlying and underlying strata are not parallel, except locally. By Late Lower Eocene time (fig. 6.29*c*), ongoing compression had tightened the syncline and produced a compression fault in its western limb. Erosion following this orogenic activity removed all but a small portion of the Paleocene (?) formation, and the deposition of Lower Eocene strata on the erosion surface produced a second angular unconformity.

Between Late Lower Eocene and Late Upper Eocene time (fig. 6.29*d*), the entire region was broadly uplifted, and stream erosion removed the Lower Eocene strata near the mountain front, exposing the compression fault. Subsequently, the region enjoyed a long respite from directed stress, during which Oligocene- and Miocene-aged sediment derived from the erosion of the mountains gradually buried the Upper Eocene erosion surface (fig. 6.29*e*). This sedimentation produced a nonconformity over the granite gneiss, angular unconformities over the pre-Medicine Bow and Paleocene formations, and a third kind of unconformity, called a *disconformity,* over the Lower Eocene strata. In a disconformity, there is an erosional break between overlying and underlying sets of strata, but the strata are parallel. Since Late Miocene time, a second episode of broad uplift has stripped most of the Upper Tertiary strata from the region (fig. 6.29*f*).

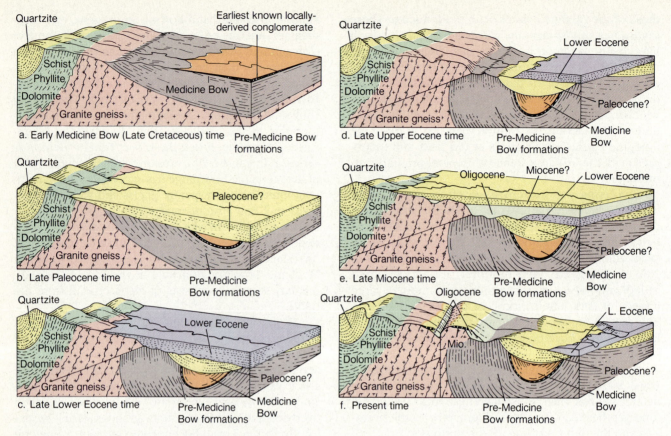

**Figure 6.29** Early Tertiary mountain-building in the Medicine Bow Range of southeastern Wyoming produced a spectacular series of unconformities separating sedimentary rock sequences of different ages.

## Summary

A **stress** is a force acting on an area. **Confining pressure** is stress that is uniform in all directions. **Directed stresses** are greatest in one direction, and can be **compressive, tensile,** or **shear. Strain** is deformation induced by stress. Hard, rigid, dry rocks tend to be brittle; soft, yielding, wet rocks tend to be puttylike. Rocks are more puttylike under increased confining pressure and temperature. When applied slowly, stress produces strain (as *creep*) without fracture. The *elastic limit* is a stress below which strain is **elastic** (temporary) and above which it is **plastic** (permanent). Brittle rocks fracture below the elastic limit; plastic rocks fracture above it (if at all). **Elastic rebound** in brittle rocks causes earthquake shocks.

Fractures along which no differential motion occurs are **joints.** Those along which motion occurs are **faults.** The **fault block** overlying a fault is the **hanging wall;** that underlying it is the **footwall.** Tensile stress prevails near divergent boundaries and backarc basins. **Extension faults** and fractures dip at about 60°. Compressive stress prevails in collision zones. **Compression faults** and fractures dip at about 30°. Shear stress prevails near transform fault boundaries. **Strike-slip faults** are vertical, and fault blocks move horizontally.

Earthquakes occur along **fault planes** when accumulated stress exceeds pressure bond strength between fault blocks. More stress is required where the plane is rough, dry, curved, or uneven. An earthquake **focus** is the point on a fault plane where slip occurs. The point on Earth's surface vertically above the focus is the **epicenter.** Shallow focus earthquakes (0–70 km) result from subduction, transform faulting, and ridge volcanism. Intermediate (70–300 km) and deep (300–700 km) focus earthquakes result from subduction only.

**Seismic waves** emanate from earthquakes in all directions. **Body waves** include **P waves** (back-and-forth motion, first to arrive) and **S waves** (side-to-side motion, second to arrive). **Surface waves** die out with depth

At 8:06 on the morning of 28 October 1983, two hunters stalking elk near Mount Borah, Idaho's highest peak, suddenly began to feel weak and dizzy. An instant later, as one of them put it, "All hell broke loose." With a deep rumbling like a giant drum roll, the road suddenly collapsed beneath them, tossing the men and their vehicle about like toys.

The Mount Borah earthquake measured 7.1 on the Richter magnitude scale. In the region around the disturbance, many changes took place. A prominent, vertical escarpment formed along the Mount Borah fault, trending northwest-southeast for 37 km. New springs burst forth from the ground, some of them as fountains of sand and water rising fifteen feet in the air. Some long-established springs as much as doubled their flow volume, whereas some others stopped flowing altogether. Mudslides cut loose from water-saturated soil. The force of the earthquake shock waves ripped trees from the ground and flung them through the air. It unseated a farmer from his tractor in a field located almost precisely at the epicenter. In the nearby town of Challis, the stone cornice of a store front collapsed, killing two children on their way to school.

None of these events was especially new and different with respect to human experience with earthquakes, but there was one particularly alarming event that was. In a brand new town called Atomic City, located 50 miles southeast of the point of maximum ground disturbance, cracks formed in two of the buildings of the Idaho National Engineering Laboratory (INEL), a nuclear research facility with 15 reactors. Prior to the construction of the INEL site, Spencer Wood, a geologist at Boise State University, had strongly advised against locating a nuclear facility anywhere in the earthquake-prone Basin and Range geologic province. Within this large area, which includes parts of Oregon, Idaho, California, Nevada, Utah, Arizona, and New Mexico, tensile stresses have been pulling the crust apart for the past 20 million years, and so earthquakes are relatively frequent there.

Wood's strong recommendation had been ignored. Idaho's Senator James McClure had championed the INEL site, and faith in its soundness had run so high prior to construction that local contractors had even managed to convince geologists at a 1980 symposium in Boise to make an indentation in the boundaries of the Intermontane seismic belt in order to exclude the site from earthquake hazard zone 3, the zone of highest risk, requiring special and expensive construction methods. The rationale was that existing faults northwest of the INEL site showed "extremely limited seismic activity" during continuous surveillance *since 1972,* an eight year period, and no single shock with a magnitude as high as 4.2. A somewhat longer surveillance would have shown a much different picture. In 1959, for example, another major earthquake had occurred at Hebgen Lake, Montana, between Mount Borah and Yellowstone Park. The Hebgen quake also produced prominent fault scarps, and it triggered a massive landslide in the Madison River canyon that buried 28 campers (see chapter 19).

In spite of the abundant and clear evidence, however, the developers of Atomic City, Idaho, and the adjacent INEL facility located their structures *directly* in line with the Mount Borah fault, a long, unstable fracture in Earth's crust. Senator McClure, observing that none of the reactors was damaged by the earthquake shocks, stated that "The recent earthquake demonstrates the integrity of the facilities at INEL, and provides important data on the dissipation of earthquake forces through the eastern Snake River plain."

Suppose, however, that the earthquake on the Mount Borah fault had measured 8.1 instead of 7.1 or that it had occurred farther southeast along the fault, both of which are quite reasonable possibilities. The recent nuclear catastrophe at the Chernobyl plant near Kiev in the Soviet Ukraine, although not fault-related, has provided the world with grim and uncontestable evidence of the consequences of a serious nuclear accident, the products of which can be wafted about the planet at the caprice of the winds. Perhaps, if Chernobyl had occurred before the Mount Borah earthquake, Senator McClure's statement might have been quite different. At least we might hope so! How much more comforting it would have been to hear: "The recent earthquake demonstrates beyond doubt that at Atomic City we have sited 15 nuclear reactors within a high-risk seismic zone. At this point, the only responsible course open to us is to dismantle those reactors as soon as possible and relocate them within an area that has been unequivocally determined by the United States Geologic Survey to be one of low seismic risk."

Unfortunately, however, reality is not always perceived the same way by one person as it is by another. All too often people see in situations what they want to see, and then act according to their misperceptions. What really counts, of course, is whose version of reality is accepted by those responsible for making decisions. In a democratic society, that should ultimately come down to you and me, unless, that is, we fail to vote, or to write to our congresspeople when they really need to be set straight!

and cause most destruction. They include *L waves* (lateral motion, third to arrive) and *R waves* (rippling motion, fourth to arrive). **Seismographs** record earthquake intensity on **seismograms.** Separation of P, S, L, and R wave arrival times increases with distance between station and earthquake. Intersection of distance circles from three stations locates the epicenter. The Richter **earthquake magnitude** scale measures logarithms of seismogram vibration amplitudes. Magnitude $n + 1$ is 10× magnitude $n$. Energy release at magnitude $n + 1$ is 31× that at magnitude $n$. The Mercalli **earthquake intensity** scale measures the destructive effects of earthquakes. P waves are excluded from a **shadow zone** between 103° and 143° from the epicenter. S waves do not occur more than 103° from the epicenter. Earth's mantle probably consists of garnet peridotite. The **core** probably contains iron, nickel, and oxygen or sulfur.

The bearing of a horizontal line within an inclined surface is the **strike** of the surface. Its **dip** is the angle the surface makes with the horizontal, and it is measured perpendicularly to the strike. A fold's **hinge** is its region of maximum flexure; its **limbs** are its sides; its **axial plane** is the surface that connects the hinges of all its strata. In *symmetrical folds,* axial planes are vertical. In *asymmetrical folds,* they are inclined. In an *overturned fold,* one limb is rotated past the vertical. In a *recumbent fold,* the axial plane is roughly horizontal. **Anticlines** are archlike folds enclosing older strata within younger. **Synclines** are troughlike folds enclosing younger strata within older. Nonplunging fold hinges are horizontal, and limbs are parallel on maps; **plunging fold** hinges are not horizontal. In plunging anticlines, limbs converge in the plunge direction; in plunging synclines, they diverge. In **domes** and doubly plunging anticlines, oldest rocks are in the center. In **basins** and doubly plunging synclines, youngest rocks are in the center. A **monocline** is a sharp flexure in horizontal strata.

Strata in **mountain belts** have been folded, broken by compressive faults, metamorphosed, and intruded by magma during **orogeny.** A **platform** is a region of young, undeformed sedimentary strata overlying the deformed, eroded basement complex of an older mountain belt. **Shields** are extensive, broadly arched regions of low relief underlain by roots of ancient mountain belts. Continents grow by *marginal accretion* of mountain belts. **Unconformities** are discontinuities that record the deformation and erosion of underlying rocks prior to the deposition of overlying strata.

## Key Terms

| | |
|---|---|
| stress | directed stress |
| confining pressure | tensile stress |
| strain | compressive stress |

shear stress
elastic strain
plastic strain
joint
fault
fault block
hanging wall
footwall
extension fault
compression fault
strike-slip fault
elastic rebound
fault plane
focus
epicenter
seismic wave
P wave
S wave
body wave
surface wave
seismograph

seismogram
earthquake intensity
earthquake magnitude
shadow zone
core
strike
dip
hinge
limb
axial plane
anticline
syncline
plunging fold
dome
basin
monocline
mountain belt
orogeny
platform
shield
unconformity

## Questions for Review

1. Distinguish between confining pressure and directed stress, and between elastic strain and plastic strain.
2. Describe the orientation of paired joint sets and the motion of fault blocks with respect to the three fundamental types of directed stress.
3. Describe the plate tectonic settings within which compression, extension, and strike-slip faulting typically occur.
4. Describe the conditions that give rise to shallow, intermediate, and deep focus earthquakes.
5. Distinguish among P, S, L, and R waves regarding travel paths, behavior, velocity, arrival time, and destructive effects.
6. How many times larger are the wave amplitude and the energy release of a magnitude 6 earthquake than one of magnitude 3?
7. Describe how you would go about measuring strike and dip.
8. Describe the attitudes of the hinges, axial planes, and limbs of a nonplunging, asymmetrical syncline and a northerly-plunging, overturned anticline.
9. Describe the fundamental structure of a continent and the manner in which it grows over time.
10. Describe the sequence of events that must occur in the production of each of three different kinds of unconformities.

# Stories in Stone: Igneous, Sedimentary, and Metamorphic Environments

## Outline

*Rocks are where you find them.*

**Anonymous rockbound**

Photo: Geyser deposits, Mammoth Hot
Springs, Yellowstone National Park.

**Figure 7.3** Pahoehoe is a ropy form of lava created by the flow of more fluid lava beneath a more viscous rind. Here, a geologist hastily captures a specimen of still-hot lava from a crack in a young pahoehoe flow in Hawaii.

Photo by U.S. Geological Survey

**Figure 7.4** Clinkery aa lava forms where more rapid flow or increasing viscosity causes the rupturing of a pahoehoe rind. Intense heat from this fresh aa flow in Hawaii incinerates a bush in the flow's path.

Photo by U.S. Geological Survey

lava at 900° C. In addition to the temperature difference, however, rhyolite is also much richer in silica than basalt; hence, it has a greater tendency to form polymerized minerals with interlinked silica tetrahedra. This renders the lava more viscous and less fluid.

The more fluid types of lava (basalt and andesite) form two common types of large-scale flow structures: clinkery lava, or **aa** (pronounced like the prelude to a sneeze, but faster: "ah-ah"), and ropy lava, or **pahoehoe** (pronounced like the sneeze itself: "pa-*hoy*-hoy!"). In more viscous types, **block lava** is characteristic. The first two terms are Hawaiian.

The surface of pahoehoe (fig. 7.3) is ropy and wrinkled because the hot interior of a lava flow moves faster than its air-chilled rind. The rough, ragged, irregular surface of aa (fig. 7.4) results when the rind of a pahoehoe flow is broken either by an increase in velocity, as when the lava passes over steeper terrain, or by an increase in viscosity, as when dissolved gases rapidly escape. Once the surface of the flow ruptures, the clots of broken rind are rotated and churned into the flow, producing the typical, ragged texture of aa.

Block lava is a jumbled mass of large lava blocks formed by rapid flow in a viscous lava, such as andesite or rhyolite (fig. 7.5). Other large-scale structures found in lava flows include *pressure ridges,* formed at, and parallel with, the margins of flows by the pressure of fluid lava against the solidifying rind of the flow, and *lava tubes,* formed by the escape of fluid lava through breaks in the rind of the flow.

**Pyroclastic Rocks.**    Mafic magma is highly fluid, because of which it usually erupts as lava flows, but more silica-rich magmas often erupt explosively. This tendency results from the greater viscosity of felsic magmas

**Figure 7.5** Block lava is characteristic of viscous and silica-rich lava flows. Here, a viscous flow of block lava overrides a human-made masonry barrier during the 1959–1960 eruption of Kilauea volcano, Hawaii.

Photo by D. H. Richter, U.S. Geological Survey

and from their higher water content. Greater magmatic viscosity causes felsic volcanoes to choke up at the vent, which in turn causes pressure to build from the rising magma beneath. Dissolved water is significant because when it *exsolves* or separates, from the rising magma, it expands into a vapor, forming bubbles in the magma and greatly increasing its volume. The process is similar to the uncorking of a champagne bottle, whereupon carbon dioxide suddenly exsolves, causing the champagne to "erupt" from the bottle.

### Doris Livesy Reynolds (Born 1899)

Early in the present century, Finnish geologist J. J. Sederholm proposed, on the basis of field observations, that the rocks he called *migmatites*—intimate mixtures of metamorphic and igneous components—represented transitional stages in the partial melting of rocks of the continental crust to form granite. Because of the impressive theoretical work of N. L. Bowen, which supported the concept that all igneous magmas are ultimately derived from a single, parent mafic magma, Sederholm's hypothesis was largely ignored by the geological community. It remained for a brilliant English geologist, Doris Reynolds, to develop the theoretical basis for the refutation of Bowen's powerful argument.

Like Bowen, Reynolds was both highly trained in physical chemistry and field-oriented. As a Research Fellow of the University of Edinburgh, Scotland, she considered the problem of the *granitization* of continental rocks. Reynold's approach was to analyze the chemistry of both migmatites and the mudstones from which they were evidently derived, as indicated by their proximity to migmatites in the field. The results of her meticulous work showed that during granitization, sodium, calcium, and silicon were introduced into the mudstones along shear planes, and that potassium, aluminum, iron, and magnesium were lost along the same planes to become absorbed within neighboring limestones and quartz sandstones.

Reynolds also discovered that at higher levels in the crust, gas exsolves from magma due to loss of pressure and flows forcefully along mineral grain boundaries, disrupting bedrock, and chemically altering and transporting the resulting fragments in a manner similar to the industrial process of *fluidization*. This contributed to an understanding of how diamonds are formed within peculiar *kimberlite pipes* that bring highly altered, fluidized peridotite to the surface from great depths in the mantle. For her work, Reynolds was awarded the coveted Lyell Medal of the Geological Society of London in 1960.

Doris Reynolds kept the very best of company. Her husband was Arthur Holmes. Their combined achievements are an eloquent testimonial to the principle that great minds stimulate one another.

---

In magma chambers, such *anhydrous* (water-free) minerals as olivine and pyroxene crystallize first, thereby enriching the remaining magma in water. At the same time, the magma is rising into cooler and lower pressure regions of the crust where its capacity to hold water is reduced, so it eventually becomes water-saturated. Further cooling and crystallization cause water to exsolve suddenly from the magma as vapor bubbles. The consequent expansion is powerful enough to fracture overlying rocks. This, in turn, results in a sudden loss of pressure, which causes further boiling of the magma as more water exsolves, whereupon a violent, explosive eruption results. The frothing magma shoots skyward from the volcanic vent, producing an enormous eruption cloud, or column, of which the lower portion (1–2 km) is propelled upward by gas pressure. The upper portion of the cloud, however (up to 55 km), is carried upward by convection because its turbulent mixture of dust, ash, magma droplets, and hot gas is, surprisingly, less dense than the surrounding air.

Pyroclastic rocks (see chapter 4) consist of fragments of a variety of sizes from *dust* (less than 1/16 mm), through *ash* (1/16–2mm) and *lapilli* (2–64 mm), to angular *blocks* and rounded *bombs* (over 64 mm). Collectively, they are known as **tephra** (fig. 7.6), and they consist of glass fragments, or *shards,* from broken magma bubbles; mineral crystals; droplets or blobs (bombs) of

**Figure 7.6** Tephra consist of fragments of wallrock, solidified lava, crystals, and glass ejected explosively from volcanic vents. In this stratified tephra deposit from a prehistoric eruption of Washington's Mount St. Helens, a variety of sizes from dust to blocks is represented.

Photo by D. R. Crandell, U.S. Geological Survey

solidified lava; and fragments of shattered wallrock, including solid lava from previous eruptions. Larger, heavier tephra follow ballistic trajectories, like mortar shells, and they accumulate near the volcanic vents that produce them. Smaller, lighter tephra are deposited by

**Figure 7.7** Successive flows of flood basalt form terraced cliffs along the walls of the Palouse River gorge in Eastern Washington state.

Photo by P. Weis, U.S. Geological Survey

**Figure 7.8** Asamayama, an active Japanese volcano, shows many features common to central eruptions: a young tephra cone topped by a crater, lava flows issuing from the flanks of the cone, an older tephra cone topped by a small caldera within which the younger cone has grown, and the stream-dissected remains of a much larger cone and caldera to the right of the newer cones.

Photo from W. K. Hamblin and J. D. Howard, *Exercises in Physical Geology*, 6/e, Macmillan Publishing Company.

fallout from the eruption cloud. Accordingly, they blanket the surrounding countryside with the coarser fragments at the bottom of the blanket and the finer at the top.

**Fissure Eruptions and Flood Basalts.** Although volcanic mountains are conspicuous and impressive, probably the greatest volumes of basalt lava have been erupted not by volcanoes but by long, open fissures in the crust. The principal example of this type of eruption is along the oceanic ridge, where basalt is erupted quietly and intermittently into the median rift as the plates slide away from the ridge. Such **fissure eruptions** have been observed during historic time in Iceland, where the oceanic ridge has been elevated above sea level, and on the flanks of the Hawaiian volcanoes. Fissure flows are also common on the continents. Figure 7.7 shows the extensive volcanic plain of the Columbia River *flood basalts* in the Pacific Northwest. These Upper Tertiary lavas erupted between 17 and 6 million years ago at a rate calculated to be at least twice that of the great Hawaiian volcanoes and at least four times that of a typical oceanic median rift. The total volume of the Columbia River basalts is estimated at about 200,000 km³. Individual flows are thin—a few meters to tens of meters—but extensive in area. The source for the lavas is thought to have been a mantle plume that is now located beneath Yellowstone National Park. Other extensive deposits of flood basalt include the Paraná basalts of eastern South America with a minimum volume of 350,000 km³, and the Karroo basalts of South Africa, which at one time covered an area of about 2,000,000 km². These two Jurassic fields represent the initial stages of rifting of the supercontinent Gondwanaland. Whether the Columbia basalts herald the imminent rifting of North America is still unclear.

**Central Eruptions and Volcanoes.** A more familiar type of volcanism is the **central eruption,** in which lava is extruded from a single *conduit,* or feeder pipe. As with fissure flows, central eruptions are far more common beneath the oceans than they are on the continents, making the basaltic submarine seamount the most abundant "landform" on Earth (see fig. 5.13).

The areal view of the Japanese volcano Asamayama in figure 7.8 illustrates the features of a typical central eruption. Several sequential events are evident in this photograph. The most recent was the formation of a symmetrical, central volcanic **cone** (*top, right of center*) with a nearly circular **crater** surrounding the central vent. A number of semiviscous lava flows issued from flank vents on the cone and flowed toward the lower left of the photograph. This cone is superimposed on an older and much larger cone whose summit collapsed following a major explosion, forming a **caldera** 1.6 km in diameter. Although craters and calderas are similar in form, craters are seldom much larger than the conduits that feed them, and they are created and maintained by the uprushing of erupted materials. Calderas, on the other hand, are created when great volumes of silicic magma have erupted explosively from near-surface magma chambers, and the overlying crust has subsequently foundered into the evacuated cavity beneath. The remains of a still larger cone and caldera can be seen on the right side of figure 7.8. Apparently, Asamayama would not say "die" after its first, or even its second, explosive evisceration!

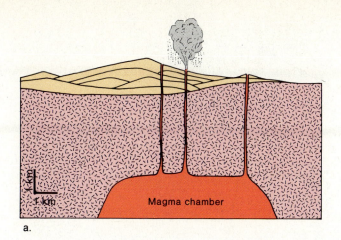

a.

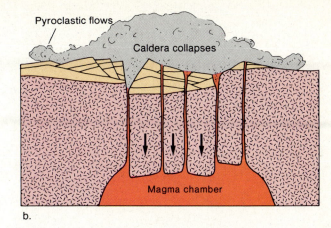

b.

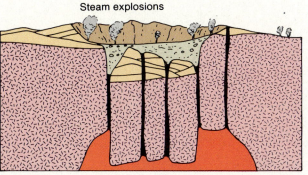
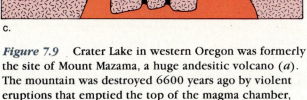

c.

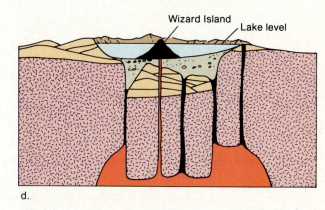

d.

**Figure 7.9** Crater Lake in western Oregon was formerly the site of Mount Mazama, a huge andesitic volcano (*a*). The mountain was destroyed 6600 years ago by violent eruptions that emptied the top of the magma chamber, causing the chamber's roof to crack, founder, and sink, forming a caldera (*b*). Steam eruptions followed (*c*). In time, lake water filled the caldera, and minor eruptions of basalt created Wizard Island (*d*).

Source: After C. Bacon, U.S. Geological Survey.

One of the best studied examples of a caldera is Crater Lake in Oregon (fig. 7.9). The terminal eruption of Mount Mazama, the huge andesitic volcano that once stood on the site of the lake, distributed ash over most of Oregon and parts of nine adjacent states and Canadian provinces. Within an area of about 300 km² adjacent to the volcano, Mazama's ash fell in a blanket over 30 cm deep.

Still larger calderas are known. Yellowstone Lake in northwestern Wyoming (fig. 7.10) lies within a caldera that measures about 50 × 65 km, an area some 80 times that of Crater Lake! Its most recent eruption occurred about 600,000 years ago, projecting over 1000 km³ of magma into the atmosphere and blanketing the entire midcontinental region of the United States from South Dakota to Texas with a layer of tephra known as the Pearlette ash. Because of its widespread and instantaneous deposition, the Pearlette ash is a useful *marker horizon* for dating strata in which it can be identified. The heat necessary to generate the felsic magma for the Yellowstone eruption is thought to have been provided by the same mantle plume that fed the Columbia River

basalts. The plume is now located farther east, however, because of North America's westward drift. New evidence indicates that felsic magma is currently rising beneath the Yellowstone plateau and will probably break out in yet another cataclysmic eruption, although the exact timing of the event remains unknown.

Depending mainly on the type of lava involved, central eruptions assume a number of different forms. The most important of these are shield volcanoes, cinder cones, composite volcanoes, and volcanic domes.

**Shield Volcanoes.** Named for their fancied resemblance to an inverted Roman shield (fig. 7.11), broad, low **shield volcanoes** are built by successive outpourings of gas-poor, fluid basalt lava from a central vent and from satellite vents on the flanks of the mountain. Typically, shield volcanoes have slopes that range from 2° to 20°. They vary in size from a few hectares to the enormous Hawaiian volcanoes with areas of hundreds to thousands of square kilometers and elevations of up to 8.9 km above the seafloor. The largest shield volcano in

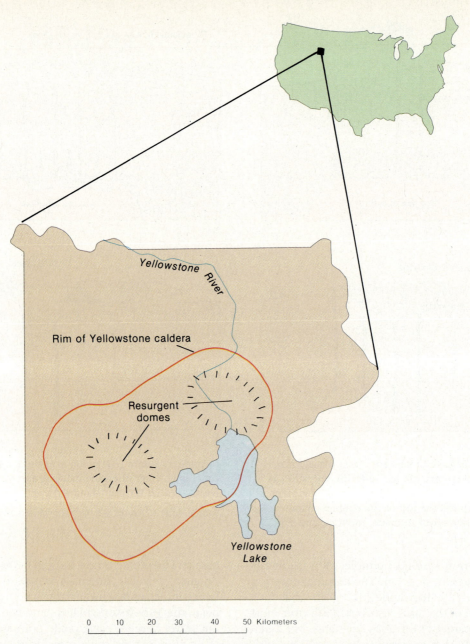

**Figure 7.10** The outline of Yellowstone caldera is shown here in relation to Yellowstone Lake. Two resurgent domes have swelled the caldera floor since its last eruption about 600,000 years ago.

the Solar System is Olympus Mons in the Tharsis region of Mars (see fig. 17.19). This huge basaltic edifice has a basal area of more than 250,000 km² and a summit elevation of 25 km!

The Hawaiian volcanoes have been studied extensively, and their manner of eruption is relatively well understood. Periodically, partial melting of the upper mantle, probably by a mantle plume, generates a batch of mafic magma at about 60 km depth beneath the main island. Within a period of about 10 years, this magma then rises buoyantly into the overlying volcanic cones, Mauna Loa and Kilauea, within which it comes to rest temporarily in shallow magma chambers. Tiltmeters installed on the flanks of Kilauea volcano show that the cone actually swells up to a meter or more over an 80 km² area around the summit while these chambers fill with magma. Eruption follows, usually from flank vents (fig. 7.12) but frequently also from a lava pit at the summit. These eruptions partially drain the magma chamber, causing the cone to deflate. Occasionally, the chamber drains underground as well, resulting in the collapse of the summit and the enlargement of the summit caldera. Because of the fluid nature of their high-temperature basaltic lava, Hawaiian eruptions are rarely

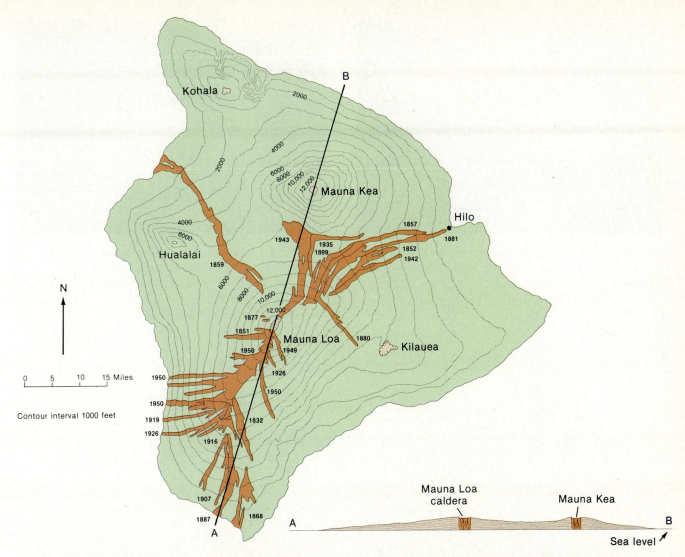

**Figure 7.11** Five basaltic shield volcanoes compose the island of Hawaii. Outlines of historic lava flows are shown, with dates. The cross section (*b*) along line *AB* illustrates the gentle slopes of the shield volcanoes, built of thin flows of fluid basalt lava. Over time, eruptive activity has shifted southeastward from now-extinct Kohala to Kilauea because of the island's northwestward drift over an underlying mantle plume.

explosive. Shield volcanoes also occur on the oceanic ridge (e.g., Iceland) and occasionally on the continents.

**Cinder Cones.** Felsic volcanoes are not the only ones that produce tephra. If basalt magma is sufficiently charged with dissolved gas, it can erupt explosively, forming steep-sided **cinder cones** (fig. 7.13) composed of tephra—mainly lava fragments that have cooled and solidified during their passage through the air. Cinder cones are usually no more than about 500 m high. This is mainly due to the relatively low energy of explosions in basalt lava. A large central crater surrounds the vent. Often the cone is asymmetrical because of the effect of the prevailing wind on the tephra as they are ejected from the vent. Cinder cones are steep-sided, with slope angles of 35° to 40°, the maximum steepness at which loose, dry, unconsolidated material will stand without slumping. They tend to occur in clusters, and basalt lava flows often issue from their bases or from low on their flanks.

**Composite Volcanoes.** Among the most imposing and majestic of Earth's mountains are the **composite volcanoes,** the internal structure of which is shown in figure 7.14. Fujiyama, located over the subduction zone of the Japan trench, is one example of this kind of volcano. Others are the peaks of the Cascade chain, including Mount Rainier and Mount St. Helens, and the volcanic peaks of the Andes, for which the rock type andesite was named. Andesite is more felsic than basalt; hence, it is more viscous, and can only flow for short

**Figure 7.12**   Behind a Hawaiian papaya field, vermilion fountains of fluid basalt lava erupt from a rift on the east flank of Kilauea volcano.

Photo by U.S. Geological Survey

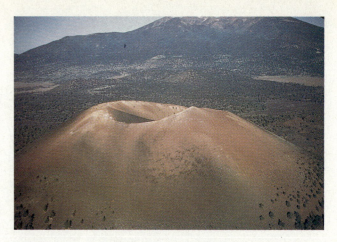

**Figure 7.13**   Cinder cones consist almost entirely of tephra. They are often asymmetric, largely due to prevailing winds, as is this one in northern Arizona's San Francisco volcanic field.

Photo by E. D. McKee, U.S. Geological Survey

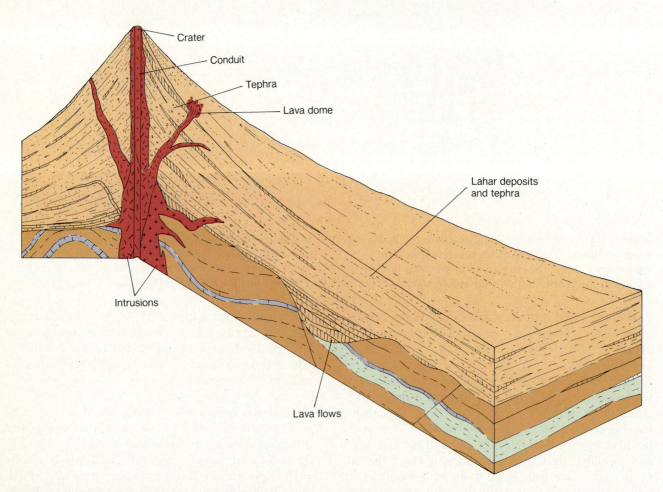

**Figure 7.14**   The immense cone of a composite volcano is built of a skeletal framework of predominantly andesitic lava flows (*vertical lines*), interstratified with tephra layers and lahar deposits (*both stippled*). The latter become more prominent with distance from the vent. Note satellitic dome of felsic lava.

**Figure 7.15** The catastrophic 1980 eruption of Mount St. Helens obliterated Spirit Lake on the north flank of the volcano and mixed its waters with exploding tephra. The resulting lahar, or volcanic mudflow, filled the valley of the North Fork of the Toutle River downstream.

Photo by R. L. Schuster, U.S. Geological Survey

**Figure 7.16** Mount St. Helens erupting, 18 May 1980.

Photo by J. G. Rosenbaum, U.S. Geological Survey

distances on steep slopes. Andesite also contains more water than basalt, which makes it more susceptible to boiling and explosive eruption. Because of these characteristics, andesitic eruptions are typically about half pyroclastic and half fluid. Fallout from eruption clouds deposits ash layers on the cone, and subsequent flows of lava from the central vents, or from flank vents, remain on the cone as coatings over the pyroclastic deposits rather than draining away to the adjacent flatlands as basalt lava would do. In time, the alternation of quiet and explosive eruptions builds an enormous cone of interlayered flows and tephra. A third important process—the volcanic mudflow, or **lahar** (an Indonesian term)—adds yet another kind of deposit to the layer cake. Lahars can be either cold, as when glacial meltwater soaks a steep portion of the cone until it collapses

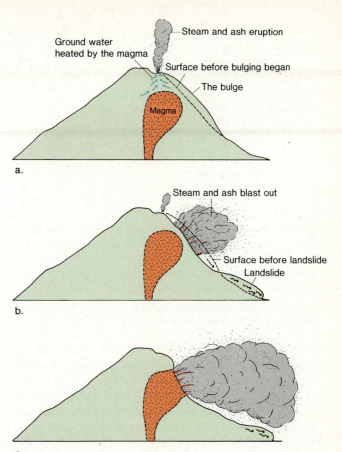

**Figure 7.17** The 18 May 1980 eruption of Mount St. Helens was preceded by the growth of a massive bulge (*a*) on the mountain's north slope as magma filled a chamber within the cone. Landsliding of the bulge (*b*) unroofed the chamber, causing magma to boil and erupt explosively (*c*).

and flows downslope, or hot, as when lava erupts beneath a glacier or a crater lake. Lahars roll down a volcano's flanks like wet concrete, depositing layers of volcanic conglomerate and often debouching on the surrounding countryside with awesome destructiveness (fig. 7.15). Loss of life and property is attributable more to lahars than to any other volcanic hazard.

The 1980 eruption of Mount St. Helens in Washington State was a classic example of a volcanic explosion due to boiling of intermediate magma (figs. 7.16 and 7.17). A small chamber of *dacite* (a quartz-rich andesite) magma had formed beneath the north flank of the volcano before the eruption. Pressure from this magma distended and oversteepened the north flank, which eventually fractured and collapsed. The resulting loss of pressure caused a sudden exsolution of gas from the magma, which in turn caused the explosion. Over 1 km³ of material was blown into the atmosphere. Geological evidence of frequent earlier eruptions in the form of lava flows, ash layers, lahars, and the structure of the cone indicates that during its 40,000-year life, Mount St. Helens has been one of the most active of the Cascade chain volcanoes, especially within the past 4500 years.

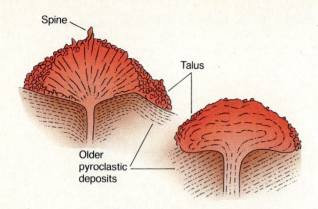

**Figure 7.18** Viscous felsic lavas are extruded from their conduits with difficulty. The most viscous are forced out as semisolid plugs or spines (*left*), which shatter easily, forming an apron of rock debris, or talus. Less viscous felsic lavas are extruded as bulbous masses (*right*).

Source: M. G. Best, *Igneous and Metamorphic Petrology.* Copyright © by W. H. Freeman and Company, New York, NY.

**Volcanic Domes.** Rhyolite, being the most felsic of lavas, is therefore the most viscous. It is also the richest in dissolved water, which makes it prone to boiling and explosive eruption; consequently, it seldom forms fluid flows. Water-poor rhyolite is almost too pasty to flow at all. Where it has flowed out at the surface, as at Mono Lake, California, and in the Yellowstone caldera, the flows are massive, thick, and of very limited extent. More typically, glassy rhyolite extrudes sparingly from vents and fissures, and piles up over them to form steep-sided **volcanic domes** (fig. 7.18). When the lava is relatively fluid, flow layers in these domes tend to be roughly concentric with the outer surface. In the case of dry, exceedingly viscous lava, the domes consist of a series of slabs or "spines" of rigid rhyolite fanning out radially from the central conduit.

Because of its high gas content and viscosity, rhyolite magma typically undergoes vigorous boiling *throughout* the magma column during major eruptions, not just in its upper part. This results in the forceful ejection of large quantities of liquid magma from the vent. As this magma spews out of the volcano, it continues to boil due to the sudden release of pressure. The boiling literally tears the viscous magma apart into furiously sizzling particles. These particles form **ash flows** that roll down the slopes of the volcano beneath glowing clouds at speeds up to 200 km/hr or more (fig. 7.19). The expulsion of gas by the individual particles keeps them suspended in the air and separated from one another as the entire mass flows downhill on a cushion of expanding gas. Needless to say, all organic matter in the path of this formidable fire avalanche is instantly incinerated. When the flow finally comes to rest, it forms a sintered mass of rhyolitic glass fragments, crystals, and ash known as *ash flow tuff.* If hot and thick enough, the tephra are actually fused together to form a *welded tuff* (see fig. 4.14*b*).

**Figure 7.19** An ash flow rolls down the south flank of Mayon volcano on Luzon Island, the Phillipines, 2 May 1968.

Photo by J. G. Moore, U.S. Geological Survey

### Igneous Intrusions

Thus far, we have been concerned with the behavior of magma as it breaks out at Earth's surface. In fact, only a minute fraction of all the magma generated actually reaches Earth's surface. The bulk of it remains below ground, and crystallizes there to form a variety of **intrusive rocks** (Latin for "thrust in").

Whereas extrusive rocks have aphanitic textures, intrusive rocks normally have phaneritic textures except where they have been intruded into cold wallrocks at shallow depths in Earth's crust. In such cases, they normally exhibit narrow, *chilled margins* with aphanitic texture.

A body of intrusive igneous rock that has crystallized beneath Earth's surface is known as a **pluton** (fig. 7.20). If the borders of a pluton are essentially parallel with the bedding of surrounding sedimentary strata or with the foliation of metamorphic rocks, it is described as **concordant.** In general, concordant plutons are forcibly intruded along bedding or foliation planes, shouldering aside the layers on either side in order to make room for themselves. Plutons whose borders truncate, or cut off, surrounding bedding or foliation are described as **discordant.** Some discordant plutons are formed by the filling of fractures. Others are intruded relatively passively as low-density magma rises buoyantly through the crust, hollowing out a space for itself by fracturing, engulfing, and digesting sections of the overlying and surrounding wallrock, a process called *magmatic stoping* (rhymes with "hoping").

**Concordant Plutons.** Sills (fig. 7.21) are concordant, tabular bodies of igneous rock intruded between adjacent strata in a series of layered rocks (igneous, sedimentary, or metamorphic). Sills range in thickness from a few millimeters to hundreds of meters and in extent from a few square meters to thousands of square kilometers. **Laccoliths** (Greek for "reservoir stone"; fig.

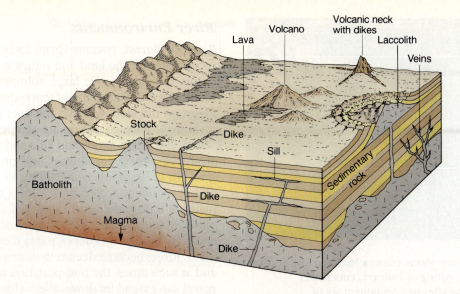

Figure 7.20 A variety of extrusive and intrusive igneous bodies. The volcanoes and lava flows are extrusive, whereas the remaining features, including the eroded volcanic neck, are intrusive. Sills and laccoliths are concordant plutons; batholiths, stocks, dikes, veins, and the volcanic neck are discordant.

Figure 7.21 High above Grinnell Lake in Glacier National Park, a dark sill traverses a cliff of horizontally bedded sedimentary rocks. Hot fluids emanating from the sill bleached the immediately adjacent rock, creating light borders above and below the intrusion.

Photo by P. Carrara, U.S. Geological Survey

Figure 7.22 A laccolith of dark Tertiary diorite arches overlying Mesozoic sandstone strata in the Henry Mountains of southeastern Utah. In this classic locality, the laccoliths occur as satellites to central conduits, and their feeder pipes are located at the edges of the plutons. In other cases, the feeder pipe is located beneath the pluton on its lower surface.

7.22) are similar in form to sills, except that the upper surface of a laccolith is arched where the intrusion has raised the overlying strata into a blisterlike uplift. Laccoliths are typically several kilometers in diameter and several hundred meters in total thickness. **Lopoliths** (Greek for "shell stone") are enormous, fractionally crystallized sills intruded into basins. Their rock types run the range from ultramafic to felsic, and they are typically thousands of square kilometers in area and several kilometers thick. The Proterozoic Bushveld lopolith in South Africa has an area of 63,000 km², and a maximum thickness of about 25 km. The origin of lopoliths is unclear, but one plausible explanation is that they represent the intrusive aftermaths of small asteroid impacts on Earth.

**Discordant Plutons.** Dikes (fig. 7.23) are tabular plutons, like sills, but they are discordant, cutting across the grain of older rocks. Of all plutons, dikes are by far the most numerous. They can occur singly, but frequently, they occur in great swarms, as in the upper part of the oceanic lithosphere, which consists entirely of dike swarms running parallel with the oceanic ridge. Dike swarms also occur in the Scottish Highlands where tensional forces associated with the rifting of Greenland from Europe in the Early Tertiary Period stretched the crust in a northeast-southwest direction. This resulted in the opening of northwest-southeast oriented fractures, which promptly filled with basaltic dikes.

**Figure 7.26** A bluff on the shore of the Bay of Fundy, Nova Scotia, reveals an inclined sequence of clastic strata from a 315-million-year-old river floodplain. The thick, upper layer consists of sand deposited in a major river channel. The older underlying strata consist of finer sand and mud deposited on the floodplain during floods. See also figure 9.16.

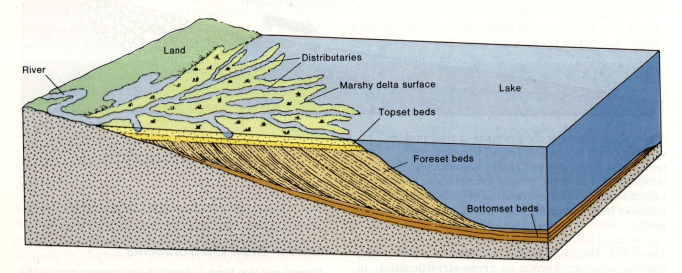

**Figure 7.27** Where rivers flow into lakes, they typically build sandy deltas outward from the shore. Coarser topset beds supply sand to dipping foreset beds, which build progressively outward over finer bottomset beds.

## Estuary and Swamp Environments

Following the last ice age, sea level has risen worldwide by as much as 100 m, except in places where shorelines have been tectonically uplifted at rates that have outstripped the rate of sea level rise. This has resulted in the drowning of the mouths of most of Earth's rivers, creating broad embayments, or **estuaries,** in the mouths of rivers that carry relatively little sediment to the sea. (Marine deltas have filled in the estuaries of rivers that carry large volumes of sediment.) Within this *estuarine environment* (from Latin *aestus,* "tide," in allusion to the strong influence of marine tides in this environment), black, organic-rich mud is deposited. If an estuary is especially rich in plant and animal life, it is called a **swamp.** Frequently, organic remains accumulate in swamp, or *paludal,* environments (pal-*oo*-dle; from Latin *palus,* "swamp") faster than decomposing organisms can consume them, leading to the formation of peat, which can eventually become lithified to coal.

### Gravity-Dominated Environments

In steep terrain, gravity is often a more effective transport agent than flowing water. Deposits of gravity-dominated environments include rockfalls, rockslides, landslides, slumps, earthflows, and mudflows (see chapter 8).

### Shoreline Environments

The incessant beating of waves on coastlines creates a high-energy environment in which relatively coarse sediment can be transported. Occasional large river floods deliver sand to the coast, and waves erode gravel from rocky headlands. These coarse sediments accumulate along the shore because of the persistent shoreward sweep of the waves. Within this high-energy *littoral environment* (*lit*-tor-ul; from Latin, *litus,* "seashore"), a variety of distinctive landforms is deposited, including beaches, spits, barrier islands, baymouth bars, and tombolos (see chapter 12). Some examples of the first three of these features are illustrated in figure 7.24. All of them show well-developed cross-stratification similar to that found in lake deltas but on a smaller scale because the laminations are usually the slip faces of migrating submarine sand ripples and sand dunes built by marine currents (fig. 7.28).

*Mudflats* usually form on shorelines in the quiet, shallow water behind spits and barrier islands, where the absence of wave action allows mud to accumulate. Although protected from waves by spits and barrier islands, in most areas mudflats are subject to the action of tides. As these flow in and out twice daily, they create ripples in the mud. When mudflats are exposed to the Sun at low tide, they often dry out and develop polygonal shrinkage cracks. Both these features can subsequently be buried and preserved as **ripplemarks** and **mudcracks,** usually separately but occasionally together (fig. 7.29). Under favorable circumstances, both can be useful in determining the original tops of subsequently deformed strata.

### Ocean Environments

Marine environments are divided into *shallow marine* (continental shelf) and *deep marine* (continental rise) because both the sediment types and the processes that deposit them are quite different in the two environments. Mud is the common denominator in both, but here the resemblance ends. In addition to mud, the shallow marine environment also comprises quartz sand, calcium carbonate, and evaporites. The quartz sand was laid down on ancient shorelines during ice ages when

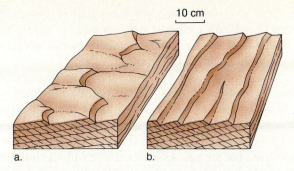

10 cm

*Figure 7.28* Cross-stratification. The cross strata (*short lines*) represent slip faces of ripples migrating toward the right under the influence of a strong current (*a*) and a weak current (*b*) flowing toward the right. Each group of cross strata is called a set, and is separated from other sets by more or less planar erosion surfaces along which one ripple overrode another. Normally, sets dip upcurrent, whereas cross strata dip downcurrent.

From R. R. Compton, *Geology in the Field.* Copyright © John Wiley & Sons, Inc., New York, NY. Reprinted by permission of John Wiley & Sons, Inc.

*Figure 7.29* An unusual, combined occurrence of ripplemarks and mudcracks in mudstone in Nova Scotia. Sharp crests of ripples and edges of cracks both face the observer, indicating that the exposed bedding plane is right side up.

sea level was lower than it is now; hence, it might more properly be classed with the shoreline environment. The mud represents all the sediment that is too fine to settle out from high-energy shoreline waters and that winds up in deeper, quieter water below the reach of wave action. Even so, continental shelf mud is often stirred up locally by storm waves and bottom currents and redeposited elsewhere.

Calcium carbonate, as I note in chapter 1, is largely a byproduct of photosynthesis and carbonate shell secretion by tropical marine organisms. Most photosynthetic organisms fare poorly in turbid (muddy) water because of low light availability. Consequently, calcium carbonate is usually deposited in places where mud is not being deposited. Because most carbonate-producing organisms require water that is constantly warm (corals, for example, will not tolerate water that is colder than 18° C), the bulk of the ocean's carbonate production occurs in the tropics.

Often, corals and carbonate-secreting algae build massive **reefs** of skeletal limestone off shorelines of tropical continents and islands. Reefs such as those shown along the edge of the continental shelf in figure 7.24 began life there when the sea level was lower during the last ice age. As sea level rose with the melting of the glaciers, the corals kept pace with that rise, building the reefs upward fast enough that their tops would stay within about 20 m of the ocean surface, a depth below which light penetration is insufficient for the growth of most coral species.

Evaporites (mainly gypsum rock and salt rock) form in restricted bays in which seawater evaporates faster than it can be diluted by inflows of fresh river water or normal seawater.

In the deep marine environment of the continental rise, two sedimentary processes are dominant: (1) the slow settling of fine mud through still water and (2) sudden submarine mudflows called **turbidity flows** (fig. 7.30). In a turbidity flow, a mass of mud on or below the edge of the continental shelf is dislodged and broken up, whereupon its clastic particles mix with seawater to form a slurry of sand, mud, and water. Because this mixture is denser than the surrounding water, it flows rapidly down the continental slope under the influence of gravity. As the flow gradually comes to rest, its energy dissipates, and the suspended material settles out of the water to form a sedimentary deposit called a *turbidite*. Turbidites come in a variety of grain sizes from coarse graywacke to fine mudstone, most of which exhibit the phenomenon of **graded bedding** (fig. 7.31) in which the coarser fragments lie at the bottom of a bed and the finer at the top. Graded bedding results from either or both of two processes:

1. Coarser clasts settle faster through the turbulent water than finer clasts
2. Progressively finer material is deposited as flow velocity decreases.

Where sediment is supplied abundantly to the outer edge of the continental shelf, a broad, thick turbidite deposit called a **submarine fan** is formed (see fig. 7.24). On these fans, the cyclic alternation of slowly deposited mud and rapidly deposited turbidites results in a monotonous interlayering of thin strata of mudstone and graywacke. In figure 1.2, Earth's two largest submarine fans are clearly visible flanking India, where they are fed by sediment-laden rivers flowing from the Himalayas.

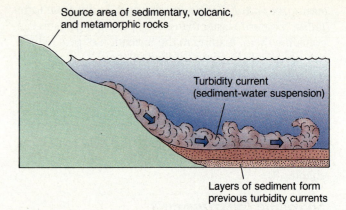

*Figure 7.30* **Composed of a turbulent mixture of sand, mud, and water, turbidity flows are much denser than seawater. Because of this, they run down submarine slopes, mixing with the overlying seawater.**

## Glacial Environments

*Glaciers* occur in regions where more snow falls in winter than can be melted in summer. This condition is met in two different settings: (1) *high latitudes,* in which thick, tabular *ice sheets* form and spread laterally on relatively flat terrain (see fig. 10.4) and (2) *high altitudes,* in which thick tongues of ice called *valley glaciers* flow down preexisting river valleys from areas of heavy snow accumulation on mountainsides (see fig. 10.5). Glacial deposits are extremely poorly sorted, exhibiting a wide range of clastic particle sizes from clay to boulders (see fig. 10.17). Those that are deposited by ice sheets blanket the land surface, whereas sediments of valley glaciers are confined to valley sides and bottoms. Each is characterized by a range of distinctive landforms illustrated in figures 10.22 and 10.23.

## Wind-Dominated Environments

In places where there is little vegetation, as in deserts, seacoasts, and glacier-margin environments, loose, unconsolidated sediment can be moved by strong wind action and deposited elsewhere. In general, the stronger the wind in this *aeolian environment* (from Greek, *Aeolus,* god of winds) and the smaller the particles, the farther those particles will be moved. Gravel is unmoved by any natural wind. Sand is moved as cross-stratified *dunes* in a bouncing, jumping motion called *saltation,* (see fig. 14.22). Silt is wafted aloft to be deposited over the landscape downwind in blankets of *loess* (see chapter 10) as wind force slackens. Clay is dispersed as fine dust in the atmosphere, settling to Earth so slowly that the only place where it can form a noticeable deposit is on the deep ocean floor, far removed from any other source of sediment.

**Figure 7.31**  Graded bedding in a siltstone. Each light band (about 3–5 cm thick) represents the base of a separate submarine mudflow. The coarser clasts settled first from the turbulent and turbid (muddy) water, leaving the finer, darker material to settle above it later as the energy of the flow abated. Graded bedding is useful in determining which direction is "up" in deformed strata.

## Metamorphic Environments of Earth's Crust and Mantle

We can observe most sedimentary processes directly, and we can infer the nature of many igneous processes from the evidence present in lavas erupted at Earth's surface. We have no convenient way to study metamorphic processes directly, however, except through deep drilling, which can only penetrate into a limited range of metamorphic environments. Consequently, most of what we know of metamorphism derives from the study of metamorphic rocks and minerals in thin section and from laboratory experiments, which rather imperfectly mimic the conditions under which metamorphic rocks form.

### The Types of Metamorphism

Some important examples from such studies are compiled in figures 7.32 and 7.33, which show metamorphic changes in minerals and rocks, respectively. Note that in both figures, pressure in kilobars (1 kb = 1000 atmospheres) is shown on the upper edges of the colored bands, whereas temperature in °C is shown on the lower edges. Vertical red lines show the temperatures at which wet granite and wet and dry basalt begin to melt. These melting temperatures place an obvious upper limit on metamorphism, but where the rocks are dry, that limit is considerably stretched. Notice that all of the melting temperatures vary with pressure.

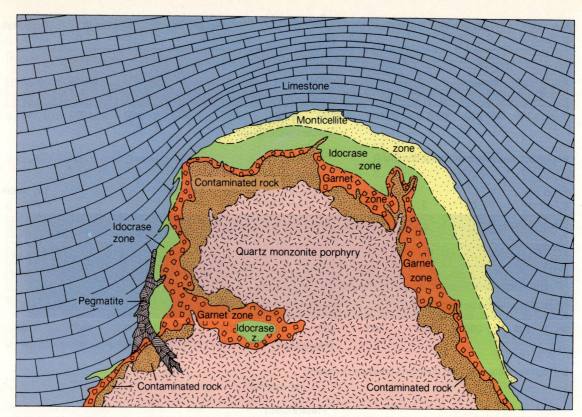

**Figure 7.34** Where igneous magmas intrude bedrock of different composition, contact metamorphism results, producing a variety of high temperature, low pressure minerals. Here, a granitic magma has intruded a limestone at Crestmore, California, producing concentric zones of garnet, idocrase (a complex silicate of calcium, magnesium, and iron), and monticellite (a rare, Ca,Mg olivine).

E. G. Ehlers and H. Blatt, *Petrology: Igneous, Sedimentary, and Metamorphic* Copyright © by W. H. Freeman and Company, New York, NY.

Metamorphic mineral changes at high pressure serve the common end of packing atoms into a more limited space. Thus, the changes from calcite to aragonite, from Ca-feldspar to garnet, from Na-feldspar to jadeite (Na,Al-pyroxene), and from pyroxene to glaucophane (Na,Al-amphibole), all involve the creation of more compact crystal structures.

### Rock Changes during Metamorphism

Figure 7.33 shows trends in the metamorphism of some major rock types, beginning with the unmetamorphosed original rock type on the left. I have described most of these rock types in chapter 4, so I leave it to the reader to review them and to use this figure and the preceding one to help develop a feel for the relationships of the various rock types to the metamorphic paths and their constituent minerals. Those rock types not described in chapter 4 follow:

1. *Greenschist,* a green, foliated rock intermediate in metamorphic grade between greenstone and amphibolite, and consisting of chlorite, epidote, green amphibole, and Na-feldspar

2. *Argillite,* a highly lithified, but nonfoliated rock, intermediate between mudstone and slate, in which there has been little growth of such platy minerals as chlorite and muscovite

3. *Granulite,* an essentially nonfoliated, high-grade metamorphic rock consisting mainly of nonplaty minerals such as quartz and feldspar

4. *Hornfels,* a hornlike, nonfoliated, aphanitic rock created by contact metamorphism and containing andalusite and a variety of other unusual, low-pressure minerals.

## Rocks and Plate Tectonics

Before the development of plate tectonic theory, the natures of the environments within which many rock types formed were obscure. Now that we have the conceptual framework of plate tectonics, however, we have a much better grasp of most rock-forming processes and the environments in which they operate. Following are sketches of some rock-forming processes within four major plate tectonic settings.

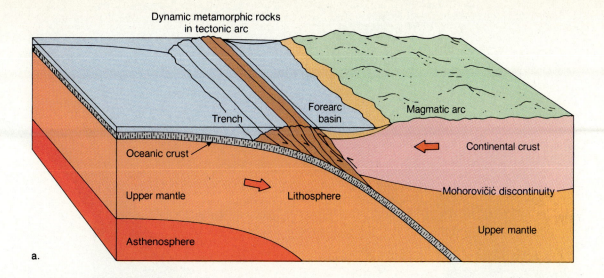

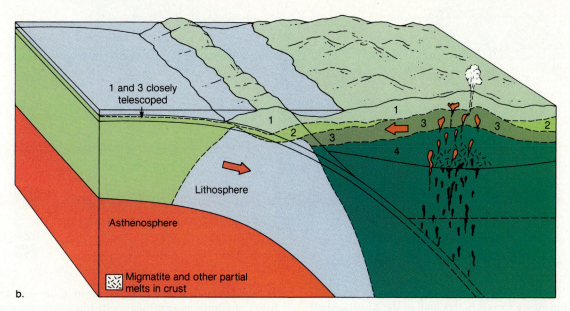

*Figure 7.35* **Metamorphic environments of an ocean-continent collision zone. Blue zone in *b* indicates mechanical metamorphism. Zones of metamorphic** intensity in the zone of regional metamorphism correspond to those in figures 7.32 and 7.33.

## Rocks of the Oceanic Lithosphere

Figure 7.37 is a diagrammatic cross section through the oceanic lithosphere at an oceanic ridge. Although there is sound evidence that the features in this cross section are approximately as depicted here, that evidence is indirect. No one has ever penetrated the seafloor to observe those features, and it is doubtful that anyone ever will. Consequently, diagrams such as this are gross oversimplifications and generalizations of a more complex reality.

The evidence on which the cross section is based is of four kinds:

1. Using the techniques of *seismic reflection and refraction profiling* (see chapter 11) geologists have generated vague images of structures beneath the seafloor from sound waves reflected and refracted from those structures.
2. Rocks of the various types shown here have been dredged from the seafloor.
3. Slivers of what are thought to be oceanic lithosphere have been caught in mountain belts between colliding plate boundaries, and the structure of these slices has been carefully studied.

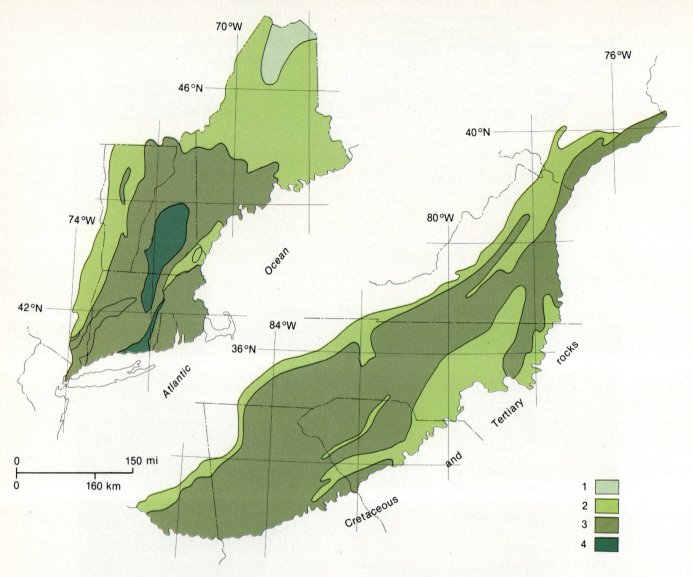

**Figure 7.36** The distributions of four zones of regional metamorphism are shown in this map of the Appalachian mountain belt. Notice that metamorphism is most intense in New England, where a coastal promontory took the brunt of continent-continent collision.

Source: M. G. Best, *Igneous and Metamorphic Petrology*
Copyright © by W. H. Freeman and Company, New York, NY.

**4.** The formation of magma beneath the ridge and its subsequent rise into the median rift is indicated by three kinds of evidence:
  a. the high elevation of the oceanic ridge
  b. earthquakes that extend from the ridge crest to a depth in the mantle of up to 70 km; and
  c. submarine volcanism.

At the base of the oceanic lithosphere is a thick layer of *undepleted peridotite* ("x" pattern in fig. 7.37). The elevated condition of the ridge and the high flow of heat from beneath it suggest that this material is being forced upward, probably by convection currents in the mantle. At a depth of about 25–30 km below the oceanic ridge, the undepleted peridotite starts to melt partially, probably because of (1) the sliding apart of the plates, which reduces pressure on the underlying hot rock; and (2) the penetration of seawater beneath the ridge, which lowers the rock's melting point. This partial melting yields an olivine-rich mafic magma which, being lighter than its surroundings, rises toward the ridge, leaving behind a residue of *depleted peridotite* ("+" pattern) about 10 km thick.

Near the top of the mantle, the rising magma pockets amalgamate to form an elongated magma chamber beneath the ridge. As the magma cools, crystals of olivine, together with some pyroxene and plagioclase, form and sink to the bottom of this chamber, where they form a relatively thin layer (0.5–2 km thick) of olivine-rich peridotite. As the plates spread apart, the remaining mafic magma crystallizes continuously on the

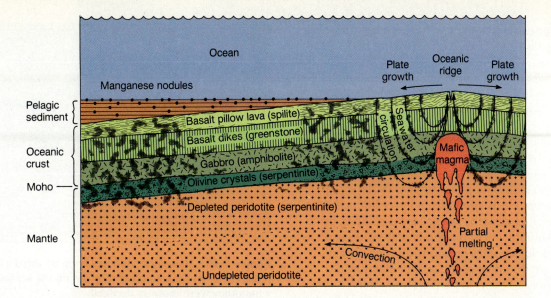

*Figure 7.37* **Rocks of the oceanic lithosphere. In this theoretical cross section of a divergent plate boundary, the effects of extension faulting have been eliminated for clarity (see fig. 5.22), but such faults provide convenient channels for the circulation of seawater through the crust and upper mantle (*arrows*). The thickness of pelagic sediments is exaggerated.**

walls of the chamber, producing a layer of gabbro of about the same thickness as the peridotite layer beneath it. The boundary between these two layers corresponds to the Moho (see chapter 1).

Next above the gabbro is a 1–4 km thick layer of vertical basalt dikes oriented parallel to the median rift at the oceanic ridge. These dikes form one at a time when seafloor spreading opens a crack in the top of the magma chamber, allowing magma to shoot upward and fill the crack. Sometimes, the previous dike is actually split down the middle by the following one! As these dikes cool and solidify, crystallizing magnetite grains become aligned with the prevailing geomagnetic field, giving rise to geomagnetic anomaly patterns, such as the one illustrated in figure 5.17.

Above the dikes is a 0.3–5 km thick layer of basalt **pillow lava,** so called because it has a distinctive structure consisting of pillowlike blobs of basalt. The individual, roughly spherical pillows range in diameter from 10 cm to 6 m. As each pillow extrudes, fed by an underlying dike or an underlying pillow, the cold seawater quickly chills its outer surface to a hard shell. Rising internal pressure soon ruptures that shell, allowing more lava to escape through the rupture and form another pillow.

Practically as soon as they are formed, the peridotite, gabbro, and basalt of the oceanic lithosphere begin to show signs of alteration to other rock types (indicated in parentheses in figure 7.37). This results from the passage of seawater through crack systems in the newly-formed rocks (mottling indicates alteration). The seawater becomes heated as it nears the magma chamber, whereupon it reacts with the minerals of the lithosphere. Nearest the surface, basalt is altered to spilite at 100–250° C. Lower in the crust, the basalt dike complex is altered to greenstone at 250–400° C. Still deeper, gabbro is altered to amphibolite at 400–500° C. In the upper mantle, periodite is altered to serpentinite.

Beginning a few tens of meters on either side of the oceanic ridge, a gradually thickening wedge of **pelagic** (deep sea) **sediments** overlies the mafic igneous rocks of the seafloor (the thickness of this wedge is greatly exaggerated in fig. 7.37). These sediments can reach a maximum thickness of 5000 m at great distances from the ridge where they have had a longer time to accumulate on older crust. They consist mainly of four components:

1. **Brown clay,** a mixture of finely powdered quartz and feldspar derived from volcanic eruptions, deserts, and glaciated regions, and blown from the continents by winds. It accumulates slowly—a few millimeters per thousand years.
2. *Siliceous and calcareous oozes.* These organically produced sediments (see chapter 4) accumulate on the order of ten times faster than brown clay.
3. *Volcanic ash* derived mainly from eruptions along the oceanic ridge.

4. **Maganese nodules** (fig. 7.38). Dense masses of manganese and iron oxides ranging in size from microscopic spheres to meter-long slabs. These deposits contain significant concentrations of such valuable metals as nickel and cobalt as well as magnesium and iron. Paradoxically, they are normally most abundant at the seafloor and become less abundant with depth in the underlying sediment, regardless of the age of the sediment. This is probably so because free oxygen is present in the uppermost layers of sediment but is absent from lower layers. In the absence of oxygen, manganese is soluble, but it is insoluble where oxygen is present. As soon as a nodule becomes buried within the oxygen-free environment of the sediment below the seafloor, it begins to be dissolved. The manganese oxide thus released is redeposited on other nodules growing on the oxygenated seafloor above.

In areas of the seafloor near continental margins, coarser, continental clastic materials can also make a significant contribution to pelagic sediments.

## Rocks of Convergent Plate Boundaries

Because of the great variety of environments that can prevail at convergent plate boundaries, an equally various array of rock types is generated in and near such boundaries. Figure 7.39 illustrates the case of an ocean-ocean convergence.

As pressure increases with depth in the subduction zone, the partially altered gabbro and basalt of the oceanic crust is progressively metamorphosed by rising heat and pressure, changing first to amphibolite and then to eclogite. Between depths of about 80 and 300 km, the pressure becomes high enough to squeeze water from cracks and pores in the rock of the subducted plate and from the crystal structures of hydrous minerals. It is thought that this water causes partial melting of both the peridotite of the overlying mantle and the top of the subducted plate, generating mafic magma, which rises buoyantly into the overlying magmatic island arc. If the magma solidifies before reaching the surface, it crystallizes as gabbro, but if it erupts in one of the islands of the magmatic arc, it solidifies as basalt lava.

Magmas derived from the deeper parts of a subduction zone (fig. 7.39, *left*) are erupted farther from the trench. These magmas typically have an intermediate composition (i.e., andesitic, see fig. 4.9*b*). Accordingly, they have a lower melting point than basalt and are richer in potassium, sodium, and silicon. At first, it was thought that different types of magma were generated at different depths, but more recently it has been suggested that mafic magma is generated at all depths

***Figure 7.38*** Manganese nodules lying on the deck of a research vessel look much as they did on the seafloor from which they were recently dredged.

Photo by Robert Hessler, courtesy of Woods Hole Oceanographic Institution.

and at about the same temperature, but because the magma generated at greater depth must pass through a greater thickness of overlying mantle rock, it is able to scavenge more potassium, sodium, and silicon during its upward passage. These contaminants alter the magma toward a more intermediate composition.

As with the more mafic magma, this intermediate magma crystallizes if it fails to reach the surface, forming diorite, but is erupted as andesite lava if it emerges in the volcanoes of the magmatic arc. During its ascent, both kinds of magma produce regional metamorphic effects at greater depths in the crust and contact metamorphic effects at shallower depths.

The tectonic arcs that normally form between trenches and magmatic arcs consist largely of chaotically sheared and jumbled breccias. Included in these **mélanges** (may-*lawzh,* French for "mixture"; fig. 7.40) are slices of ocean floor, pelagic sediment, and clastic sediment from the magmatic arc, all of which have been thrust against the leading edge of the overriding plate in a process called *offscraping*.

Within the forearc basin, between the tectonic and magmatic arcs, clastic sediment derived from the erosion of both arcs accumulates, mainly as interstratified mudstone and graywacke. Occasionally, lahars flush coarse gravel down from the magmatic arc. When lithified, this gravel becomes volcanic conglomerate. Mudrock, graywacke, and volcanic conglomerate can also accumulate in the backarc basin behind the magmatic arc.

Where a subduction zone coincides with a continental margin (fig. 7.41), as on the west coast of South America, the patterns of rock formation differ in some respects from those of an ocean-ocean convergence. The principal differences between figures 7.39 and 7.41 can

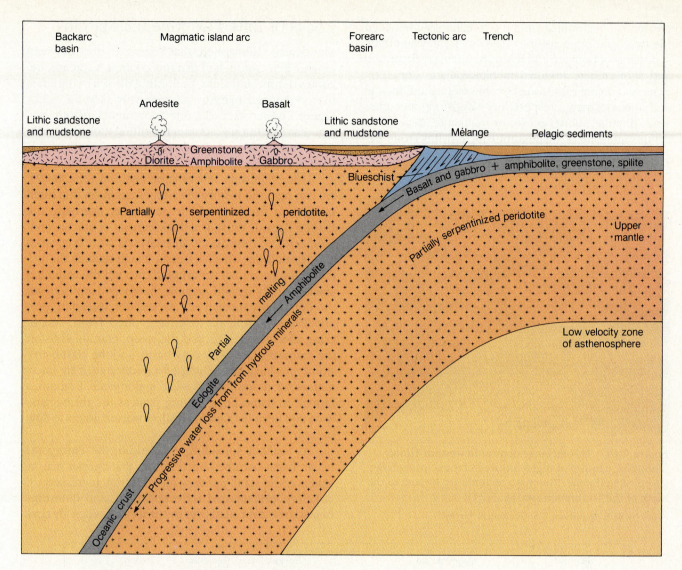

*Figure 7.39*  **Rocks of ocean-ocean convergence zones.**

be seen to the left of the tectonic arc. Where the over-riding plate carries continental crust, magmas rising from the subduction zone either assimilate and react with felsic crustal rocks or (more probably) simply provide enough heat to cause partial melting of such rocks. As with intermediate magma, this felsic magma can either crystallize at depth as a phaneritic, plutonic rock (granite), or it can rise to the surface and be erupted as aphanitic lava (rhyolite).

On the side of the forearc basin adjacent to the continent, conglomerate, feldspathic and quartz sand-stones, mudstone, and limestone are deposited. Conglomerate is deposited mainly in the mountains; sandstones mainly near the shoreline; mudstone mainly in quieter, offshore water; and limestone mainly where the ocean water is clear and warm and where there are no rivers, mudflows, strong winds, glaciers, or other transport media capable of delivering clastic sediment to the sea. The continental backarc basin can also accumulate any of these deposits, although backarc limestones are uncommon.

If the ongoing subduction of the oceanic plate should eventually bring a magmatic arc or continent into the trench (fig. 7.41, *right*), the tectonic arc is likely to be forced against the continental margin. Such an event crushes the tectonic arc and the forearc basin between the two colliding coastlines (see fig. 5.27). The clastic sediments of both arc and basin are intensely deformed and subjected to regional metamorphism. Conglomerate and quartz sandstone become metaconglomerate and metaquartzite, respectively. Mudstone becomes slate, phyllite, schist, gneiss, or migmatite, depending on the intensity of metamorphism, and limestone becomes marble.

**Figure 7.40** This mélange deposit in western Timor, Indonesia, consists of highly sheared clay enclosing large slivers of limestone. Timor is a tectonic arc located to the north of the Java trench (see fig. 5.21).

Photo by W. B. Hamilton, U.S. Geological Survey

## Rocks of Rifted Continental Margins

The oceanic lithosphere that begins to grow between newly rifted continental margins cools as it spreads, becomes denser, and sinks, permitting the sea to inundate the continental margins progressively (see fig. 5.22). Figure 7.42 shows such a rifted continental margin several tens of million years after the rifting took place. The east coast of North America is a good example of this type of continental margin.

Using Stensen's law of superposition (chapter 2), we can read the history of sedimentation on this rifted margin. At the base of the sequence, overlying the faulted oceanic crust, is a thick deposit of arkosic conglomerate and sandstone interlayered with evaporites and basalt lavas. This rock assemblage records the initial rifting event. Basalts erupted through the widening cracks as they tapped magma chambers deep beneath the embryonic median rift. As seafloor spreading progressed, widening the overlying rifted basin, weathering and erosion attacked the steep, granitic cliffs on the margins of the widening basin, and the clastic erosional products were dumped rapidly into it by intermittent floods forming arkosic conglomerate. Evaporites formed in natural evaporating pans as the rift became wide and deep enough for shallow marine waters to flow in.

As seafloor spreading progressed, the embryonic ocean basin became large enough that deeper marine waters could stand permanently within it. Increased circulation compensated for evaporation, and limestone began to be deposited over the older evaporites. By now,

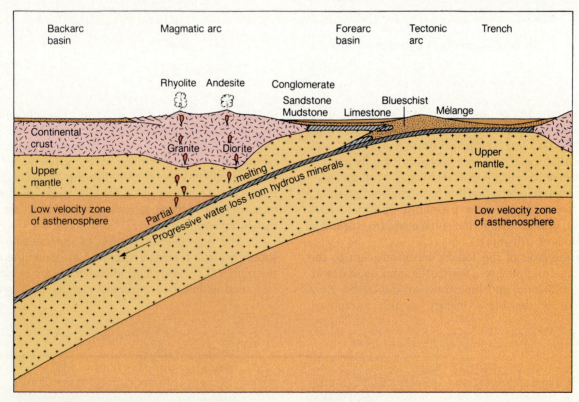

**Figure 7.41** Rocks of ocean-continent convergence zones. Note that the magmatic arc and the backarc basin is underlain by continental crust and that a continental margin is approaching the convergence zone from the right.

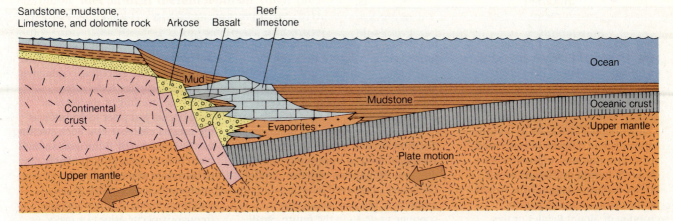

**Figure 7.42** Rocks of rifted continental margins. The continental shelf overlies the slightly submerged edge of the continent; the continental rise overlies the faulted boundary between continental and oceanic crust, and the abyssal plain overlies the oceanic crust of the deep ocean floor.

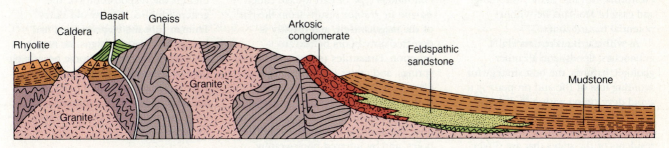

**Figure 7.43** Rocks of continental interiors.

the marine environment was of normal salinity (about 3.5% salt), so limestone-reef-building organisms, such as calcareous algae and corals, could thrive in the tropical waters. In the back-reef environment toward the continent, evaporites, mudstone, and arkose could still accumulate. Eventually, a rapid sea level rise or subsidence of the crust submerged the reef environment faster than it could grow upward, killing the reef-building organisms. The fossil reefs then become buried by thick deposits of mudstone and graywacke, forming a massive prism underlying the continental rise. Toward the continent, limestone, dolomite rock, mudstone, and sandstone were deposited on the continental shelf.

Looking ahead, at some future time, a new subduction zone could form offshore from the continental margin, producing a covergent plate boundary environment like that of figure 7.41.

## Rocks of the Continental Interiors

Generally, continental interiors are stable regions composed of the deeply eroded root zones of ancient mountain belts that have been accreted over time to a continental nucleus. Occasionally, however, even these ancient stable regions are broken by mountain-building forces and magmatic activity, as when a mantle plume develops or an increase in seafloor spreading rate causes

a flattening of the angle of subduction at an ocean-continent convergence (fig. 7.39 shows slow subduction, and fig. 7.41 shows rapid subduction). Such a change affected western North America during the Late Mesozoic and Early Cenozoic Eras, resulting in the uplift of the central and southern Rocky Mountains in the stable continental interior. Figure 7.43 shows a schematic cross section through an uplift of the Rocky Mountain type. Here, a great fault block has been uplifted along high-angle faults (*arrows*). Note that the massive, uplifted block has sagged slightly under its own weight, deforming the western boundary fault outward and away from the uplift.

If the boundary faults extend deeply enough, they can serve as conduits for the rise of mafic magma from the underlying subduction zone. Vast eruptions of rhyolite and rhyolitic ash-flow tuff have also occurred in the central and southern Rocky Mountains. This felsic volcanism has produced sizable bodies of granitic rock types in the southern and central Rockies, and in places it has also produced huge calderas.

Close to the mountain uplifts, arkose can be deposited if the bedrock of the mountains is granitic. Rugged uplifts can yield deposits of conglomerate. At greater distances from mountain uplifts, arkose, quartz sandstone, and mudstone are laid down as the transporting streams lose their energy in flatter terrain.

**Figure 8.5** The roots of a large paper birch have penetrated the joints and bedding planes of a Devonian sandstone exposed along Trout Brook near Mount Katahdin, Maine. Although plant roots are much softer than rock, their slow growth is capable of quarrying enormous blocks.

Rates of mechanical weathering depend on the physical properties of the rock, especially the presence or absence of fractures and bedding planes, and on geographical factors. Figure 8.6 indicates the susceptibility of five different categories of landforms to both physical weathering and chemical weathering, measured in terms of the amounts of clastic and dissolved materials removed from these landforms annually by individual rivers. Note that because of the enormous range of values represented in figure 8.6, the data must be shown on a logarithmic scale, in which even divisions represent powers of ten (1, 10, 100, 1000, etc.). For example, tropical mountain rivers erode anywhere from 135 to 2250 tons of suspended clastic sediment from each square kilometer of the drainage area they serve each year.

Several important generalizations can be made on the basis of this information. The suspended sediment yield of tropical mountain rivers is about 10 times greater than that of rivers that drain tropical plateaus (25–190 tons/km²/yr). This suggests that mechanical weathering is more effective in tropical mountains than on tropical plateaus, probably because freeze/thaw cycles prevail year-round in the mountains. Rivers in temperate mountain regions, where freeze-thaw cycles occur only during the summer, yield much less sediment (5–105 tons/km²/yr). Tropical plateaus are somewhat more productive than temperate mountains, probably because of the greater influence of chemical weathering in loosening clastic grains, whereas both tropical and temperate plains yield the least suspended sediment (2–65 tons/km²/yr and 2–20 tons/km²/yr respectively).

**Figure 8.6**
Mechanical and chemical weathering rates in different geographic settings are reflected, respectively, in the amounts of suspended clastic sediment (*brown*) and dissolved ions (*blue*) transported annually by rivers draining such settings. Note logarithmic scale.

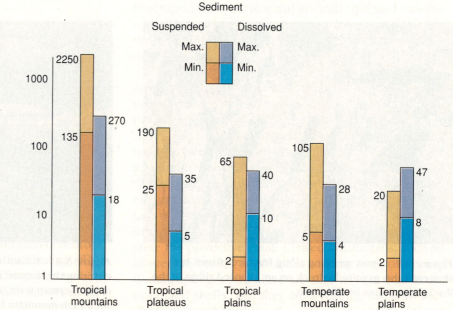

# Chemical Weathering

While these various agents of mechanical weathering shatter the bedrock, a more subtle kind of weathering is also at work. Chemical weathering agents present in the atmosphere, the hydrosphere, and the biosphere selectively attack certain rock minerals in a variety of ways, breaking down their crystal structures and releasing soluble ions.

The simplest form of chemical weathering is **solution,** the dissolving of a mineral in water without chemical reaction. Ionic-bonded minerals, such as halite and gypsum, are subject to this kind of weathering. Where salt rock or gypsum rock lie at or near Earth's surface, the solution of these rock types by groundwater can present major problems for landowners because of the frequent occurrence of sinkholes and the contamination of drinking water by the dissolved salts.

Instead of simply dissolving mineral matter, water often reacts with it chemically, a process known as **hydrolysis** (Greek for "water loosening"). Among the simplest of hydrolysis reactions is the slow corrosion of calcite in limestone (or of dolomite in dolomite rock) by rainwater or groundwater in which dissolved carbon dioxide is present (as it usually is), rendering it mildly acidic. In the case of calcite, this reaction produces calcium bicarbonate in the form of dissolved calcium and bicarbonate ions:

$CaCO_3 + H_2O + CO_2 = Ca^{+2} + 2HCO_3^-$
(calcium carbonate + water + carbon dioxide = calcium ion + 2 bicarbonate ions)

If carbon dioxide is subsequently removed from the water, the calcium and bicarbonate ions will recombine to form calcite. (This can be seen by reading the equation backward.) The hydrolysis of carbonate rocks can be as much a source of vexation for landowners as a solution, as it, too, can create sinkholes capable of swallowing houses.

Hydrolysis also affects silicate minerals. In chapter 1, I discuss the hydrolysis of Ca-feldspar by carbonic acid rain, producing kaolinite clay and calcium bicarbonate:

$CaAl_2Si_2O_8 + 3H_2O + 2CO_2 \rightarrow Al_2Si_2O_5(OH)_4 + Ca^{+2} + 2HCO_3^-$
(Ca-feldspar + 3 water + 2 carbon dioxide → clay + calcium ion + 2 bicarbonate ions)

In this case, the original mineral is structurally altered in the solid state from a framework silicate to a sheet silicate (clay). Consequently, it is not possible to reconstitute it out of solution. Another example is the hydrolysis of Na-feldspar. This also produces kaolinite, but instead of calcium bicarbonate, it produces sodium bicarbonate, or baking soda. A third product, silicic acid ($H_4SiO_4$), is also formed because Na-feldspar is richer in silicon than is Ca-feldspar. Some of this dissolved silicic acid is deposited as cement within sediments, and most of the rest is used by one-celled marine plants and animals to make microscopic shells of silica or by certain sponges to make internal skeletal parts.

Structurally, hydrolysis involves the replacement of large, phaneritic silicate crystals with aggregates of aphanitic clay crystals. As this happens, the clay aggregates expand as much as 40%, forcing adjacent mineral grains aside and greatly reducing the strength of the rock. Granite, for example, is about one third quartz—one of the most resistant minerals—and about one third K-feldspar—another fairly resistant mineral—but the remaining third is plagioclase, which is much less resistant to weathering. In addition, granite often contains such nonresistant accessory minerals as biotite and amphibole. Just as a chain is no stronger than its weakest link, a rock is no more resistant than its least resistant mineral. As these "weak links" disintegrate, they weaken the fabric of the rock, which then rapidly disintegrates.

As with other weathering processes, hydrolysis takes place on the exposed surface of the bedrock, or along fracture surfaces within the bedrock. When hydrolysis produces an insoluble residue, as it does in the weathering of most silicate minerals, that residue accumulates as a weathering rind of altered, or "rotten" rock material, called **saprolite** (see figs. 1.10 and 8.7), that can be as much as 70 m thick in tropical regions. Because many of the diagnostic properties of rocks are destroyed or obscured during weathering, it is standard geological field practice to trim off the weathering rind on hand specimens with a few well-directed blows of the rock hammer.

Weathering and saprolite formation proceed relatively slowly along a joint surface. They proceed more rapidly where two joint surfaces intersect, forming an edge, and most rapidly of all where three joint surfaces intersect, forming a corner (fig. 8.8). In such homogeneous rock types as granite and basalt, these differences in weathering rates often produce spherical or rounded forms. This phenomenon is known as **spheroidal weathering.**

In 1938, S. S. Goldich proposed a hypothesis to explain the susceptibility of silicate minerals to hydrolysis. During an extensive microscopic study of mineral alteration in thin sections, Goldich found that the highest minerals in the Bowen reaction series (see fig. 3.15), such as olivine and Ca-feldspar, are the most readily and rapidly altered in the weathering environment, and that the lowest minerals in the series, such as quartz and muscovite, are the least affected. In other words, the greater the difference between a given silicate mineral's environment of formation and Earth's surface environment, the more susceptibile that mineral will be to hydrolysis.

**Figure 8.7** The penetration of acidic rainwater into joints in this impure limestone in central Vermont has produced a thick weathering rind of rotten rock, or saprolite, around a central core of dark, unweathered bedrock.

In the case of essentially monomineralic rock types, such as quartz sandstone and olivine peridotite, Goldich's principle is quite obvious. Quartz sandstone formations are almost always resistant to weathering, forming prominent landforms, whereas peridotite usually is found in valleys or subdued landforms.

**Oxidation** is a chemical weathering process that often accompanies hydrolysis and that acts mainly on iron-bearing minerals, such as olivine, pyroxene, amphibole, and biotite. The oxidation of Fe-olivine ($Fe_2SiO_4$), for example, produces hematite and silicic acid:

$$2Fe_2SiO_4 + 4H_2O + O_2 \rightarrow 2Fe_2O_3 + 2H_4SiO_4$$
(2 Fe-olivine + 4 water + oxygen → 2 hematite + 2 silicic acid)

Note that this is also a hydrolysis reaction. The silicic acid is washed away in solution, but the insoluble hematite remains, usually imparting a rusty red color to

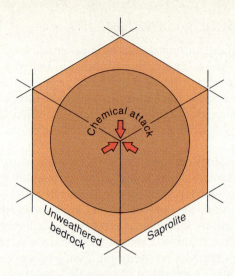

**Figure 8.8** Chemical attack from two sides at edges of joint-bounded bedrock blocks and from three sides at corners results in the phenomenon of spheroidal weathering, which produces rounded boulders from originally angular blocks.

the saprolite. Oxidation is responsible for the brilliantly colored "red beds" of sandstone and mudstone that impart such scenic splendor to much of the western United States and to many other parts of the world (see fig. 2.10).

A fourth weathering process is that of **hydration,** the combining of a mineral with water to form a different mineral. In hydrolysis, the water molecule is destroyed by the reaction, but in hydration, water enters the crystal structure of the new mineral without losing its identity. Hematite, for example, is often hydrated to Goethite ($Fe_2O_3 \cdot H_2O$), and *anhydrite* ($CaSO_4$) is hydrated to gypsum ($CaSO_4 \cdot 2H_2O$). Both these reactions, especially the latter, result in expansions in volume, which can disrupt the structure of rocks in which these minerals are abundant.

As is the case with mechanical weathering, the effectiveness of chemical weathering is dependent on environmental factors that vary with latitude and altitude. Figure 8.6 shows that mountain rivers of tropical regions transport chemical weathering products at a relatively rapid rate (18–270 tons/km²/yr), whereas the rate for temperate mountain rivers is much lower (4–28 tons/km²/yr). Inasmuch as chemical reactions tend to go faster at higher temperatures, this seems reasonable. Similarly, chemical weathering is somewhat more effective on temperate plains (8–47 tons/km²/yr) than in temperate mountains, which again is reasonable, because at a given latitude, plains are normally warmer than mountains.

In the case of tropical mountains and tropical plains, however, the situation is reversed. The explanation for this apparent paradox is that the higher rate of mechanical weathering in warm mountain regions provides a

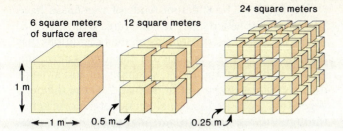

**Figure 8.9** Comminution, or breaking up, of rock increases the surface area exposed to chemical weathering. Each time the dimensions of the cubes are cut in half, the total surface area is doubled.

much greater surface area on which chemical weathering can take place. Figure 8.9 shows how the mechanical disintegration of a block of rock increases its surface area. Each time the dimensions of a fragment are cut in half, its surface area doubles.

High temperature, high precipitation, and extensive vegetative cover all favor high rates of chemical weathering, as is illustrated in figure 8.10. High evaporation, on the other hand, has the opposite effect. Weathering is most intense in the tropics where the first three factors are most influential, moderately active in temperate regions (coniferous and deciduous forest) where the same factors are less intense, and least active in the subtropical desert and polar regions. In the polar regions, low values of all three factors limit chemical weathering. In deserts, low precipitation, high evaporation, and sparse vegetative cover are the limiting factors.

## Soil: The Interface of Land, Water, and Life

The products of weathering have, over geologic time, accumulated in a blanket of unconsolidated material overlying bedrock at Earth's surface. This blanket, which ranges in thickness from zero to many tens of meters, is called **regolith** (Greek for "blanket stone"). If it consists of saprolite that has formed in place and has not been eroded or transported, it is *residual regolith.* If it consists of clastic material that has been transported by some sedimentary process (e.g., landslide, stream deposition, glaciation, wind action, etc.), it is *transported regolith.*

Under the influence of a number of factors, of which climate and living organisms are the most significant, regolith is slowly transformed into **soil,** a mundane but vital mixture of five essential components:

1. Weathered and unweathered, loose mineral matter
2. Altered, decay-resistant organic material, or **humus**
3. Air
4. Water, including dissolved gases and ions
5. A variety of living organisms from bacteria to bears.

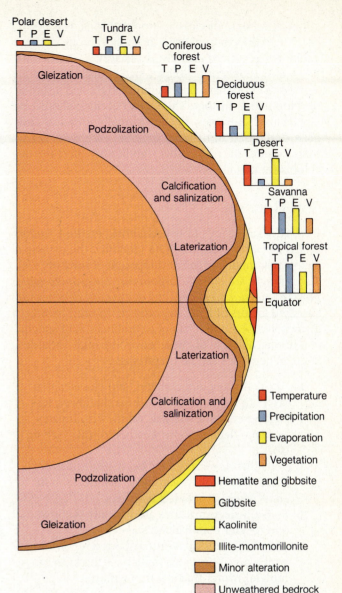

**Figure 8.10** In this cross section, Earth's soil mantle has been greatly thickened to illustrate changes in depth of weathering and saprolite formation with latitude. Intensities of three influential climatic factors and of vegetation are shown as colored bars. Soil-forming processes are shown within the zone of unweathered bedrock.

Soil that develops on residual regolith is known as **residual soil,** and soil that develops on transported regolith is known as **transported soil.**

Without soil, life on land would be impossible. It is the medium within which practically all plant life must begin. Soil is also indispensible to all nonmarine animals because plant life forms the base of the food pyramid on which all animal life rests. Soil is far from being just a simple mixture, however. It is a highly evolved and complex system whose five components interact in systematic ways to produce over 7500 known *soil types.* Many attempts have been made to muster soil types into a logical scheme of classification with practical utility, but due to the complexity of the subject, none of these attempts has proved entirely satisfactory.

A basic concept of soil science is that soils within a given category have had a similar origin and a similar developmental history. The word "similar" should be emphasized, however, as often virtually identical soils have developed from quite different materials. Let us now consider the factors that significantly affect the development of soils.

## Soil Genesis

In 1941, Hans Jenny published a book, entitled *Factors of Soil Formation,* in which he identified the principal factors that influence the development of a mature soil. Those factors are *climate, organisms, topographic relief, parent material, time,* and any of a variety of other unspecified factors of local importance. In other words, a given soil type is the result of a combination of specific conditions of each of these environmental variables. Some generalities can be emphasized:

1. Cold climates, dry climates, and climates that are both cold and dry produce shallower, less weathered soils, whereas warm, wet climates produce deeper, more weathered soils (see fig. 8.10).
2. Soils that develop under heavy vegetation are more organic-rich than those that develop in areas where vegetation is sparse.
3. Soils on steep hillsides are typically shallower and rockier than flatland soils.
4. The type of residual or transported regolith on which a soil develops influences the kind of mineral matter that is found within it.

Figure 8.11 illustrates an example of the combined effects of the first four of Jenny's variables on the development of a soil in three adjacent microenvironments: a sunny slope (*right*), a valley bottom (*center*), and a shaded slope (*left*). Figure 8.11*a* shows the effects of parent material. A residual regolith of sand has formed on the sunny slope from weathering of the sandstone, and a regolith of clay has formed on the shady slope from weathering of the limestone (of course, these two rock types do not always correspond to sunny and shady slopes!).

In figure 8.11*b,* the effects of climate are modified by those of parent material and topography. The porous sand allows rainfall to infiltrate at a high rate. This, combined with a high rate of evaporation on the sunny slope, results in a warm, dry regolith with little runoff or erosion. The nonporous clay, on the other hand, allows only limited infiltration, but what water does infiltrate is tightly held by the clay. These factors, combined with a low evaporation rate on the shady slope, result in a cool, moist regolith with considerable runoff and erosion, which tends to steepen the shady slope.

In figure 8.11*c,* different types of vegetation have become established in the different microenvironments, each suited to the environmental conditions that prevail there: sparse shrubs on the dry, sunny site; broadleaf trees on the moist, shaded site; and swamp-loving conifers in the wet bottomland. These different vegetative types cycle elements and organic material through the regolith (*circular arrow,* fig. 8.11*c*), converting it to soil. On the warm, sunny site, because of the sparseness of the vegetation and rapid loss of organic material to decomposing organisms, only a thin soil is able to develop. The lusher vegetation and slower rate of decomposition on the cool, shady site favor a thicker, richer soil, whereas the thickest soil forms in the waterlogged valley bottom where decomposition is slower still.

The characteristics of the soil that develops under the influences of all these combined, interactive factors are shown beneath each of the microenvironments at the bottom of figure 8.11. The effect of time is not represented, but in general, it can be said that older soils tend to be deeper and to have higher contents of (1) organic material, (2) clay and oxides of iron and aluminum from weathering, and (3) calcium carbonate from the evaporation of groundwater, especially in arid regions.

## Soil Profiles and Horizons

Figure 8.12 is a generalized representation of a **soil profile,** a vertical section through a body of soil showing its internal structure. Most evident in this figure are the two principal sources of soil material: bedrock beneath and vegetation on top. Within the soil body, these two sources blend to form a sequence of **soil horizons,** each with distinctive and recognizable properties, inputs, storage, and outputs; hence, each is a unique *system* by the definition given in chapter 2. Specific inputs to these systems are fresh organic material, unweathered rock material, and air and water containing various gases and dissolved materials. Outputs are decomposed organic material, weathered mineral matter, and air and water with altered contents of gases and dissolved materials. The horizons of a mature soil run parallel to one another and to the overlying ground surface. Unlike sedimentary strata, however, they develop in place as a result of the alteration of existing material rather than being transported in and deposited, layer on layer. From top to bottom, the principal horizons are designated simply A, B, and C.

The **C horizon** consists of unconsolidated regolith, either residual or transported. The C horizon in a residual soil consists of saprolite, and it sometimes retains ghostly traces of bedding, foliation, and other structures of the original bedrock. Similarly, the C horizon of a transported soil can retain traces of sedimentary structures produced by the transport medium. Silica, carbonates, and even some soluble salts can accumulate within the C horizon as a result of weathering reactions.

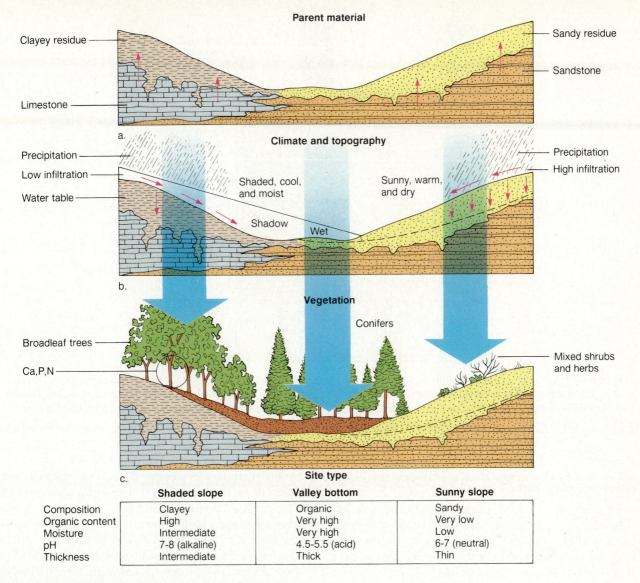

**Parent material**

Clayey residue

Limestone

Sandy residue

Sandstone

a.

**Climate and topography**

Precipitation

Low infiltration

Water table

Shaded, cool, and moist

Shadow

Wet

Sunny, warm, and dry

Precipitation

High infiltration

b.

**Vegetation**

Conifers

Broadleaf trees

Ca,P,N

Mixed shrubs and herbs

c.

**Site type**

| | Shaded slope | Valley bottom | Sunny slope |
| --- | --- | --- | --- |
| Composition | Clayey | Organic | Sandy |
| Organic content | High | Very high | Very low |
| Moisture | Intermediate | Very high | Low |
| pH | 7-8 (alkaline) | 4.5-5.5 (acid) | 6-7 (neutral) |
| Thickness | Intermediate | Thick | Thin |

**Figure 8.11** Influences of parent material, climate, topography, and vegetation on soil development within three adjacent microenvironments.

The **B horizon** consists of highly weathered parent material from which all traces of bedrock or sedimentary structure have been eradicated. It usually contains substances, including clay, humus, and iron and aluminum oxides, that have been *leached* or dissolved out from the overlying A horizon by downward percolating rainwater that has been rendered acidic by carbon dioxide and plant acids. These leached materials bind the soil particles of the B horizon into larger grains, imparting to this horizon a characteristic **soil structure.** Soil structure, which comprises a variety of forms, including *crumbs, granules, blocks, prisms, columns,* and *plates,* is essential to the efficient flow of gases, water, and nutrients through the soil.

The **A horizon,** overlying the B, is more or less impoverished in the clay, humus, and iron and aluminum oxides that have been flushed into the B horizon by percolating acid rainwater. Where this leaching action is intense, it bleaches the base of the A horizon, which consequently assumes the color of the mineral grains of the parent material. Humus can amount to as much as 30% of the soil in this horizon. Often, an organic-rich **O horizon** of plant litter overlies the A horizon at the surface.

In arid and semiarid climates, downward-percolating carbonated rainwater usually is unable to penetrate to or below the base of the soil profile before it evaporates or is taken up by plants. Because of this, calcium carbonate dissolved in the water in small amounts is deposited in the soil, where it accumulates gradually.

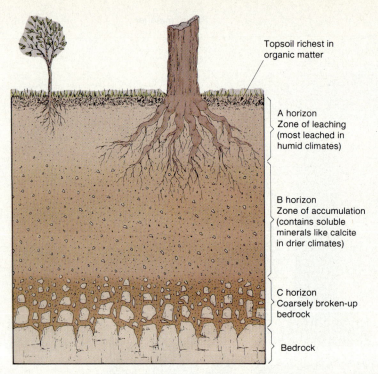

Topsoil richest in
organic matter

A horizon
Zone of leaching
(most leached in
humid climates)

B horizon
Zone of accumulation
(contains soluble
minerals like calcite
in drier climates)

C horizon
Coarsely broken-up
bedrock

Bedrock

*Figure 8.12*    Soils are systems with inputs of bedrock and organisms, throughputs of gases, water, and energy, and outputs of residual and dissolved mineral and organic matter. C horizons show dominant influence of parent rock, A horizons show dominant influence of organisms (especially vegetation), and B horizons show maximum development of soil characteristics. A residual soil is illustrated here.

Eventually, such accumulations can constitute more than 50% of the soil by weight. In the southwestern United States, these hard, white carbonate deposits are known as *caliche* (kah-*lee*-chay). Under certain conditions, accumulations of gypsum, soluble salts, humus, iron oxide, silica, and clay can also form, especially within the B horizon. If uncemented, clay layers are called *claypan*. If cemented, they are called *hardpan*.

## The Physical and Chemical Properties of Soil

The suitability of soil for various purposes, such as supporting forest growthbearing loads, or preventing erosion, is dependent on its physical and chemical properties. Soil structure is one such property. **Soil texture** is another, measured and described in terms of the proportions of sand, silt, and clay in the soil. The triangular diagram of figure 8.13 represents the range of soil textures, each apex representing 100% of a given component, and the opposite base representing 0% of that component. The term **loam** is used to designate a mixture of all three components in which clay does not exceed 40%. For example, a soil having 10% sand, 60% silt, and 30% clay would be a silty clay loam.

The moisture-retaining capacity of the soil is of critical importance to plants, and it is dependent on soil texture. Sand is poorest and clay is best at retaining water.

This is because smaller voids retain water more tenaciously than larger ones do. Water thus bound in soil is unavailable to plants; hence, for a given amount of rainfall, sandy soils tend to yield more water to plants than clay soils, a fact well known to the Hopi people, who grow corn in little hills of sand in their dry, Arizona homeland. Furthermore, sand, unlike clay, permits a free flow of air to plant roots. Its main agricultural disadvantage is that in comparison with clay, it is relatively sterile, and it lacks the ability of clay minerals to exchange cations with its environment. The most productive soils are loams, which combine the advantage of large pore spaces with that of a high capacity to exchange cations.

*Soil color* is dependent mainly on the concentration and chemical state of organic matter and iron in the soil. An increase in organic content corresponds with an increase in the blackness of the soil. Red color (due to hematite) indicates a highly oxidized, well-drained soil, yellow color (due to Goethite) an oxidized, but wetter soil. Gray to blue colors indicate reducing conditions (if iron is present). White can indicate either extreme leaching, as in the base of an A horizon, or the deposition of calcium carbonate or some other salt.

The *colloid chemistry* of soils determines their fertility and to some extent their moisture retention, texture, and structure. **Colloids** are particles larger than the largest molecules but invisible to the unaided eye when

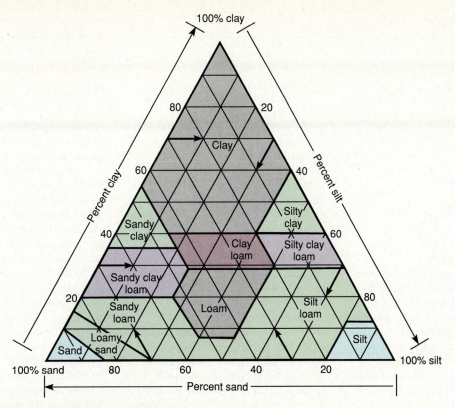

**Figure 8.13** Soil textures are shown by points on this triangular diagram. Each apex represents 100% of clay, silt, or sand, the opposite base, 0%. A soil with equal amounts (33⅓%) of the three constituents would be a clay loam.

Source: Soil Conservation Service.

suspended in water. They range in diameter from 0.005 to 0.25 $\mu$m (*micrometers* $= 10^{-6}$m), and they include clay minerals and organic matter. Clay minerals form tabular crystals whose flat sides are negatively charged because they are composed largely of the electron clouds of closely packed oxygen atoms. Cations are attracted to these negative charges, and they form films of positive charge surrounding the crystals. Organic colloid particles behave similarly, except that their positive charge films are more than twice as dense as those of the clay minerals. These cations are the source of the mineral nutrients required by plants. There is a sort of "pecking order" among them such that some can replace others by a process known as **cation exchange.** Hydrogen ion heads this pecking order, followed by calcium, magnesium, potassium, and sodium ions. Because hydrogen ions are extremely abundant in *acid rain* (see chapter 15), it is clear why this form of pollution is so threatening. Large concentrations of hydrogen ions in soil waters displace and leach the nutrients from soil colloids, ultimately rendering the soils sterile.

## Soil Types and Their Formative Processes

Classification is the cornerstone of any science. In some sciences, such as chemistry, physics, and biology, the subject matter lends itself readily to categorization. In others, including soil science, in which categories are elusive and ill-defined, classification can be a nightmare. Many soil classification schemes have arisen, each with its own unique approach and purpose. In the United States, C. F. Marbut developed a classification system in 1927 based on soil-formative processes, and it was used for many years by the U.S. Soil Conservation Service (SCS). In 1960, a newer and vastly more complicated classification was developed by the SCS, based on observable characteristics of soils and using frightening conglomerations of Latin and Greek prefixes and suffixes that sound more like the names of Teutonic kings than the names of soil types (e.g., Umbrept, Cryoboroll, and Eutroboralf). In this chapter, I use a simplified version of Marbut's classification that emphasizes the major soil-forming processes. A more detailed classification and map of world soils is given in appendix 7.

### Curtis Fletcher Marbut (1863–1935)

**D**emocrat, Unitarian, Renaissance man, freethinker, and from Missouri, C. F. Marbut had a temperament that was well suited to the considerable challenge of developing a new science, especially one that deals with so complex a subject as soils. Undergraduate study at the University of Missouri and graduate work at Harvard University, from which he received a masters degree in 1895, further prepared him for this task, but interim fieldwork with the Missouri Geological Survey probably

contributed as much background as his academic studies. Following his graduation from Harvard, Marbut returned to the University of Missouri to work at the Institute of Geology and Mineralogy until 1897, when he obtained an assistant professorship. In 1899, he became a full professor and curator of the Geological Museum, a post he held until 1913.

Marbut's official work in soils began in 1905, when he became director of the Soil Survey of Missouri. In 1909, he assumed the additional post of special agent with the Bureau of Soils of the U.S. Department of Agriculture. One year later, he was promoted to Chief of the Soil Survey, in which capacity he continued until his death. While in that position, he engaged in an ambitious and successful project to map the soils of the United States. Concurrently, in collaboration with other workers, he developed the

science of *pedology,* or soil science. In acknowledgment of his debt to the pioneering work of Russian soil scientists, he translated the German edition of a Russian text entitled *The Great Soil Groups of the World and Their Development* (1927), a work that contributed both terminology and methodology to the emerging science of pedology in the United States.

In addition to his work on domestic soil, Marbut also had more than one occasion to set foot on foreign soil. He was sent on an expedition to Central America by President Wilson and on another to Brazil by Secretary of Commerce Hoover. In 1935, the year in which his major work, *Soils of the United States,* was published, he went to China at the request of the Chinese government to study the soils of that country. Before he was able to begin the project, however, he died of pneumonia in Harbin, Manchuria.

---

The primary division within the Marbut classification is twofold, corresponding to two major terrestrial environments: humid and arid. In humid climates, where precipitation exceeds evaporation, calcium and other cations are flushed out of the soil by the downward percolation of excess water. Under these conditions, a category of soils known as **pedalfers** develops. The term combines the Greek root "ped" (ground) with "alfer" for aluminum and iron (Latin *ferrum*), reflecting the tendency in such soils for these two elements to be leached from the A horizon into the B horizon. In arid climates, where evaporation exceeds precipitation, calcium carbonate is precipitated within the soil profile as water rises through the soil and evaporates. Soils in which this occurs are known as **pedocals** in reference to the accumulation of calcium carbonate within them.

Figure 8.14 shows how the pedalfers and pedocals are distributed in the United States. Within these two major categories, five main subcategories can be identified, three within the pedalfers and two within the pedocals. Figure 8.15 shows how the subcategories are distributed worldwide. Let us briefly consider each of them.

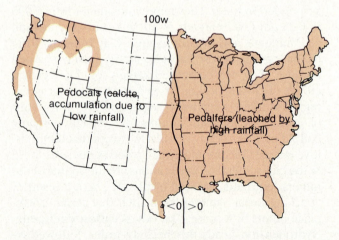

**Figure 8.14**  The boundary between humid soils (pedalfers) and arid soils (pedocals) in the United States lies between the 100th meridian and the 0 cm water surplus contour (precipitation minus evaporation). Annual water surplus exists to the east of the latter boundary, annual deficit to the west.

Source: U.S. Department of Agriculture.

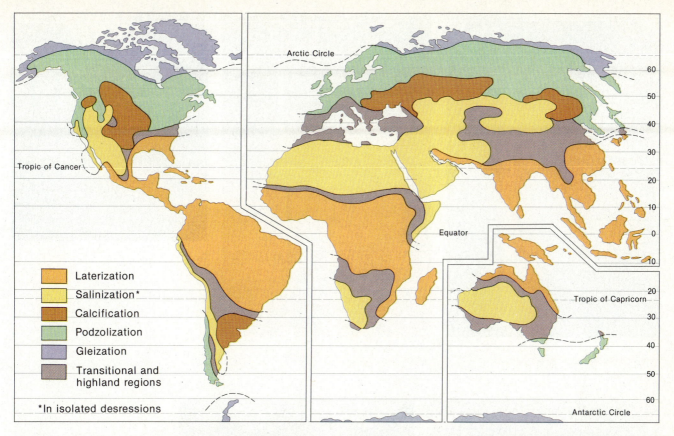

**Figure 8.15** The global distribution of five major soil-forming processes reflects the influences of climate and vegetation. Poor drainage and cold temperature favor gleization, the accumulation of clay and humus in Arctic soils. Cool, moist climates favor podzolization with strong acid leaching. Hot, wet climates favor laterization with oxidation and intense neutral leaching. In semiarid climates, calcium carbonate is deposited in soils by calcification. In arid climates, soluble ions are deposited by salinization.

From A. N. Strahler and A. H. Strahler, *Introduction to Environmental Science.* Copyright © 1974 Hamilton, a division of John Wiley & Sons, Inc., New York, NY.

## Pedalfers

**Podzolization** is a process in which cations, clay, humus, and iron and aluminum oxides are leached from the A horizon by organic acids. These acids are produced by the slow decomposition of organic material under a forest cover in a cool, humid climate. *Podzol soils* (Russian for "ash soil;" a *spodosol* in the new classification) are best developed in Canada, the northeastern United States, and northern Europe and Asia.

The podzol from northern New York illustrated in figure 8.16a exhibits the essential features of podzolization. A whitish, leached layer underlies a moderately acidic, humus-rich layer in the A horizon. The B horizon is thick, dense, and colored yellowish- to reddish-brown by iron oxide and humus leached from the A horizon by acidic water.

In **laterization,** subtropical to tropical warmth and humidity promote the extremely rapid decay of organic material, which leaves soil waters neutral or slightly acid. Under these conditions, no leaching of iron and aluminum oxides occurs. Consequently, these minerals accumulate as rock-hard, brick-red nodules called laterite (from the Latin *latericus:* brick) mainly within the thick, deep red B horizon.

In the Hawaiian *latosol* (or *oxisol*) of figure 8.16b, the A horizon is reddish-brown due to iron oxide and a minor amount of humus. Continuous leaching by neutral water beneath a heavy forest cover has robbed the soil of virtually all its silica and nutrients. The forest has adapted to this condition with a host of mechanisms that keep nutrients constantly cycling through the vegetative cover. Now that tropical rain forests are being extensively cleared, however, the nutrients are also being cleared and forever lost, because there is no way for the soil to reclaim them. A further difficulty is that latosols have a tendency to lithify promptly, becoming brick-hard as they dry out. This further diminishes the potential (and the sense) of converting tropical rain-forest into cropland and pasture.

**Gleization** (glay-*zay*-shun, from Russian, *glei:* clay) is a process whereby a sticky, blue-gray accumulation of clay and humus, called *gley,* forms within cold, poorly-drained soils of polar regions. Low temperatures preclude much decomposition of the organic material, and layers of *permafrost,* or permanently frozen ground (see chapter 10), prevent the draining of water from the regolith; hence, this organic-rich clay remains in a highly reduced (i.e., unoxidized) condition, which accounts for its bluish-gray color.

c.

a.

b.

e.

d.

*Figure 8.16* Representative soil profiles exhibit major trends in soil-forming processes. *a.* Podzol, northern New York; *b.* Latosol, Kauai, Hawaii; *c.* Tundra soil, northern Alaska; *d.* Chernozem soil, southeastern South Dakota; *e.* Red desert soil, central Arizona.

Reproduced from the C. F. Marbut Memorial Slide Set, 1968, by permission of the Soil Science Society of America, Inc.

In the northern Alaskan *tundra soil* (or *inceptisol*) of figure 8.16*c,* a thick, dark, humus-rich accumulation at the base of the A horizon overlies a mottled, blue-gray gley horizon. Such soils are characteristic of the vast expanse of the arctic tundra in northern Russia and Canada.

### *Pedocals*

In semiarid and arid climates where evaporation equals or exceeds precipitation, the process of **calcification** replaces leaching. In the United States, this condition generally prevails west of the 100th meridian (see fig. 8.14). In the grassland and desert soils of this region, water has a greater tendency to move upward in the soil profile and evaporate from the surface than to percolate downward through the profile. Consequently, the cat-ions released from mineral grains during weathering are not leached out of the soil. Under these conditions, cal-cium ions tend to combine with bicarbonate ions in soil water to form deposits of calcium carbonate.

In the *chernozem soil* (or *mollisol*) from south-eastern South Dakota (fig. 8.16*d*), a whitish accumula-tion of calcium carbonate is evident. Less obvious carbonate deposits occur throughout both the B and the C horizons, along with clay in the B horizon. A dense grass cover lies above this profile, contributing to its highly fertile, nutrient- and organic-rich A horizon.

In extremely arid climates where weathering rates exceed rates of leaching by rainwater, soluble ions ac-cumulate within the soil profile, a process called **sal-inization.** The process is most active in desert basins with internal drainage, in which floodwater evaporates, depositing its dissolved salts in the soil instead of car-rying them in rivers to the sea. Only salt-tolerant veg-etation, immune to the toxic effects of chloride ions, can survive on salinized soils.

The *red desert soil* (or *aridisol*) from central Ari-zona (fig. 8.16*e*) totally lacks an A horizon. Not enough vegetation is produced here to provide a significant amount of organic material to the surface layers of the soil. The B horizon is reddish, indicating waters of neu-tral pH, and some importation of clay from the A ho-rizon has occurred. Within the lower B and C horizons, white flecks of calcium carbonate (caliche) have formed.

### *Immature and Unusual Soils*

In some areas, local conditions produce soils in which profiles are poorly developed or absent. The most common causes of such soils are floods and mudflows, landslides, dune migration, lava flows and ashfalls, and glaciation, all of which provide fresh mineral regolith on which transported soils can develop. Transported soils are typically highly fertile, as percolating water and acids have had little time to leach them of their nu-trients.

Where standing water persists at the ground sur-face, peat- and gley-rich *bog soils* develop. Special soils develop on certain rock types, such as limestone and serpentine, and they usually support distinctive vege-tative communities.

## Mass Wasting: The Role of Gravity

Both regolith and bedrock are subject to the influence of gravity, which tends to remove detached masses of either of these materials from higher positions and transport them to lower ones. The term **mass wasting** comprises a wide variety of erosional and transporta-tional processes in which gravity is the prime mover but in which water can also play a significant role. There is no clear boundary between what constitutes stream transport and what constitutes mass wasting, but a water content of 60% by weight is commonly considered a maximum for mudflows, the most fluid of all mass wasting phenomena.

A slope of about 33°–37° is required to maintain the downslope sliding of loose, dry clastic material. On gentler slopes, the force of friction exceeds the force of gravity and loose rock material will not slide unless it is lubricated by water or actively moved downslope by flowing water. This equilibrium slope is called the **angle of repose** for clastic material. This, incidentally, ex-plains why most hillslopes and stream gradients are gentler than about 34°. On slopes steeper than this, re-sidual regolith produced by weathering is swept away by gravity or flowing water as soon as it is released from the bedrock. Hence, most slopes that are steeper than about 34° are developed on bedrock, usually by glacia-tion or rapid downcutting by streams.

## Why Slopes Fail: The Causes of Mass Wasting

Regardless of the slope angle, mass wasting only occurs on a slope when the *cohesiveness* (Latin for "sticking together") of the slope materials is exceeded by the pull of gravity in the downslope direction. Depending on a number of factors, this can happen on slopes ranging from 1° or 2° to 90°. Let us consider first those factors that reduce slope cohesion and then those factors that increase the effectiveness of gravitational stress.

### *Factors that Decrease Slope Strength*

The most effective agent in reducing the cohesiveness of a slope is weathering. Exfoliation is also effective in that it permits the mass wasting of large slabs of granite and other rock types. Other mechanical and chemical weathering processes release smaller clasts from the bedrock, rendering them more susceptible to the action of gravity.

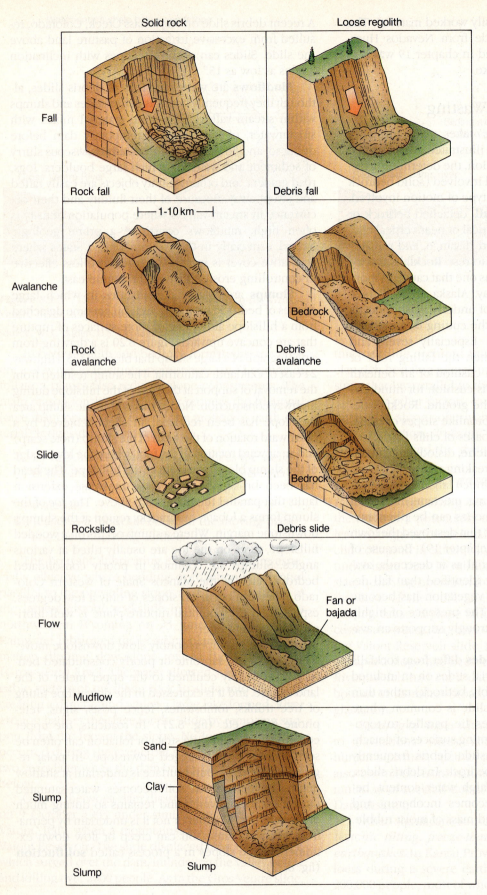

**Figure 8.18** In this classification of rapid mass wasting processes, the horizontal rows represent mode of transport and the vertical columns represent the type of material being transported.

From B. J. Skinner and S. C. Porter, *Physical Geology.* Copyright © John Wiley & Sons, Inc., New York, NY. Reprinted by permission of John Wiley & Sons, Inc.

Solid rock

Loose regolith

Fall

Rock fall

Debris fall

├─ 1-10 km ─┤

Avalanche

Rock avalanche

Debris avalanche

Bedrock

Slide

Rockslide

Debris slide

Bedrock

Flow

Fan or bajada

Mudflow

Slump

Sand

Clay

Slump

Slump

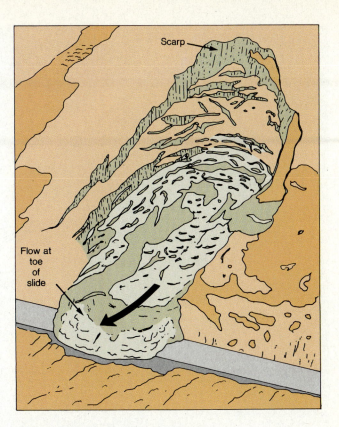

**Figure 8.20** Some features of a slump.

**Figure 8.19** Boulders of granite weathered from an overlying, glacially eroded cliff have formed a partially vegetated talus slope above the wildflower meadows in Garnet Canyon, Grand Teton National Park, Wyoming.

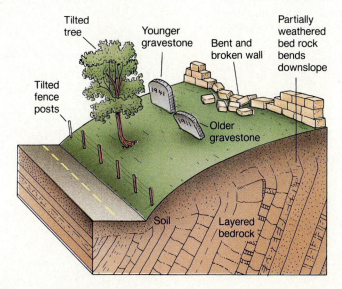

**Figure 8.21** Hillside creep disturbs trees, human artifacts, soil, and underlying bedrock as gravity slowly drags all surface materials downslope.

From Plummer/McGeary, 1988, after C. F. S. Sharpe.

*Figure 8.22*  Solifluction over permafrost has produced a wavy, lobate surface on this barren terrain in northwest Greenland. Note the tendency of boulders to become concentrated in hollows.

Photo by R. B. Colton, U.S. Geological Survey

## Food for Thought

Suppose you owned a hundred commercially zoned acres of prime chernozem soil in the midwest near an urban center. Faced with the question of whether to build and operate a truck farm on your land or to sell it for ten million dollars to a developer who intends to build a shopping center on it, which would you do? Assuming that you would do what most Americans in such a position would do, Earth's food-producing capacity would be diminished by one hundred acres of prime agricultural soil.

This kind of decision is made many times each day and with increasing frequency. The soil that is stripped away to make room for commercial developments required over a thousand years to mature. Even if the decision were to be reversed before construction began, the soil could not be regenerated once it has been stripped.

Are such decisions reasonable? We know that Earth's resources, including its soil, are finite, yet we seem to be unwilling to control the human population that depends on them. Our socioeconomic traditions applaud individuals who make decisions that lead to personal economic gain. In the Soviet Union, the state is praised for similar decisions. The *mercantile* tradition in America and much of the rest of the world, including the Soviet Union, has its origin in a debate between two of our founding fathers, Alexander Hamilton and Thomas Jefferson. Faced with the challenge of defining a socioeconomic policy for our new nation, these two men engaged in vigorous contention over the direction that policy should take.

While Hamilton argued for mercantilism, Jefferson championed **usufruct,** the practice of caring for resources in such a way as to maintain or to increase their health and productivity for future generations. Mercantilism placed the monetary value of resources above the resources themselves, which Hamilton considered expendable, an attitude possibly defensible at a time when America promised vast, and seemingly "inexhaustible" frontiers. Usufruct, on the other hand, places a greater value on the resources themselves. Jefferson envisioned a nation whose strength would lie not in a system of commercial interdependency that trades resources for currency, but in a system of independent, self-sufficient family farms in which a person's richness is measured by the health and productivity of the land he or she owns and cares for.

Given the influences of gold's unfailing glitter and the apparent vastness of the American frontier, it is little wonder that Hamilton's bold concept won the favor of our democratic nation over Jefferson's more conservative one. As long as there was still a frontier to fuel the consumptive fires of mercantilism, the concept flourished with amazing success. Since Hamilton first ignited them, those fires—measured by the gross national product—have been raging and consuming resources at an ever-increasing rate. Today, our nation stands at the farthest horizon of the once-promising American frontier, sensing trouble but not yet fully acknowledging it, far less making a solid commitment to avert it. Due to the concern of some farsighted

individuals like Gifford Pinchot, Walter. C. Lowdermilk, and Stephen T. Mather, some institutions based on the concept of usufruct have arisen in America, among them the United States Forest Service, the Soil Conservation Service, and the National Park Service. These government-based institutions are, however, a far cry from Jefferson's idea of usufruct as a sense of responsibility toward the land that is sanctioned by law, administered by government, but vested in the landowner.

In terms of the depletion of our soil resource, the legacy of the mercantile ethic in America is sobering. Studies indicate that erosion rates have tripled since North America was colonized by Europeans. Under equivalent rainfall regimes, cropland can erode up to 100 times faster than forest soil. In the United States, sediment is eroded from the land at a rate of 250 tons/km²/yr. The global average is 180 tons/km²/yr. Figure 8.23 shows the cumulative effects of soil erosion in the United States. When we add to this picture the prospect of acid rain sterilizing what soil remains (see fig. 15.33), the concept of single-valued striving for short-term gain begins to take on the appearance of gross irresponsibility. Those who are dedicated to the mercantile ethic might rush to its defense with the argument that "we train and pay Earth scientists to solve these kinds of problems for us," but what happens when the Earth scientists run out of answers? There are two possibilities: replace the Earth scientists with public relations experts or develop national policies that show greater responsibility toward the natural resources on which we depend.

*continued*

## Summary

Plutonic rocks and minerals high in Bowen's reaction series are physically and chemically unstable in the weathering environment. **Mechanical weathering** is the disintegration of rock by such agents as **unloading** (which produces **exfoliation domes**), **frost wedging,** *salt crystal growth, fire, root wedging,* and *animal biting and burrowing.* **Chemical weathering** disintegrates rock by **solution** (the dissolving of ionic minerals), **hydrolysis** (reaction with water), **oxidation** (reaction with oxygen), and **hydration** (addition of water). The higher a mineral is in Bowen's reaction series, the more susceptible it is to chemical weathering. **Saprolite** is rotten rock, the result of hydrolysis and/or oxidation. **Spheroidal weathering** results from chemical weathering along joint planes. **Regolith** is unconsolidated

# The Work of Surface Water and Groundwater

## Outline

*There is nothing under Heaven*

*that is more yielding than water,*

*Yet it attacks that which is strong and durable.*

*It is not able to do this, however,*

*without suffering change.*

*The weak overcomes the strong.*

*The soft overcomes the hard.*

*Everyone knows this, yet none can act accordingly.*

*Lao Tzu*

**Tao, Te Ching, Chapter 78**

Photo: The inner gorge of the Colorado River from Toroweap Point, Grand Canyon, Arizona.

# The Interaction of Land, Water, and Gravity

In chapter 1, I stress that Earth's surface, unlike that of the Moon, is a dynamic, ever-changing one, whose form adjusts to the forces of gravity and flowing water that act continually upon it (see fig. 1.13). Over time, these forces slowly reshape the land surface toward that ideal form which provides for the most efficient removal of precipitation and sediment. This reshaping involves erosion, transportation, and deposition, and it operates according to the laws that govern equilibrium systems (see chapter 2). In the first part of this chapter, I consider the behavior of flowing water and the landforms that develop as a result of that behavior: stream channels, stream networks, drainage basins, and valley sideslopes.

## Dynamic Fluids: The Behavior of Flowing Water

Like any other object located within Earth's gravitational field, water tends to reduce its potential energy by moving toward the center of that field (i.e., toward Earth's center). Frustrated by the hard reality of continental bedrock, water flows not straight down, but slantwise from highlands in rivers to the sea. Why should it flow in rivers rather than as thin, continuous sheets over the land surface? The answer lies in the greater efficiency of channeled flow.

### Sheet Flow and Channel Flow

The tendency of water to flow in channels can be understood from a simple analogy. If paint is slopped over the rim of a can, it begins to run slowly down the side as a curtainlike **sheet flow.** Soon, however, drips form on the lower edge of the curtain of paint, and these run much more quickly to the bottom of the can, effectively halting the downward progress of the curtain by draining paint from it. Obviously, this "drip flow" is much more efficient than sheet flow in reducing the potential energy of the paint. The reason for this is that the *internal* friction experienced by paint in flowing against itself within a drip is much less than the *external* friction it experiences in flowing as a sheet against the can label.

The same thing is true of water flowing on land. **Channel flow** (equivalent to drip flow) is a much more efficient means of moving water downslope than is overland sheet flow. This is evident in figure 9.1, which shows that the velocity of water flowing in a straight channel is lowest immediately adjacent to the channel bed and banks and highest along the centerline of the channel and about one third of the way down from the water surface to the bed (arrows). From this efficient, high-velocity core, flow velocity decreases slightly upward, due to surface tension and friction against the

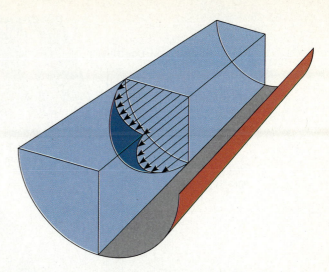

**Figure 9.1** In a straight stream channel, flow velocity is lowest near the bed and highest along the centerline about a third of the way from the water surface to the bed. Flow is turbulent and not straight (as shown) except immediately adjacent to the bed. Arrows show average motion.

air, and strongly downward and sideward due to friction against the channel bed and banks. Within the thin film of water next to the mineral surface of the channel bed, the flow velocity drops almost to zero. As this film of slowly moving water is essentially equivalent to overland sheet flow, it is easy to see why channel flow is so much more efficient.

The straightness of the arrows in figure 9.1 gives a misleading impression of the actual behavior of water flowing in natural channels, however. In reality, only the very low-velocity flow immediately adjacent to the stream bed and banks is smooth, tranquil, and straight. As flow velocity increases toward the core region of the stream, the flow becomes highly disturbed, or *turbulent.* The straight arrows, then, simply indicate the average motion of water molecules at particular points within the flow. A second unrealistic aspect of figure 9.1 is the straightness of the stream channel. In reality, stream channels are seldom straight, but tend to be either sinuous or braided. I discuss the reasons for the occurance of these forms in a following section.

The velocity of channeled flow depends on its efficiency, and this, in turn, depends on the *slope, depth,* and *roughness* of the channel. In general, the steeper, deeper, and smoother the channel, the faster the flow. Conversely, the gentler, shallower, and rougher the channel, the slower the flow. Of the three factors, roughness is the most influential, and slope is the least influential. This relationship further clarifies the difference in behavior between overland sheet flow and channel flow. Sheet flow is both rougher and shallower than channel flow and is therefore slower and less efficient.

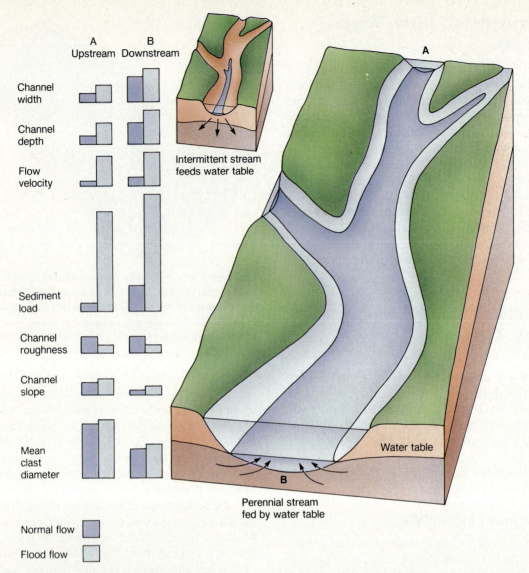

A B
Upstream Downstream

Channel width

Channel depth

Flow velocity

Sediment load

Channel roughness

Channel slope

Mean clast diameter

Normal flow

Flood flow

Intermittent stream feeds water table

A

Water table

B

Perennial stream fed by water table

**Figure 9.2** Perennial streams (*right*) receive water as input from high water tables, whereas intermittent streams (*top*) yield water as output to low water tables. Physical properties of perennial stream channels (*left*) are affected by downstream increases in discharge (from *A* to *B*) under a given flow, and by increases in discharge at a given station (from *dark blue* to *light blue*) during floods.

## Stream Channels as Systems

Stream channels are dynamic equilibrium systems. Their inputs are (1) *water* in the forms of groundwater, precipitation, and melted snow; and (2) *sediment,* both suspended and dissolved, derived from the weathering and erosion of bedrock and regolith. Outputs occur as evaporation of water and as seepage of water into the ground but principally as the **discharge** of water and sediment down the channel. Discharge is defined as the volume of water and sediment passing by a given point in a given time interval.

As long as outputs do not exceed inputs for any extended period, water and sediment are present in storage within a stream channel. In humid regions, a stream is normally kept flowing by inputs of water all along its length from a high water table. Such streams are described as **perennial.** The larger block diagram in figure 9.2 illustrates a perennial stream. In arid regions, however, water tables are typically too deep to feed streams and are fed by them instead. In some cases, such as that of the mighty Colorado River of the American southwest, inputs of water from mountain tributaries are sufficient to keep such streams perennial, even in the arid lowlands where water is lost rapidly into the dry ground. If the rate of such inputs is less than the output rate, however, the streams quickly lose whatever water they have in storage within their channels and become **intermittent,** flowing only during heavy rainfall or snowmelts. The smaller block diagram in figure 9.2 illustrates an intermittent stream.

Discharge normally increases regularly from the headwaters to the mouth of a given stream because of increasing downchannel inputs from groundwater and incoming tributaries. Variations are imposed on this regular downstream increase in discharge by floods, which further increase flow, and by droughts, which decrease it. The forms and dimensions of stream channels tend to adjust in response to such variations in streamflow. In figure 9.2, two extreme stages of discharge are shown for the perennial stream, a normal flow in dark blue and a flood flow in lighter blue. The bar graphs at the left of the figure show how several features of the stream channel change in response to changes in discharge. On the larger diagram in figure 9.2, *A* is an upstream station, and *B* is a downstream station. Comparing the two stations under low flow conditions (dark blue bars), we find that the first four features—*channel width, depth, flow velocity,* and *sediment load*—all increase downstream from *A* to *B*. The same observation holds in times of flood (light blue bars). We also find that at a given station, either *A* or *B,* these same four features increase from low flow to high flow (dark blue bars to light blue bars).

By far the most impressive of the at-a-station changes is the dramatic increase in sediment load from low flow to high flow. This explains the muddiness of streams in flood, and it also illustrates the most significant single factor in stream erosion: *the power of a stream to erode its bed and to carry sediment depends on its flow velocity.* This factor is so influential that a doubling of the flow velocity results in approximately a 20-fold increase in sediment load! Similarly, an increase in flow velocity at a given station results in a corresponding (though much less dramatic) increase in the *mean clast diameter* of the sediment that a stream can transport (bottom bar graph in fig. 9.2).

The maximum sediment load that a given flow can transport is called the flow's **capacity,** and the maximum (*not* the mean!) clast diameter it can transport is called its **competence.** Strictly, these two properties refer only to *bed load,* or sediment moving along the channel bed, and not to sediment in suspension. Together, they measure the erosive power of a given stream flow. Both capacity and competence increase with flow velocity, and this underscores another important principle: that *most significant stream erosion takes place during large, but infrequent floods.* This is one of the main arguments against Lyell's simple interpretation of Hutton's principle of uniformitarianism (see chapter 2).

In view of the foregoing discussion, it might come as a surprise that mean clast diameter *decreases* downstream under both normal flow and flood flow, in spite of the downchannel *increase* in flow velocity. There are two reasons for this: (1) clasts are broken up into smaller pieces during transport (see fig. 7.25), and (2) the force

of gravity is less effective on the decreased channel slopes downstream, placing more of the burden of moving clasts on stream flow.

*Channel roughness* remains relatively constant downstream under a given flow, due largely to vegetation on streambanks. During floods, however, it decreases at a given station due to the smoothing effects of suspended sediment and increased depth of flow. *Channel slope* decreases downstream under both normal flow and flood flow, but increases at a given station during floods. Slope decreases downstream because as channels become deeper, efficiency of flow increases; hence, a gentler slope is adequate to move the water. This is the reason for the typical, smooth, concave-up profiles of most stream channels (see fig. 9.4). The increase in slope at a given station during floods results because water that is heavily laden with sediment at high discharges is more viscous than slower, cleaner water; hence, steeper slopes are required to move it at a given velocity. Because of this, streams tend to steepen their channels by depositing sediment during periods of high, sediment-choked flow, and to erode their channels to make them less steep during periods of low flows that are relatively free of sediment.

As complex as these relationships appear to be, the situation is rendered even more complex by interactions among these seven characteristics, some of which are alluded to in the foregoing discussion. Stream channels are indeed among the most complicated dynamic equilibrium systems in Nature!

## The Geometry of Streams

The preceding sections highlight certain generalities about streams:

1. They start small, steep, and clean, and become large, gentle, and dirty.
2. Their velocities increase only slightly along their lengths, but increase markedly at a given station during floods.
3. They tend to be more sinuous than straight.

Some other generalizations are evident in figure 9.3, which illustrates the **drainage system** of the combined Mississippi, Missouri, and Ohio Rivers. This great network of streams is contained within a **drainage basin** (area enclosed by dashed line), from which all its flowing waters are derived. The drainage system is branched like a tree: "twigs" merge into larger "branches," which, in turn, lead to a "trunk," whose base (the rivermouth) is branched into the rootlike *distributary channels* of the Mississippi delta (see fig. 12.31).

Two of James Hutton's many insightful conclusions drawn from his observations of the European countryside were that (1) streams carve the valleys in which they flow, and (2) except in areas that have recently undergone faulting or valley glaciation, stream tributaries are

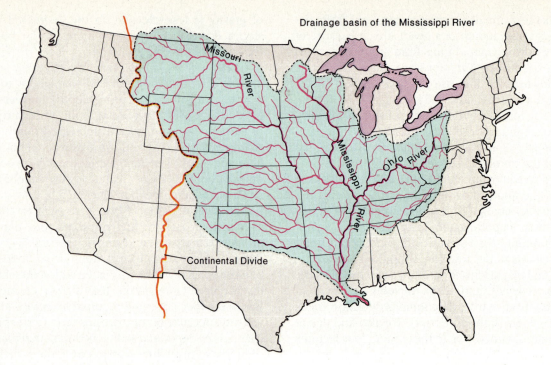

**Figure 9.3** The drainage system of the Mississippi River illustrates the features of a typical stream system: intricately branched tributaries collect water and sediment from rain and snowmelt and feed them into a main stream, which transports them to a distributary system at the stream mouth, which in turn disperses them into an ocean or lake.

almost always *accordant,* i.e., the elevation of the mouth of a tributary is the same as the elevation of the larger stream into which it flows. If this were not the case, the mouths of many tributaries would have either a waterfall or a lake at the point of entry into the larger stream. The concept of accordance of stream junctions later led to the concept of the **base level** of a stream, or that elevation below which the stream can not continue to cut its channel. A stream's base level is simply the elevation of the water body into which the stream empties. The word "level" is actually a misnomer, because streams must always maintain a slope in order to keep water flowing. A stream that has cut its channel down to its base level is called a **graded stream.**

The slope, or *gradient,* of a graded stream varies with the prevailing climate. Figure 9.4 shows the longitudinal profiles of two small, headwater streams, Dothan Brook in Norwich, Vermont, and Camp Foster Creek in Carbondale, Colorado. Both these profiles have smooth, almost mathematically perfect, concave-upward shapes that are steepest in their upper reaches and gentlest in their lower reaches. In the humid climate of Vermont, stream profiles are more concave and less steep, whereas in the semiarid climate of Colorado, profiles are less concave and steeper. The reason for the difference is that in Colorado, streams carry more sediment

than they do in Vermont, and so they need consistently steeper channel slopes in order to move that sediment.

The tributaries of a drainage system branch out from the trunk stream to create a fairly even distribution of channels over the land surface (see fig. 1.13). Such a distribution is most efficient for removing water and sediment from the land. Beyond this basic tendency, the patterns assumed by stream systems are diverse, and they are largely determined by the topography and the structure of the bedrock. Figure 9.5 illustrates four common types of *drainage patterns.*

The treelike branching pattern of **dendritic drainage** develops on homogeneous bedrock or on sedimentary rocks that lack such strong structural features as bedding, foliation, or fracture systems. In **trellised drainage,** the controlling factor is inclined stratification in the bedrock. Longer tributaries erode valleys parallel to the regional strike of dipping strata (NE–SW in fig. 9.5). Shorter tributaries, oriented transversely to the regional strike, then develop to drain the slopes of these valleys and have carved an intricate drainage system on the impervious mudstone strata in the northwest (area *A*). On the porous sandstone strata in the southeast (area *B*), a lower-density drainage system has developed.

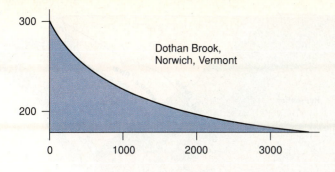

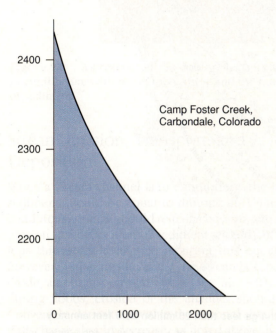

Dothan Brook,
Norwich, Vermont

Camp Foster Creek,
Carbondale, Colorado

**Figure 9.4** Stream profiles often exhibit near-mathematical precision, the forms of which are determined largely by climate. Elevations and distances are in meters. Vertical exaggeration is 10x.

**Parallel drainage** develops on relatively homogeneous bedrock or sediment, the surface of which has a strong regional slope (toward the southeast in figure 9.5). Often, parallel drainage forms where uplift or a lowering of sea level has freshly exposed portions of the continental shelf. **Deranged drainage** develops in recently glaciated regions, where the shape of the landscape is determined more by glacial scouring than by stream erosion. Deranged drainage is characterized by an abundance of lakes, as in Canada, Minnesota, and New England. Such lakes gradually disappear as the drainage becomes better organized.

In many cases, it is possible to interpret the topography, the geologic structure, and even the dominant bedrock types of a given region simply through studying the drainage patterns present in that region. This is especially true in humid regions, where bedrock is often poorly exposed.

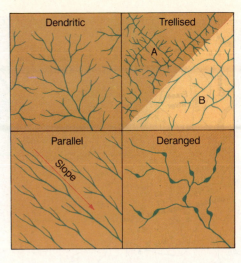

**Figure 9.5** The patterns assumed by drainage systems reflect the topography and geologic structure on which they develop.

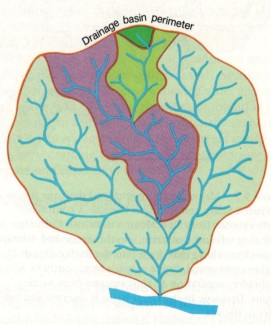

**Figure 9.6** Stream systems are contained within drainage basins, from which they draw all their water and sediment. A given drainage system comprises a hierarchy of subbasins nested within basins nested within superbasins. One basin of each order is highlighted.

## Drainage Basins: Their Forms and Functions

As the dendritic drainage system in figure 9.6 reveals, a drainage basin collects and drains precipitation from a clearly defined geographic area. The *perimeter* of the basin is a ridge that encloses all the various tributaries of the stream system and separates them from those of adjacent basins. This ridge increases in elevation from the mouth of the stream toward the headwaters, but it is, of course, everywhere higher than the immediately

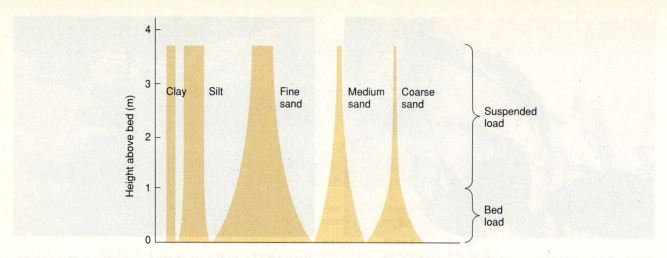

*Figure 9.12* Distribution of sediment with depth in a stream channel (Missouri River at Kansas City, January 1930), shows the distinction between bed load and suspended load. Flow velocities high enough to erode sediment are also high enough to suspend it only if grains are medium sand size or smaller. Coarser particles settle to form the bed load.

*Figure 9.13* Clastic sediment eroded by glaciers from Alaska's Copper River drainage basin is delivered to the trunk stream at such a high rate that the normal stream discharge lacks sufficient competence to flush it away. Instead, the flowing water merely shoulders it aside, resulting in a sediment-choked, braided stream channel. Note the small lake (*left center*) dammed by glacial till and a large sand bar.

Photo by E. Brabb, U.S. Geological Survey

sediment to local streams. The highly braided South Platte River in eastern Colorado and Nebraska is one such stream whose headwaters lie within coarse-grained, Precambrian granites of the Colorado Front Range. Braiding of the South Platte has been intensified by modern withdrawals of water for human use.

In arid mountainous regions, streams are often intermittent, only flowing after heavy rains or when snow melts in spring. Their relatively straight channels descend from the mountains at steep gradients, which are sharply reduced where the streams break out onto the plains beyond the mountain front. At this abrupt break in slope, a fan-shaped deposit of coarse alluvium called an **alluvial fan** is typically present (fig. 9.14). Intermittent streamflow can deliver great volumes of discharge and sediment down the mountain channel, but little water or sediment is carried much farther into the plains than the outer edge of the alluvial fan. There are

**Figure 9.14** Alluvial fans, such as these in Death Valley, California, result from an abrupt loss of competence as streams flow from steep, impervious mountain channels onto the gently sloping, permeable materials of the valley floor. On the fan, flow becomes braided, and it wanders freely across the shallowly conical, concave-upward fan surface.

Photo by W. B. Hamilton, U.S. Geological Survey

two main reasons for this. First, whereas mountain channels are often cut within relatively impermeable bedrock, alluvial fans are highly permeable; hence, they quickly soak up water, thereby reducing discharge. Second, as a stream emerges from a mountain front, its channel changes abruptly from steep, smooth, and deep to low-angle, rough, and shallow, all changes that reduce flow velocity. Both of these effects sharply reduce capacity, as a result of which the stream's suspended and bed load sediment is abruptly dumped on the alluvial fan. This "sediment graveyard" is coarsest, thickest, and steepest at its apex where it joins the mouth of the mountain canyon that feeds it, and finest, thinnest, and least steep at its outer edge where it blends into the valley floor. In time, the alluvial fans of adjacent canyons can grow and merge into a broad clastic apron, called a **bajada** (ba-*ha*-da; Spanish for "slope") bordering the mountain front (see fig. 9.22).

Notice that the channel pattern on the surface of the fan in figure 9.14 is braided, and it spreads out laterally in a 180° sweep, encompassing the full breadth of the fan. This pattern does not imply that the flow covers the entire fan all at once, but rather that the channel migrates freely over the fan surface.

## The Dynamics of Floodplains

The processes of stream erosion, transport, and deposition are nowhere more evident than on a **floodplain,** the most characteristic feature of a graded stream in equilibrium (figs. 9.15 and 9.16). During floods, when streamflow spills out of the channel and onto the surrounding land, alluvial deposition occurs on a massive scale. As floodwater begins to move away from the channel, its flow velocity is abruptly reduced. This, in turn, reduces its capacity and competence, forcing it to deposit sediment. Coarser sediment settles out on the tops of the stream banks where the flow is swiftest, raising the banks into low ridges called **natural levees.** Finer sediment settles at increasing distances from the channel on the floodplain, where the flow velocity is less. Natural levees present an obstacle to tributary streams, which often must run downvalley for several kilometers before they are able to break through the levee and join the main stream. Such deflected tributaries are known as **yazoo streams** after the Yazoo River in Mississippi, which bears this relationship to the Mississippi River.

On most floodplains, graded streams develop bends or **meanders.** The outer, concave bank of a meander is both steep and deep (note locations of pools in the meandering channel in fig. 9.8). This is because flow velocity (hence competence) is highest along the outer margin of a curved channel. Hydraulic erosion is, therefore, the prevailing process on such banks, which are appropriately known as *cutbanks* (fig. 9.16). In contrast, the inner, concave bank of a meander is gentle and shallow because flow velocity is lowest along the inner margin of a curved channel; therefore, deposition occurs

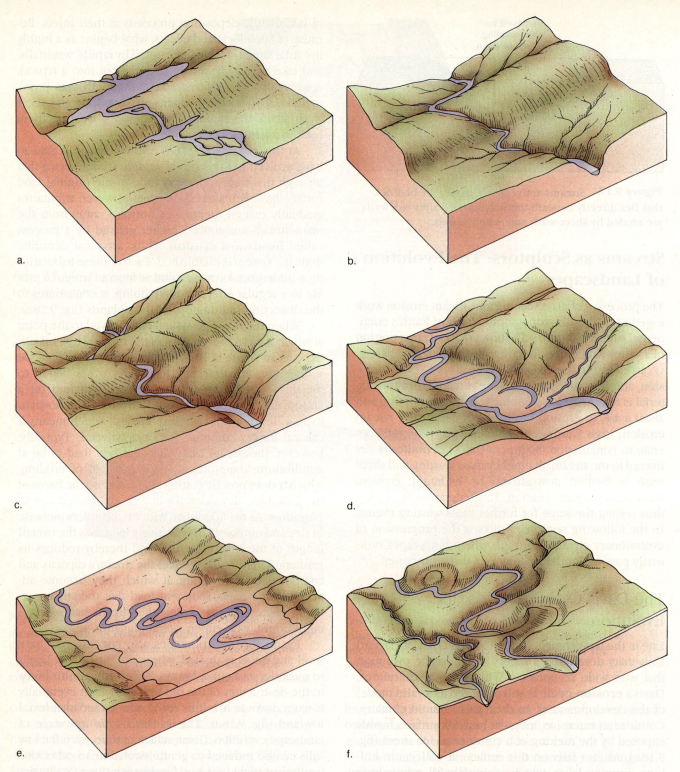

**Figure 9.18** The evolution of a glaciated landscape in a humid region. Youthful landscapes (*a, b, c*) consist mostly of original upland, some valley side slopes, and little river bottom land. Mature landscapes (*d*) consist mostly of valley side slopes and have little original upland but have more bottom land. Old landscapes (*e*) have no original upland and consist predominantly of bottom land. Subsequent uplift (*f*) can rejuvenate the landscape.

Copyright © 1971, by Arthur N. Strahler. Reproduced by permission.

a.

b.

c.

d.

e.

f.

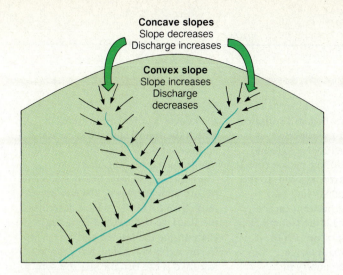

**Figure 9.19** Suitable adjustments in discharge and slope angle allow mature landscapes in equilibrium to maintain uniform competence over the land surface, essential for removing the products of erosion. Arrows indicate sheet flow.

time onward the drainage system establishes its influence over the entire landscape. Consequently, all landforms that develop from maturity through old age are *equilibrium landforms,* whose shapes are continually readjusted to assure the most efficient transfer of water and regolith off the land. Under this concept, the shape of a given hill or valley is a reflection of the influences of a number of factors, including bedrock type, grain size of regolith, climate, vegetation, rainfall, sheet flow, and channel flow.

Figure 9.19 illustrates this concept in terms of the question, "What specific combinations of slope and discharge are required to erode and transport sediment of a given grain size?" As sheet flow gathers into hollows on a hillside (*upper right* and *left*), the flow becomes concentrated, resulting in increased discharge. This, in turn, results in increased erosion, which accentuates the hollows, deepening their bottoms and reducing the channel gradient downstream. Reduction of channel gradient reduces competence until sediment begins to be deposited, whereupon equilibrium is established. Equilibrium is maintained on *convex slopes* (fig. 9.19 *center*) because sheet flow is dispersed on such slopes, resulting in a decrease in discharge and competence. In order to keep sediment moving on such slopes, the slope angle must increase downward, as it indeed does on a convex slope.

## Disturbances to the Davis Cycle

Although some peneplains have been identified beneath unconformities in the stratigraphic record, peneplanation seldom has time to happen before some plate tectonic event uplifts the landscape, or base level falls, forcing the drainage system to begin downcutting

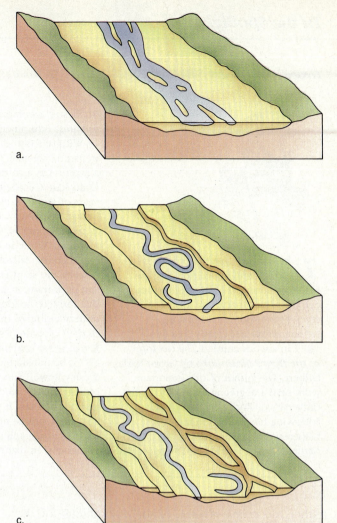

**Figure 9.20** Changes in climate, glacial activity, and human land use can cause a graded stream either to build up its floodplain (*a*) or to erode it and form alluvial terraces (*b*). Multiple terrace sets (*c*) result from cyclic changes in these factors.

until it reestablishes a graded condition. This process is known as **rejuvenation.** Such a rejuvenating event has occurred in figure 9.18*f.* Notice that as the stream system has cut down through the bedrock, it has preserved its meandering form in a series of **incised meanders.** Following rejuvenation, the erosion cycle begins anew and continues until the next tectonic disturbance or drop in base level.

Another kind of disturbance, even more common than rejuvenation, involves changes in the various factors that determine equilibrium in stream systems. Figure 9.20 shows a graded stream that has experienced a change from a surplus of sediment to a deficit. The braided stream channel in figure 9.20*a* indicates that the sediment that is being fed into the stream system is

### Grove Karl Gilbert (1843–1918)

*G.K. Gilbert*

The concept that the processes of erosion and deposition reshape landscapes toward ideal equilibrium configurations that allow for the most efficient transport of water and sediment was first clearly stated in 1877 in a work entitled *The Geology of the Henry Mountains* by Grove Karl Gilbert. The author of this important work held a degree in Classics from the University of Rochester, New York. Following his graduation in 1862, Gilbert went to work for the Ward Natural Science Establishment in Rochester, a company that still supplies scientific equipment to schools.

With this introductory background in natural science, Gilbert then secured a position as a volunteer assistant to the Ohio State Geological Survey. Two years of work in that position earned him an assignment with the George M. Wheeler Survey west of the 100th meridian. While in that post, he explored the canyon country of Arizona and southern Nevada by boat and pack train, making observations that formed the foundation of our understanding of the geologic structure of the Colorado Plateau and Basin and Range physiographic provinces. In 1875, he transferred to the John Wesley Powell Survey, with which he remained until 1879, studying the geology of Utah. It was during this appointment that he gathered field data for his monograph on the Henry Mountains. In that report, he described the concordant igneous plutons of which the Henry Mountains consist, and named them "laccolites," a term that has since become altered to "laccolith." Also from this period came his monumental work, *The Bonneville Monograph* (1890), which describes the vast lake that existed in central Utah during the last ice age, a freshwater body of which the Great Salt Lake is the last, meager remnant.

In this modern age of specialization, it hardly seems possible that one person could have contributed as substantially to as many diverse subdisciplines of geology as G. K. Gilbert did during his long, energetic, and productive life. Gilbert was one of the six senior geologists chosen to staff the newly-formed United States Geological Survey in 1879. He became head of the Survey's Appalachian Division in 1884 and its chief geologist in 1889.

---

either too coarse-grained or too voluminous for the stream's competence or capacity. This situation typically arises downvalley from melting glaciers from which large volumes of clastic sediment are being flushed by meltwater.

In figure 9.20*b*, the stream has cut beneath the original floodplain surface, forming a new, narrower floodplain at a lower level, and it has begun to meander, both indications that its competence and capacity have increased. This could result from either an increase in discharge or a decrease in sediment supply, or both. The complete disappearance of the glacier from the valley head during an interglacial interval could produce both these effects. Notice that remnants of the higher floodplain have been preserved against the valley sides. Such remnants are known as **alluvial terraces.**

In figure 9.20*c*, a subsequent episode of downcutting has occurred, possibly reflecting a second glacial/interglacial fluctuation, and another set of alluvial terraces has developed below the first and above the most recently developed floodplain. Multiple sets of paired alluvial terraces are commonly seen within drainage systems of the western United States. Many of these are attributable to changes between glacial and interglacial conditions, but some have certainly resulted from changes in climate or in human land use (see Food for Thought). For example, a change to a wetter climate would favor erosion by downcutting and a more concave-upward stream profile, whereas a change to a drier climate would favor the deposition of alluvium and a less concave profile.

## Modifications of the Davis Cycle

Such factors as climate, ruggedness of the original terrain, rock types, and vegetation types strongly affect the course of fluvial landscape evolution, often producing landforms that differ markedly from the "normal," rolling, concave-convex hillslopes so characteristic of mature fluvial landscapes in humid regions. For example, flat valley side slopes should not be common in Nature, because they require inefficient overland sheet flow. In fact, however, flat slopes do commonly occur in forested valleys, where they are maintained by the interwoven network of forest tree roots. In grassland, shrubland, and desert, where the vegetation does not produce laterally extensive root nets, flat slopes are less common because channel flow can readily erode them into valleys and ridges with concave and convex slopes.

*Figure 9.21* **Hogback ridges of resistant sandstone warped upward by the rising of the Rocky Mountains to the west (*right*) present a spectacular example of differential erosion.**

Photo courtesy of Colorado Geological Survey

Bedrock types and structures are of great significance to the course of fluvial landscape evolution. Old age comes much more quickly to river systems developed on soft, easily erodible rocks, such as mudstone, than it does to those developed on resistant rocks, such as sandstone. In regions of complex geologic structure, variations in the resistance of different geologic formations result in **differential erosion,** which etches the land surface so that the more resistant formations stand out as highlands and ridges, whereas the less resistant formations form lowlands and valleys. A classic example of differential erosion south of Denver, Colorado, is illustrated in figure 9.21, which shows a series of resistant sandstone ridges formed by the edges of strata that were upwarped by the rise of the Rocky Mountains to the west (*right*) at the end of the Mesozoic Era. The softer shale formations between the sandstones have been readily eroded by intermittent streams flowing east from the mountains during cloudbursts, leaving the sandstones standing in prominent *hogback ridges.*

Arid regions, such as the Great Basin in Nevada and adjoining states (see fig. 9.7), are characterized by *internal drainage,* meaning that they have no outgoing drainage to the sea. Streams in the Great Basin start in the high mountains, where groundwater, precipitation, and snowmelt are adequate to maintain them, but they soon lose these sources and become intermittent as they descend to the arid plains, where they eventually dry up (note, for example, the Humboldt River in northern Nevada). In addition to these unusual drainage conditions, the Great Basin also has an unusual geologic structure. In combination, these two unusual conditions have produced an unusual kind of landform evolution that has been called the **arid erosion cycle** (fig. 9.22).

In the initial stage (fig. 9.22*a*), extensional faulting breaks the crust into raised, elongate fault blocks called **horsts,** or Hörste (*her*-stuh; German for "eagle nests"), and parallel sunken blocks called **grabens,** or Gräben (*gray*-bn; German for "trenches"). In time, the relief of the landscape is greatly reduced by intense erosion of the horsts and rapid accumulation of the erosional products as alluvial fans on the floors of the grabens (fig. 9.22*b*). During moist episodes, runoff becomes ponded in the grabens as short-lived **playa lakes.**

Instead of hillslopes becoming gentler over time, as they seem to do in humid landscapes, slopes in arid regions maintain a constant angle and retreat parallel to themselves as they erode. The reason for this is that intermittent streams in arid regions lose water downchannel. In order to keep their sediment loads moving despite this loss of discharge, stream channels must maintain steep, fairly straight gradients, like that of Camp Foster Creek (see fig. 9.4). As a slope retreats, a gently sloping erosion surface called a **pediment** develops on the bedrock at its base (fig. 9.22*b* and *c*). Meanwhile, the coalescence of alluvial fans along the mountain fronts produces bajadas. Eventually, the growing pediment eliminates most of the upland topography of the horsts except for a few, scattered remnants. If renewed movement along the extension faults occurs (fig. 9.22*d*), the stage is set for a repetition of the arid erosion cycle.

## Groundwater Systems

Of the precipitation that falls on continental surfaces, only about 30% runs off directly as stream discharge, while another 57% is evaporated. The remaining 13% is intercepted by vegetation and surface litter, whereupon it percolates slowly through soil profiles and into underlying regolith or permeable bedrock, where it becomes **groundwater.** According to recent estimates, the total amount of groundwater held in storage within Earth's crust is almost 7000 times the amount of water that is present at any given time within the channels of

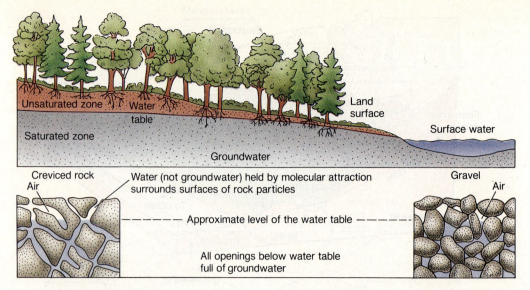

**Figure 9.24** In the saturated zone below the water table, interconnected pores in rock, saprolite, and soil are filled with water. In the overlying unsaturated zone, pores can be dry but are usually partially water-filled. Water from rainfall and snowmelt percolates down through these pores to the water table. Where the water table intersects the ground surface (*right*), water bodies appear at the surface.

Source: "Ground Water," 1986 (491–402/04), U.S. Geological Survey.

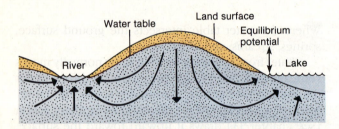

**Figure 9.25** The form of the water table usually mimics that of the overlying topography, but with lower relief. Groundwater moves along curved flow paths in response to local equilibrium potentials (*double arrow*).

the water table drops. Because this also reduces the height of the equilibrium potential, streamflow is also reduced.

## Groundwater Reservoirs

Not all ground is equally suited for the storage of groundwater. Two factors, porosity and permeability, determine the suitability of Earth materials as groundwater reservoirs. **Porosity** is defined as the percentage of the total volume of a rock occupied by empty space. Porosities of rocks range from 0% in the case of massive, unfractured igneous rocks and some limestones and well-cemented sandstones to 43% in Cretaceous chalk from Great Britain. Some soft, unconsolidated muds have porosities as high as 90%. Figure 9.26 shows examples of porosity in six different materials.

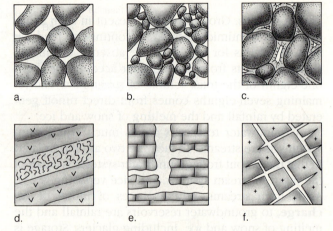

**Figure 9.26** Porosity, the percentage of a rock, sediment, or soil occupied by empty space, is high in well-sorted sand (*a*), and low in poorly sorted sediment (*b*). Porosity can be highly variable in cemented sandstone (*c*), buried lava flow tops (*d*), jointed and hydrolyzed carbonate rock (*e*), and fractured bedrock (*f*).

If the pore spaces of a rock are not connected with one another, or if there is great resistance to the free flow of water through the rock, even the most porous rock can be a poor producer of water. The property that determines the capability of Earth materials to transmit water is called **permeability.** It is dependent on the size and shape of pores and on the size, shape, and extent of the interconnections among them. Rock formations (including unconsolidated sediments) that have high permeabilities are called **aquifers.** Those

**Table 9.1** Relative Porosities and Permeabilities of Various Earth Materials in Approximate Order of Decreasing Value as Aquifers

| Material | Porosity | Permeability |
|---|---|---|
| 1. Gravel | Medium | High |
| 2. Medium to coarse sand | Medium | High |
| 3. Dune sand | Medium | Medium |
| 4. Sand and gravel | Medium | Medium |
| 5. Glacial till | Medium variable | Medium variable |
| 6. Fine sand | Medium variable | Medium variable |
| 7. Sandstone and conglomerate | Low variable | Medium |
| 8. Loess | High | Low variable |
| 9. Silt | High | Low variable |
| 10. Volcanic tuff | High variable | Low variable |
| 11. Clay | High | Low |
| 12. Glacial outwash | Medium variable | Low variable |
| 13. Unjointed limestone | Low | Low variable |
| 14. Lavas | Low | Low variable |
| 15. Granite and granodiorite | Low | Low |

Source: T. Dunne and L. B. Leopold, *Water in Environmental Planning.* Copyright © 1978 by W. H. Freeman and Company, New York, NY.

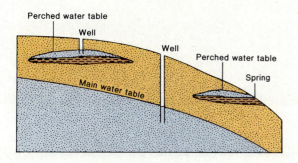

**Figure 9.27** Perched water tables often form above local aquicludes within a larger aquifer. Here, shale lenses (dashed) perch water within sandstone (stippled).

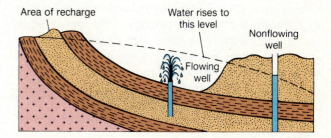

**Figure 9.28** Artesian water systems can develop where an inclined aquifer, such as a sandstone formation, lies between overlying and underlying aquicludes, such as shale strata. Wells drilled into such an aquifer flow freely if their upper ends lie below an artesian pressure surface (*dashed line*) that slopes with the aquifer gently away from the recharge area.

with low permeabilities are called **aquicludes.** Table 9.1 lists some rock types and unconsolidated Earth materials in decreasing order of porosity and permeability.

## Perched Water Tables and Springs

Aquifers and aquicludes are interlayered and juxtaposed irregularly in Earth's crust. **Perched water tables** can arise when aquicludes are located above the main water table within an aquifer (fig. 9.27). Downward percolating water becomes ponded on the upper surface of such local aquicludes. If an aquiclude intersects a hillslope, the water can then flow laterally and emerge as a line of **springs.** Springs can also occur (1) along fault planes where an aquiclude has been brought into contact with an aquifer and (2) where layers of porous limestone or lava intersect a hillslope.

## Artesian Systems

A special situation, called an **artesian system,** arises when an alternating series of aquifers and aquicludes dips in the same general direction as the regional topographic slope. In figure 9.28, which illustrates such a system, an eastward-dipping sandstone aquifer is enclosed between mudstone aquicludes. The aquifer crops out at the surface in the foothills of a mountain uplift, providing a surface recharge area. There has been enough precipitation to fill the aquifer to capacity. The heavy, dashed line indicates the equilibrium potential of this artesian system, or the level to which the water within the sandstone aquifer theoretically would rise if

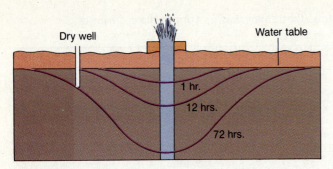

**Figure 9.29** High rates of pumping result in a drawdown, or lowering of the water table, which creates a cone of depression. Here, the effects of pumping over a 72 hour period are shown. Drawdown can affect neighboring wells, causing them to go intermittently or permanently dry.

it were not confined above by the mudstone. Because pressure is lost through underground leaks within the system, the line slopes gently toward the right from the level of the water table in the recharge area. Wherever an open fracture or well penetrates the upper aquiclude, water from the aquifer rises to the level of the artesian pressure surface above, creating an *artesian spring* or *well,* respectively. If the top of the fracture or well is located above the pressure surface, no water flows from it unless it is pumped, but if the top is located below the pressure surface, water flows from it spontaneously.

## Pumped Wells and Their Problems

Most wells have to be pumped before they yield water. If the depth to the water table in a well is 10.3 m or less, water can be drawn from the well by means of a suction pump installed at the surface. In wells in which the water table lies at depths exceeding 10.3 m, however, water must be forced upward by a pump that pressurizes the well because suction is unable to support a column of water that rises higher than 10.3 m above the water table. The principle is the same as in the case of the mercurial barometer (see fig. 14.1). A column of mercury 76 cm high, a column of water 10.3 m high, and a column of air from ground level to the top of the atmosphere all weigh the same if they all have the same cross-sectional area.

A few water wells pumping at moderate rates in a given area normally have a relatively insignificant effect on the local water table, but a large number of wells pumping at moderate rates or even a few wells pumping at high rates can cause substantial drawdowns of the water table. Figure 9.29 shows the development, over 72 hours of continuous pumping, of a deep **cone of**

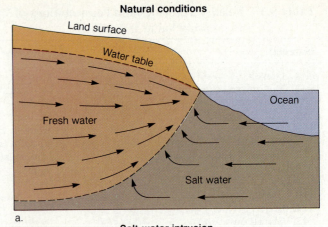

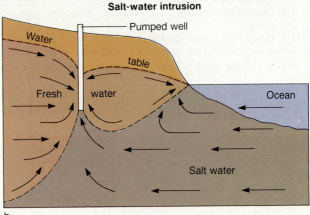

**Figure 9.30** Salt-water intrusion is a serious problem in wells of seacoast communities but one that can be controlled by the use of recharge wells.

Source: "Ground Water," 1986 (491–402/04), U.S. Geological Survey.

**depression** in the water table around a high-discharge well, such as is often found in agricultural areas. Drawdowns can affect neighboring wells, and if they are sustained enough, they can eventually lower the water table regionally to such an extent that all water wells in the region must be drilled to greater depths to keep their lower ends in the saturated zone.

Wells drilled in seashore areas have special problems (fig. 9.30). Rainwater percolating down through coastal land surfaces does not mix with underlying salt water, but forms a low density wedge of fresh water that floats on the denser seawater beneath (fig. 9.30*a*). Wells drilled into this fresh water wedge create both the normal cone of depression in the top of the water table and an inverted cone of *salt-water intrusion* below (fig. 9.30*b*), which can contaminate domestic water supplies. The only practical method for eliminating intrusion cones is to sink recharge wells into the ground near the pumping wells and force treated, fresh waste water down them.

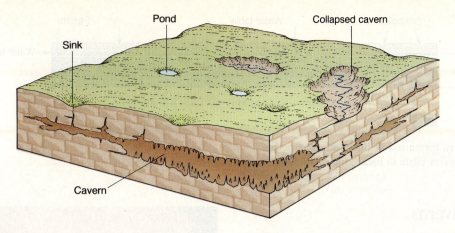

**Figure 9.31**  Karst topography results from the hydrolysis of a landscape underlain by limestone bedrock.

Labels: Pond, Sink, Collapsed cavern, Cavern

# The Soluble Landscape: Karst and Caverns

The acidity of water is measured by the *pH scale,* a logarithmic scale on which neutral water has a value of 7.0. Lower values are acidic, and higher values alkaline. Normal, unpolluted rainwater contains enough dissolved carbon dioxide to give it a pH of about 5.6. When this slightly acid water percolates down through fracture systems in limestone bedrock, it reacts with calcium carbonate by hydrolysis, producing soluble calcium and bicarbonate ions and leaving voids in the rock in their place. This hydrolytic process is thought to take place principally in the upper part of the saturated zone, just below the water table. In time, it can produce vast systems of underground caverns and even reduce great limestone plateaus to knobby lowlands.

## Karst Topography

In regions where the hydrolysis of limestone bedrock is the principal agent of erosion, a distinctive kind of landscape called **karst topography** results, named for a limestone region in northwestern Yugoslavia. Figure 9.31 shows some characteristic features of karst topography. Karst regions seldom have major, throughgoing, perennial streams because such streams have abundant opportunities to sink below the surface into channels that have been hollowed out by hydrolysis along fractures and bedding planes in the limestone. Frequently, solution proceeds far enough to open up large, underground caverns. The collapse of cavern roofs where the overlying bedrock is thin often produces an abundance of **sinkholes** at the surface. Sinkholes constitute a prime geologic hazard because they can, and often do, swallow houses, cars, and other human effects (see fig. 19.25).

**Figure 9.32**  Fantasyland or real landscape? In this painting of the Emperor Ming Huang's journey to Szu Ch'uan, the anonymous Chinese artist has not greatly exaggerated the spectacular karst topography of one portion of the journey.

In more advanced stages of karst development, the coalescence of sinkholes results in the formation of deep valleys and intervening ridges. These are often of bizarre and spectacular form, well portrayed in many Chinese paintings (fig. 9.32).

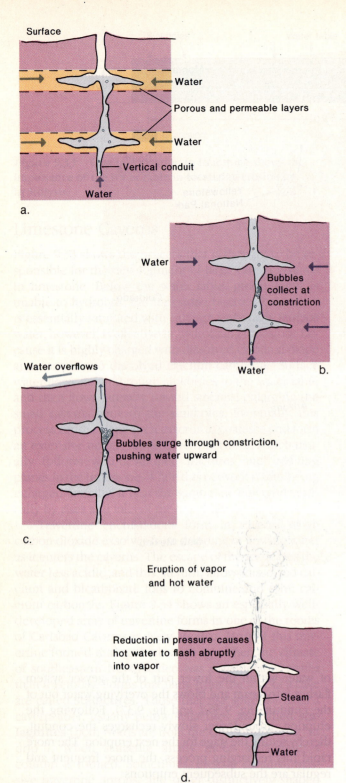

**Surface**

Porous and permeable layers

Water

Water

Vertical conduit

Water

a.

Water

Bubbles collect at constriction

Water

b.

Water overflows

Bubbles surge through constriction, pushing water upward

c.

Eruption of vapor and hot water

Reduction in pressure causes hot water to flash abruptly into vapor

Steam

Water

d.

***Figure 9.36*** The effect of a steep geothermal gradient on groundwater is spectacularly illustrated by geysers. In these illustrations, the volume of space within the geyser conduit has been greatly exaggerated.

Source: U.S. Geological Survey Bulletin 1347.

***Figure 9.37*** Roughly once every 65 minutes, Old Faithful in Yellowstone National Park hurls about 40,000 liters of hot groundwater up to 45 meters in the air. Each eruption lasts about 4 minutes.

Photo by J. R. Stacy, U.S. Geological Survey

notably the one at The Geysers, California, which, at 1180 megawatts, is designed to be the world's largest producer of geothermal energy. In addition to the use of steam for the generation of electricity, hot water from thermal springs can be used directly for the heating of houses and buildings.

An interesting new concept that is still in the experimental stage is the drilling of wells where hot, dry rocks lie at shallow depths in the crust. Cold water would be pumped down one well and retrieved as hot water from another, adjacent well. The approach could even be used (with caution!) in the drilling of wells into active or dormant volcanoes. Potentially, such schemes could tap prodigious sources of energy, but even so, it has been estimated that only between 1% and 20% of United States' energy needs in 2000 A.D. could be met by geothermal resources. Moreover, the development of geothermal power presents some significant practical problems, including the release of such noxious and corrosive gases as sulfur dioxide to the atmosphere and the deposition in pipes of encrusting substances present in geothermal water and steam. Here, then, is yet another indication that the wisest solutions to our long-term energy needs lie in the directions of conservation and population control.

**I**n a remote section of the Navajo Indian Reservation in northeastern Arizona, the Tsegi drainage system has incised itself deeply within flat-lying, red Triassic and Jurassic sandstones, forming vertical-walled box canyons. A thick alluvial fill has been deposited in the canyon bottoms, and this fill has been deeply incised by two arroyo-cutting events. Following the first of these, which has been dated to about 1300 A.D., a new alluvial fill was deposited within the older arroyo. In 1883, the second arroyo was cut within this younger fill. As a result, there are two alluvial terraces within the Tsegi canyons: an older, higher one, and a younger, lower one (fig. 9.38).

While working as a park ranger in the Tsegi country, I began to suspect that the earlier erosional event might have been caused by a combination of climatic fluctuation and human land abuse. Evidence from tree rings indicated that a prolonged drought had immediately preceded the erosional episode. Nevertheless, tree rings also indicated that the canyons had endured several other such drought episodes before and since without suffering from severe erosion. Anthropological evidence revealed, however, that the population density within the Tsegi canyon system had reached its maximum immediately before the earlier arroyo cutting. This suggested that intensive corn farming might have progressively stripped the alluvial bottom lands of their natural, protective vegetation, an idea supported by evidence of prolonged sheet wash erosion just prior to the arroyo cutting. This would have rendered the alluvial fill unusually susceptible to erosion by heavy cloudburst floods. A prolonged dry spell would have reduced the protective effect of vegetation still further. When arroyo cutting finally occurred, it lowered the water table so drastically that the land was no longer suitable for corn farming, whereupon the inhabitants moved south to the Hopi Mesas in central Arizona.

Evidence for the human origin of the second arroyo cutting episode is even clearer. Topographic maps of the Tsegi area made around 1880 show

**Figure 9.38** Severe stream erosion has occurred twice in the Tsegi canyon system of northeastern Arizona. The prominent, sagebrush-covered terrace was once the canyon floor until flash floods cut a deep arroyo down to bedrock about 1300 A.D. Aggradation then built a new, lower floor of alluvium within the arroyo, but flash flooding occurred again in 1883. All that now remains of this second floor are a few, narrow terrace remnants.
Photo by U.S. National Park Service

evidence of the first gully, but not of the second. In fact, the maps show an abundance of swamps, lagoons, and dense vegetation within the drainage. These undoubtedly were the source of the Tsegi's other name, Laguna Creek. A topographic map made in 1883, however, shows the modern arroyo in place of the swamps, lagoons, and vegetation. Between 1880 and 1883, some drastic event had clearly taken place, but what was it?

In the years prior to 1883, Navajo shepherds had been running increasingly larger flocks of sheep on the Tsegi alluvium. Of all grazing animals, sheep are the most destructive

to rangeland. When Chief Ranger Harold Timmons and I asked some of the older people in the district about the possibility that overgrazing had caused the erosion, however, they flatly denied it. It was most emphatically *not* their fault. Their explanation? *"The canyons were bewitched!"*

Riding our ponies back downcanyon toward Park Service headquarters, Timmons and I pondered this emphatic excuse, and agreed on one thing: if somehow human beings could learn to accept responsibility for their own actions, they just might become willing to learn how to stop changing their homes from paradises into wastelands.

# 10

# The Cold Carver: Ice on the Land

## Outline

*When I revealed to Harris the fact that the passenger part of this glacier—the central part—the lightning—express part, so to speak—was not due in Zermatt till the summer of 2378, and that the baggage, coming along the slow edge, would not arrive until some generations later, he burst out with: "That is European management, all over!"*

*Mark Twain*

***A Tramp Abroad***

Photo: Victoria Glacier near Lake Louise, Alberta, Canada.

**Figure 10.1** An erratic boulder rests in a shallow glacial groove in Devonian schist on a mountain ridge near Gorham, New Hampshire.

## Flood or Freeze? The Meaning of "Drift"

The realization that glaciers have been far more influential in modifying Earth's surface than their present distribution would suggest was slow in coming. Prior to 1795, the traditional wisdom about the presence of enormous, rounded boulders in highly unlikely locations (fig. 10.1) was that they had been dropped from icebergs drifting on the Noachian deluge—the biblical flood. Consequently, such boulders and the sediments associated with them were called **drift,** and the boulders themselves were called **erratics.** As late as the 1830s, no less august a scientist than Charles Darwin confidently explained a deposit of such erratics in Tierra del Fuego, South America, as the product of flood-borne icebergs. Not surprisingly, it was the insightful Scotsman James Hutton who first suggested in 1795 that a more probable source of erratics was nearby mountain glaciers in a greatly expanded condition, but the idea did not catch on. Twenty years later, a Swiss civil engineer named Ignaz Venetz-Sitten (*Ven*-ets-*Zit*-n) paid a visit to a mountain village where the terminus of a local glacier was depositing gravel ridges and scratched and faceted boulders within its valley. A local peasant easily convinced the engineer that identical materials far downvalley had been deposited at some time in the past by that same glacier. Venetz-Sitten was sufficiently impressed with the evidence to conduct further investigations, on the basis of which he proposed in 1829 that all of northern Europe had at one time been covered by glacier ice.

Understandably, this revolutionary idea generated a lot of controversy. Among its opponents was a young Swiss biologist named Louis Agassiz (Ah-gah-see), who went to the French Alps with the intent of disproving the concept, but was soon as impressed with the evidence as Venetz-Sitten had been. With the zeal of a convert, Agassiz then wrote eloquently and prolifically about Venetz-Sitten's "ice age," thereby paving the way for the acceptance of the idea by the geological community. Agassiz's subsequent appointment to a professorship at Harvard University further increased his credibility and lent further strength to the concept of continental glaciation. Today, we have a detailed picture of the growth of great ice sheets over vast northern areas of North America, Europe, and Asia, and their persistence over tens of thousands of years (fig. 10.2). Moreover, we have clear evidence that these ice sheets formed, advanced, retreated, and disappeared at least four times and possibly more during the 2 million years, more or less, of the Pleistocene Epoch. I discuss the causes of this remarkable, cyclic phenomenon in a following section, but first we must consider the processes of glacier growth and motion; glacial erosion, transport, and deposition, and the highly distinctive effects of glaciers on our modern-day landscapes.

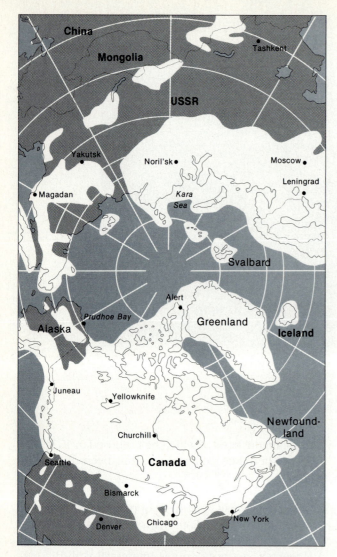

**Figure 10.2** The extent of ice cover in the Northern Hemisphere during the Pleistocene Epoch. In addition to the major continental ice sheets, mountain ice sheets also covered northwestern North America, the Alps, the Caucasus Mountains, the Himalayas, and the east Asian highlands.

Source: "Our Changing Continent," 1986 (181–404–155/40006), U.S. Geological Survey.

## Snow, Firn, and Ice: Glacial Origins

In most places where snow falls, it falls in winter and disappears in spring. In some places, however, notably high mountain regions that are shaded from the Sun or sheltered from wind, or both, some of the snow that falls in winter is able to survive the summer despite the effects of warm rains, warm winds, dust accumulation, and direct solar radiation. Such snow is known as *firn* (German for "of last year").

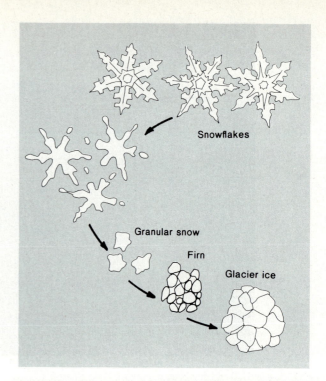

**Figure 10.3** The transition from snowflakes to glacier ice involves the progressive elimination of air space and the welding of dense, rounded firn grains into a compact, crystalline mass.

Firn is quite a different substance from the freshly fallen snow from which it is derived. That snow typically consists of feathery, hexagonal crystals (fig. 10.3), aggregates of which can have a specific gravity as low as 0.05 and a porosity as high as 95%. Over the course of several weeks, snowflakes in fallen snow change from feathery crystals to rounded grains by the evaporation of molecules from their projections and the redeposition of molecules nearer their centers. As this happens, the snow becomes more compact. Grains are pressed together, and at the points of highest pressure, melting occurs. The water thus formed refreezes in voids, thereby reducing the porosity of the snow to about 50% and raising the specific gravity to about 0.5. Most firn is of about this consistency.

If firn accumulates to depths ranging between about 30 m in warmer climates and about 150 m in the polar regions, its lowermost layers are sufficiently compressed to be converted to ice (specific gravity 0.92). If the thickness of the deposit continues to increase, this basal ice layer likewise continues to thicken until it begins to flow downslope under the influence of gravity. At this point, the entire mass of ice and overlying firn and snow is considered to be a **glacier.**

*Jean Louis Rodolphe Agassiz
(1807–1873)*

Seemingly insignificant events in early childhood can often subtly chart the course of a lifetime. Rose Mayor, the wife of the Protestant pastor of Môtier-en-Vully on the shores of Lake Morat, Switzerland, was a lover of Nature. She imparted that passion to her son, Jean Louis, along with the means of communicating it to others. This kind gift to the world probably did more to promote public understanding and appreciation of Nature than any other factor in human history.

Louis Agassiz pursued studies in natural history and medicine at the Universities of Zürich, Heidelberg, and München, for which he received degrees of doctor of philosophy and doctor of medicine in 1830. In the preceding year, at age 22, he published his first major work, entitled *Selecta Genera et Species Piscium* (*Selected Genera and Species of Fish*), the fruits of a study of Brazilian fish that he had inherited three years previously on the death of J. B. Spix, who had made the collection. During that effort, Agassiz developed a lasting passion for the study of fish, which was to manifest itself later in a large number of landmark papers.

At the time of Agassiz's graduation, the Mecca for scholars of natural history was Paris. In 1832, he moved to Paris, where he met several eminent naturalists, among them Baron Cuvier and Baron von Humboldt. The latter secured him a professorship at the university in Neuchâtel, Switzerland, a stone's throw from his home town. He remained there until 1846, gathering a large following of students and associates and setting up his own private publishing business for the circulation of developing concepts in natural history. While at Neuchâtel, he became aware of Venetz-Sitten's ice age hypothesis. To test that idea, he built a hut, which he called the "Hôtel des Neuchâtelois," on the ice of the Aar glacier. From this "hôtel," he and his associates conducted the investigations that convinced Agassiz, and soon the world, that Venetz-Sitten's idea had merit. From this effort came Agassiz's monumental work, *Études Sur les Glaciers* (1840), in which he depicted the vast ice sheet that had formerly covered all of Switzerland and much of its neighboring nations.

In 1846, Agassiz was invited to give a series of lectures for the Lowell Institute in Boston, Massachusetts. The success and popularity of this and another series of lectures in Charleston, South Carolina, earned Agassiz an appointment to the chair of natural history in the Lawrence Scientific School at Harvard University. Thus began an association with Harvard that was to last for the rest of his life. In addition to producing many publications on the zoology and geology of the United States, Agassiz used his Harvard years to design the Museum of Comparative Zoology and to develop a tradition of Nature study that has become the ideal in American schools (albeit seldom attained). Among his outstanding philosophical statements on that tradition are: "If you study Nature in books, when you go out-of-doors, you cannot find her . . . The book of Nature is always open . . . Strive to interpret what really exists."

In the midst of such a shrine to the written word as Harvard University, such a belittling of books might well have been taken as heresy, but true to its motto, *Veritas* (Truth), the university recognized the possibility that in the search for Truth, the written word might not be the last word. It could hardly object, then, when this sensible Swiss turned the eyes of the world toward the picture that is worth a thousand million words: Nature itself.

## Glaciers: Ice in Motion

Flowing ice can be as effective an agent of erosion and deposition as flowing water. This should come as no surprise, as both these fluids have a number of properties in common:

1. They flow faster and more efficiently in channels than in sheets.
2. They flow faster as their thickness (or depth) increases and as the steepness of the slopes on which they flow increases.
3. They pluck loosened Earth materials from the surfaces over which they flow and transport them.
4. They use clastic materials as abrasives to erode underlying surfaces.

## Ice Sheets and Valley Glaciers

There are two fundamentally different kinds of glacier, each with several varieties. The simpler kind is the **continental ice sheet,** of which the great ice masses covering Greenland and Antarctica are modern examples (fig. 10.4). Ice sheets form by the accumulation of snow

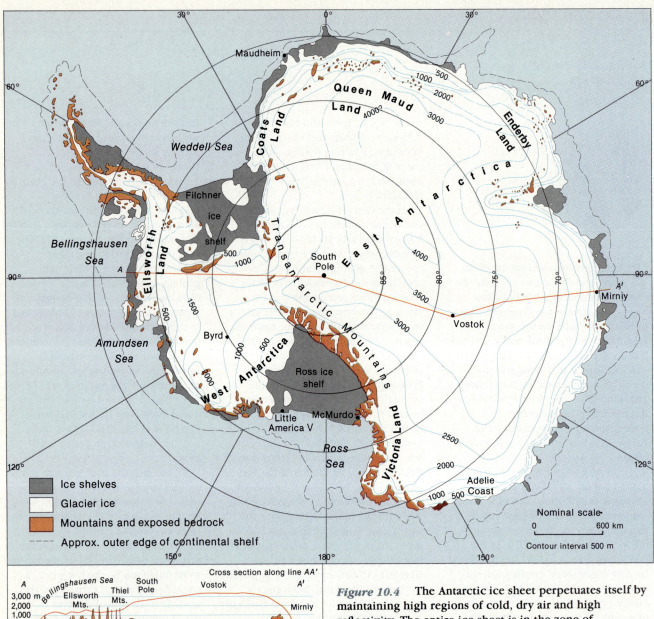

**Figure 10.4** The Antarctic ice sheet perpetuates itself by maintaining high regions of cold, dry air and high reflectivity. The entire ice sheet is in the zone of accumulation (see fig. 10.7). Outlet glaciers pass through mountain barriers to the sea, carving deep fjord valleys. Floating ice shelves form where ice extends beyond the shoreline.

From R. F. Flint, *Glacial and Quaternary Geology*. Copyright © John Wiley & Sons, Inc., New York, NY. Reprinted by permission of John Wiley & Sons, Inc.

in high latitude areas of low to moderate elevation and low topographic relief, and they flow outward from these areas in a manner analogous to the sheet flow of water. They are thickest in their central regions, from which their smooth, flat, and essentially featureless surfaces slope gently away toward the thinner, outer margins. Where those margins overlie mountainous coasts, *outlet glaciers* flow outward from the main ice sheet through mountain passes. These often coalesce and extend beyond the margin of the continent into the ocean, where they are bouyed up by ocean water to form

floating *ice shelves,* such as the Ross and Filchner ice shelves of Antarctica. Large sections of the outer edges of these ice sheets continually break off, producing massive, tabular *icebergs,* which can have surface areas in excess of 30,000 km². Ice shelves and icebergs both illustrate the principle of isostasy (see chapter 1), as they both float within the slightly denser water medium with nine-tenths of their volume submerged.

In contrast to continental ice sheets, elongate ice bodies called **valley glaciers** (fig. 10.5) form within former stream valleys in cold regions of high topo-

**Figure 10.5** Three small valley glaciers flow from a high ridge into a valley in Alaska's Copper River region. Prominent marginal moraines, or till ridges, mark the glaciers' farthest advances, from which they have since receded due to warmer conditions.

Photo by E. Brabb, U.S. Geological Survey

graphic relief. Valley glaciers move by channel flow rather than sheet flow. They occur at all latitudes, but with increasing proximity to the equator they are found at higher altitudes. Those of east Africa's Mount Kenya on the equator, for example, lie at altitudes above 4900 m.

In their early (and late) stages of development, valley glaciers are small, short, and mostly confined to bowl-shaped depressions at the heads of valleys and on high mountain slopes protected from wind (fig. 10.6). These depressions, created by the erosive action of the glaciers they contain, are called **cirques.** As valley glaciers grow, they extend downvalley, and often coalesce with other valley glaciers, progressively burying the mountains within a *mountain ice sheet.*

## Glaciers as Systems

Of all natural equilibrium systems, glaciers are among the most clear cut. Figure 10.7*a* and *b* shows sections through a valley glacier and a continental ice sheet. Notice the position of the **annual snowline,** which is the lower limit of unmelted snow and the upper limit of bare, exposed glacier ice at the end of the summer. Between the annual snowline and the head of a valley glacier, or the center of a continental ice sheet, lies the **zone of accumulation,** marked by a clean, white, snow-covered surface (see also fig. 10.5). Within this zone,

snow accumulates as input, contributing to the storage of snow, firn, and ice within the glacier. Input also occurs in the form of rock material, some of which is eroded by both valley glaciers and ice sheets from their beds, and some of which falls on a valley glacier's surface from adjacent slopes.

A glacier's **terminus** is its outer, or lower, margin. Between the annual snowline and the terminus lies the **zone of ablation** (from Latin *ablatus,* "taken away"), marked by exposed and usually rather dirty ice. Within this zone, ice melts or evaporates, yielding water vapor, glacial meltwater, and rock material as output.

Clearly, inasmuch as ablation can only take place below the snowline, there must be some mechanism that moves ice from the zone of accumulation into the zone of ablation. That mechanism is *glacier flow,* and it is shown by the long arrows, or *flow lines,* in figure 10.7. The flow of valley glacier ice is similar to that of stream flow, as is revealed by a comparison between figures 9.1 and 10.8. In both cases, velocity is highest in mid-channel and drops to near zero at the bed. Furthermore, in both cases, flow velocity increases with both slope angle and the depth of flowing water or ice, although the effects of these two variables are much more pronounced in the case of glacier flow than they are in the case of stream flow.

**Figure 10.6** Prominent basins on the north side of Colorado's Wichoacan (or Mount Sopris) are classic examples of glacial cirques. Ice-filled during glaciations, these cirques presently hold active rock glaciers (see fig. 10.20).

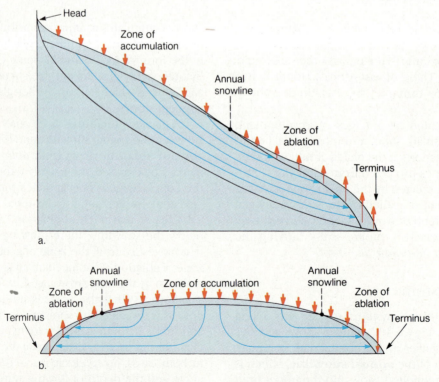

**Figure 10.7** Sections through a valley glacier (*a*) and a continental ice sheet (*b*) show net accumulation (*downward-pointing arrows*) above the annual snowline and net ablation (*upward-pointing arrows*) below it. Flow lines (*long arrows*) show the motion of ice within the glacier. If accumulation equals ablation, the terminus remains stationary. If accumulation exceeds ablation, the terminus advances. If ablation exceeds accumulation, the terminus recedes.

From R. F. Flint, *Glacial and Quaternary Geology.* Copyright © John Wiley & Sons, Inc., New York, NY. Reprinted by permission of John Wiley & Sons, Inc.

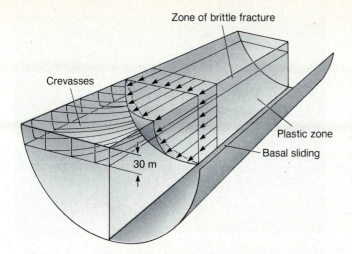

**Figure 10.8** As with stream flow, the velocity of flow in glacier ice increases with distance from bed and banks and with slope angle. At depths below 30 m, ice flows plastically, whereas fracturing and crevasse formation prevail in the overlying, rigid zone of brittle fracture. During surges, glaciers move by basal sliding.

The increase in flow velocity with ice thickness actually places a practical limit on the thickness an ice sheet can attain. Beyond a thickness of about 4000 m, natural snowfall is unable to accumulate fast enough to compensate for the lowering of the glacier surface caused by increased flow rate. Accordingly, even though the great Antarctic ice sheet has over seven times the area of the Greenland ice sheet, the two glaciers have about the same thickness.

In addition to internal flow, a certain amount of *basal sliding* of glacier ice on the underlying bedrock also occurs (fig. 10.8). This type of motion is sporadic and difficult to study quantitatively. It is favored by accumulations of water beneath the glacier and by the buildup of unusual thicknesses of ice upvalley. Taken together, internal flow and basal sliding produce flow rates that vary from 0 to over 120 m/day, although rates of 25 to 50 cm/day are more typical. The highest rates are characteristic of outlet glaciers on the margins of continental ice sheets flowing through steep, narrow passes in bordering mountain ranges. Following periods of relative inactivity, these and other glaciers often experience *surges,* in which a slow wave passes downvalley, thickening the glacier by as much as 60 m as it passes and greatly increasing its rate of movement, both by shearing flow and by basal sliding. The causes of surges are still obscure.

Notice, in figure 10.8, that whereas the lower portion of a glacier is a zone of plastic flow, the upper 30 m or so compose a rigid zone of brittle fracture. Within this brittle upper crust, deep **crevasses** form as a result of the uneven flow (fig. 10.9). Where a glacier flows over a break in slope, its crevasses open widely, often forming a maze of impassable ice towers. When the slope becomes gentler again, most of these crevasses close.

**Figure 10.9** Flowing rapidly down a steep, uneven gradient, the Margerie Glacier in Alaska's Glacier Bay National Monument exhibits a maze of crevasses crisscrossing its brittle upper layer.

Photo by U.S. Geological Survey

## Glacial Budgets: Advances and Retreats

As with any other equilibrium system, the amount of material in storage within a glacier at any given time depends on the rate of input. If input increases, storage and output rate also increase; if input decreases, storage and output rate also decrease. If the total accumulation and the total ablation at the end of the summer (expressed as centimeters of water) are equal in a given year, output was equal to input, and the system was in equilibrium. Under such conditions, the position of the glacier terminus remains fixed.

If, on the other hand, accumulation is significantly greater than ablation, the amount of ice in storage within the glacier increases. Under these circumstances, the glacier's thickness increases, which in turn increases the flow velocity. The net effect of thickening the glacier within the zone of accumulation, therefore, is to *lengthen* the glacier downvalley. Thus, the glacier terminus moves forward in what is known as a **glacial advance.**

Conversely, if ablation is greater than accumulation, the glacier decreases in thickness, resulting in a lower flow velocity and a *shortening* of the glacier upvalley. Such a shortening is called a **glacial recession** rather than a retreat because the ice does *not* actually retreat upvalley! Glacier ice always flows downvalley, but when ablation exceeds accumulation, the rate at which the terminus recedes upvalley exceeds the diminished rate of downvalley flow; hence, it is only the *position* of the terminus that retreats.

**Figure 10.10** A plunging anticline revealed by glacial erosion on an extensive peneplain in the Canadian Arctic. A succession of continental ice sheets has stripped the regolith from the underlying bedrock.

Photo by the Geologic Survey of Canada

## Glacial Erosion, Transport, and Deposition

Glaciers, like streams, are efficient producers, movers, and depositors of clastic sediment. Although there are many similarities between the processes of fluvial erosion, transport, and deposition and their glacial counterparts, there are also many important differences. Let us now explore the most significant of those differences.

### *Glacial Erosion*

The exact mechanisms whereby glaciers erode the surfaces over which they flow are not well known because it is difficult to observe these mechanisms directly. Most hypotheses of glacial erosion are based on studies of glacially eroded ground from which glacier ice has long since ablated. The abundance of fresh, unweathered bedrock exposed in heavily glaciated regions (fig. 10.10) shows that ice sheets have no difficulty in removing soil and saprolite from the surfaces over which they flow. Furthermore, the broad, troughlike forms of glaciated stream valleys (fig. 10.11) show that valley glaciers and mountain ice sheets actively erode the bedrock surfaces

**Figure 10.11** The valley of Lost Lakes Creek, Colorado, shows a troughlike, U-shaped cross profile characteristic of glaciated valleys.

over which they flow. Beyond these general observations, it is clear that the range of glacial erosion is extremely broad. In many localities on the Canadian shield, for example, it appears that glacial erosion was negligible, whereas in some Norwegian fjords (see fig. 12.20b), the apparent depth of erosion exceeds 2400 m!

The following factors are responsible for this wide range in the effectiveness of glacial erosion:

1. Glacier thickness
2. Flow rate
3. The hardness, angularity, coarseness, and abundance of rock tools embedded in the sole of the glacier
4. The relative resistance of the underlying bedrock to erosion by such tools

Other factors being equal, glacial scour is invariably deepest in valleys and mountain passes, within which the ice flow is channeled. This suggests that flow rate is the most influential of the four factors (as it is in stream flow). In the Canadian shield, the lowest rates of erosion correspond with the hard, smooth, sparsely-jointed surface of Pre-Phanerozoic crystalline rocks. In the softer platform sediments of the northeastern United States, however, glacial erosion was locally much greater. The basins of the Great Lakes and those of the Finger Lakes in New York, for example, were scooped out of weak mudstone formations.

The effects of glacial erosion are more evident in hard rocks than in soft. One especially characteristic feature is **stoss and lee topography** (*Stoss* means "push" in German; fig. 10.12), in which the upstream sides of bedrock outcrops, from knobs to mountains, have been abraded and smoothed by the passage of a glacier, whereas the downstream sides have been plucked, or quarried, by the ice so that their surfaces are rough and irregular. Stoss and lee forms are among the most reliable indicators of the direction of glacier

**Figure 10.12** An outcrop of Devonian rhyolite in Baxter State Park, Maine, displays well-formed stoss and lee topography. Smoothly abraded surfaces (*right*) and rough, quarried surfaces (*left*) indicate that glacier flow was from right to left. Quarrying is the more effective of the two erosional processes.

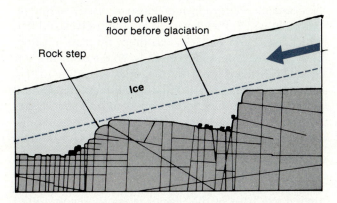

**Figure 10.13** Stairstep profiles in some glaciated valleys result from more effective quarrying of bedrock where joints are closely spaced.

Source: After F. E. Matthes, 1930, U.S. Geological Survey.

**Figure 10.14** Striations on this glacially polished outcrop of Devonian schist at the summit of New Hampshire's Mount Kearsarge indicate that the glacier flowed in a direction roughly parallel with the line of sight in the photograph.

**Figure 10.15** Glacial grooves in Devonian schist near the summit of Mount Kearsarge, west-central New Hampshire.

flow. Careful studies have shown that the volume of rock quarried by glaciers greatly exceeds that which is abraded. Furthermore, quarrying is more effective where joints in the bedrock are closely spaced than where they are far apart. Figure 10.13 shows the longitudinal profile of a glaciated valley in which basins have been eroded from densely jointed bedrock, leaving the more resistant, sparsely jointed bedrock to form steps in between.

On a finer scale, glaciated bedrock shows a number of characteristic features. **Glacial polish** (fig. 10.14) is a smooth—sometimes even mirror-smooth—surface that is best developed on hard, fine-grained rocks by very fine clastic particles embedded in the sole of a glacier. Polished surfaces usually also show **glacial striations** made by clasts larger than those which produced the polish. Where glaciers pass over relatively nonresistant bedrock, such as limestone and mudstone, striations can become enlarged into deep *grooves* (fig. 10.15). Under exceptional circumstances, these grooves can attain

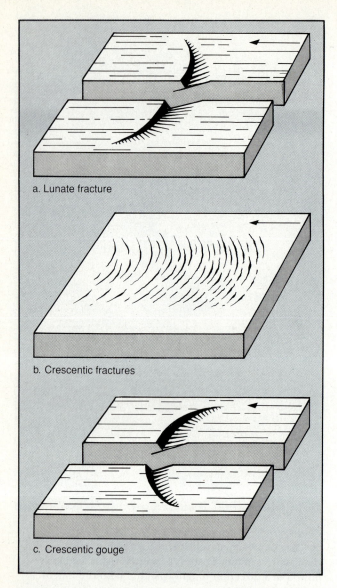

a. Lunate fracture

b. Crescentic fractures

c. Crescentic gouge

**Figure 10.16** Lunate and crescentic fractures (*a, b*) are concave in the direction of glacier motion, whereas crescentic gouges (*c*) are convex. Arrows indicate flow direction.

depths of 30 m and lengths of 1500 m, but few exceed 2 m in depth and 100 m in length. It is thought that glacial grooves grow by repeated chance engravings along, or close to, the same path. As the groove enlarges, ice is forced into it, favoring continued abrasion.

When large clasts, such as cobbles and boulders, are embedded in the sole of a glacier, their passage is often recorded by curved fractures and gouges in the bedrock (fig. 10.16). Experiments conducted with steel balls and glass suggest that fractures that are concave downstream are caused by sliding clasts, whereas those that are convex downstream are caused by rolling clasts.

**Figure 10.17** A spectacular deposit of glacial till in the valley of the Roaring Fork near Aspen, Colorado, illustrates the extreme range of particle sizes, from clay to boulders, found in some tills.

## Glacial Transport

Like streams, valley glaciers transport clastic sediment: both that which they have eroded themselves and that which has been delivered to them by mass wasting. Here, however, the similarity ends. Whereas streams sort their clastic loads into size categories according to flow velocity, glaciers do not. **Till,** the sediment transported by glacier ice, is very poorly sorted, and all size classes are represented in it (fig. 10.17). Instead of being distributed as bed load and suspended load, as it would be in streams, till is carried mainly at the bottom, sides, and top of the flowing ice, and very little of it is actually embedded within the ice (fig. 10.18). That which is carried at the base of the glacier is appropriately called *basal till.* Basal till is deposited under high pressure beneath the sole of a glacier. Accordingly, it is highly compact, and it often contains rounded, polished, striated, crushed, or faceted pebbles, cobbles, and boulders, which are frequently oriented with their long dimensions parallel to the direction of glacier flow.

At the sides of of the glacier, basal till blends into *lateral till.* Talus accumulates on the margins of the glacier from mass wasting of the valley sides, contributing to lateral till. Where two glaciers join, their adjacent lateral till bands blend to form a band of *medial till* along the suture between the two glaciers. Figure 10.18 also shows that if a tributary glacier is much smaller than the main glacier, its base is correspondingly much higher than that of the main glacier. Some large glaciers have so many tributaries that their upper surfaces are candy-striped with medial till bands (fig. 10.19). These surface deposits of both lateral and medial till are usually thick enough that they insulate the underlying ice from solar radiation, thus protecting it against melting. As the cleaner parts of the glacier surface are lowered by ablation, the till ridges increase in height (up to 40 m or more) allowing the talus of the ridges to slide off and

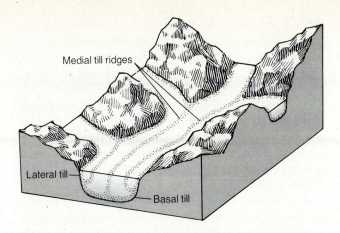

**Figure 10.18** Basal till is carried within and beneath the sole of a glacier and blends sideways into lateral till, the upper part of which consists of rockfall on the glacier's lateral margins. Medial till forms where a lateral till ridge of a tributary glacier blends with the adjacent lateral till ridge of the main glacier. Note that the upper surfaces of all tributaries are level with that of the main glacier regardless of the thickness of the ice in the tributary.

**Figure 10.19** The candystriped surface of Alaska's Barnard Glacier reveals how a new ridge of medial till is added with the entry of each new tributary. Lateral sliding of till has begun to create a blanket of ablation till on the downvalley portions of the ice surface (*bottom*).

Photo by Bradford Washburn/Boston Museum of Science.

spread out, forming a more even cover of till on the surface of the glacier downvalley. This cover is known as *ablation till*. Ablation till is less compact than basal till, and its clasts show few or none of the characteristics listed for basal till. In addition, some of its finer materials have frequently been selectively washed away by rain and meltwater.

Near the terminus, ablation often reduces the thickness of the glacier until it is too thin to flow, so a certain length of the glacier's lower end becomes stagnant. Continued pressure from the actively flowing ice upvalley, however, causes the glacier to thrust itself up and over its stagnant terminus along a closely-spaced set of shear planes. This thrusting brings basal till upward to mix with the till on the top of the glacier. The lower portions of some glaciers are so liberally covered with the resulting mixture of basal and ablation till that to a casual observer, they do not even appear to be glaciers.

A somewhat enigmatic extension of this trend is the **rock glacier,** a thick, lobate deposit of large talus blocks, usually located within a cirque (fig. 10.20). Rock glaciers flow slowly downslope due to the annual freezing and thawing of interstitial ice. Some geologists hypothesize that rock glaciers are the remnants of ice glaciers that stagnated and shrank beneath their own till, but others doubt this.

### Glacial Deposition

Glacial deposits can be divided into two broad categories. **Moraine** (a colloquial French term) is a deposit of till that has been laid down directly by moving ice and that has not been modified by the action of glacial meltwater. Although there is some confusion in usage,

**Figure 10.20** A large rock glacier flows slowly northward from one of the cirques of Wichoacan (see fig. 10.6). Fed by rockfall from the cirque walls, the rock glacier flows under the influence of gravity and the freezing and thawing of interstitial ice.

the term *moraine* should only be applied to glacier-built landforms, whereas the term *till* should be applied to the sediment composing those landforms. In contrast to till, **stratified drift** composes landforms in which the deposit has been more or less washed and sorted by meltwater. Let us consider each of these categories in turn.

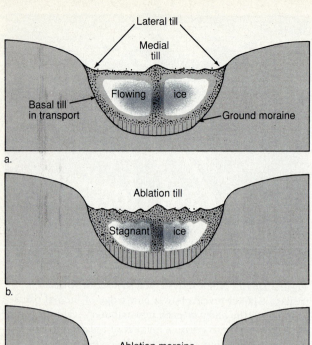

**Figure 10.21** The deposition of a till sheet begins with the laying down of basal till under high pressure beneath a flowing glacier (*a*). As the glacier stagnates and melts, ablation till is laid down on top of the basal till (*b, c*).

From R. F. Flint, *Glacial and Quarternary Geology.* Copyright © John Wiley & Sons, Inc., New York, NY. Reprinted by permission of John Wiley & Sons, Inc.

**Moraine.** There are several varieties of moraine, and each has a different origin (figs. 10.21 to 10.23). **Ground moraine** is an extensive, sheetlike deposit composed mainly of basal till but often with a "frosting" of ablation till on top (fig. 10.21*c*). In many glaciated regions, including much of the northeastern United States, southern Canada, and northern Europe, extensive blankets of ground moraine cover the landscape. These vast blankets vary considerably in thickness. In central Ohio, for example, the average depth of the ground moraine is 29 m, but it thickens to as much as 231 m in buried stream valleys. In the Spokane Valley between Idaho and Washington, the ground moraine is 180 to 400 m thick, whereas in New Hampshire, it averages 10 m in thickness.

Because it is formed mainly beneath glaciers, ground moraine has little topographic relief. Moraines formed at glacier margins, or at glacial sutures, on the other hand, normally stand as ridges above the surrounding terrain (fig. 10.22). **Lateral moraines,** composed of lateral till, often form ridges or terraces along valley sidewalls, whereas **medial moraines,** composed of medial till, typically form ridges that extend downvalley from bedrock spurs at tributary junctions. **Recessional moraines,** composed of ablation till, are crossvalley moraines formed at the glacier terminus under equilibrium conditions when the till-charged ice flowing downvalley is melting at such a rate as to hold the terminus fixed at one position; hence, all the till in transport is dumped at that position. During the course of an extended recession of a glacier, any number of recessional moraines can be built by such stillstands of the terminus. Subsequent readvances of the terminus can override and destroy all recessional moraines as far downvalley as the readvance extends.

A glacier terminus often remains for some time at the position of its farthest downvalley excursion. Accordingly, a moraine ridge formed at this position is called a **terminal moraine.** Often, this is the only kind of cross-valley moraine left by a vanished glacier. Both recessional and terminal moraines frequently blend laterally into lateral moraines.

Because of their geometry, continental ice sheets lack lateral moraines, but they do have all the other varieties of moraine already mentioned with the exception of the medial moraine. The latter is replaced by its near-equivalent, the **interlobate moraine,** shown, along with other features of a retreating ice sheet margin, in figure 10.23. Near the right margin of this figure, another distinctive landform appears that is often produced in ground moraine beneath continental ice sheets. Long thought to be the products of ice flow, but recently reinterpreted as the products of the sudden escape of meltwater from beneath stagnant glaciers, these low, streamlined, elongated hills, called **drumlins,** can consist entirely of till, or they can contain a core of bedrock at their upstream ends. A field of drumlins is shown in figure 10.24.

**Stratified Drift.** As its name indicates, stratified drift usually exhibits a more or less stratified structure, but this can be rather rudimentary. The most significant feature of stratified drift is that it has been sorted, however crudely, by the action of flowing water. The clasts composing the drift tend to be large close to the glacier terminus or where slopes are steep, discharges are high, or both. Clasts tend to be small far from the terminus or where slopes are gentle, discharges are low, or both. The more extensive the reworking of the drift by flowing water, the better the sorting (compare figs. 10.25*a* and 10.25*b*). As you might expect, stratified drift is laid down more abundantly during glacial retreats, when meltwater is plentiful, whereas moraine is laid down more abundantly during glacial advances.

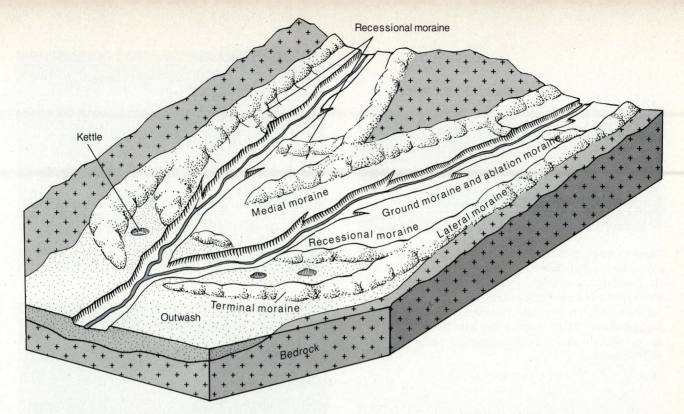

**Figure 10.22** Depositional landforms laid down by two merging valley glaciers. Two lateral moraines have coalesced into a medial moraine (*center*). A terminal moraine marks the farthest downvalley position of the terminus. Two recessional moraines mark stillstands during recession. Ground moraine blankets the valley floor, and outwash has filled the valley below the first recessional moraine. Kettles have formed in the outwash behind the terminal moraine.

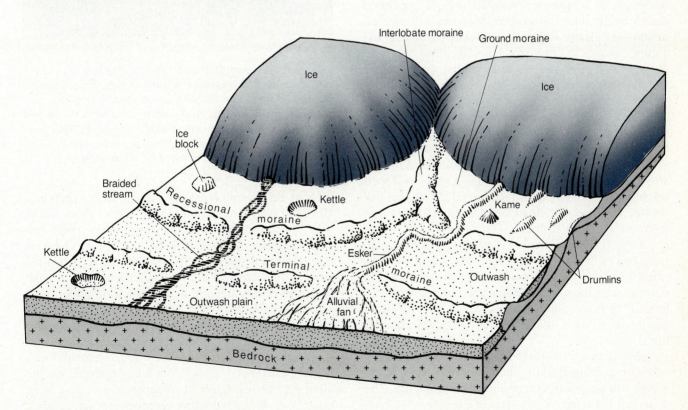

**Figure 10.23** Depositional landforms at the terminus of a continental ice sheet. Compare with figure 10.22.

**Figure 10.24** Fields of drumlins, such as this one in Saskatchewan Province, Canada, have long been interpreted as having been formed as flowing continental ice shaped the underlying ground moraine into streamlined forms. Recently, however, it has been suggested that these features might have been formed by catastrophic floods released suddenly from beneath a melting glacier.

Photo by W. G. Pierce, U.S. Geological Survey

There are two kinds of stratified drift. *Ice-contact stratified drift* is, as the name indicates, laid down in contact with glacier ice. **Glacial outwash,** on the other hand, is deposited downvalley from the glacier terminus by meltwater streams (see figs. 10.22 and 10.23). Inasmuch as ice-contact stratified drift is deposited mainly by irregular and sporadic flow events during the melting of a glacier, it typically exhibits poor sorting and abrupt changes in dominant grain size from one layer to the next (fig. 10.25*a*). Glacial outwash, on the other hand, is deposited by relatively constant streamflow; therefore, it is much better sorted (fig. 10.25*b*).

Figure 10.23 shows the more common forms assumed by stratified drift in the vicinity of the stagnant, wasting terminus of a continental ice sheet and the processes that create them. Downslope from the terminus and the outermost terminal moraine, a well-developed **outwash plain** has been formed by braided streams that emerge from ice tunnels in the melting glacier as well as from the glacier surface.

While the glacier margin is melting, detached, outlying pieces of it may persist for a while as ice blocks (fig. 10.23 *left*). These become progressively buried in outwash and ablation till. When the blocks melt, they form deep holes called **kettles,** which fill with water and become ponds in regions where the water table is high.

Meltwater streams often flow in ice tunnels beneath or within melting glaciers. There they deposit sand and gravel, and lesser amounts of silt and boulders. After the surrounding ice melts this sediment is preserved as long,

a.

b.

**Figure 10.25** Stratified drift shows the effects of reworking by flowing water. A kame terrace deposit near Aspen, Colorado (*a*), displays crude stratification and poor sorting. A sand pit on the shore of Cross Lake, Maine (*b*) exposes glacial outwash in a lake delta that displays much better sorting, smaller grain size, and cross-stratification. Flow direction was from left to right.

sinuous ridges called **eskers** (Irish, *eiscir:* ridge; figs. 10.23 and 10.26). As the containing walls of its ice tunnel melt, the sediments on the margins of an esker collapse. Because of this, the strata on the sides of eskers often slope steeply away from the horizontally bedded strata of the center of the deposit. Eskers have a range in height of about 2–200 m, in breadth of about 5 m–3 km, and in length of about 100 m–500 km. In general, eskers follow topographic depressions, but sometimes they rise as much as 250 m over drainage divides.

Commonly, depressions between the margin of a melting glacier and the adjacent valley sidewall become filled with stratified drift. Such deposits, which are often quite large, are called **kame terraces** (see fig. 10.25*a*). In glaciated regions, kame terraces are important commercial sources of sand and gravel. *Kames* are small

*Figure 10.26* A large esker snakes across the barren peneplain of the Canadian shield.

Photo by the Geological Survey of Canada

*Figure 10.27* Varved clays are deposited in ice-marginal lakes. Each varve consists of a light, silty, summer layer overlain by a dark, clay-rich, winter layer.

Photo by E. Brabb, U.S. Geological Survey

bodies of crudely stratified drift deposited within sink-holes in glacier ice. When the glacier melts, the kame is left as an isolated hillock standing on the underlying ground moraine (see fig. 10.23, *right*).

Temporary lakes often form at an ice margin as it recedes. As with all lakes, these accumulate deposits of clastic sediments in the forms of small deltas (fig. 10.25*b*) and fine, laminated silts and clays, which typically show an alternation of light and dark layers (fig. 10.27). Each pair of such layers is called a **varve,** and the sediments composing it are called *varved clays.* Varves record seasonal cycles in which coarser, lighter sediment (chiefly silt) is deposited relatively rapidly in spring and summer following high runoff events; and then finer, darker sediment (chiefly clay) settles out at a much slower rate during the low runoff season of winter. Because varves are expressions of an annual cycle, they can be used as a dating tool. In Sweden, Gerhard de Geer developed a varve chronology that extends back from 1900 A.D. for almost 17,000 years. Of this span, the last 12,000 years are considered reliable. The chronology was made by correlating the varve sequences of lakes whose life spans overlapped. As long as de Geer compared lakes that are separated by a distance of no more than 1 km, he found that the pattern of wide and narrow varves matched quite well from one lake to another.

## The Evolution of Glaciated Landscapes

Many of the erosional and depositional effects already mentioned are distinctive and can easily be recognized and interpreted by anyone who is aware of Venetz-Sitten's theory of glaciation. There are many other, broader-scale effects that are not as immediately obvious, however, yet they have profoundly influenced the land-scapes on which they operated. Let us turn now to these effects, considering first those of continental glaciation and then those of alpine glaciation.

## Effects of Continental Glaciation

Continental ice sheets develop mainly on subdued topography and spread by inefficient sheet flow. Because of this, their modifications of the landscape are relatively modest, except where easily erodible rocks, such as shale, crop out at the surface. A chain of large lakes, including the Great Lakes, (see fig. 6.24) was produced when ice sheets eroded the edges of the soft, midcontinental platform strata where they lap onto the hard, crystalline rocks of the Canadian shield. One might think that where ice sheets move into regions of high relief, such as New England, the mountains would be severely worn down by the overriding ice, but the evidence does not support such a conclusion. Rather, the ice tends to mold itself around mountain masses and flow through the valleys between them. The erosive power of ice sheets is greatly enhanced within these valleys because of increased flow rates. Because of this, the overall relief of mountainous terrain is actually often increased during continental glaciation.

In both shield and mountain areas of the world, continental glaciation has deposited extensive sheets of ground moraine. Careful field studies in the northeastern and north-central United States and Europe have revealed that in most localities there are several such till sheets superimposed on one another (fig. 10.28). Before the end of the last century, it was realized that each of these till sheets was laid down by a different

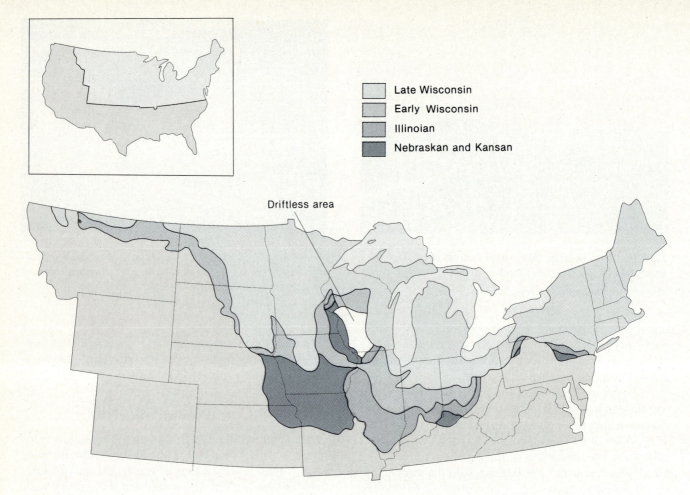

Late Wisconsin
Early Wisconsin
Illinoian
Nebraskan and Kansan

Driftless area

**Figure 10.28** Drift borders of different ages show the extent of Pleistocene glaciation in North America. Wisconsin drift, with two separate advances, is youngest.

The "Driftless area" in southwestern Wisconsin results from a high bedrock plateau in northeastern Wisconsin that diverted ice flow to the west and south.

**Figure 10.29** Hummocky, boulder-strewn terrain north of Mount Wichoacan, Colorado, is typical of marginal moraine.

continental ice sheet during a series of successive episodes of glaciation that were separated by warmer, interglacial episodes. The most recent of these glacial episodes, the *Wisconsin,* began some 70,000 years ago and ended about 10,000 years ago. At least four major glaciations have been identified by till sheets, but there might have been many more since the beginning of the Pleistocene Epoch about 2 million years ago. There might even have been others before that, and we have little reason to hope that there will be no more in the future.

In the interior platform of the north-central United States, the topography developed on ground moraine is usually quite subdued. Except for the local occurrence of drumlins and broad, shallow, parallel grooves sometimes visible from the air, but seldom from the ground, the country is typically flat. Where marginal moraines underlie the surface, however, the topography becomes hummocky, with rolling and often stony, hills interspersed with hollows (fig. 10.29). This "knob and kettle" topography is a favorite setting for English country estates and their American counterparts.

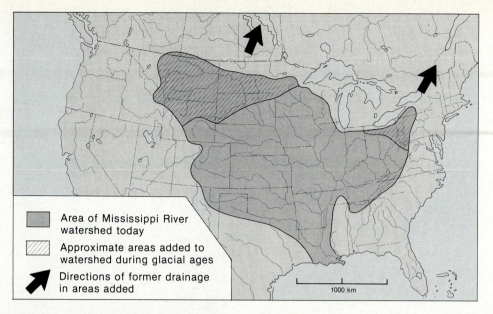

Area of Mississippi River watershed today

Approximate areas added to watershed during glacial ages

Directions of former drainage in areas added

1000 km

**Figure 10.30** Before the Pleistocene glaciations, the headwaters of the Missouri and Ohio Rivers drained northward to Hudson Bay and northeastward to the St. Lawrence River, respectively. Glaciation redirected the flow of these areas (*lined*) southward into the Mississippi River and deranged Canada's formerly well-integrated drainage system.

From R. F. Flint, *Glacial and Quaternary Geology*. Copyright © John Wiley & Sons, Inc., New York, NY. Reprinted by permission of John Wiley & Sons, Inc.

The deposition of ground and marginal moraines by southward spreading ice sheets has significantly altered the preglacial drainage system of the midcontinental United States. Figure 10.30 shows the extent to which glacial advances have deflected formerly northward-flowing drainage networks southward into the Mississippi River system.

## Crustal Subsidence and Rebound

When continental crust is loaded with a 3–4 km thickness of glacier ice, it subsides by isostasy (see chapter 1) to accommodate the increased load. This occurs because an amount of rock material equal in weight to the ice load flows slowly away from beneath the ice-covered area. When the ice load melts, this displaced rock material gradually flows back into place, and the crust slowly rises again in a process called **isostatic rebound.**

Studies of raised beaches in Europe and North America have revealed the amount of rebound that has occurred since the melting of the Wisconsin continental ice sheets. The contours in figure 10.31 represent uplift of northeastern North America in meters above the sea level of 6000 years ago. Notice that there are four centers of maximum rebound, two of which exceed 100 m. These two centers probably correspond to the areas of maximum ice thickness, and presumably, because of this they were the last to become ice-free. By the same reasoning, they were probably also the first to accumulate ice at the beginning of the Wisconsin glacial episode.

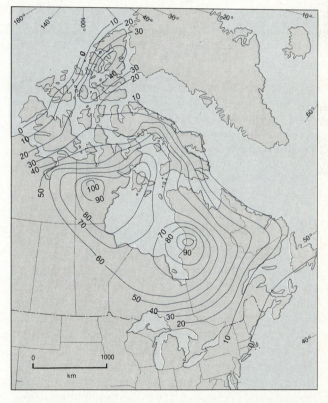

km

**Figure 10.31** Contours of equal rebound in meters reveal three principal centers of uplift in Canada. Those exceeding 100 m probably represent areas both of maximum ice thickness and of earliest growth of ice sheets.

From R. F. Flint, *Glacial and Quaternary Geology*. Copyright © John Wiley & Sons, Inc., New York, NY. Reprinted by permission of John Wiley & Sons, Inc.

The Cold Carver: Ice on the Land    **269**

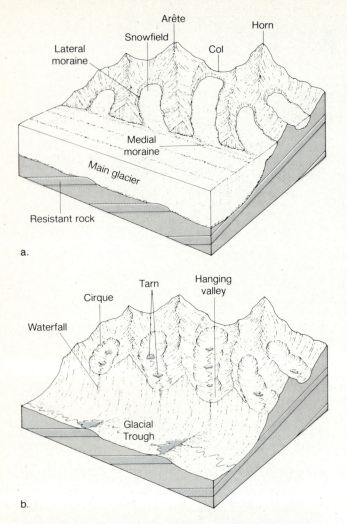

a.

b.

**Figure 10.32** The dominant effects of valley glaciation in a fluvial mountain landscape are to widen and deepen the valleys, to sharpen and truncate interstream ridges, and to produce cirques on northeast slopes (southeast in the Southern Hemisphere).

## Effects of Mountain Glaciation

When glaciation occurs in previously unglaciated mountain terrain, the general effects are a deepening and widening of stream valleys and a narrowing and sharpening of peaks and ridges between streams. Figure 10.32 shows a mountain landscape that has undergone at least three successive cycles of mountain glaciation, each of which was slightly less extensive than the previous one, as is attested by the nesting of smaller glaciated valleys within larger ones. The most obvious characteristic of these valleys is their troughlike shape. Glacial abrasion broadens the bottoms of fluvial valleys into this trough shape, oversteepens the valley sides, and removes support from their bases. This results in rapid mass wasting, which sharpens the overlying ridges into jagged, knifelike **arêtes** (a colloquial French term). Often, the lower ends of arêtes become cut off (as here) as the trough of the main glacier is enlarged by glacial

**Figure 10.33** The Grand Teton, or Mount Hayden, viewed across a small lake contained by the terminal moraine of one of the vanished valley glaciers that carved the peak into a horn.

abrasion. Thus modified, arêtes are known as *truncated spurs.* Following deglaciation, steep alluvial fans and talus aprons soon form on glacially oversteepened valley sidewalls, and these in time restore the former V-shaped cross-profiles of typical alluvial valleys.

Mountain peaks are similarly oversteepened and sharpened into **horns** (a colloquial German term) by valley glaciers, three or more of which often converge upon a peak, eroding their way headward in the same manner as tributary streams. The Swiss Matterhorn is a well-known example of a horn and the Grand Teton in Wyoming (fig. 10.33) is another. Notches, or passes, between adjacent horns are known as *cols* (another colloquial French term).

Recall that the bases of tributary glaciers lie at higher levels than the base of the main glacier (see fig. 10.18). Following deglaciation (fig. 10.32*b*), it can be seen that this difference has produced **hanging valleys,** from which tributary streams now enter the main stream by means of waterfalls and cascades. Both these tributaries and the main stream are *underfit,* meaning that they appear to be too small to have carved the valleys within which they flow. The basins between bedrock steps in the valley bottoms (see fig. 10.13) are often occupied by small lakes, and the deeply scoured floors of the cirques also contain small lakes, called **tarns** (a colloquial Icelandic term).

The depositional features of valley glaciers are summarized in figure 10.22. In common with the deposits of continental ice sheets, these include terminal, recessional, lateral, medial, and ground moraines, kame terraces, and kettles. Drumlins, kames, and eskers, however, are normally absent from glaciated valleys, probably because valley glaciers, with their steep gradients, normally maintain downvalley flow, and with it erosion, until they waste away. Continental ice sheets, in contrast, because of their extremely slight gradients, normally stagnate as they melt, thereby preventing the erosion of such features.

## Periglacial and Transglacial Phenomena

The effects of glaciation are far from limited to the glaciated region itself. Within the **periglacial zone** immediately adjacent to glacier margins, various phenomena characteristic of frozen and unvegetated ground occur. Beyond this expectable margin of influence, however, major changes occur during glaciations that suggest, by their breadth and scope, that few areas of the world (with the probable exception of the highly stable tropics) remain unaffected by the process of glaciation. There is, at present, no name for the region beyond the periglacial zone within which the effects of continental glaciation are felt, nor is there a general, categorical name for the effects themselves, but I shall use the term **transglacial** in that sense for the purposes of this text.

### Periglacial Phenomena

Along the drift border that marks the maximum advance of the Wisconsin ice sheet in North America (see fig. 10.28), various frost action effects are evident within a band that varies in width from a few tens of kilometers in lowland areas to more than 600 km in places in the Appalachian Mountains. These effects include block fields, patterned and frost-cracked ground, and solifluction.

As I note in chapter 8, high-frequency cycles of freezing and thawing create the intense frost wedging conditions under which block fields form. **Patterned ground** (fig. 10.34) is another effect of freeze-thaw cycles. Because solid rock conducts heat about ten times more efficiently than loose saprolite or soil, large stones cool more rapidly than the finer materials that surround them. Consequently, groundwater first freezes on the underside of such stones. The resulting ice grows at the

a.

b.

*Figure 10.34* The formation of stone rings or polygons (*a*) is thought to involve the migration of frost heaved stones into a network of cracks caused by the shrinkage of frozen soil in periglacial regions. On sloping ground (*b*), these polygons become stretched out downhill due to solifluction.

expense of soil water, causing the stone to be moved upward in a process known as *frost heaving.* In time, a layer of such stones forms at the ground surface. These stones gradually become distributed into rings or polygons, especially where the surface is underlain at shallow depths by **permafrost,** or permanently frozen ground. The reason for this is not entirely clear, but it is probably due to the shrinking and cracking of the soil, either as a result of drying in summer or sudden chilling in winter, or both, resulting in the development of a fairly regularly spaced network of cracks.

In permafrost regions, such cracks often fill with *ice wedges,* which can grow to widths of up to 5 m at the top and to depths of as much as 10 m. Because ice wedges are unable to grow if they melt in summer, their presence (or evidence of their former presence) indicates that the ground was underlain by permafrost at the time of their formation. Another piece of evidence for underlying permafrost is solifluction, discussed in chapter 8. In places, solifluction has moved several meters of soil and saprolite down slopes as gentle as 2° in the periglacial zone.

## Transglacial Phenomena

Under ice age conditions, global marine and atmospheric circulation patterns are greatly altered. As a result, many things happen, even in places far from the glacier margins, that do not happen in those places under normal conditions. One important effect of piling ice on the land is that the level of the sea, from which the ice was derived, drops. Fossil shells of organisms that live only at sea level have been found in growth positions at various depths on the continental shelves. Radiocarbon dating of these shells has shown that sea level dropped about 100 m and then rose again over the past 30,000 years, as the great ice sheets of the second phase of the Wisconsin glacial age first grew and then melted.

During such low stands of the world ocean, vast expanses of the continental shelves are laid bare and are quickly invaded by plant and animal communities, including human beings. Evidence from submerged beach sand deposits, radiocarbon dated organic remains, and bones of extinct elephants indicates that during the Wisconsin glacial maximum, sea level dropped enough to expose practically all of the continental shelf. Perhaps the Atlantis fable is a folk memory of the loss of such borrowed land when it was reclaimed by the sea. If the ice in existing glaciers were to melt, sea level would continue to rise, perhaps by as much as 35 m. This would be inconvenient for a human civilization that has had sufficient faith in the constancy of sea level to build hundreds of major cities at the existing shoreline.

Although sea level is lowered during a glacial episode, the opposite effect occurs in water bodies of many land areas. Figure 10.35 shows the outlines of the large

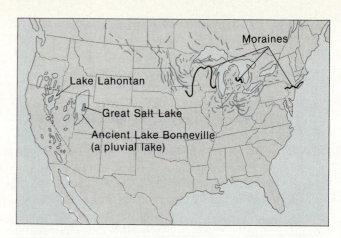

***Figure 10.35*** Some glacial and pluvial features of the United States.

Sources: After C. S. Denny, U.S. Geological Survey, the Geological Map of North America, the Geological Society of America, and The Geological Survey of Canada.

**pluvial lakes** (from Latin, *pluvialis,* "of rain") that formed in the western United States during the Wisconsin glacial age. Note that the area shown is one of internal drainage (see fig. 9.7). Having no outlets to the sea, the topographic basins in this area fill with water when the climate changes from its normally arid state to a more humid regime. That such a change occurred during the Wisconsin is shown not only by the extent of the pluvial lakes but also by evidence of much lusher vegetation, remains of which have been found in (among other places) the fossilized nests of pack rats throughout this region.

There are two likely reasons for this more humid, pluvial regime in western North America. First the polar jet stream appears to have been brought southward by an expansion of the total volume of cold, polar air over the glaciated north. Second, the cold California current (see fig. 12.11) was apparently replaced by a warm surface current, which would have supplied more moisture to the air flowing eastward across the North American continent. With the melting of the continental ice sheets, the climate returned to its present circulation patterns, and the pluvial lakes dried up except for a few highly saline remnants. The largest and best known of these is the Great Salt Lake in northern Utah. The surrounding landscape retains traces of the vanished pluvial lakes in the form of dry lake beds, such as the Bonneville salt flats, and such shoreline features as beaches, wave-cut platforms, spits, and deltas (see chapter 12).

Both loess and windblown sand are abundant in peri- and transglacial environments and are derived from both till and outwash, but primarily from the latter. Figure 10.36 shows the distribution of loess and windblown sand in the United States. There is a close genetic relationship between these deposits and local

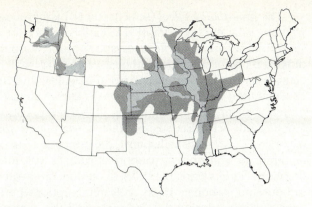

**Figure 10.36** Derivation of loess and aeolian sand from glacial outwash deposits is suggested by their relation to the drainage systems of the Mississippi, Missouri, and Columbia rivers.

Sources: U.S. Bureau of Reclamation, 1960, and various others.

drainage systems and the coarser material is concentrated around the headwaters of the drainages. The location of the sediments on the eastern sides of the drainage channels reflects the prevailing westerly wind direction.

The presence of so much windblown material in this region raises the question whether, unlike the far west, the midcontinent region was arid during glaciation. Studies of fossil pollen from this region suggest, however, that the climate was considerably wetter than it is today, as it apparently supported forests of spruce, birch, alder, and tamarack. It seems, then, that the presence of abundant outwash in the drainages was all that was needed to generate so much aeolian sediment even under a more humid climate.

## The Causes of Glaciation

Although we know a great deal about *what* has happened during the Peistocene glacials and interglacials, we still do not know for certain *why* it happened. We have a number of facts, or boundary conditions, however:

1. From the evidence of tillites and glacial striations in the stratigraphic record we know that major earlier glaciations occurred on Earth in the Early Proterozoic, the Ediacarian, the Late Ordovician, and the Late Paleozoic.
2. From the evidence of fossil plants and temperature-sensitive marine shells we know that global climate has steadily cooled by about 15° C over the past 50 million years.
3. We have been able to trace the gradual build-up of glacier ice on the Antarctic continent starting as long as 36 million years ago.
4. We know that within the past 2 million years or so (the Pleistocene Epoch) there have been at least six climatic fluctuations that have resulted in the repeating glacial-interglacial cycle.

5. Study of gases trapped in the ice of the Greenland and Antarctic ice sheets has revealed that during the last ice age, global temperatures were up to 10° C cooler than they are today, and that the atmosphere contained about one-third less carbon dioxide, about 50% less methane, and about 30 times more dust than it does today.

Any hypothesis that attempts to explain the occurrence of ice ages must take all these facts into account. There are at least ten current hypotheses that do this, but to date, none has won general acceptance, partly because of the large number of possible causative factors and partly because of the lack of adequate evidence regarding the behavior of many of these factors in the past. The principal factors evoked so far are:

1. Long-term changes and cyclic variations in the Sun's energy output.
2. Long-term changes and cyclic variations in Earth's reflectivity.
3. Horizontal and vertical displacements of Earth's crust (mountain building, broad uplifts, and continental drift).
4. Long-term changes and cyclic variations in the greenhouse effect.
5. Cyclic variations in Earth's orbital motions.
6. Changes in the patterns of atmospheric and oceanic circulation.

Before attempting to evaluate each of these six factors as to its probable influence on glaciation, I should emphasize that there is not just one problem to be solved, but two: the gradual, global cooling trend since the Cretaceous Period, and the phenomenon of cyclic glaciation. These problems are quite different; hence, it is likely that at least two causative factors are responsible and possibly many more. Let us consider the long-term cooling trend first and then the pattern of glacial-interglacial fluctuation.

## Global Cooling

Variations in the Sun's energy output are appealing as an explanation for global cooling because of the obvious, simple, and direct effect they would appear to have on global temperature. It seems logical that a long-term decrease in solar energy output should result in a cooling trend on Earth. We know that the value of that output has fluctuated by as much as 1.5% over the past several decades. Unfortunately, however, we have no basis for reconstructing the variations in solar energy output over geologic time; therefore, the hypothesis is untestable. Furthermore, in theory, the Sun should actually be getting hotter over time; hence, its energy production rate should have increased (see chapter 18). Finally, there are theoretical reasons to suspect that variations in solar energy input to Earth are effectively countered by variations in the greenhouse effect.

An increase in reflectivity would serve to cool the planet by reflecting more incoming solar energy. This could come about in a number of ways: an increase in cloud cover; an increase in atmospheric dust from volcanic eruptions, the growth of deserts, forest fires, or Earth's passage through interstellar dust clouds; or from an increase in the ratio of land area to ocean (land surfaces are two to three times more reflective than ocean water). The volcanic dust source seems unpromising, as we know that both global volcanic activity and its cause, seafloor spreading, have *decreased* during the Cenozoic. The proliferation of deserts and fire is likewise unsupported in the geologic record. An increase in cloud cover is also an unlikely cause of global cooling because in theory, it should result from *increased* global temperature (favoring evaporation). Interstellar dust, like a decrease in solar output, would interfere with the transmission of solar energy to Earth, but it also has the same problem of untestability.

Of all the reflectivity factors, the least problematic is an increase in the ratio of land area to ocean. A slowing of seafloor spreading rates since the Cretaceous has resulted in a generally cooler, and therefore lower, seafloor, which has resulted in a general lowering of sea level. Further lowering of sea level over the past 15 million years has occurred because of the withdrawal of the water that is now locked into the Greenland and Antarctic ice sheets. Together, these effects have significantly increased the ratio of land area to sea.

Another consequence of seafloor spreading has been the drifting of a large continent, Antarctica, over the South Pole about 90 million years ago. This excluded warm ocean currents from the south polar region and created a large, ice-covered area with extremely high reflectivity. We know that during each of the Phanerozoic ice ages, a glaciated continental landmass has been similarly situated over the South Pole.

During the Late Ordovician and Permo-Carboniferous ice ages, sea level was relatively high, and glaciation was apparently restricted to the south polar continent. During the Ediacarian Period, however, the continents stood well above sea level, as they do today, and *widespread* global glaciations occurred. This strongly suggests that *lowered sea level* is an important factor in the initiation of glaciation outside the polar regions. The main reason for this is that air temperature decreases with altitude, and moist air blowing inland from the ocean is more likely to condense if it is forced to rise onto a highland. Furthermore, when precipitation occurs on a highland, it is more likely to do so in the form of snow. A second reason is that oceanic circulation is more restricted during times of lowered sea level, and this increases the opportunity for large, polar water bodies, such as the Arctic Ocean, to become isolated from warm, tropical currents.

The *greenhouse effect* (see chapter 13) is a process whereby solar energy is trapped at Earth's surface for a short time before it is able to escape to space. Carbon dioxide is one of several greenhouse gases (all of which contain more than two atoms per molecule), and its concentration in Earth's atmosphere is known to have played a major role in regulating global temperature (some Earth scientists suspect that it has played *the* major role). As I note in chapter 1, the principal source, or input, of atmospheric carbon dioxide is volcanic eruptions, and its principal sinks, or outputs, are hydrocarbons and carbonate rocks. Volcanic eruptions have been declining during the Cenozoic with the general slowing of seafloor spreading and subduction rates. If there has been no change in the rate at which carbon is stored in hydrocarbons and carbonate rocks, the concentration of carbon dioxide in the atmosphere should have gone down, resulting in a lowering of global temperature.

The rate of carbon storage might not have held steady, however. Recently, Tyler Volk of New York University has suggested that the rate of carbon dioxide removal by organisms might actually have *increased* during the Cenozoic, further intensifying the cooling effect! In chapter 2, I mention the hypothesis that the first of Earth's ice ages was caused by the sudden proliferation of carbon dioxide-consuming cyanobacteria. Volk's hypothesis ascribes a similar role to the evolution of *angiosperms,* or flowering plants, including most deciduous trees. He bases his concept on the fact that bedrock weathers some two to four times faster under angiosperm forests than under gymnosperm (or "conifer") forests. This is because angiosperm roots deliver carbon dioxide to soil about ten times faster than gymnosperm roots do. Therefore, when angiosperms began to proliferate, weathering increased, and bicarbonate ions were released and delivered to the oceans at an accelerated rate, whereupon marine photosynthesis and calcium carbonate secretion and deposition also increased (see fig. 1.22). This accelerated drain on atmospheric carbon dioxide resulted in a decrease in the greenhouse effect and, according to Volk's computer model, a global temperature decrease of up to 12° C, more than enough to cause continental glaciation.

Beyond suggesting a possible mechanism for the onset of glaciation, Volk's model also suggests that it would be wise to be cautious about cutting large tracts of angiosperms, as is happening worldwide on an accelerating schedule, because these trees appear to be one of the planet's principal defense mechanism against overheating due to the greenhouse effect.

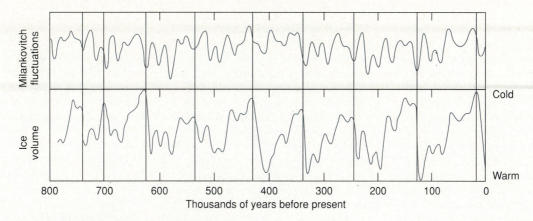

**Figure 10.37** A curve representing cyclic variations in Earth's orbital motions (*top*) corresponds closely to a curve representing ice volume estimated from fluctuations in sea surface temperatures estimated in turn, from the ratio of heavy oxygen to normal oxygen in shells of foraminifera found in dated deep sea cores (*bottom*). Vertical lines indicate ends of glacial ages.

## Cycles of Glaciation

As we have seen, long-term cooling can be adequately explained by the combination of a south polar continent, increased land area, lowered sea level, decreased volcanism, and angiosperm evolution, but we still need to identify some mechanism that can produce glacial-interglacial cycles.

The mechanism that seems to be most favored by glacial geologists today is one that was first proposed in the 1860s by James Croll, a university janitor in Glasgow, Scotland. In Croll's model, glacial-interglacial cycles are caused by a complex system of cyclic variations in Earth's orbital motions. First described in 1920 by a Yugoslavian meteorologist named Milutin Milankovitch, that system comprises three components:

1. Variation in the eccentricity of Earth's orbit (93,000-yr cycle)
2. The *precession*, or cyclic wandering, of Earth's axis (25,800-yr cycle)
3. Variation in Earth's axial tilt (41,000-yr cycle)

In the upper part of figure 10.37, these three components have been combined into a single curve that shows deviations from Earth's mean orbital motions for the past 800,000 years. The lower curve in this same figure shows fluctuations in the total volume of glacier ice over the same interval. It is a reconstruction based on the ratio of a heavy oxygen isotope ($^{18}O$) to normal oxygen ($^{16}O$) in microfossil shells found in deep sea drill cores. This ratio is high during glacial ages, when the temperature of the sea surface (where the organisms live) is low, and low when sea surface temperature

is high. Comparison of the curves with respect to the vertical lines, which represent glacial maxima, reveals two significant trends:

1. Both curves display a roughly 100,000-year periodicity that is well-correlated between the two curves. The periodicity drops to about 41,000 years prior to 600,000 years ago.
2. Deglaciations were abrupt, and they frequently corresponded to sudden increases in amplitude of Milankovitch fluctuations.

Proponents of the Croll/Milankovitch model have argued that these fluctuations would affect Earth's climate by causing corresponding fluctuations in *insolation,* or the rate at which solar energy is received by the Earth system (see chapter 13). The strong correlations between the two curves support the model, but the model fails to account for the marked drop in atmospheric carbon dioxide and methane and the rise in dust during glaciations, noted earlier. It also fails to explain why the growth and shrinking of glaciers has always occurred *evenly and simultaneously worldwide.* Inasmuch as seasonal insolation effects are reversed in the Northern and Southern Hemispheres, this is a serious, if not fatal, flaw in the model. Probably, as so often happens in Nature, the answer lies in a combination of two or more effects. In this chapter's Food for Thought section, I explore one possible candidate for the additional effect.

## Food for Thought

*I*n spite of nearly a century and a half of speculation about the cause of glacial-interglacial cycles, that speculation still constitutes one of the great unanswered questions in Earth science. In this chapter, I note the recent discovery of evidence in deep ice cores from the Greenland and Antarctic ice sheets showing marked increases in atmospheric carbon dioxide and methane accompanying the abrupt rise in temperature at the transitions between glacial and interglacial conditions. That evidence should help to clarify the question, but to many researchers, it is less than welcome because it seems to make little sense in terms of the currently favored Croll-Milankovitch model of cyclic glaciation. According to that model, the rise in temperature at the glacial-interglacial transitions should be due chiefly to increases in insolation. The discovery that the temperature rises correspond to increases in the concentration of atmospheric carbon dioxide raises the possibility that interglacial conditions might have been brought on not by increased insolation but by greenhouse warming.

Some defenders of the insolation hypothesis have countered this dark thought with the idea that the temperature increase might have stimulated increased production of carbon dioxide through increases in such processes as microbial respiration. Even that hope now appears to have faded, however, with the recent finding that the increases in carbon dioxide *preceded* the temperature increases. This leaves us with a serious dilemma: the glacial-interglacial cycles are clearly in phase with the Milankovitch curve, but it now seems probable that the carbon dioxide increase, and not the Milankovitch effect, is the chief cause of the interglacial warming. How can these two observations be reconciled?

Certainly the most conservative and prudent approach to this problem is to look for an explanation that takes both these factors into account. Because it is highly unlikely that both would act independently with the same periodicity and in phase with one another, it is further necessary to assume some sort of causal relationship between the two. It makes no sense to assume that the carbon dioxide

periodicity could cause the Milankovitch cycle, so that leaves us with the converse. Is it reasonable to suppose that the Milankovitch cycle could somehow cause major fluctuations in Earth's atmospheric carbon dioxide concentration?

Here, an important clue emerges in core number 386-64 of the Deep Sea Drilling Project (see chapter 5), which shows a rhythmic alternation of black, oxygen-free strata and greenish-gray, oxygen-rich strata in Lower Cretaceous marine mudstone with a roughly 41,000-year periodicity, one of the dominant rhythms of the Milankovitch curve. A likely interpretation of this core is that the dark strata were deposited during times of submarine volcanism that rendered the ocean water toxic. Those times would have corresponded to active seafloor spreading.

A second major clue emerges in another observation resulting from the Deep Sea Drilling Project. Before considering that clue, I should first explain that the concentration of dissolved carbon dioxide in seawater increases with depth (see chapter 11). There is a certain depth, called the

## Summary

Sedimentary deposits resulting from glaciation are called **drift. Erratics** are huge boulders moved by glaciers. Snow that survives the summer becomes compact *firn,* which, if thick enough, compacts further into ice. A **glacier** is ice moving under the influence of gravity. **Continental ice sheets,** analogous to sheet flow in water, form on high latitude uplands, whereas **valley glaciers,** analogous to channel flow, create, and grow within, **cirques** in middle to low altitude mountains, and thence flow down former stream valleys. The **zone of accumulation** lies above the **annual snowline,** the **zone of ablation** lies below it. **Crevasses** form in the brittle upper layer of a glacier, especially when it flows over a break in slope. In glaciers in equilibrium, the **terminus** neither advances nor recedes. Net accumulation results in **glacial advances,** net ablation in **glacial recessions.** Glacial *surges* involve the sliding of a glacier on its bed.

*lysocline,* at which the carbon dioxide concentration becomes great enough to dissolve calcium carbonate as fast as it is deposited. Studies of the calcium carbonate content of a vast number of cores from the Deep Sea Drilling Project show that at the glacial-interglacial transitions, the lysocline rises abruptly worldwide by as much as 1000 m. Such a rise suggests a sudden influx of massive quantities of dissolved carbon dioxide into the waters of the deep ocean. If it did, then what might be the source of such an influx?

As I note in chapter 1, carbon dioxide is a major constituent of volcanic gases, and as I also note in that chapter, the median rift of the oceanic ridge is the site of the most extensive volcanic activity on Earth. It is evident, however, that the volcanic activity along the oceanic ridge is not continuous, but is in fact an intermittent, catch-and-go affair, as is the jockeying of lithospheric plates about the globe. We know that seafloor spreading occurs in episodic spurts, that new rifts propagate themselves as old ones die out, and that the directions of plate motions change

almost capriciously. We also know that *tidal flexing* is the cause of the extreme volcanic activity of Jupiter's closest moon, Io (see chapter 17). It seems reasonable to suspect that a greater than normal amount of tidal flexing occurs within Earth's lithosphere during the high-amplitude phases of the Milankovitch cycle. It also seems reasonable to suspect that times when such flexing is in effect would be the most favorable times for the occurrence of seafloor spreading and ridge volcanism, which would inject massive amounts of carbon dioxide into the deep ocean.

At the same time, it is likely that the associated ridge volcanism would warm the deep ocean waters sufficiently to change the circulation patterns of the ocean. Recently, Edward Boyle of the Massachusetts Institute of Technology found geochemical evidence that during the last ice age one of the key subsystems of present oceanic circulation was not in operation. One of the modern ocean's great migratory water masses, the North Atlantic Deep and Bottom Water (see fig. 12.19), stirs the ocean from top to bottom, forming what Wallace

Broecker of Columbia University has called the "oceanic conveyor." This vast, northward-flowing current also transfers large amounts of heat to northern Europe. Without that heat transfer, northern Europe would have a climate similar to that of Labrador (northeastern Canada), which lies at the same latitude. As a result of the shutdown of the oceanic conveyor during the last ice age, two things happened: northern Europe went into a deep freeze, and most of the exchange between sea surface water and the deep ocean was cut off.

That cutoff is significant because the concentration of carbon dioxide in the atmosphere is controlled by its concentration in the surface waters of the ocean. At present, the oceanic conveyor mixes gas-poor surface waters and gas-rich bottom waters throughout the world ocean, as it probably also did in past interglacial ages. Because of this stirring, the carbon dioxide concentration of surface waters and the atmosphere should be higher during interglacial ages than they are during glacial ages, when the conveyor is not in operation. This is exactly what we see happening in the ice cores.

**Stoss and lee topography** and **glacial striations** in bedrock indicate the prevailing direction of glacier flow. Glaciated bedrock usually exhibits **glacial polish.** **Till** is deposited directly by glacier ice. **Stratified drift** is more or less sorted and transported by glacial meltwater. A **moraine** is a landform composed of till. **Ground moraine** comprises basal till and overlying ablation till. **Lateral moraine** is a till-talus ridge deposited at the side of a valley glacier. **Medial moraine** is a ridge formed by the coalescence of two lateral moraines downvalley from a glacier junction. A **recessional moraine** is a cross-valley ridge formed during a temporary stillstand of the terminus in a glacial recession when the glacier is in equilibrium. A **terminal moraine** is a cross-valley ridge formed during a stillstand at the farthest downvalley excursion of the terminus. An **interlobate moraine** is a ridge formed between two lobes of a continental ice sheet.

**Rock glaciers** are slowly-flowing accumulations of talus, ablation moraine, or both, with varying amounts of interstitial ice. **Drumlins** are streamline molded forms in ground moraine. **Glacial outwash** is deposited on an **outwash plain** downvalley from a glacier terminus by braided meltwater streams. **Kettles** are depressions formed by the melting of ice blocks buried in moraine or outwash. **Eskers** are sinuous ridges composed of the bedloads of subglacial streams. **Kame terraces** are large deposits of stratified drift that formed between a melting glacier and a valley sidewall. A **varve** is an annual layer of lake bottom sediment composed of a thicker, lighter lamina of warm-season silt and a thinner, darker overlying lamina of cold-season clay.

After a glacier melts, the unloaded crust undergoes **isostatic rebound.** Widening of valley sidewalls by glacial abrasion results in mass wasting of the overlying ridges, which sharpens them into **arêtes.** The down-valley ends of arêtes are often abraded by main glaciers, forming *truncated spurs.* **Horns** form by the coalescence of headward-eroding cirques on mountain peaks. Greater abrasive valley deepening by a main glacier than by its tributaries results in **hanging valleys. Tarn** lakes form in glacial cirques. Frost effects, such as **patterned ground** and solifluction, appear typically above **permafrost,** within the **periglacial zone.** Altered climate effects, such as loess and **pluvial lakes,** appear within the **transglacial zone.**

Reduced volcanism, the drifting of Antarctica over the South Pole, lowered sea level due to slower seafloor spreading, and angiosperm evolution can account for Cenozoic global cooling. Cyclic glaciations are correlated with the Milankovitch curve of variations in Earth's orbital motions, and with reduced atmospheric $CO_2$ concentration. Possibly, stresses on Earth due to Milankovitch fluctuations stimulate plate tectonic activity, which in turn activates the "Atlantic conveyor," resulting in the mixing of gas-poor surface seawater and gas-rich deep water. This would raise atmospheric $CO_2$ concentration, thereby terminating glaciation.

# Key Terms

| | |
|---|---|
| drift | medial moraine |
| erratic | recessional moraine |
| glacier | terminal moraine |
| continental ice sheet | interlobate moraine |
| valley glacier | drumlin |
| cirque | glacial outwash |
| annual snowline | outwash plain |
| zone of accumulation | kettle |
| terminus | esker |
| zone of ablation | kame terrace |
| crevasse | varve |
| glacial advance | isostatic rebound |
| glacial recession | arête |
| stoss and lee topography | horn |
| glacial polish | hanging valley |
| glacial striations | tarn |
| till | periglacial zone |
| rock glacier | transglacial zone |
| moraine | patterned ground |
| stratified drift | permafrost |
| ground moraine | pluvial lake |
| lateral moraine | |

# Questions for Review

1. Distinguish among drift, till, stratified drift, and moraine.
2. Describe the evolution of snow into glacier ice.
3. In as much detail as possible, compare and contrast continental ice sheets and valley glaciers.
4. Explain glacial advances, stillstands, and recessions in terms of accumulation and ablation.
5. Describe the erosional forms produced by continental glaciation, and explain how some of them can be used as indicators of glacier flow direction.
6. Name six different types of moraine and describe their individual modes of origin.
7. Name three different types of stratified drift and describe their individual modes of origin.
8. What principal characteristic of a vanished continental ice sheet is reflected in the pattern of isostatic rebound?
9. Name five different mountain landforms produced by valley glaciation, and describe their individual modes of origin.
10. Compare and contrast the periglacial and transglacial environments.
11. Describe four likely causes of Cenozoic global cooling and one possible cause of glacial-interglacial cycles.

# Part Three

# Earth's Oceanic Systems

Photo: A rocky coastline at Granite Creek, near Big Sur, California.

# 11

# Water, Salt, and Basalt: The Structure of the World Ocean

## Outline

*W*ith whisper of her mellowing grain,

*With treble of brook and bud and tree,*

*Earth joys for ever to sustain*

*The bass eternal of the sea.*

*Roden Berkeley Wriothesley Noel*

**Beatrice**

Mafic lavas of a magmatic arc stand dark behind a beach of coral sand in the West Indies.

# Exploring the Seas

As so often happens in most areas of scientific inquiry, serious study of the world's ocean had to await the development of economic incentives. When the fate of fishing fleets, merchant ships, and naval vessels became a matter of common concern, the motivation soon arose for understanding the behavior and hazards of the oceans. Ferdinand Magellan's circumnavigation of the globe from Spain (1519–1522) demonstrated the continuity of the world ocean. His unsuccessful attempt to hit bottom with a cannonball tied to a 730 m line demonstrated the "unfathomable" depth of the Pacific Ocean. In 1770, Benjamin Franklin and Timothy Folger, a ship captain from Nantucket Island, published the first map of the Gulf Stream. Their map explained for the first time why ships sailing east to Europe made better time than did those sailing west to America. In 1855, a crippled U.S. Naval officer named **Matthew Fontaine Maury** published his landmark work, *The Physical Geography of the Sea*. Maury had gathered all reliable data from ship's logs, including some whose captains he had cajoled into making special sojourns into "unknown territories." From these data (singular *datum;* Latin for "that which is given"), he was able to construct detailed and elaborate charts of winds and currents for all the world's oceans (fig. 11.1). Maury also compiled systematic sounding data that allowed him to produce the first bathymetric (depth-contoured) map of the Atlantic Ocean in 1854.

Great Britain's contributions to *oceanology,* the science of the sea (also called oceanography), have been formidable. Between 1768 and 1777, scientist and navigator James Cook sailed the British *Endeavour* through most of the "seven seas," vastly increasing our knowledge of marine geography, astronomy, and navigation. From 1831 to 1836, the ten-gun British brig *HMS Beagle* sailed from England round Cape Horn to the Pacific to circumnavigate the globe, carrying a brilliant young naturalist named Charles Darwin. The extraordinary insights that Darwin gleaned from this voyage led biological science from a dark age of half truth and speculation to a new understanding of how Nature works on its own terms.

## In the Spotlight

### Matthew Fontaine Maury (1806–1873)

Personal misfortune, however unwelcome, can sometimes open avenues to achievement that might otherwise have remained closed. Midshipman M. F. Maury was crippled for life when he was involved in a stagecoach accident during a tour of southern harbors in 1839. As a result of his injury, he was removed from active service and placed in charge of the United States Naval Depot of Charts and Instruments, the precursor to the United States Naval Observatory and Hydrographic Office.

Maury launched himself into his new career with an ambitious project to compile data on global winds and currents. To facilitate the data collection process, he developed special logbooks and distributed them to ship captains. The success of this venture inspired an international conference on navigation and meteorology in Brussels, Belgium, in 1853. The results of Maury's productive work at the Naval Depot were published in his *Physical Geography of the Sea* in 1855. Acknowledged as the first textbook of oceanology, the work went through eight editions, and for many decades it served as the standard international reference on the physical aspects of the global marine environment. Included in its scope were charts showing atmospheric circulation; seasonal winds, currents, and temperatures; sources of atmospheric moisture; bathymetry of the North Atlantic Ocean; and the storm track of a major hurricane.

When the Civil War broke out in 1861, Maury resigned from the Naval Depot and joined the Confederate forces, as befitted his heritage as a native Virginian. While serving as head of coast, harbor, and river defenses, he developed an electric torpedo. The need to secure supplies for that project took him to Great Britain in 1862, where he remained as Agent for the Confederacy until 1865. At the war's end, he sailed for Mexico, where he became Imperial Commisioner of Immigration to the Emperor Maximilian. While in this position, he attempted to establish a colony for Confederate veterans, a project that was initially endorsed by Maximilian, but abandoned in 1862, whereupon Maury returned to England. In 1868, he accepted a position as professor of meteorology at the Virginia Military Institute, where he taught until his death in 1873.

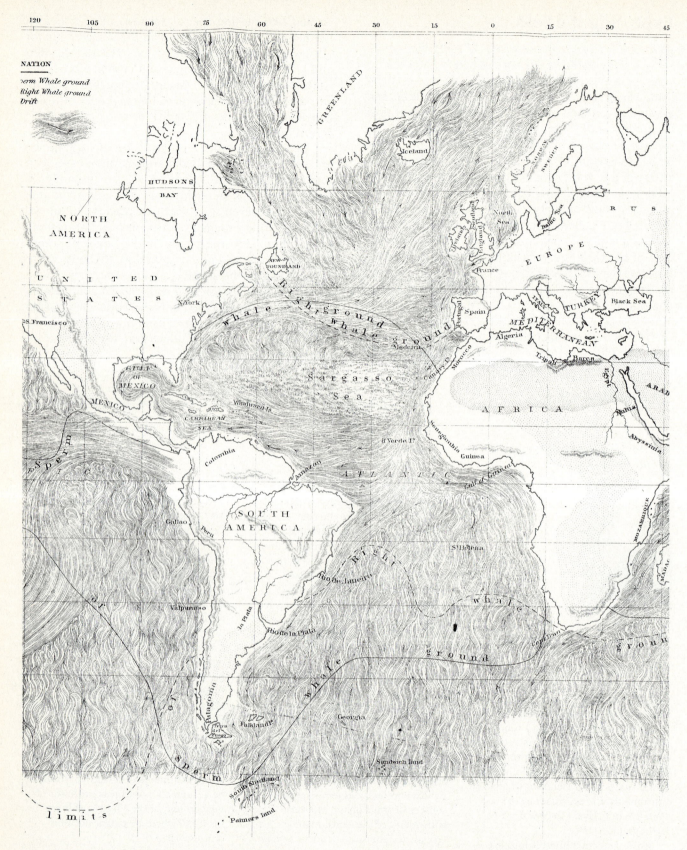

**Figure 11.1** Using data from ships' logs, Matthew Fontaine Maury carefully constructed charts, such as this one showing "sea drift" (ocean currents), and the ranges of two species of whale. Published in 1855 in his pioneering treatise on *The Physical Geography of the Sea,*

Maury's charts proved of great value to both science and commerce but was not of much benefit to whale populations.

Source: M. F. Maury, *The Physical Geography of the Sea,* 1855.

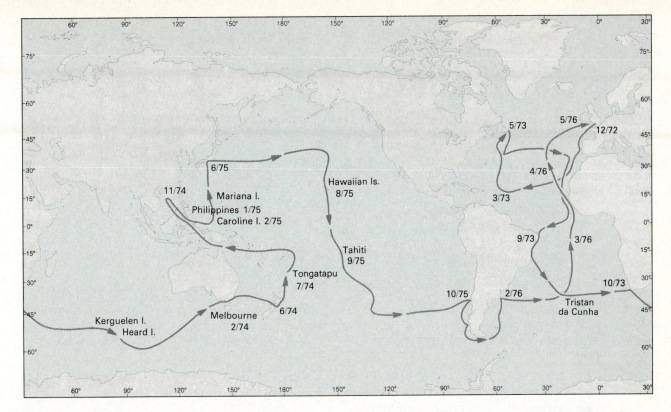

**Figure 11.2** The five-year voyage of the British steam corvette *H.M.S. Challenger* (1872–1876) covered the Atlantic Ocean intensively and the Pacific and Indian Oceans less intensively, gathering scientific data that would forever change our ways of thinking about the world ocean.

From W. J. Cromie, *Exploring the Secrets of the Sea,* © 1962, p. 38. Prentice-Hall, Inc., Englewood Cliffs, NJ.

In 1872–1876, Great Britain undertook an ambitious and highly successful oceanological expedition with a single vessel, **HMS Challenger** (fig. 11.2), directed by Sir Charles W. Thomson. The *Challenger's* circumnavigation of the globe produced a far greater body of fundamental knowledge about the oceans and their inhabitants than had all previous expeditions combined. The data collected and analyzed by the *Challenger's* six scientists fill 50 volumes, which are still regarded as definitive works in oceanology. At each of 362 oceanological stations, the following observations were routinely made:

1. Water depth
2. Water temperature at depth intervals and at the top and bottom
3. Water samples at depth intervals, including the bottom
4. Sediment samples from the seafloor
5. Dredge samples of bottom fauna (animals)
6. Net samples of surface and near-surface organisms
7. Speed and direction of surface currents (and occasionally of subsurface currents)
8. Meteorological observations

The expedition provided the first solid evidence of the consistently great depth of the world ocean, including one incredible sounding of 8180 m in what is today known as the Challenger Deep of the Mariana trench, south of Japan where the Pacific plate is being subducted beneath the Phillipine plate.

The *Challenger's* was a hard act to follow. In the late 1800s and early 1900s, several less ambitious oceanological voyages were undertaken. The most productive of these were probably the United States expeditions of Alexander Agassiz (son of Louis) in the *Blake* and the *Albatross* to Florida, the Caribbean, and the Pacific (1877–1904), and the Norwegian expeditions of Fridtjof Nansen in the *Fram* to the Arctic Ocean (1893–1896).

In this century, emphasis in oceanological research has shifted considerably from marine biology and the physical and chemical characteristics of ocean water to the exploration of the seafloor. Although many vessels from many nations have been involved in this ambitious undertaking, the **Glomar Challenger** and its successor, the **JOIDES Resolution,** stand out among them (figs. 5.14 and 11.3). During its 15 year lifetime, the Deep Sea Drilling Project has drilled holes more than 1700 m deep within the oceanic crust and in water depths of over 7000

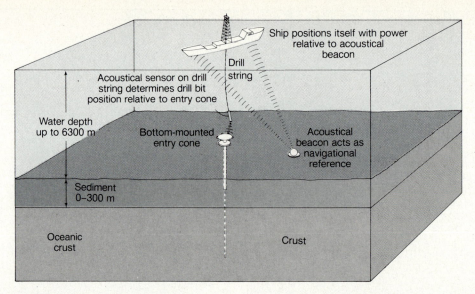

**Figure 11.3** The drill ship *Glomar Challenger* was specially designed to hold its position over a drill hole in the seafloor as much as 6300 m below the surface. Bow and stern thrusters and the main propellers are computer controlled. Reentry into the hole, following retrieval of the drill string to change bits, is accomplished by means of a large funnel and an array of sonar positioning devices.

m. Drill cores have been taken from over 600 localities. The information presented in the oceanic portions of the world map of figure 1.5 is a distillation of the results of the Deep Sea Drilling Project.

## Oceanological Instruments

The exploration of the oceans presents difficulties of the same kind as those which face geophysicists and astronomers: the object of investigation is often inaccessible to close-range observations. In studying the topography of the deep ocean floor, for example, we are limited by a number of factors. There is no light on the ocean floor below a depth of about 100 m, and the pressure due to the weight of overlying water exceeds 1000 atmospheres (atm) in the deeper parts of the ocean. Neither oceanologists nor their instruments fare very well under such conditions.

Three main strategies are used to overcome these difficulties:

1. *Protection* of investigators against hostile marine environments so they can observe undersea phenomena firsthand
2. *Direct* measurement or *sampling* of undersea phenomena by instruments let down to the sampling point on cables or drill strings
3. *Remote sensing* of submarine environments from oceanological vessels using electronic, photographic, and sonic means.

## Suits and Submarines

The first of these strategies takes either of two forms: special suits that provide air or oxygen and protect the wearer from pressure and cold, and submarine research vessels within which the operators are free to move about. Early diving suits were cumbersome contraptions with large, spherical headpieces. Air intake and exhaust hoses led from the headpiece to the surface. In 1942, Jacques Cousteau modernized this archaic device, and the result was the *Aqualung,* of which the modern *SCUBA* (Self-Contained, Underwater Breathing Apparatus) is a more recent version (fig. 11.4). This device permits divers to carry their air supply in tanks strapped on their backs. Often, an insulating wet-suit is worn in colder waters. Divers using SCUBA are restricted by water pressure to the upper 45 to 90 m of the ocean. Because most of the ocean's biomass lives in this topmost layer, SCUBA is used extensively by marine biologists.

Submarine research vessels range from small and relatively uncomplicated to large and complex. The earliest known "submersible" was built by a Dutchman, Cornelis Drebbel, about 1620. It was essentially a leather-covered submarine rowboat in which he reportedly transported his friend, England's King James, across the Thames River at a depth of 4–5 m. The boat was propelled by 12 oarsmen, and could remain submerged at that depth for several hours. Modern research submersibles, such as the French-built *Trieste,* are equipped to withstand the great pressures of the ocean depths. In 1960, the *Trieste* descended nearly 11 km into

**Figure 11.4** A marine scientist equipped with SCUBA studies a living reef with the aid of a light ring. Note cylindrical sponges (*left*) and net-veined fan coral (*right*).

Photo by D. E. Kesling, NOAA

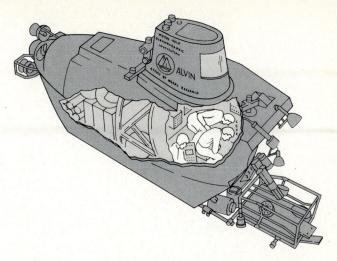

**Figure 11.5** The American research submersible *Alvin*. A mechanical sampling arm and an instrument package extend forward (*right*). Spheres, of which only two are visible in this cutaway view, are rubber ballast bags filled with oil. When oil is pumped from these bags into reservoir tanks, the bags collapse, and the surrounding space is filled with heavier water, permitting *Alvin* to sink.

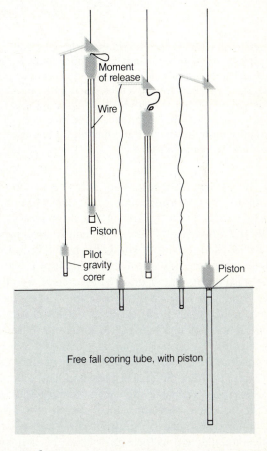

**Figure 11.6** In this coring device, a gravity corer (*left*) is used as a trigger for a larger piston corer. When the gravity corer strikes the seafloor and embeds itself, the piston corer is released. The internal piston rises, allowing the corer to penetrate still farther into the sediment.

the Mariana trench, the deepest part of the ocean. Submersibles are equipped with a variety of devices for examining the seafloor, including cameras, closed circuit television, and mechanical sampling arms that can be extended from the vessel. The small submersible *Alvin* (fig. 11.5) was used in the French-American Mid-Ocean Undersea Study (FAMOUS) to survey a portion of the North Atlantic section of the oceanic ridge in the mid-1970s.

## Probing the Depths

The second strategy has evolved a variety of devices for probing deep ocean waters and the seafloor. In additon to core drilling, simpler and less expensive devices are available for sampling the seafloor at shallow depths. These **corers** are of two types: *gravity corers,* which are driven by their own weight into the soft sediments of the seafloor and *piston corers,* in which the coring tube fits over a piston that is pulled upward by tension on the cable, to which it is attached, creating suction when the corer drives into the mud. Although gravity corers are limited to cores of about 1 m length, piston corers can retrieve cores up to 30 m long (fig. 11.6).

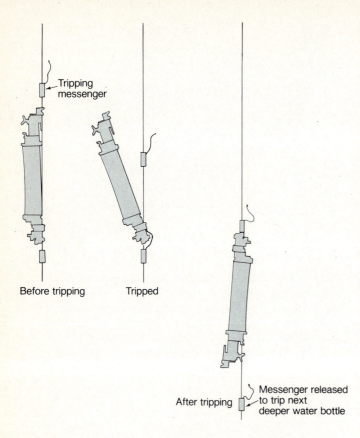

Tripping messenger

Before tripping

Tripped

After tripping

Messenger released to trip next deeper water bottle

**Figure 11.7** The Nansen bottle is a tethered water sampler. Before tripping, the bottle is open and seawater can flow through freely. When a messenger weight slides down the wire and trips the bottle, the upper end is released, and the bottle's valves close, trapping a seawater sample as it inverts. Inversion breaks the mercury columns within the attached thermometers, thereby recording the temperature.

**Figure 11.8** An oceanologist lowers a secchi disk over the side of a research vessel.

Photo courtesy of NOAA

The **tethered samplers** are a diverse group of instruments for collecting samples or measurements of seawater or sediments. One of the most basic is the *Nansen bottle,* named for its inventor, the explorer Fridtjof Nansen (fig. 11.7). This is an open, tubular bottle fastened permanently to a cable at its lower end and temporarily at its upper end. When a steel weight, or "messenger" is dropped down the cable, the upper end is freed, whereupon the bottle inverts itself, closing and trapping water as it does so. Two reversing thermometers attached to the bottle simultaneously record the temperature at the sample point. One of these thermometers is shielded against pressure. The difference between the readings of this thermometer and the unshielded one can be compared by means of a formula to give a good estimate of depth. Usually, several Nansen bottles are attached to the same cable. As each one is tripped by a messenger from above, it releases another messenger, which in turn trips the bottle below. A more modern and sophisticated instrument is the *STD probe,*

which measures salinity, temperature, and depth electronically. Unlike the Nansen bottle, it has the ability to take continuous samples as it is lowered on its cable.

The *secchi disk* (fig. 11.8) is a white or black and white disk, 30 cm in diameter, that is lowered on a wire to a depth at which it disappears from view. This is called the *secchi depth,* and it is a measure of the optical clarity of the water. *Transmissometers* and *nephelometers* are used to measure the clarity and turbidity (cloudiness) of deeper water electronically. *Dredges* and *grab samplers* are used to scoop samples of sediment and benthic (bottom-dwelling) organisims from the ocean floor (fig. 11.9). *Plankton nets* and *trawls* are used to collect **plankton** (floating organisms, such as jellyfish), and **nekton** (swimming organisms, such as fish).

Ocean currents are measured by anchored or tethered *current meters. Drogues* are free-floating devices used to track the paths of ocean currents. *Wave rider buoys* float at the surface and record the action of waves. *Tide gages* record the passage and height of tidecrests. These can consist of *pressure transducers* installed on the seafloor, or of various *wave staffs* that record tide crests electronically.

**Figure 11.9** An oceanologist, protected by a "belt belay," adjusts a grab sampler prior to a bottom sampling project.

Photo courtesy of Woods Hole Oceanographic Institution

## Remote Sensing

Prior to 1912, less than 6000 soundings of the ocean had been made by wire or rope in depths greater than 1000 fathoms (1800 m). Using some of these data, Maury produced a **bathymetric** (depth) **map** of the North Atlantic in 1854, showing depth by 1000 fathom contour lines. Because of the sparseness of the data available to Maury, his map was inadequate to delineate the oceanic ridge.

In 1922, the *U.S.S. Stewart* employed for the first time a new device called an **echo sounder,** which was soon to change dramatically our conception of seafloor topography. The principle is simple (fig. 11.10). A sound pulse travels downward from a research vessel, hits the seafloor (or some other object, such as a school of fish) and bounces back. The depth to bottom (or to fish) in meters is then one-half the speed of sound in water (1460 m/sec) times the length of time in seconds that the signal takes to return to the research vessel. A typical echo sounding profile is shown in figure 11.11. A little over a decade later, the German *Meteor* expedition to the South Atlantic made over 70,000 echo soundings during 14 oceanic crossings. From these and subsequent soundings, the modern picture of the seafloor's

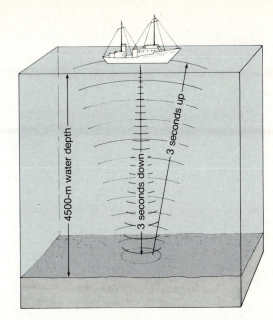

**Figure 11.10** The principle of echo sounding. Depth of water is calculated from the time required for a sound signal to travel from ship to seabed and back again.

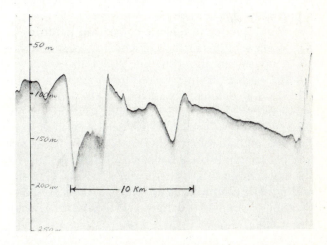

**Figure 11.11** An echo sounding profile. Note that the vertical dimension in this profile is exaggerated 73 times.

Photo courtesy of A. C. Duxbury and A. B. Duxbury

rugged topography shown in figure 1.2 gradually emerged. More recently, a side-scanning, radarlike device called *sonar* has been developed which sweeps the ocean bottom with a rotating "beam" of sound pulses. Sonar can produce images of the seafloor that are of almost photographic quality.

In chapter 5, I mention the role of *magnetometers* in accessing the hard evidence that finally forced acceptance of the concepts of seafloor spreading and plate tectonics. *Gravimeters,* or gravity meters, are also routinely used to measure density variations, which can be used to infer the rock types and geologic structures that underlie the seafloor.

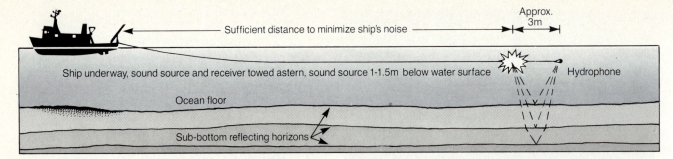

**Figure 11.12** In the technique of seismic reflection profiling, loud sound signals are reflected from prominent discontinuities beneath the seafloor, such as bedding, unconformities, and faults.

From J. B. Hersey, "Continuous Reflection Profiling" in *The Sea,* M. N. Hill (editor), Vol. 3:47. Copyright © 1963 John Wiley & Sons, Inc., New York, NY. Reprinted by permission of John Wiley & Sons, Inc.

Perhaps the most powerful of all the remote sensing techniques is **seismic reflection profiling** (fig. 11.12). Like sonar, this is a refinement of echo sounding, but the sound signal used is powerful enough to penetrate the seafloor and reflect from discontinuities at various depths within the crust as well as from the seafloor itself. This technique gives a fuzzy, but usually interpretable picture of the subsurface structures within the oceanic crust in addition to the overlying seafloor topography (fig. 11.13).

The technique of **seismic refraction profiling** is a refinement of this method in which a research vessel picks up signals from another vessel that were *refracted,* or bent, as they passed from one layer to another of a different consistency. The concept is analogous to deducing Earth's deep structure by observing how earthquake waves behave as they pass through, and are refracted by, layers of differing densities and rigidities (see chapter 6). Similarly, the refracting of sound signals by layers of various densities and rigidities in the oceanic crust can be analyzed to determine geologic structures and rock types beneath the seafloor.

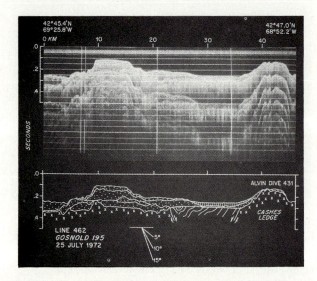

**Figure 11.13** An east-west seismic reflection profile of the ocean floor east of Cape Cod, Massachusetts shows a number of prominent reflecting horizons, which have been interpreted in the geologic cross section (below).

Photo courtesy of Woods Hole Oceanographic Institution

## Physiography of the Ocean Basins

Figure 11.14 gives a general impression of a fact that became increasingly evident as more soundings of the world ocean became available: Earth has a split-level surface. The land surfaces form plateaus with a mean elevation of about 840 m, whereas the ocean basins form submerged plains with a mean depth of about 3750 m. This arrangement makes sense in terms of the principle of isostasy (see chapter 1), whereby less dense continental crust stands higher than denser oceanic crust. The continental platforms, including the continental shelves, constitute the 29% of Earth's surface area that lies between 1000 m above and below sea level, whereas the ocean basins constitute the 53% that lies between 3000 and 6000 m below sea level. Only about 9% is higher than 1000 m (mountains and plateaus), and only 1% is deeper than 6000 m (trenches). The remaining 8% that lies between 1000 and 3000 m depth represents the transitional zones between the margins of continents and ocean basins (the continental slopes and rises).

The generalized physiography of the North Atlantic ocean floor, greatly exaggerated in the vertical dimension, is shown in figure 11.15 (compare with fig. 5.12). Let us now explore the principal features of that once dark and mysterious realm, newly illuminated for us by the devices of echo sounding, profiling, and drilling.

## The Oceanic Ridges

In chapter 1, I describe the general form and function of the oceanic ridge. In chapter 5, I do the same for the transverse fracture zones along which the trend of the

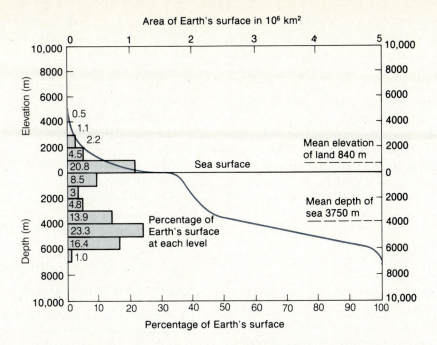

**Figure 11.14** The distribution of Earth's surface elevation with respect to sea level is shown by the bar graph at left and by the curve that results from adding all the bars sequentially from top to bottom. The high and low ends of the curve represent mountains and trenches, respectively. The steep section to the left of center represents continental slopes and the broad plateaus centered on 840 m and −3750 m represent continental platforms and ocean basins, respectively.

From H. U. Sverdrup, M. W. Johnson, and R. H. Fleming, *The Oceans*, © 1942, renewed 1970, pp. 18, 19, 241, 242, 245. Prentice-Hall, Inc., Englewood Cliffs, NJ.

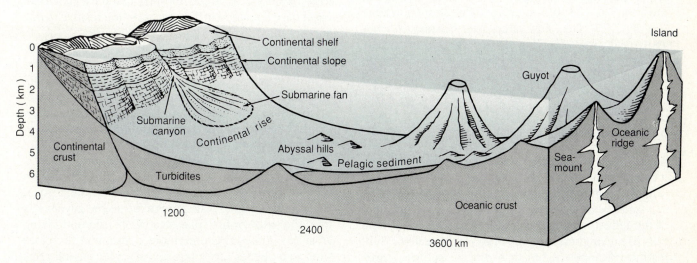

**Figure 11.15** Seafloor morphology from the mid-Atlantic ridge to the coast of North America.

ridge is offset. The mid-Atlantic section of the ridge, which lies along a line approximately equidistant from the east and west coasts of the Atlantic Ocean, averages about 1000 km in width but narrows in some places to less than 500 km and broadens in others to over 2000 km. It rises about 3000 m above the adjacent deep seafloor, but in some places, such as Iceland and the Azores, west of Spain and Portugal, the ridge or its volcanic peaks rise above sea level.

Compared with other sections of the oceanic ridge, the mid-Atlantic ridge is relatively rugged and prominent. This results from its relatively slow rate of seafloor spreading: 1–2 cm/yr for each plate, compared with 3–8 cm/yr for the east Pacific section of the ridge. As the lithosphere cools, it becomes denser and sinks to a lower elevation. The same amount of sinking occurs farther from the rift in the fast-spreading east Pacific ridge (or rise, as it is often called) than in the more slowly spreading mid-Atlantic ridge.

The rise of magma from the underlying mantle is indicated by unusually high heat flow over the oceanic ridges. Another indicator of rising magma is the median rift, marked by its narrow, deep valley 1–50 km wide and up to 200 m deep. In the east-Pacific section, this trough is missing. Where present, the median rift is probably caused by subsidence of the ridge crest following the extrusion of basaltic lava. The great fracture zones that separate the offset segments of the Pacific ridge can be as much as 3500 km in length and can drop as much as 450 m from one side of the fault to the other. Oceanic ridges underlie 32.7% of the total area of the world ocean.

## The Oceanic Trenches

Trenches (not shown on fig. 11.15, but see figs. 1.2 and 5.21) mark the "end of the line" for spreading seafloors. They vary in depth from 6662 m (Middle America) to 11,034 m (Mariana); in width from 55 km (Tonga) to 120 km (Kurile); and in length from 900 km (Japan) to 5900 km (Peru-Chile). In cross section, trenches are normally V-shaped with walls that slope from 4° to 45°. Their floors are typically narrow, no more than a few kilometers wide, but those located near plentiful sources of clastic sediments can accumulate appreciable thicknesses of mud and turbidites. The growth of a tectonic arc over the trench during subduction tends to displace the trench axis laterally away from the subduction zone (see fig. 5.25). Heat flow values over oceanic trenches are typically low, reflecting the descent of cold oceanic lithosphere within the subduction zones beneath trenches.

## The Deep Ocean Floor: Abyssal Plains and Hills

As the ocean floor spreads away from the oceanic ridge, it carries with it topographical evidences of its volcanic origin. Foremost among these is the **abyssal hill.** This is a small basaltic extrusion, the submarine equivalent of the terrestrial shield volcano. By definition, abyssal hills rise no higher than 1000 m above the seafloor. Submarine volcanoes higher than this are known as **seamounts.** Figure 5.13 shows abyssal hills and seamounts on a portion of the Pacific Ocean floor. Sedimentary deposits that might bury and obscure such features are thin or absent on the Pacific floor because the unbroken line of trenches on its northern and western boundaries efficiently traps continental clastic sediment before it can reach the deep ocean floor.

Both seamounts and abyssal hills are also present on the Atlantic Ocean floor, but the margins of the Atlantic are practically free of oceanic trenches. As a result, continental clastic sediments have inundated the deep seafloor, burying most of its topographic features beneath deep blankets of mud and turbidites. Only the crest and flanks of the oceanic ridge remain relatively free of this sedimentary cover. Near its continental sources, that cover can reach thicknesses of 3 km. The broad **abyssal plain,** formed by a combination of such continental sediments and pelagic sediment, lies at depths of 3000–6000 m below sea level, has slopes of less than 1 m/km, and is by far the most extensive and featureless plain on Earth. Abyssal plains underlie 41.8% of the area of the world ocean. Heat flow values on the deep ocean floors are intermediate between oceanic ridges and trenches, but closer to the latter.

## Conveyor Belt Mountains: Seamounts and Guyots

On the right side of figure 11.15, several large volcanic structures are evident, rising above the more abundant abyssal hills. The one with a peaked summit is a seamount, whereas those with flat summits are called **guyots** (*gee*-ohs, with a hard g; named for the Swiss-American geologist Arnold Guyot).

What is the source of seamounts and guyots? Their chemistry suggests that they are nourished by deep-seated, persistent mantle plumes. These plumes appear to be immobile, as all the hot spots in the Pacific Ocean basin have remained in the same position relative to one another for at least the past 80 million years.

Why do guyots have flat tops? The favored explanation is that they are seamounts whose summits were planed off by wave erosion. This might seem implausible because the tops of many guyots lie at depths of up to 1500 m, far below the reach of wave erosion. As it has done with so many other apparent enigmas, however, plate tectonic theory has provided an answer. When a volcanic seamount is formed at a ridge or a hot spot, it can project above sea level as an island, whereupon wave erosion can plane off the summit, forming a guyot (fig. 11.15). In time, plate growth carries the guyot away from the ridge or hot spot. Meanwhile, the underlying crust has cooled, become more dense, and therefore sunk to a deeper level. In tropical regions, coral reefs often grow on the summits of slightly submerged guyots and persist as long as their growth can keep pace with the rate of subsidence of the seafloor. Many coral **atolls** originated in this way (fig. 11.16). Deep drilling through hundreds of meters of reef coral down to the underlying basalt foundation has confirmed this interpretation of sinking seamounts, first proposed by Charles Darwin in 1837.

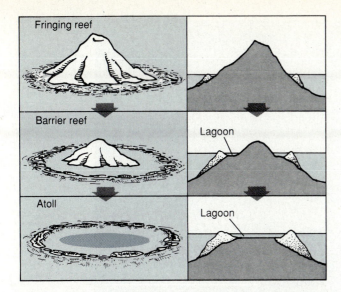

**Figure 11.16** Charles Darwin's theory of evolution of coral atolls.

Seamounts and guyots provide a convenient means of determining rates of seafloor spreading, as their basaltic rock can be dated radiometrically. Distance of the seamount or guyot from its point of origin divided by its age yields its average rate of movement.

## The Continental Margins

It would be satisfying if the shorelines of the world ocean coincided with the outer edge of the continental shelf, the approximate boundary between oceanic and continental crust. At present, however, the ocean is too big for its basins, and it has spilled over onto the continental margins. Nearly 15% of the world ocean's total area lies over the *continental shelves,* the gently-sloping, submerged outer edges of the coastal plains where quartz sandstone, limestone, and mudstone accumulate (see chapter 7).

Continental shelves can be absent in places where the continental margin is actively rising or where strong ocean currents remove clastic sediments faster than they can accumulate (e.g., southern Spain; see fig. 15.12). In places where the continental margin is subsiding, however, or where large rivers have built extensive deltas, the continental shelf can be as wide as 1500 km (e.g., the New England-Newfoundland area and the Gulf Coast area of the southern United States). Slopes on the continental shelf average only 0.2%, or a drop of 2 m/km. Such slopes appear essentially level to the human eye. The outer edge of the continental shelf ranges in depth from 20 to 550 m.

Much of the area of the continental shelves was exposed during glaciation because the conversion of seawater to glacial ice during the Pleistocene Epoch lowered sea level by about 100 m worldwide. It has been estimated that the total land area was thereby increased by about 6%. The legend of Atlantis could well be a folk memory of the drowning of inhabited coastal plains as sea level rose at the end of the last major glaciation, from about 15,000 to 10,000 years ago.

Beyond the *shelf break* at the outer edge of the continental shelf, the continental margin slopes away more steeply toward the ocean floor (fig. 11.15). This *continental slope* (see chapter 5) is inclined at an average of about 4°, or 70 m/km, but ranges from 1° (18 m/km) off deltaic coasts up to 25° (466 m/km) off faulted coastlines. One puzzling feature of continental slopes is the presence of deep **submarine canyons** cut into their surfaces (e.g., east of Nova Scotia and north of the mouth of the Amazon River; see fig. 5.12). The most reasonable explanation for their origin is that they were at first subaerial canyons on rifted continental margins. As those margins drifted progressively farther away from their median rift, they gradually sank to positions well below sea level. At present, they serve as channels for turbidity currents.

Although some mud and turbidite is deposited on the continental slope, by far the largest volume of clastic sediment ends up in a thick prism between the base of the continental slope and the adjacent abyssal plain, known as the *continental rise* (see chapter 5). This feature varies in width from 100 to 1000 km, and its slopes average about 0.5° (about 9 m/km). It lies between 1400 and 5100 m below sea level. When involved in continent-continent or continent-arc collisions, its sediments are compressed and metamorphosed into broad belts of slate and schist, such as the one that extends from southern New Brunswick through central Maine and into southern New Hampshire (see figs. 4.35, 4.37, and 6.14).

## The Origin of Seawater

In chapter 1, I mention the hypothesis that Earth's atmosphere was produced by volcanic eruptions. Of the gases erupted by modern volcanoes, water makes up about 70%. Much of this water condenses soon after eruption and falls as rain (fig. 11.17). Considering the amount of volcanic activity that has occurred since Earth has been cool enough to retain an atmosphere (roughly the last 4000 million years), this seems an adequate source for the world ocean.

At present, water is being released through *subaerial* (above sea level) volcanoes at a rate of about 66 $\times$ 10$^{15}$ g/yr and through submarine volcanoes at an unknown, but perhaps much higher rate. Surprisingly, the subaerial rate alone is roughly 165 times greater than necessary to account for all the water present in the world ocean! One might infer from this that volcanism

**Figure 11.17** Volcanic eruptions deliver vast quantities of water to the atmosphere. Here, lightning flashes through the eruption cloud of Galunggung volcano in West Java, Indonesia, during an eruption on 29 September 1967.

Photo by Ruskia Hadian, Volcanological Survey of Indonesia, via U.S. Geological Survey

is far more active today than in the past, but we know this to be untrue. Volcanic gases could have contained less water in the past, although that is unlikely. Furthermore, water is lost from the oceans by the hydration of minerals in pelagic sediments and the seafloor and by subduction beneath oceanic trenches. Most significant of all is the recent research of geologist Donald Wise, which indicates that the height of continents above sea level has remained virtually constant for the past 3800 million years. This implies that the volume of the world ocean has also been essentially constant for that length of time. If so, then the average amount of water added annually by volcanism must be closely balanced, over time, by the amount withdrawn by subduction and seafloor hydration—a steady state situation! If we divide the amount of water present in the world ocean ($1.4 \times 10^{24}$ g) by the production rate of volcanic water we arrive at an average **residence time** for water molecules in the world ocean of about 21 million years (this would probably be at least halved if the rate of submarine volcanism were included). This, then, is the average minimum length of time that an individual water molecule spends in the ocean.

## Water: The Wonder Chemical

Like most substances, hydrogen oxide ($H_2O$ or $OH_2$) can exist in any of three states: *solid, liquid,* and *gas* (fig. 11.18). Hydrogen oxide is unique, however, in being the only common substance on Earth that simultaneously exists in all three of these states. If it were not for this happy circumstance, neither life on Earth, nor most of the geological, oceanic, and atmospheric phenomena described in this book would be possible. The three states of $H_2O$ differ from one another in the arrangement of their molecules. In *ice*—a solid—$H_2O$ molecules are packed together in an orderly crystal structure. In *water*—a liquid—this order is disrupted, and the molecules are free to shift in relation to one another, but they remain in mutual contact. In *water vapor*—a gas—the molecules move freely and independently of one another except for frequent, brief collisions. Note, however, that water vapor and *steam* are not the same thing. Whereas steam is a suspension of fine water droplets in air, water vapor is *invisible*. In a typical classroom, at 25° C, there can be as much as 4 liters (about 1 gallon) of water vapor dissolved in the air.

## Changes of State

Figure 11.19 shows how variations in two factors, temperature and pressure, together determine whether hydrogen oxide exists as ice, water, or vapor. Pressure is shown by the vertical axis (using a logarithmic scale, which compresses the diagram's upper end). Temperature is shown by the horizontal axis (using a square root scale, which expands the diagram's central portion).

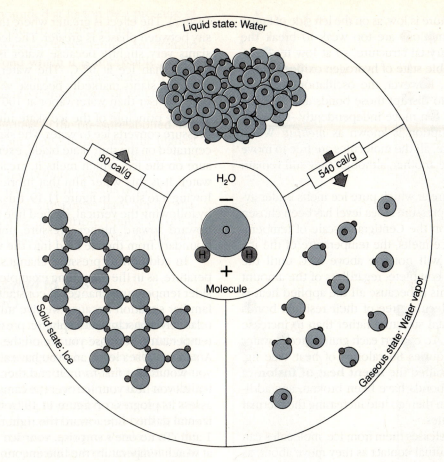

Liquid state: Water

80 cal/g

540 cal/g

H₂O

O

H H

−

+

Molecule

Solid state: Ice

Gaseous state: Water vapor

**Figure 11.18** In the lopsided hydrogen oxide molecule, H₂O (*center*), two hydrogen atoms are unsymmetrically bonded to an oxygen atom, creating a negative "end" (*top*) and a positive "end" (*bottom*). At low temperatures (*lower left*), H₂O assumes the ordered structure of the solid state (ice). At higher temperatures (*top*) bonds break, allowing molecules to move about in the more chaotic liquid state (water), but they remain in mutual contact. At still higher temperatures and low pressures (*lower right*), molecules fly apart into the vapor state. Broad arrows indicate the amount of heat required to convert 1 g of H₂O from solid to liquid at 0° C, and from liquid to gas at 100° C.

Of these two factors, temperature is the more effective in inducing changes of state in hydrogen oxide. Temperature is simply a measure of the average speed at which the atoms or molecules of a substance are oscillating or vibrating. At **absolute zero** (−273.16° C), there is no molecular motion. If the substance is exposed to a flow of energy, such as sunlight, however, some of that energy is stored temporarily within the molecules, causing them to move faster and raise the temperature of the substance. The subsequent release of this stored energy of molecular motion is what we call *heat*. The standard unit of heat in the metric system is the **calorie,** the amount of energy required to raise the temperature of one gram of water by one degree Centigrade.

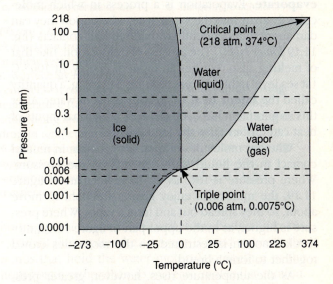

Critical point
(218 atm, 374°C)

Water
(liquid)

Ice
(solid)

Water
vapor
(gas)

Triple point
(0.006 atm, 0.0075°C)

Pressure (atm)

Temperature (°C)

**Figure 11.19** Prevailing pressure and temperature conditions determine the state(s) in which hydrogen oxide can exist. Conditions within fields permit only one state (e.g., only ice at 1 atm and −100° C); conditions along boundaries permit two (e.g., water and ice at 1 atm and 0° C), and conditions at the triple point permit all three. Above the critical point, water and water vapor are indistinguishable. Note distorted scales.

overcome it. Mountain climbers and visitors to such high-elevation localities as Denver and Mexico City experience this principle as they find that it takes longer to boil eggs and noodles at such altitudes than it takes at lower ones.

## Nature's Heat Sponge

Substances differ from one another in their ability to absorb heat. It takes 1 calorie to raise the temperature of 1 g of water by 1° C, but most other substances require far less heat per gram to raise their temperatures by the same amount. This difference is measured by the **specific heat** of a substance, or the amount of heat required to raise the temperature of one gram of a substance by a certain number of degrees compared with the amount of heat required to raise the temperature of one gram of water by the same amount. Thus, a substance having the same heat absorbing capacity as water would have a specific heat of 1.00 which is about twice that of ice and an unusually high value for a natural substance. The specific heats of rocks range from 0.17 (mudstone) to 0.26 (sandstone). Because of this difference, the same quantity of solar energy falling on equal areas of land and ocean will heat up the land area much faster than it can heat up the ocean area. The higher specific heat of water allows the oceans to soak up more of the heat energy for a given rise in temperature. Oceans and lakes, in other words, are excellent "heat sponges," capable of absorbing large quantities of heat without experiencing a correspondingly rapid rise in temperature.

## Why Ice Floats

Unlike almost all other substances, hydrogen oxide is less dense in its solid phase than in its liquid phase. Ice has a density of 0.917 g/cm³. The density of water is 0.99987 g/cm³ at 0° C, rises to 1.00000 g/cm³ at 3.98° C, and falls to 0.95838 g/cm³ at 100° C. Throughout the temperature range in which $H_2O$ is a liquid, then, ice remains between 4.3% and 8.3% less dense than water. This is why ice always floats on water. This fact that we take so much for granted is a great blessing, because if water became more dense on freezing, all the ice that has ever formed on the oceans or that has flowed from glaciers into them would have sunk to the bottom, and the oceans would have been frozen solid, except for a shallow, tropical belt, since some time in the Cryptozoic Eon.

Figure 11.21 shows why ice is less dense than water. In the crystal structure of ice, each hydrogen atom (small spheres) is located between two oxygens (large

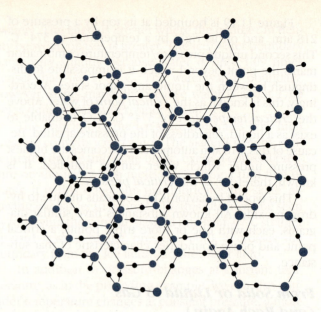

**Figure 11.21**   In ice, the orientation of lopsided $H_2O$ molecules with respect to one another creates an open, honeycomblike structure with a repeating hexagonal pattern. That structure is greatly expanded here for clarity.

spheres), to which it is attracted by its opposite charge. As water cools toward 0° C, its molecules oscillate and vibrate less and less, allowing residual bonding forces to jockey them into the positions shown in figure 11.21. The resulting structure is hexagonal because the two hydrogens of each $H_2O$ molecule are oriented at an angle of 105° from one another. This is close to 120°, the angle between adjacent sides of a hexagon, so it is easiest for the molecules to arrange themselves in a honeycomblike hexagonal pattern. This structure leaves large holes among the molecules, which impart to ice its low density. The perfect hexagonal shapes of the minature ice crystals we call snowflakes reflect this internal geometry.

As noted, water is most dense not at 0° C but at 3.98° C, and the density drops toward both lower and higher temperatures. This odd behavior results from the tendency of water to form extremely short-lived "clusters" of $H_2O$ molecules that have the crystal structure of ice. These clusters last only $10^{-10}$ to $10^{-11}$ seconds on average, but they form more frequently and persist longer as the temperature decreases. Accordingly, water expands slightly as it cools toward the freezing point. At the same time, water also contracts with falling temperature as molecular motions decrease. As water cools to 3.98° C, the expansive effect of increasing cluster formation overshadows the contractive effect of cooling, so water expands slightly on cooling below that temperature.

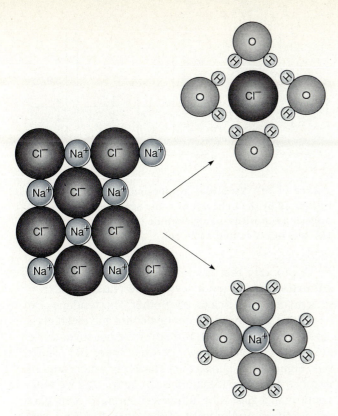

*Figure 11.22* Oriented H₂O molecules form hydration halos around sodium and chloride molecules freed from a dissolving salt crystal.

## The "Universal" Solvent: Powers of a Lopsided Molecule

Hydrogen oxide is a **polar molecule** in which one side is positively charged, whereas the other is negatively charged (see fig. 11.18). The side bearing the two hydrogen atoms is the positive side because of the relatively exposed proton in each of the hydrogen nuclei. The opposite side of the molecule has a negative charge because the electron orbits of the oxygen atom are concentrated on that side, away from the positive hydrogen nuclei.

Because of its polar nature, the water molecule is attracted both to charged particles, such as ions, and to other polar molecules, such as sugars, alcohols, and ammonia. Such substances readily dissolve in water because water molecules are able to insert themselves between adjacent ions or particles of the substance, and destroy its crystal structure (if it is a solid). Figure 11.22 shows water molecules dissolving a crystal fragment of sodium chloride and forming *hydration halos* around the sodium and chloride ions. In the hydration halo of a sodium ion ($Na^+$), the water molecules are oriented with their hydrogens pointing outward, whereas in the halo of the chloride ion ($Cl^-$), the hydrogens point inward. Hydration halos around ions in aqueous (water) solution prevent those ions from recombining into crystal structures. This explains why it is so difficult to

induce a dissolved substance, or **solute,** to crystallize from a solution in a dissolving liquid, or **solvent,** unless the solution is saturated (i.e., no more solute will dissolve). Even then, crystallization does not readily occur unless "seed crystals" of the solute are dropped into the solution.

Although water has been called the "universal solvent" due to its ability to dissolve polar and ionic substances, its solvent powers are actually far from universal. Such nonpolar, nonionic substances as wax, oil, fat, gasoline, quartz, and clay do not dissolve readily in water. Certain rock types, such as limestone, marble, gypsum rock, and salt rock, whose constituent minerals are ionic, are soluble to some extent (see chapter 8).

A final point about water as a solvent concerns the effects of solutes on the vapor pressure and the freezing and boiling points of water. The presence of dissolved molecules or ions, such as $Na^+$ and $Cl^-$, that are not able to evaporate reduces the vapor pressure of a liquid by an amount that is proportional to the concentration of the solute. (*Volatile* dissolved substances *increase* the vapor pressure, however.) This effect, known as *Raoult's* (Raw-*ooze*) *Law,* is independent of the chemical identity of the dissolved substance.

At the same time as dissolved substances lower vapor pressure, they also raise the boiling points of solutions, inasmuch as these are directly related to vapor pressure. A liquid boils when its vapor pressure becomes equal to the atmospheric pressure surrounding it. Under 1 atm pressure, pure water boils at 100° C. If the vapor pressure is lowered by solutes, however, the water must be heated above this temperature until its vapor pressure again equals 1 atm before it can boil. This is why soup boils at a slightly higher temperature than pure water.

The freezing point of a solvent is nearly independent of vapor pressure, but it, too, is affected by the presence of solutes. In fact, it is more strongly affected than is the boiling point, but for a different reason. The dissolved solute molecules interfere with the assembling of solvent molecules into crystal structures, so a lower temperature is required to slow their motions enough to permit those few crystal nuclei that do form to persist and grow. This is why you add antifreeze to the water of your car's cooling system to prevent it from freezing in winter.

The role of water as an efficient solvent and the effects of dissolved substances on ocean water are of great significance to the Earth system. Many ionic species and polar organic molecules are dissolved in seawater. These are derived originally from the weathering of rocks, from terrestrial and submarine volcanic eruptions, and from the decay of organisms. These ions and molecules enter into a great variety of chemical reactions that influence virtually every aspect of Earth's surface environment and

**Figure 11.23**  A water strider takes advantage of surface tension to exploit an ecologic niche occupied by few other species.

Photo © BioPhoto Assoc./Photo Researchers, Inc.

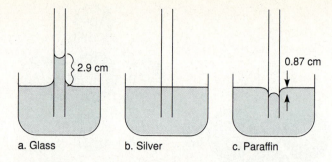

a. Glass    b. Silver    c. Paraffin

**Figure 11.24**  Variations in adhesion and capillarity are illustrated by the behavior of water standing in tubes. Adhesion is positive in a glass tube (*a*), neutral in silver (*b*), and negative in paraffin (*c*).

much of the subsurface to a depth of 700 km. The presence of dissolved substances also affects the behavior of seawater, depressing its freezing point and creating density differences that influence the motions and the locations of water masses (see chapter 12).

## Sticking Together: Cohesion and Adhesion

Because of the electrically dipolar nature of $H_2O$ molecules, water is a highly **cohesive** liquid, that is, its molecules tend to stick together. If you have ever tried to separate two panes of glass held firmly together by a thin film of water, you have experienced this truth directly. Another proof of the cohesiveness of water is its unusually high boiling point for a liquid of such low molecular weight.

The water strider in figure 11.23 is taking advantage of the cohesion of water in a form known as **surface tension.** Because the surface molecules of a liquid are surrounded on all sides except the upper side by other molecules, they experience a net attractive force downward that serves to shrink the water surface to a minimum area. There is no actual film on the surface despite the dimples produced by the insect's feet. The dimples are simply equilibrium forms that express a compromise between the insect's weight and the surface tension. That compromise tends to produce a smooth surface of minimum area. Another expression of surface tension is the spherical form of raindrops (they are *not* tear-shaped!), because a sphere is the form of least area for a given volume. On Earth, water surfaces are flat because gravitational forces are much more powerful than the forces of surface tension. In space craft, however, where there is no gravity, liquids tend to assume a spherical form. Measurements of the surface tension of water show that it is twice to three times as high as that of most organic liquids.

A property that is closely related to cohesion is **adhesion,** or the tendency of different substances to stick to, or not to stick to, each other. This tendency can be postive, as between water or alcohol and clean glass; zero, as between water and silver; or negative, as between mercury and glass or between water and paraffin wax. Figure 11.24 shows the behavior of water standing in tubes of different materials. In figure 11.24*a,* water standing in a glass tube adheres so strongly to the glass that it climbs up the walls of the tubes, forming a *meniscus,* or curved liquid surface and drawing the water beneath it upward. This effect is called **capillarity.** In figure 11.24*b,* water in a silver tube shows no tendency to adhere more strongly to the silver than to itself, so it forms no meniscus, and the water level in the tube remains the same as the level in the vessel beneath the tube. In figure 11.24*c,* water in a paraffin tube is so repelled by the paraffin that it forms a convex meniscus, which depresses the water level below that of the water in the vessel.

## The Salt Sea

For a long time, geologists accepted two popular ideas about the saltiness of the sea: that it is due to rivers washing the ionic products of weathering into the oceans, and that, like Utah's Great Salt Lake, the oceans have been getting saltier as time goes on. Now that we have gained a better understanding of the oceans, neither of these ideas is still tenable. In the following sections, I discuss the saltiness of the sea, where the salt comes from, and where it goes.

## The Composition of Seawater

In 1865, J. C. Forchammer analyzed hundreds of surface samples of seawater from different localities and confirmed a hypothesis that had first been proposed in 1819: in all his samples, the ratios of the major dissolved ions to one another were virtually constant. *Forchammer's principle* was subsequently upheld by C. R. Dittmar of the British *Challenger* expedition, who showed that the generalization held not only for surface waters, but also for waters of the ocean depths. What is not constant, however, is the total **salinity,** or the percentage of seawater represented by dissolved ionic solutes. Normal seawater ranges between 3.3% and 3.7% salt, or to use the preferred expression for salinity, between 33‰ and 37‰ ("per mil," or parts per thousand). In rivermouth estuaries, however, salinity can approach 0‰, and in shallow embayments with no freshwater input and with restricted circulation from the ocean, evaporation can raise salinity as high as 60‰, as it has in the Hamelin Pool Basin at the head of Shark Bay, western Australia (see fig. 2.7). Salinities of some isolated brine pools can rise to several hundred per mil.

Figure 11.25 shows the principal ingredients of seawater. The diagram assumes a salinity of 35‰, and the composition of the dissolved salt is shown in percentages. Chloride and sodium ions are dominant, and sulfate and magnesium are rather distant runners-up. Calcium and potassium each accounts for only about 1%, and the chemically important bicarbonate ion accounts for only 0.4%. All other species combined amount to less than 0.25% of the total. These include many biologically important nutrients, such as phosphate and nitrate.

## Dissolved Gases

The principal gases dissolved in seawater are oxygen, nitrogen, and carbon dioxide. In contrast to dissolved solids, the proportions of dissolved gases in the oceans are extremely variable. The following factors influence the concentration of a gas in seawater:

1. The concentration of the gas in the atmosphere
2. The relative activity of processes that generate the gas in, or remove the gas from, seawater
3. The temperature of the water
4. The salinity of the water.

More gas will dissolve if its atmospheric concentration increases, if marine gas-generative processes are active, if water temperature decreases, and if salinity decreases. Figure 11.26 shows the major gas exchanges that affect the ocean.

The first of the above factors is not likely to affect the concentration of oxygen or nitrogen, because their atmospheric concentrations are constant. Local variations in the concentrations of atmospheric carbon dioxide do occur, however, as a result of: (1) subaerial volcanism, organic decay, respiration (breathing) and both natural and human-caused oxidation (including combustion), all of which increase the concentration of $CO_2$, and (2) weathering, photosynthesis, and carbonate secretion, which decrease it. The main processes that generate gases in seawater are submarine volcanism, which produces nitrogen and carbon dioxide; photosynthesis, which produces oxygen and removes carbon dioxide; and animal respiration and organic decay, which remove oxygen and produce carbon dioxide. Blue-green algae have the ability to fix nitrogen by converting it to ammonia. Oxygen and carbon

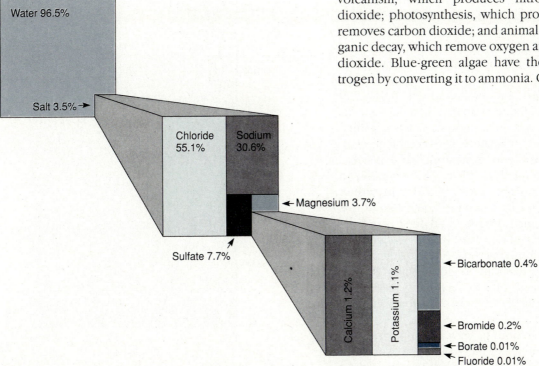

*Figure 11.25*  The salinity of seawater (*upper left block*) averages 3.5%, or 35‰. Percentages of the total salt content accounted for by seawater's ten most abundant ions and are shown in the expanded projections in the center and lower right. These percentages are constant in both space and time.

From D. A. Ross, *Introduction to Oceanography,* p. 199.
© Prentice-Hall, Inc., Englewood Cliffs, NJ.

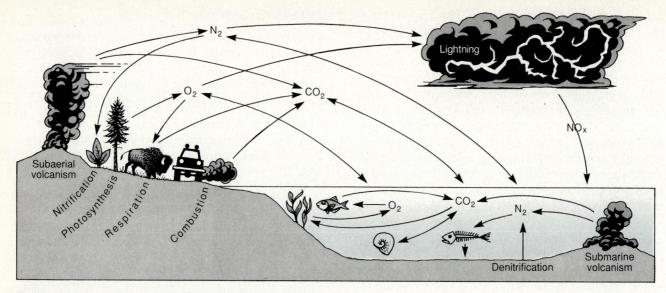

**Figure 11.26** Oxygen, nitrogen, and carbon dioxide, the three principal gases dissolved in the oceans, follow different paths as they enter and leave the oceanic reservoir.

dioxide dissolved in seawater react with minerals in rocks of the oceanic crust as it circulates through them, producing oxides and hydroxides of iron, and manganese, and carbonates of calcium, iron, and magnesium.

Figure 11.27 is a vertical profile of dissolved oxygen and carbon dioxide from the tropical east Pacific Ocean. The upper hundred meters of the ocean are rich in oxygen due to algal photosynthesis and the mixing of water with air by the action of wind and waves. At the same time, they are relatively poor in carbon dioxide because that gas is used in photosynthesis. Below 100 m depth, however, mixing is drastically reduced, as is photosynthesis due to loss of sunlight, whereas animal respiration continues unabated. This combination of factors creates an **oxygen minimum zone** that extends downward as much as 1 km. The reduction in photosynthesis also results in a rapid rise in carbon dioxide into the oxygen minimum zone. Below this zone, oxygen concentrations rise again due to the presence of well-oxygenated water that has descended from the polar regions beneath the surface water of the tropics (see chapter 12). In this same region, carbon dioxide concentrations increase due to decreasing water temperature and presumably due also to submarine volcanism.

Oxygen concentrations in seawater range from 0 to about 9 ml/l (milliliters per liter). Nitrogen concentrations are less variable, ranging from 8 to 15 ml/l. Carbon

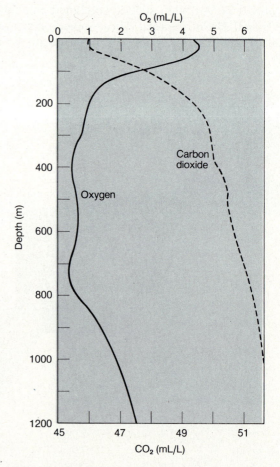

**Figure 11.27** Profiles of dissolved oxygen and carbon dioxide in the upper 1200 m of the ocean.

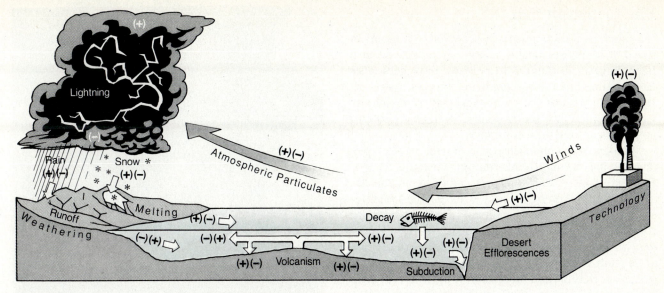

**Figure 11.28** The sources of salt in the ocean are more numerous than was formerly assumed. Ions are added not only by river water, but also by glaciers, submarine volcanism, dust, rain, snow, lightning, and the decay of organic material.

dioxide concentrations, between 45 and 54 ml/l, are the least variable of all because they are regulated by the various reactions involving calcium bicarbonate (see chapter 1). Under this **carbonate buffer system,** an increase in dissolved $CO_2$ tends to dissolve calcium carbonate deposits and convert them to calcium bicarbonate, whereas a decrease in dissolved carbon dioxide reverses the process, stimulating both the chemical precipitation and the biological secretion of calcium carbonate.

## The Sources of Salinity

The old assumption that the source of the salts in the world ocean is ions weathered from continental rocks and carried to the sea by rivers is correct, but only partially so. Today, we know that this is only one of many sources, of which the most significant are illustrated in figure 11.28. One clue that the weathering of rocks, is, by itself, an inadequate source is the abundance of chloride ion in seawater. Chlorine is a rare element in most rocks (except salt rock), but it is abundant in volcanic gases. The same observations apply to sulfur, which, as sulfate ion, is also a major constituent of seawater.

The major *cations* in seawater, however, including sodium, magnesium, calcium, and potassium, are abundant in terrestrial rocks. Clearly, therefore, rock weathering is a principal source of these cations. Another source of cations is dust, which can originate in volcanic eruptions, deserts, river floodplains, or forest fires and is subsequently blown out to sea by strong winds.

Dust can also be carried to the ocean by rain and snow, inasmuch as both raindrops and snowflakes begin life by condensing on dust nuclei.

Improbable as it might seem, lightning can also be a source of salinity. Electric discharges in the atmosphere combine nitrogen and oxygen into oxides of nitrogen, some of which form nitrite ($NO_2^-$) and nitrate ($NO_3^-$) ions when dissolved in water. These anions are then flushed into the oceans by rain.

A more significant source of dissolved substances is the death and decay of organic material. In addition to both cations and anions, decaying organic matter also yields a variety of organic molecules, some of which form soluble complexes called *chelates* (*key*-lates) with normally insoluble metal ions, such as iron, copper, zinc, manganese, and cobalt, thereby greatly increasing their availability to biological processes in the marine environment.

The most recently identified salinity source is submarine volcanism in median rifts and active seamounts. Undersea eruptions charge the overlying water with metal oxides and sulfides, and with chloride, bromide, and bicarbonate anions. Magmatically heated seawater passes through hot crustal rocks and leaches out heavy metal cations, such as manganese, iron, cobalt, nickel, copper, zinc, silver, and gold. These metalliferous brines are eventually discharged into the ocean as hot springs (fig. 11.29). Most of the cations precipitate out on the seafloor as colloidal (jellylike) oxide and sulfide ores, but enough remains in solution to qualify such submarine hot springs as a major, and perhaps the chief, source of the heavy metal cations in seawater.

## The Constant Sea

Inasmuch as the chemistry of seawater is constant in space, it seems reasonable to ask whether it has also been constant in time. Until recently, most geologists have assumed that it has *not* been constant. Although the available data are still somewhat equivocal, the prevailing opinion now seems to have swung more toward approximate long-term constancy.

Studies of the chemistry of sedimentary rock types reveal some definite trends through time (fig. 11.30). Dolomite rock is the dominant carbonate in the Archaeozoic and Proterozoic Eras, whereas limestone becomes dominant in the Phanerozoic. Because dolomite contains magnesium and calcite does not, this change *appears* to reflect a decrease in magnesium ion relative to calcium ion in seawater. Similarly, younger mudstones are rich in montmorillonite clay, whereas older mudstones are rich in illite clay. Again, this *appears* to represent increases in magnesium, sodium, and calcium ions relative to aluminum and potassium ions. This trend was regarded by many as an indication of change over time in the composition of seawater.

Several recent studies have shed light on this and other ambiguous chemical evidence. Analysis of mudstones in the Gulf Coast region has shown that the trend from illite to montmorillonite represents not a change in the kinds of clay being deposited then and now but the slow *conversion* of montmorillonite to illite as these mudstones exchange ions with the aqueous solutions that have been percolating through them ever since they were deposited. Furthermore, it has been shown that magnesium-rich brines convert limestone to dolomite at marine shorelines. These discoveries strengthen the case for constancy because they imply that chemical changes following deposition can account for the difference observed between modern and ancient sediments, so there is no need to assume original differences in the sediments.

In 1972, H. D. Holland of Harvard University published a paper in which he considered the chemical environments within which certain sedimentary rock types must have been produced. His results indicate that throughout its history, the chemistry of the oceans has deviated little from its present condition. Specifically, he found that none of the seven major components of seawater (sodium, magnesium, calcium, potassium, chloride, sulfate, and bicarbonate ions) could ever have been more than twice or less than half as concentrated in seawater as they are today. Furthermore, he concluded that the acidity of the oceans must have remained within the range of pH 7 to pH 9 (neutral to mildly alkaline). Their present acidity is pH 8.1.

A few years earlier, another piece of compelling evidence had been discovered by R. C. Reynolds of Dartmouth College, who analyzed the boron content of

**Figure 11.29** Heavy metal sulfides leached from underlying crustal rocks are being discharged into seawater by this dark "smoker," a hot spring located on the east Pacific ridge at a depth of about 3 km. The water temperature in such hot brines exceeds 350° C.

Photo courtesy of NOAA

illite clay in mudstones and limestones as old as 2700 million years. His results showed no significant variation in boron content over time. Because the concentration of boron in illite is directly correlated with its concentration in seawater, Reynolds concluded that the salinity of marine waters has remained virtually constant.

As is so often the case in science, these conclusions beg a question. We know that new ions are constantly entering the oceans from a variety of sources. If the chemistry of marine waters really is in a steady state, then all these ions must somehow be removed as quickly as they are added. The question is, how?

## Where the Salt Goes: Ionic Reservoirs

Ions, like water, have residence times in the sea, and these differ widely. Attempts have been made to calculate residence times for the various ions in seawater by dividing the total mass of each ion by its annual input rate (the output rate would work just as well). Unfortunately, these calculations are of little value because they are based on ions dissolved in river water and ignore inputs from submarine volcanism, dust, and other sources. In general, however, it can be said that the more concentrated an ion is in seawater, the more likely it is

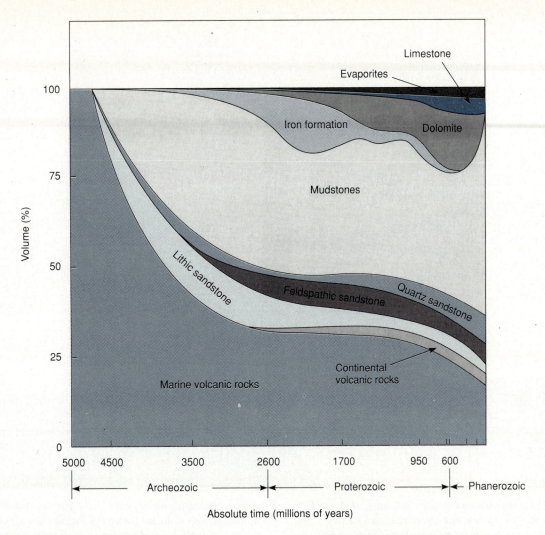

**Figure 11.30** Changes over time in the relative percentages by volume of sedimentary rock types. Because the total volume of sedimentary rocks has increased markedly through time, the diagram gives a false impression of the abundance of volcanic rocks in the Archeozoic.

that it has a long residence time. Ions that are essential to life processes, such as bicarbonate and phosphate, have relatively short residence times, whereas those that have little value in biological systems, such as sodium and chloride, have relatively long residence times. Let us consider these biologically unappealing ions first.

Buried deeply beneath the state of Michigan lie thick and extensive beds of evaporite, and similar deposits occur in many other parts of the world. These evaporites are reservoirs for five of the most abundant ions in seawater: sodium, potassium, calcium, chloride, and sulfate. Limestone and dolomite rock are reservoirs for calcium, magnesium, and bicarbonate ions. The lion's share of potassium is accounted for by the gradual conversion of montmorillonite clay to potassium-rich illite in mudstones. A third known reservoir for potassium is the green leaf-silicate *glauconite,* which forms on shallow seafloors by the alteration of biotite.

To these cold water chemical reactions, we must add the alteration of basalt and gabbro to spilite and am-

phibolite by hot, circulating seawater (see chapter 7). Taken together, these chemical interactions of minerals with cold and hot seawater amount to an efficient "ion sponge" for removing certain ions from seawater, notably iron, potassium, sodium, and sulfate. At the same time, seawater can acquire other ions in exchange, especially calcium. Curiously, magnesium is added to seawater at low temperatures but removed from it at higher temperatures.

Finally, we must again acknowledge the powerful influence of marine organisms, most notably the algae and cynaobacteria, in removing from seawater calcium, bicarbonate, and phosphate ions, together with lesser amounts of many other ions. In chapter 1, I take a brief look at the carbon cycle, which governs such removals (see fig. 1.22). This vital mechanism, essential to life on Earth, thoroughly pervades the plate tectonic, rock, and hydrologic cycles. Now, with a little background behind us, we can examine the carbon cycle in greater depth.

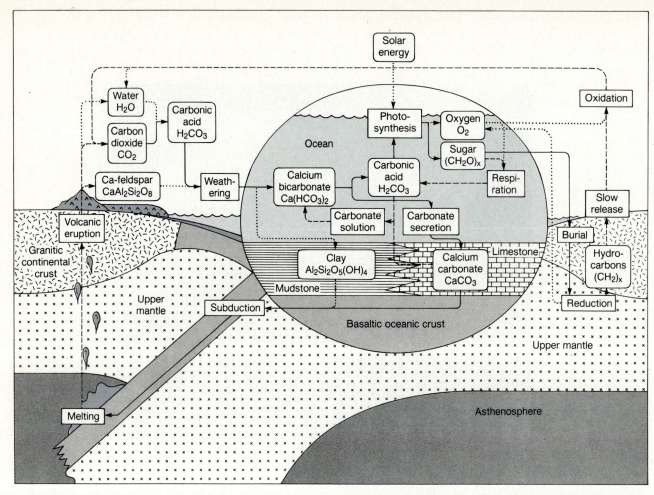

**Figure 11.31** Pervading the plate tectonic, rock, and hydrologic cycles, the carbon cycle modifies Earth's nonliving systems to the advantage of its living ones. Rounded boxes represent materials, rectangular boxes represent processes. Solid arrows indicate pathways that remove carbon from the atmosphere and ocean, dashed arrows indicate pathways that have the reverse effect, and dotted arrows indicate pathways that do not involve carbon. Chemical symbols are H, hydrogen; O, oxygen; C, carbon; Ca, calcium; Al, aluminum; and Si, silicon. The cycle begins with volcanic eruption (*left*).

## The Carbon Cycle

Before walking through the more detailed version of the carbon cycle in figure 11.31, you would probably find it helpful to review the simpler version in figure 1.22. As in that figure, the solid arrows in figure 11.31 indicate pathways that decrease the greenhouse effect and the acidity of seawater by removing carbon dioxide from the atmosphere and ocean; the dashed arrows indicate pathways that reverse those effects by returning carbon dioxide to the ocean and atmosphere; and the dotted arrows indicate pathways that do not involve carbon.

The carbon cycle begins as water vapor and carbon dioxide combine to produce carbonic acid, which weathers Ca-feldspar by hydrolysis (fig. 11.31, *left*), yielding calcium bicarbonate and clay. The clay settles

to the seafloor, where it becomes lithified into mudstone, whereas the calcium bicarbonate is dissolved in the ocean as calcium and bicarbonate ions. Marine algae then extract carbonic acid from the dissolved calcium bicarbonate during photosynthesis, leaving a residue of calcium carbonate that ultimately is deposited on the seafloor as limestone, although much of it is incorporated into shell material along the way. Note that both limestone and mudstone can be *subducted* (solid arrows at *bottom* of fig. 11.31), and ultimately *melted,* reconstituting basaltic magma (dashed arrow at *left*), and the cycle can then repeat.

This fundamental, cyclic pattern is further complicated by two other subcycles shown in the upper right and lower center of the inset oval in figure 11.31. As indicated by the presence of dashed arrows, both these subcycles return carbon to the ocean and atmosphere.

# Food for Thought

Although we seem to have reasonable answers to the question of "where the salt goes," there is one troublesome aspect of the problem that still remains to be solved, and that has to do with the nature of the steady state in which the ocean seems to be maintaining itself. In chapter 2, I note an important difference between equilibrium and cybernetic systems. Equilibrium systems equalize inputs and outputs *passively* by varying the amounts of matter and energy in storage within the system. Cybernetic systems on the other hand, maintain desired amounts of matter and energy in storage by *actively* regulating input or output. I also note that in Nature only living systems, or systems created by or regulated by living systems, are cybernetic.

In this chapter, I review much of the evidence that has led to a new view of the oceans as relatively *constant in time* with respect to salinity, acidity, and composition. I also point out that many of the inputs and outputs of dissolved ions involve processes that can be extremely variable in time. The formation of large, shallow-marine evaporite basins depends on tectonic processes that are episodic and irregularly distributed in both space and time. Rates of weathering, erosion, and river flow also depend on tectonic processes; nevertheless, no significant fluctuations in the chemical or physical steady states of the sea have yet been found.

The world ocean, in other words, is behaving very much as if it were alive or part of something alive. The thought of a living ocean is a bit bizarre, perhaps, but the thought of the ocean being part of a living system is not a wholly outrageous one. Human blood is not "alive," but it is a vital part of living human beings. In a provocative book entitled *Gaia: A New Look at Life on Earth,* a British biochemist named James Lovelock reviews and evaluates the evidence in favor of a view of our planet as a single, living system of which the oceans are the lifeblood. Lovelock's idea has met with a deal of eyebrow raising, although

scarcely with the cries of outrage earned by Charles Darwin's theory of natural selection! Possibly, much of the objection to the "Gaia hypothesis" (Gaia is the Earth goddess of ancient Greece) stems from the understandable misconception that it postulates an intelligent or at least a sentient being, but such a characterization of our planet is dearly not Lovelock's intent.

The idea of a living "Earth-being" is, rather, simply an extension of the ecosystem concept, in which various living organisms are functionally integrated into a larger system that has an identity and a characterizable behavior of its own. I doubt that many biologists would maintain that ecosystems are sentient or intelligent, but the concept that they are living systems, by the definition given in chapter 2, is a thoroughly respectable one.

The analogy between seawater and human blood is not as far-fetched as it might sound. Blood plasma, like seawater, is mostly water (roughly 90%), and it contains dissolved ions and gases in addition to proteins, organic nutrients, nitrogenous wastes, and special products in transport. With the additon of phosphate, the ions are, in fact, the same ones as the major ions in seawater and in roughly the same proportions! Furthermore, the dissolved gases in blood plasma are also the same three as are present in seawater: oxygen, nitrogen, and carbon dioxide. Perhaps it should come as no surprise that we carry in our veins and arteries this liquid legacy of the constant sea. Life began in the sea, and seawater serves the function of blood for all unicellular animals and many of the less advanced multicellular ones as well. Only when organisms became so complex that direct access to seawater was cut off for internal cells and organs did it become necessary to have a cirulatory system to carry vital ions, gases, and nutrients to those cells and organs, and to flush away their metabolic wastes.

Can a similar function for the sea with respect to the whole planet be envisioned for Earth? If so, we raise a

plethora of new questions, of which the principal one is, what are the controls? If the chemistry of seawater is being stabilized by a worldwide network of biocybernetic feedback systems, what are they, where are they, and how do they function? The biologically mediated carbonate buffer system is one prime candidate, but that is only a beginning. Many more basic data and much more research are needed before we can develop hypotheses regarding the extent of biological control over the steady state of marine ionic chemistry.

If the "Gaia hypothesis" proves to be valid, it should stimulate some significant rethinking about our relationship to our planet. We might, for example, become more concerned over the welfare of "Earth's lifeblood" under the increasing impact of human use and abuse of the marine environment. Curiously, however, Lovelock himself sounds a note of optimism in this regard. Just as our human physiology can withstand a staggering degree of abuse by stimulants, depressants, narcotics, carcinogens, and other systemic poisons, so, perhaps, the cybernetic oceans are more capable of cleansing themselves of human abuses than they would be if they were simple equilibrium systems. Although this view might relieve us from some of the doomsday predictions of environmental extremists, it should not be taken as a license to pollute and pillage our marine environment on the assumption that it will take care of itself. Until we have enough new information to be able to raise intelligent hypotheses about the nature of controls on the marine system, we should regard Lovelock's concept as a "working hypothesis," which could be a reasonable guide to management practices for the years ahead.

An astute student of mine recently suggested what seems to be a sensible and workable approach to the problem: "Why quibble over a definition? If Earth *acts* as if it were alive, then we ought to be treating it as if it *were* alive."

The first of these paths is the process of **respiration,** which reverses that of photosynthesis by combining sugar and oxygen to yield carbonic acid. Respiration is the process whereby all organisms—both plants and animals—obtain energy for bodily functions by "burning" carbohydrate food with oxygen within their bodies. This is, in fact, the reason why we eat carbohydrates and breathe oxygen, and it stands behind every concerned parent's admonishment: "Eat so you'll have energy!"

Normally, respiration exactly balances photosynthesis. If respiration should exceed photosynthesis, as it occasionally does, an excess of carbonic acid will be formed, rendering the ocean more acidic. As noted previously, volcanism has the same effect. This potentially life-threatening trend is reversed by the second subcycle, representing *carbonate solution.* In this wholly inorganic buffering process, the excess carbonic acid is neutralized as it reacts with calcium carbonate to form calcium bicarbonate.

A solid arrow leaves the oval on the right, showing that some carbohydrate (in the form of dead organic matter) is not respired but is buried in seafloor sediment, where it is heated and stripped of its oxygen, a process called *reduction.* Ultimately, reduction alters organic matter to **hydrocarbons,** the principal ingredients of petroleum. The stripped oxygen enters the ocean and atmosphere. Over time, these buried hydrocarbons also leak slowly into the atmosphere where they combine with oxygen (see the "oxidation" box in fig. 11.31 *right*) to form water and carbon dioxide (dashed arrow, *top*). Because respiration balances photosynthesis, our atmosphere actually owes its oxygen content not to photosynthesis but to this slow burial of organic matter. Therefore, if all Earth's buried petroleum were extracted and burned, it would consume all the oxygen in the atmosphere, a sobering thought as our oil-hungry society strives to produce and burn more petroleum every year!

# Summary

Notable pioneers in the science of *oceanology* include Magellan, Franklin, Folger, **Maury,** Cook, Darwin, Thomson, Agassiz, and Nansen. Notable vessels include *HMS Endeavour, HMS Beagle,* **HMS Challenger, Glomar Challenger,** and **JOIDES Resolution.** The Aqualung and SCUBA are self-contained, underwater breathing apparatuses that can take divers as deep as 90 m. Gravity **corers** can retrieve about 1 m of sediment, piston corers up to 30 m of sediment. Other **tethered samplers** include the *Nansen bottle, secchi disk* (for measuring *secchi depth*), *transmissometer, nephelometer, dredge, grab sampler, plankton net* (for collecting **plankton**), *trawl* (for collecting **nekton**), *current meter, drogue, wave rider buoy, tide gage, and wave staff.* **Bathymetric maps** show depth contours. **Echo sounders** measure seafloor topography from travel time of reflected sound waves. *Sonar* is similar, but uses a rotating, side-scanning sound signal. *Magnetometers* and *gravimeters* remotely sense magnetic and gravitational anomalies, respectively, on the seafloor, giving clues to rock type and structure. **Seismic reflection profiling** uses a powerful sound signal to reveal structures beneath the seafloor. **Seismic refraction profiling** determines such structures from sound waves refracted by them.

Nine percent of Earth's surface is above 1000 m above sea level, 29% between 1000 and −1000 m, 8% between −1000 and −3000 m, 53% between −3000 and −6000 m, and 1% below −6000 m. *Oceanic ridges* underlie 32.7% of the ocean. **Abyssal hills** are small (<1000 m) seafloor volcanoes; **seamounts** are larger. **Guyots** are seamounts with flat, wave-eroded tops. Fringing coral reefs often grow upward as fast as a guyot sinks, forming a circular **atoll. Abyssal plains** are thick, continent-derived, sedimentary blankets on the deep seafloor. The *shelf break* is at the top of the *continental slope.* **Submarine canyons** probably are sunken river valleys formed at divergent margins during early rifting. The *continental rise* is a thick prism of continental sediment seaward of the continental slope. Volcanic exhalations can account for Earth's water. Sea level has been approximately constant in time. The **residence time** of an ion, or of water, in the ocean is its mass in storage divided by its input or output rate.

At **absolute zero,** molecular motion stops. The **calorie** is the amount of heat required to raise the temperature of 1 g of water 1° C. Increased temperature raises molecular motion until it breaks the crystal bonds of the solid state, causing melting into the liquid state at higher pressures or *sublimation* and **evaporation** into the vapor state at lower pressures. The **latent heat of fusion** of ice is 80 cal/g; the **latent heat of evaporation** of water is 540 cal/g. At the *triple point,* solid, liquid, and gas phases coexist stably. Past the *critical point* (above the *critical temperature* and *pressure*), the liquid phase does not exist. The **vapor pressure** of a solid or liquid is that portion of the adjacent atmospheric pressure due to the evaporation of high-energy molecules of the substance. **Saturation vapor pressure** is attained when the *condensation* rate of molecules equals the evaporation rate. The *boiling point* of a liquid is that temperature at which its vapor pressure equals the atmospheric pressure; hence, it varies with the atmospheric pressure. The **specific heat** of a substance is the heat required to raise the temperature of the substance by a certain amount relative to the heat required to raise an equal mass of water by the same amount. Most substances have a lower specific heat than water.

Because water, the "universal" **solvent,** consists of **polar molecules,** it dissolves **solutes,** forming *hydration halos* around their ions. By *Raoult's Law,* solutes raise the boiling points of solvents. **Cohesion** is the property whereby a substance sticks to itself. **Surface tension** is a measure of cohesion at a free liquid surface. **Adhesion** is the property whereby one substance sticks to another. It is expressed by **capillarity,** the tendency of a liquid to stand higher or be depressed in a tube. A concave-up *meniscus* indicates positive adhesion; no meniscus, neutral adhesion; a concave-down meniscus, negative adhesion.

*Forchammer's principle* states that ionic ratios in the ocean are constant in space. Normal **salinity** ranges from 33‰ to 37‰. More gas dissolves in seawater with increases in its atmospheric concentration and gas generation processes, and with decreases in temperature and salinity. The **oxygen minimum zone** extends from about 100 m to as much as 1 km depth and is due to respiration and lack of mixing and photosynthesis. The biologically controlled **carbonate buffer system** regulates oceanic acidity. Sources of oceanic cations include weathering, organic decay, and dust from volcanism, deserts, floodplains, and forest fires. Sources of anions include volcanism and lightning. Many otherwise insoluble metals are rendered soluble as organometallic complexes called *chelates.*

Changes in sedimentary rocks and minerals over time (e.g., dolomite to limestone, illite to montmorillonite) are due to alteration, not to original deposition. Geochemical studies show that ocean chemistry has been constant in time. Evaporites store $Na^+$, $K^+$, $Ca^{+2}$, $Cl^-$, and $SO_4^{-2}$; carbonates store $Ca^{+2}$, $Mg^{+2}$, and $HCO_3^-$; clay minerals and glauconite store $K^+$. Alteration of basalt to spilite stores $Fe^{+2}$, $K^+$, $Na^+$, and $SO_4^{-2}$. In the *carbon cycle,* weathering, photosynthesis, and carbonate secretion transform lava, water, and carbon dioxide into clay, limestone, carbohydrate, and oxygen, thereby reducing oceanic acidity and the greenhouse effect. Subduction, **respiration,** *carbonate solution,* and the oxidation of buried **hydrocarbons** reverse this trend. The ocean resembles human blood plasma in its constant chemistry, and behaves as if it were part of a living system.

## Key Terms

Matthew Fontaine Maury
*HMS Challenger*
*Glomar Challenger*
*JOIDES Resolution*
corer
tethered sampler
plankton
nekton
bathymetric map
echo sounder

seismic reflection
   profiling
seismic refraction
   profiling
abyssal hill
seamount
abyssal plain
guyot
atoll
submarine canyon

residence time
absolute zero
calorie
latent heat of fusion
evaporation
latent heat of
  evaporation
sublimation
vapor pressure
condensation
saturation vapor
  pressure
specific heat

polar molecule
solute
solvent
cohesion
surface tension
adhesion
capillarity
salinity
oxygen minimum zone
carbonate buffer system
respiration
hydrocarbon

## Questions for Review

1. Identify the chief contributions to oceanology made by Franklin and Folger, Maury, the *Challenger* expedition, and the *Glomar Challenger.*
2. Describe the three main strategies for accessing data about the oceans and ocean floors. Give examples of each, and explain their uses.
3. What are the differences among echo sounding, sonar, seismic reflection profiling, and seismic refraction profiling? Explain the function of each.
4. What are the distinguishing characteristics of abyssal hills, seamounts, guyots, and atolls? Explain the origin of each.
5. Explain the origins of continental shelves, continental slopes, continental rises, and abyssal plains. Why are abyssal plains rare in the Pacific Ocean?
6. Explain how increases in temperature and pressure affect the state of hydrogen oxide.
7. What is the difference between a triple point and a critical point?
8. What is vapor pressure? What condition is necessary for the attainment of saturation vapor pressure?
9. Would you expect dry rock or wet rock to heat up faster on a sunny day? Why?
10. Water slowly dissolves gypsum and limestone but will not dissolve sandstone at all. Why?
11. Name four factors that would favor the exsolution of carbon dioxide from seawater.
12. Identify the main sources and reservoirs for the major oceanic cations and anions.
13. List each of the various processes of the carbon cycle under one of the following headings as appropriate: decreases oceanic acidity and the greenhouse effect; increases oceanic acidity and the greenhouse effect; does not affect oceanic acidity and the greenhouse effect.

# 12 Waves, Currents, and Battered Shores: Processes of the World Ocean

## Outline

*When you do dance, I wish you*

*A wave o' the sea, that you might ever do*

*Nothing but that.*

*William Shakespeare*

*The Winter's Tale*

Photo: Waves battering a headland on the coast of Maine.

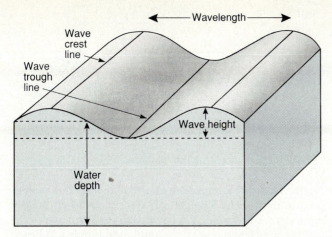

**Figure 12.1** Anatomy of a water wave. Most natural waves have crests that are somewhat narrower, and troughs that are somewhat broader, than this mathematically ideal sine curve.

## Shapes on the Wind: Water Waves

When wind blows across an open water surface, it exerts a frictional drag on that surface, transferring energy to it. Out of this energizing of water surfaces, a familiar type of closed system arises: the **water wave.** In a water wave, inputs of kinetic wind energy are stored as potential energy in the raising of a wave crest and are then released in the form of kinetic and heat energy as the crest collapses. As long as the input of kinetic energy from wind continues, the water wave system continues to oscillate in this fashion. The amount of energy input is proportional to the square of the height of the resulting waves; hence, a wave with a height of 10 m, for example, has 100 times the energy of a wave with a height of 1 m.

The shape of water waves is similar to a mathematical figure called a *sine curve* (fig. 12.1), although the crests of most natural waves are actually narrower, and the troughs broader, than the crests and troughs of sine waves. The perpendicular distance between the crests of adjacent waves is called the **wavelength** ($L$). *Wave height* ($H$) is the vertical distance between the bottom of a wave trough and the top of the adjacent crests, and this generally increases with wavelength. The *water depth* ($D$) below a wave is measured from mean sea level (i.e., from one-half the wave height).

A wave's motion is measured by its **period** ($T$), the time required in seconds for two successive crests to pass a stationary point, and the *velocity* ($V$) with which the crests are moving. Velocity and period are related to wavelength by the expression, $L = VT$, a form of the familiar "distance equals rate times time" relationship. As wavelength increases, both velocity and period increase as well; hence, longer waves are faster than shorter ones, and they also take longer to repeat.

## Deep-Water Waves

Waves on the open ocean vary enormously in wavelength, period, velocity, and height. The smallest have wavelengths of less than 2 cm, periods of less than 0.1 sec, velocities of about 0.2 m/sec, and heights of a few millimeters. The largest storm waves can have wavelengths of over 350 m, periods of over 15 sec, velocities of over 25 m/sec, and heights of over 30 m. The maximum size of the waves produced by a given storm wind depends not only on the energy of the wind, but also on how long it lasts and on the *fetch,* or the length of free ocean surface over which the wind blows. This distance is generally limited to about 250 km, or about one-fourth the circumference of a typical cyclonic storm system (see chapter 15).

During a storm at sea, a broad spectrum of waves is generated. Because these waves differ in wavelength and veolcity, they soon become sorted out as they travel away from the storm center, a phenomenon known as *wave dispersion.* Longer waves travel ahead of shorter ones because of their greater velocity. This explains why the sea surface beneath a storm center is chaotic, with waves of all sizes, but becomes more orderly with increasing distance from the storm as the waves sort themselves into groups of more uniform size. As the waves move away from the influence of the storm wind, they become smoother and more regular and are known as *swells.*

It is a common misconception that water moves bodily along with a moving wave. Figure 12.2 shows the path actually followed by a cork floating on the sea surface as a wave passes through the underlying water. Notice how the cork is first lifted by the oncoming wave crest (fig 12.2*a* and *b*) and then carried forward by the crest (fig. 12.2*c* and *d*) until it finally slides down the back side of the wave into the oncoming trough (fig. 12.2*e* and *f*). In so doing, the cork follows a circular path whose net movement in the direction of wave advance is zero, and whose diameter is equal to the wave height. The water molecules at the surface of the wave all behave identically to the cork; hence, the water does not travel forward with the wave form, but simply oscillates in a vertical, circular orbit as the wave form passes through it.

In figure 12.3, these orbits are shown in various stages of progression throughout a single wavelength and at various depths. Notice that the diameter of the circular orbits decreases exponentially with depth below the surface. At a depth of ⅑ the wavelength below mean sea level, the orbits have only ½ their surface diameter, and at a depth of ½ the wavelength, their diameters have been reduced to only ¹⁄₂₃ of the surface value. This is a sufficiently small amount of motion that a depth of half the wavelength is regarded as **wave base,** the level below which the wave energy is no longer significant.

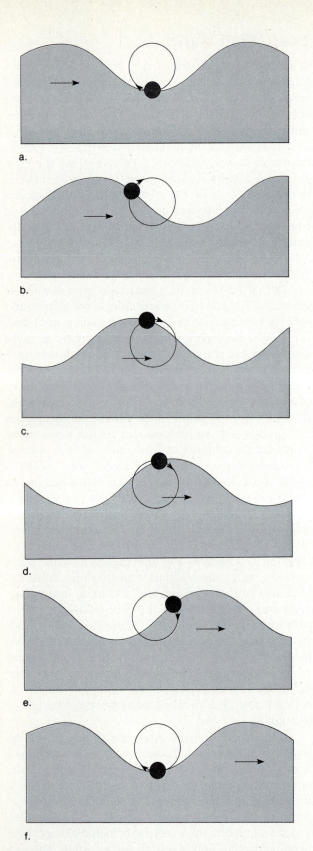

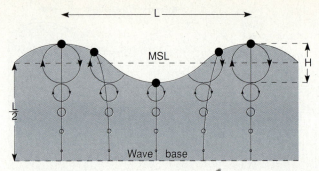

**Figure 12.3** The amplitude of orbital motion in a wave is equal to the wave's height (*H*) at mean sea level (*MSL*), but it decreases exponentially with depth. At a depth of half the wavelength (*L/2*), the amplitude is only 1/23 its surface value; hence, this depth is regarded as the base of the wave. Lines connecting equivalent points on all circles of a given column suggest that water wave motion is similar to that of wheat stalks waving in the wind.

It follows that for long wavelengths, wave base is deeper than for shorter wavelengths. Waves whose bases lie above the seafloor are known as **deep-water waves.**

## Shoaling Waves and Wave Refraction

As deep-water waves approach a coastline, they move into progressively shallower water until the wave base contacts the bottom, whereupon they are known as **shoaling waves.** Shoreward of this point, the circular orbits of the water molecules in the waves become compressed into horizontal ellipses that become more elongate as the water gets shallower (fig. 12.4). At the same time, the wavelength is shortened, the troughs are flattened, and the waves become steeper as friction from the seafloor reduces their velocity.

Figure 12.5 illustrates an important effect of shoaling on the behavior of waves. On irregular coastlines, the seafloor remains relatively deep off bays, but rapidly *shoals,* or becomes shallow, off headlands. Because of this, the wavelength and velocity of waves are reduced sooner off headlands than they are off bays. This results in a bending of the wave crests (solid lines in fig. 12.5)

**Figure 12.2** A cork floating on the sea surface illustrates the orbital motion of the water surface as wave forms pass through it. Motion of the top of the orbit is the same as the direction of wave advance.

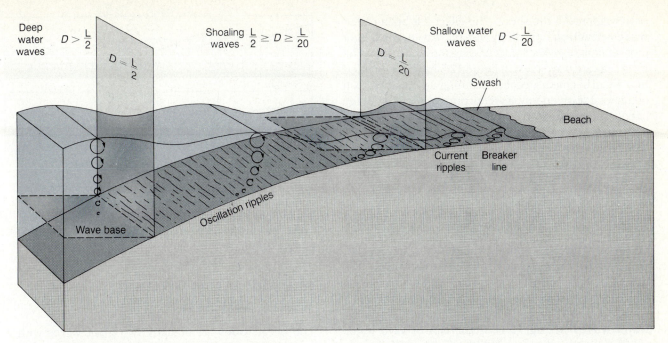

**Figure 12.4** As the wave base of shoaling waves is compressed against the rising seabed, wavelength and velocity are reduced, wave orbits flatten, and wave height and steepness increase. This increasing interaction of wave base with the bottom is reflected in ripples on the seafloor, which begin as symmetrical oscillation ripples but become increasingly asymmetrical current ripples toward shore. At the breaker line, the depth falls below 1.28 times the wave height. Here, waves become unstable, and they collapse and rush up the beach face as swash.

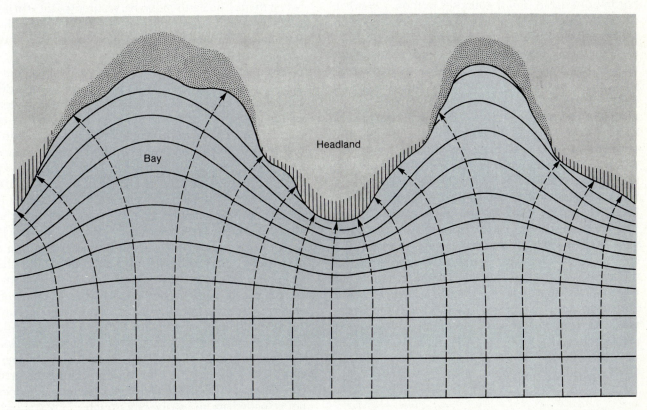

**Figure 12.5** Waves shoal off headlands before they shoal in bays, resulting in a bending, or refraction, of the wave crestlines. Equal wave energy parcels, shown as squares at bottom, become concentrated on headlands and dispersed in bays.

as they approach the shore. The effect has been called **wave refraction** by analogy with the way in which light rays are bent or *refracted* on passing through a lens. The most seaward wave crest in figure 12.5 has been divided into equal segments, each containing an equal portion of the total wave energy. As refraction occurs, the energy becomes concentrated on the headlands but dispersed in the bays. This has important consequences for shoreline erosion and deposition, which I discuss in a following section.

## Water on the Move: Shallow-Water Waves

Shoaling waves still continue to behave essentially like deep-water waves for some time after their wave base has hit bottom, but when the water depth approaches $\frac{1}{20}$ the deep-water wavelength, their behavior changes markedly. At this point, they have become **shallow-water waves** (see fig. 12.4). In this type of wave, water tends to move shoreward with the wave form. As shoaling waves approach the shoreline, friction against the rising bottom causes the upper parts of the orbital paths to overshoot the lower parts. Because their crests have become crowded together, the waves become steeper, but only up to a point. When the angle between the two sides of a wave narrows to 120° or less, the wave becomes unstable, and it collapses to form a *breaker*. Normally, this happens where the water depth falls to 1.28 times the wave height. Because of this relationship, the breaker zone migrates toward deeper water offshore when waves are large and high, and returns inshore when waves are small and low. Breakers can be either of the *spilling* type, in which collapse occurs gradually and progressively as the wave approaches shore (fig. 12.6*a*), or of the *plunging* type, in which the crest tends to shoot suddenly forward and then to roll down, enclosing a tubular air space in front of the collapsing wave (fig. 12.6*b*). Waves that are steep in deep water are likely to become spilling breakers, whereas low swells are more likely to be "plungers."

The farthest shoreward advance of a shallow-water wave is the **swash,** or watersheet, that rushes up the beach face. When waves approach the shore head-on, this uprush occurs perpendicularly to the shoreline, but when waves approach the shoreline obliquely, the uprush occurs at an angle. In either case, the **backwash,** or return flow of water, occurs at a right angle to the shoreline (fig. 12.7).

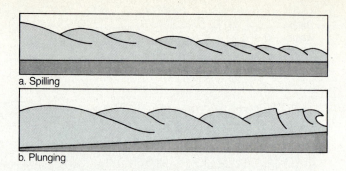

a. Spilling

b. Plunging

*Figure 12.6* Short, steep waves tend to be spillers (*a*), because they are too shallow to be compressed much by the seafloor during shoaling. Long, gentle swells, on the other hand are usually plungers (*b*) because of their greater depth.

## The Tides: The Greatest Waves on Earth

The ocean **tides** are actually two giant waves with a wavelength equal to half Earth's circumference (fig. 12.8). The tides are generated not by the wind, but by the gravitational pulls of the Moon and Sun, and driven by the relative motions of the Moon and Earth. Strange as it may seem, the Moon does not actually revolve around Earth. Rather, both Earth and Moon revolve around a *barycenter,* or common center of gravity, between them. The Earth-Moon barycenter (point *B* in fig. 12.8) is located approximately 1700 km below Earth's surface along a line connecting the centers of the two bodies.

George Darwin (son of Charles Darwin) studied these forces and their effect on Earth's oceans and developed the theory on which we base our understanding of the tides. The simplest interpretation of Darwin's mathematically complex theory is as follows. The Moon's gravitational force decreases with distance from its center. Accordingly, that force is about 6.5% stronger on the side of Earth that faces the Moon than it is on the opposite side. As a result of this, Earth is "stretched," or elongated slightly in the direction of the Moon. Because water is more easily stretched than rock, most of that stretching occurs in the oceans, raising a prominent "water mountain," or *tidal crest,* on the side facing the Moon (point *P* in fig. 12.8). A second tidal crest is raised on the opposite side, where Earth is literally pulled away from the ocean by the Moon's gravity. Sea level is depressed within the broad, intervening band of Earth's surface, giving the water surface a shape that is reminiscent of a football in the highly exaggerated view of figure 12.8. The band separating the tidal crests is referred to as a *tidal trough,* even though it does not actually have the shape of a trough.

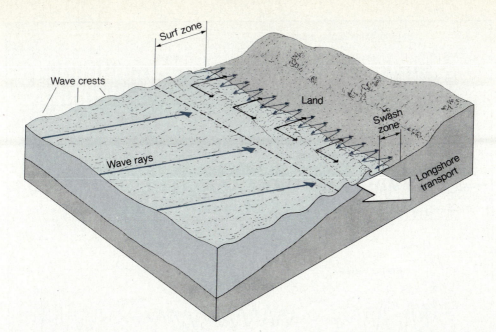

**Figure 12.7** Breakers often approach a shoreline obliquely, causing swash to rush up the beach face at an angle, but backwash returns perpendicularly to the shoreline. In the surf zone, a longshore current (*broad arrow*) develops and flows in the direction opposite to that from which the waves are approaching. Swash and backwash move sediment along the beach face while the longshore current moves it below the water level.

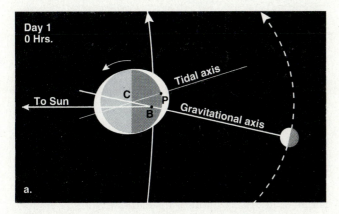

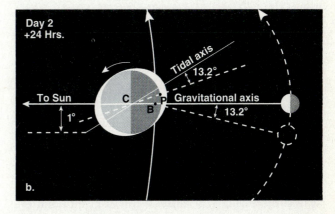

**Figure 12.8** Earth and Moon both orbit a common center of gravity, or barycenter (point *B*), which in turn orbits the Sun. The unequal pull of lunar gravity on opposite sides of Earth raises the tidal crests, but Earth's rapid easterly rotation (*curved arrow*) skews the tidal axis strongly away from the gravitational axis. Twenty-four hours later (*B*), the barycenter has moved about 1° of arc along its orbit while the Moon and tidal axis have moved through 13.2°. Earth, however, must still rotate for an additional 50 minutes to bring point *P* directly beneath the tidal crest, as it was 24 hours earlier.

The tidal crests move slowly eastward (counterclockwise) at the same rate as the Moon's revolution in its orbit (fig. 12.8). At the same time, Earth rotates *rapidly* eastward (note that *rotation* means turning on an axis, whereas *revolution* means following an orbit). The oceans, of course, rotate with the planet, moving through the tidal crests and troughs as river water moves through a stationary wave form in a rapid. Waves of this kind, that remain stationary while water moves through them, are known as *standing waves*. They are the reverse of water waves, in which the water remains stationary and the wave form moves through it. Because of this, the tidal crests and troughs actually travel *westward* with respect to the ocean surface as it passes through them.

In figure 12.8, the tidal axis is not aligned with the Earth-Moon gravitational axis, but lies at a considerable angle to it. This apparent paradox is simply a result of Earth's rapid eastward rotation, which drags the ocean's water forward as the Moon's gravitational force pulls it

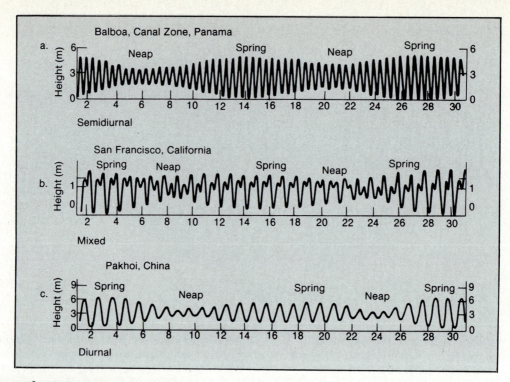

**Figure 12.9** The shapes of tidal curves reflect the relative effects of the two tidal crests as they pass. In Balboa (*a*), both crests are equally effective, producing a semidiurnal tide. In San Francisco (*b*), one crest is dominant, creating a mixed tide. In Pakhoi (*c*), only one crest is effective, resulting in a diurnal tide.

up into the tidal crests. If Earth's surface were free of landmasses, and if the ocean basins were at least 22.5 km deep at the equator, this drag effect would not occur. As we have seen, however, ocean depths average only about 3.8 km; hence, the tidal crests with their enormous wavelength (about 20,000 km at the equator) are actually huge shallow-water waves, slowed by friction against the seafloor. Because of this, a given point on Earth's surface (such as point *P* in fig. 12.8) passes beneath the Moon several hours before it passes through the tidal crest. The duration of this so-called *lunitidal interval* varies from place to place as a function of latitude, ocean depth, the shape of the tidal crest, and the coastal geometry of any landmasses that might interrupt the passage of the tidal crest.

In figure 12.8, it is clear that any given point on Earth's surface will pass beneath a tidal crest twice during a complete rotation of the planet. The time of the point's passage beneath a crest is called **high tide.** In contrast, **low tide** occurs when the point passes beneath a tidal trough. If the Moon were stationary with respect to Earth, the time interval between successive high (or low) tides would be exactly 12 hours as Earth rotated beneath the Moon. Because the Moon moves

eastward in its orbit, however, Earth must rotate for an additional 13.2° each day in order to catch up with it (fig. 12.8*b*). This takes about 50 minutes, so high and low tides occur about 50 minutes later than they did on the preceding day. Because there are two tides each day, the *tidal period,* or the time interval between the passage of point *P* beneath one tidal crest and the next, is about 12 hours, 25 minutes instead of exactly 12 hours.

Other factors further complicate this already complex situation. The plane of the Moon's orbit is inclined at an angle of about 5° to the ecliptic, the plane of Earth's orbit (see figs. 16.15 and 16.16). Furthermore, the Moon's orbit is somewhat eccentric, and the form of the ocean basin is highly irregular. These factors vary the patterns of tidal fluctuation such that tides range in height from about 0.5 m in the equatorial regions to over 15 m at higher latitudes in restricted bays such as the Bay of Fundy, Nova Scotia. In some localities (fig. 12.9*a*) both tidal crests are equally effective, and a **semidiurnal** (twice daily) **tide** results with two subequal high tides and two subequal low tides occurring each day. In other localities (fig. 12.9*b*), one crest is more effective than the other, and a **mixed tide** results, in which successive high tides alternate between high and moderate. **Diurnal tides** (fig. 12.9*c*), in which only one tidal crest is effective, are relatively rare.

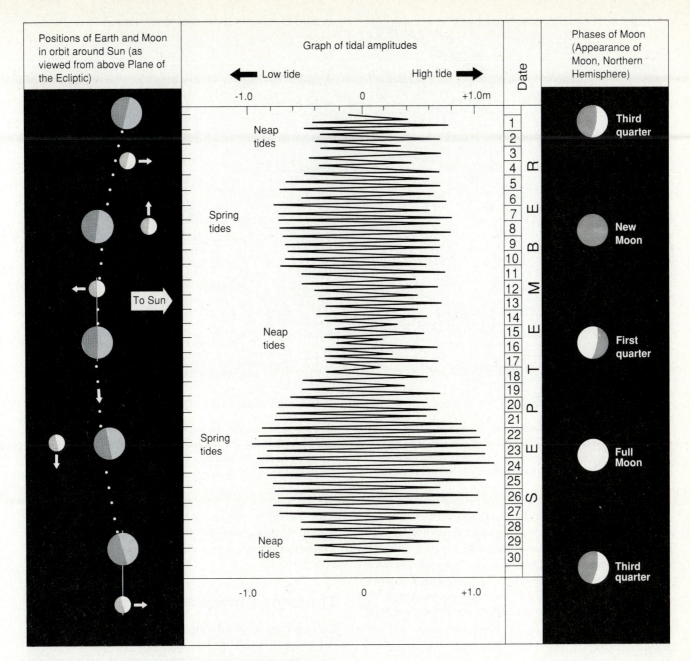

**Figure 12.10** High, spring tides occur when Earth, Moon, and Sun are aligned, as at the new and full Moons. Low, neap tides occur when the three bodies form a right angle, as at first and third quarters.

The amplitudes of the tidal waves in figure 12.9 exhibit a cyclic variation, which is due to interference from a second tide-producing force: that of the Sun. Despite the fact that the Sun is 27 million times more massive than the Moon, its tide-producing effect is only 45% as great as the Moon's because it is 389 times farther away from Earth. Figure 12.10 shows how this interference works. The tides of highest amplitude, called **spring tides,** occur when Earth, Sun, and Moon are all in a line as at the new and full Moons. In these positions, the tidal effects of Moon and Sun are additive. In contrast, when the three bodies form a right angle as at first and third quarters, the tidal effects of Sun and Moon tend to cancel each other, producing tides of low amplitude, called **neap tides.** Secondary variations within the fundamental pattern of spring and neap tides are caused by orbital eccentricities and inclinations.

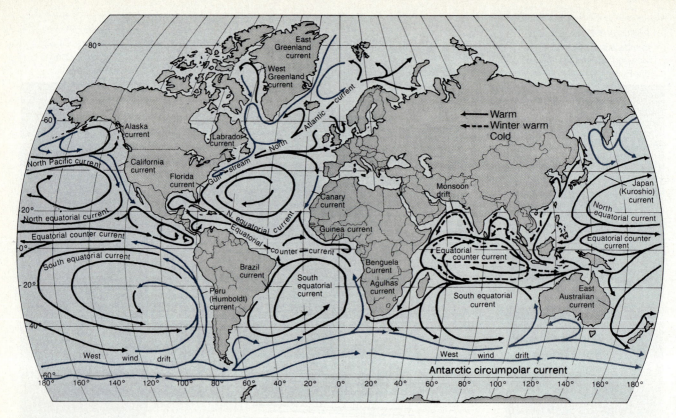

**Figure 12.11** Surface currents of the world ocean exhibit many regularities. The gyres of the major ocean basins circulate clockwise in the Northern Hemisphere and counterclockwise in the Southern Hemisphere, and they are separated by equatorial countercurrents. The unbroken flow of the Antarctic circumpolar current is replaced in the Arctic by cold currents flowing mainly south between landmasses.

As the tide rises, it produces a landward flow of water in rivermouth estuaries and in tidal inlets through off-shore barrier islands. This **flood tide** lasts for 6 hrs 13 min (on average). Following a brief period of *slack water,* the tide reverses and becomes a seaward-flowing **ebb tide** that also lasts for an average of 6 hr 13 min. In estuaries, the flow of river water downstream strengthens the ebb tide and weakens the flood tide.

## Ocean Currents

A glance at the pattern of surface currents of the world ocean in figure 12.11 suggests that there are few truly motionless regions of the sea. What causes this incessant circulation? Another glance at figure 14.6 should suggest an answer. The similarity between the patterns of atmospheric and ocean surface circulation in the summer hemisphere is no coincidence. Summer winds are, in fact, the main driving force for the surface currents of the ocean (winter winds are too variable), but a more careful look should uncover a technical difficulty. Although the patterns are *generally* the same, current directions are rotated as much as 45° to the right of wind directions in the Northern Hemisphere and as much as 45° to the left of them in the Southern Hemisphere. How can this be explained? The answer lies in a rather enigmatic phenomenon known as the Coriolis effect.

## Throwing Curves: Why Currents Turn

Consider an artificial satellite orbiting Earth in a circular, north-south, polar orbit (fig. 12.12). The satellite's orbit is fixed in position in space by its rotational inertia (except that as a part of the Earth system, it also orbits around the Sun). Because the satellite's orbit is circular, its speed is constant. Let us assume that it moves through 1° of arc in 120 sec. At this *angular velocity,* it sweeps from a point above the North Pole to a point above the equator (90° of arc) in 10,800 sec, or 3 hr. Meanwhile, however, Earth turns eastward with a constant angular velocity of 1° in 240 sec, so in the same time interval, the planet has rotated eastward through an angle of 45°.

Notice, however, what this easterly rotation of the planet does to the apparent path of the satellite. For each 10° that the satellite moves southward, the planet's surface moves 5° eastward; thus, with respect to Earth's

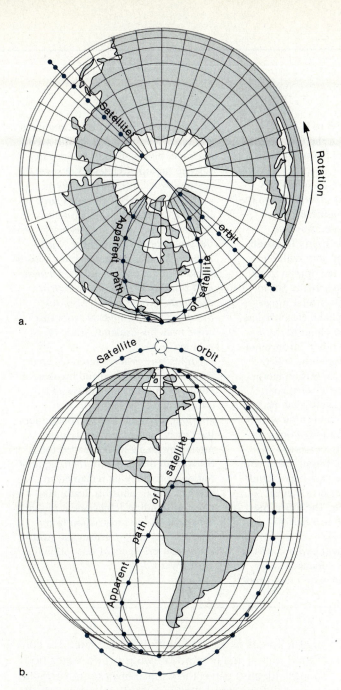

**a.**

**b.**

***Figure 12.12*** An artificial satellite in polar orbit
illustrates the Coriolis effect. As Earth turns eastward, the
satellite's circular orbit appears to be deflected to the
west. The satellite moves 5° west for each 10° of
southerly travel. The spreading of meridians southward in
the Northern Hemisphere gives the satellite's path an
apparent right curvature. Convergence of the meridians in
the Southern Hemisphere gives the appearance of left
curvature. The path appears straight at the equator, where
the meridians are parallel.

surface, the satellite has a southwesterly course instead
of a due southerly one. Furthermore, the course appears to curve toward the right in the Northern Hemisphere. The course straightens out again over the
equator, but then it curves toward the left in the Southern
Hemisphere.

The reason for these peculiar effects is evident from
the dots on the satellite's apparent path in figure 12.12.
In this figure, **parallels,** or east-west latitude lines, and
**meridians,** or north-south longitude lines, are drawn
at 10° intervals. Each dot is located 10° south, and 5°
west, of the one above it. Notice, however, that whereas
the spacing of the parallels remains constant, the meridians diverge from the poles and only become parallel at the equator. Therefore, if you follow the
succession of dots south from the North Pole, each 5°
westward shift covers a greater distance until you arrive
at the equator. Then, in continuing past the equator
toward the South Pole, the effect reverses as each 5°
shift covers a shorter distance. This accounts for the reversing curvatures of the satellite's apparent path. Notice
that the effect is strongest at the poles, where the path's
apparent curvature is greatest, and drops to zero at the
equator, where the path "becomes" straight.

Long before the advent of artificial Earth satellites,
scientists and others had recognized that fluids moving
on our planet's surface have a tendency to curve toward
the right of their path of travel in the Northern Hemisphere and toward the left in the Southern Hemisphere.
This caused considerable vexation, as it appeared to vi-
olate **Newton's first law of motion,** which states that
an object at rest or in uniform, rectilinear (straight-line)
motion must persist in that state unless acted on by some
external force. The nineteenth century French mathematician Gaspard Gustave de Coriolis first analyzed this
unruly behavior of fluids and found that "his" effect increased with velocity of the moving fluid. In his honor,
this behavior of currents has been named the **Coriolis
effect.**

Some scientists regard the Coriolis effect as a *force,*
because, in classical mechanics, a deviation from rectilinear motion is an acceleration, which can only be
generated by a force acting on a mass, a rule expressed
in **Newton's second law of motion,** $F = ma$ (force
equals mass times acceleration). This raises a serious
problem, however, because there is no evident energy
source to generate such a force. A simple but profound
observation of Albert Einstein's comes to our rescue,
however. In his concept of *relativity,* Einstein recognized that the laws of classical mechanics tend to lose
their grip in situations where there is more than one
frame of reference. In the case of the Coriolis effect,
there are two frames of reference: (1) space, in which
objects obey Newton's first law (note that the satellite's
*curved* path results from the *external* force of Earth's

## Fridtjof Nansen (1861–1930)

Probably no one better exemplifies the value of field study than the Norwegian explorer Fridtjof Nansen. In order to test his hypothesis that Greenland was covered by an ice sheet, he skied across it in 1888 with five companions. The trip took 40 days. Five years later, he raised funds and built a special research vessel, the *Fram* (Norwegian for "Forward"), that was designed to (1) survive freezing in the pack ice of the Arctic Ocean, which had crushed the American exploring ship *Jeanette* in 1881, and (2) test the hypothesis that the pack ice would carry the *Fram* eastward from Siberia, between Alaska and the North Pole, and southward east of Greenland, as it had done with the remains of the *Jeanette*. With Otto Sverdrup—another famous Norwegian explorer—in command of the *Fram,*

Nansen sailed in 1893 to the Novosibirsk Islands in the East Siberian Sea, where the vessel became firmly lodged in pack ice.

Fortunately, Nansen's hunch was right. The *Fram* duly emerged from the ice three years later north of Spitzbergen, but Nansen was not aboard to witness the event. In March 1895, he had jumped ship with one companion and 28 sled dogs in quest of a glory that would actually fall to Admiral Robert Peary of the United States Navy 14 years later. Nevertheless, Fridtjof Nansen came closer to the North Pole than any other European of record had been before him. When forced to turn back, he was no more than 320 km from the elusive goal. Returning southward across the pack ice, Nansen and his companion were held up by a combination of open water and the coming of winter. Therefore, they built a stone cabin with a roof of walrus hide on a large island and passed the winter on a diet of walrus and polar bear. The following spring, they returned to a hero's welcome in Norway.

Nansen's subsequent explorations, although perhaps less dramatic, were equally significant. In addition to proof

of continental glaciation in Greenland and of the eastward drift of the Arctic pack ice, his chief contributions to science include the design of several oceanological instruments and the refinement of existing ones; a large body of research on the nature of Arctic Ocean waters; the discovery of the mechanism whereby the North Atlantic deep and bottom water mass is formed; and a treatise on *The Structure and Combination of Histological Elements of the Central Nervous System,* a classic medical work that arose from his academic study of zoology.

In his later years, Nansen became increasingly interested and involved in concerns of humanitarianism and statesmanship. In 1906, he was appointed first minister in London for the newly-established Norwegian monarchy. In 1920, he headed the Norwegian delegation to the League of Nations, of which he remained an active member until his death. His most outstanding achievement in that capacity was the repatriation from Russia of 427,886 German and Austro-Hungarian prisoners of war following World War I.

---

gravity); and (2) our rotating planet, on whose surface rectilinear motions are impossible except over very short distances. In short, the apparent deviation from rectilinear motion is an artifact of the interaction of the two different frames of reference; hence, there is no actual acceleration. The Coriolis effect is, therefore, only an effect and not a true force.

### The Ekman Spiral

In 1902, V. W. Ekman made an elegant mathematical analysis of a phenomenon observed by the Norwegian Arctic explorer/scientist Fridtjof Nansen from the deck of his research vessel, the *Fram,* while it lay frozen in the Arctic pack ice. Nansen noted that the ice was drifting consistently 20° to 40° to the right of the wind direction. Ekman postulated that the stress of wind on the ocean surface is balanced by both the Coriolis effect

and the forces of internal friction in the water. He concluded that a steady wind blowing over a water body should (ideally) cause the surface layer to move 45° to the right of the wind direction in the Northern Hemisphere, and 45° to the left in the Southern Hemisphere. Furthermore, he concluded that the lower water layers are influenced by the surface layer in the same way as the surface layer is influenced by the wind, but with declining effect at progressively lower levels.

The **Ekman spiral** (fig. 12.13) is a diagrammatic representation of Ekman's concept. Notice that between the ninth and tenth arrow from the top of the spiral, the flow of water has turned a full 180° from that of the surface layer. At this depth—typically about 100 m—the flow rate is only about 1/23 (4.4%) of the surface flow. This level is regarded as the effective base of the Ekman spiral.

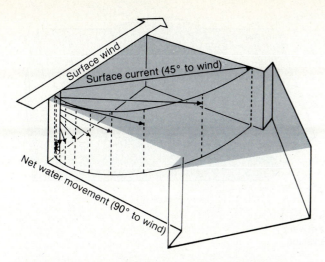

**Figure 12.13** The Ekman spiral illustrates the effect of wind on the surface layers of the ocean. Surface currents flow 45° to the right of the wind direction. Successively lower water layers are in turn driven farther to the right, but with less force. The average drift of the entire water column is at 90° to the right of the wind (*arrow box*). The Ekman spiral for the Southern Hemisphere is identical, but reversed.

What is the net effect of the Ekman spiral on the entire body of water that lies above its base? In order to determine this, we must average the flow vectors (arrows) down to this depth. (Note that the flow pattern in fig. 12.13 is actually continuous from top to bottom and not discrete as the arrows might suggest.) The result is a net transport at 90° to the wind direction (mega-arrow), to the right in the Northern Hemisphere, and to the left in the Southern Hemisphere. Hence, whereas a cork floating on the sea surface will drift at a 45° angle to the right of the wind (Northern Hemisphere), the entire body of water above the base of the Ekman spiral is actually moving at a 90° angle to the right. This motion is known as **Ekman transport,** and it explains why surface ocean currents are rotated with respect to the winds that cause them and why these currents are restricted to the upper 100 m or so of the ocean surface.

Figure 12.14 illustrates one important consequence of Ekman transport. In figure 12.14*a*, surface winds generate currents that flow toward the equator along west coasts (e.g., the California, Peru, Canary, and Benguela currents; see fig. 12.11), or toward the poles along east coasts (e.g., the Gulf Stream, Kuroshio, and East Australian currents). Such currents produce an Ekman transport of seawater away from the continent. This creates a zone of low surface water pressure along the coast, which draws up cold water from below. Coming from depths at which biological productivity is low, these upwelling waters are rich in dissolved organic compounds and in nutrients, such as phosphate, nitrate, and silica.

As a result, such coasts enjoy high biological productivity, and are prime resource areas for the world's fishing industry.

In figure 12.14*b*, surface winds generate currents that flow away from the equator along west coasts (e.g., the Alaskan and North Atlantic currents), or toward the equator along east coasts (e.g., the East Greenland current). Ekman transport from such currents is directed *toward* the shore, resulting in a pileup of water against the coastline, which forces coastal waters to sink.

### Geostrophic Flow

The various currents of the North and South Atlantic, the Western and Southern Pacific, and the Indian Oceans flow in great swirling patterns that have been called **gyres**. In the Northern Hemisphere, these gyres rotate clockwise, and in the Southern Hemisphere they rotate counterclockwise (see fig. 12.11). Given Ekman's theory, we might predict that surface water should be flowing toward the centers of the gyres and piling up. This does indeed happen, and as a result, the sea surface in these regions stands up to 2 m higher than in surrounding areas (although the pileup is offset to some extent by evaporation of seawater into the dry air of the high pressure cells that lie above the gyres). Both the piled up sea surface and the sinking of relatively dense, high-salinity seawater beneath it create a region of high water pressure, from which a **pressure gradient** radiates outward in all directions. One might think that water sinking beneath a high pressure zone would flow radially outward along this pressure gradient, and indeed it does, but not for long!

Figure 12.15 shows a pressure gradient represented by concentric lines of equal pressure, or *isobars*. High pressure can be caused by a raised water surface, as here, or by concentrations of denser, more saline water, or by both. Here, driven by the *pressure gradient force, P* in figure 12.15, a small "parcel" of water has begun to flow away from the high pressure at point 1. As it moves, however, this parcel is influenced by the Coriolis effect, *C,* which acts at 90° to the right of the path of travel in the Northern Hemisphere. The water parcel, therefore, veers to the right (point 2) until its velocity and direction are such that the Coriolis effect is both equal and opposite to the pressure gradient force, as at point 4. By now, the water parcel is travelling at a right angle to its former direction, and its path is parallel to the isobars. In the Southern Hemisphere, the result is the same, except that the water parcel is deflected to the left instead of to the right. This kind of flow, running parallel to the contours of a pressure gradient, is known as **geostrophic flow,** meaning "turned by Earth." Geostrophic flow is instrumental in driving not only the oceanic gyres, but also the circulation of the atmosphere (see chapter 14).

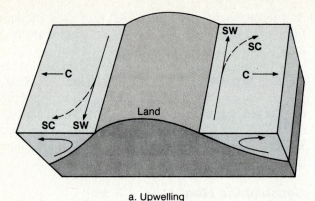

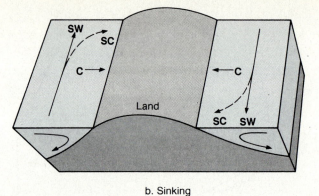

a. Upwelling    b. Sinking

SW = Surface winds
SC = Surface current
C = Coriolis vector

*Figure 12.14*  Surface winds (*SW*) blowing parallel to a shoreline generate surface currents (*SC*) that become deflected toward the right (Northern Hemisphere) by the Coriolis effect (*C*). If the winds blow equatorward along a west coast or poleward along an east coast, as in *a*, the Ekman transport is directed offshore, resulting in upwelling of deeper waters. If the winds blow in the opposite directions, as in *b*, the Ekman transport is onshore, resulting in sinking.

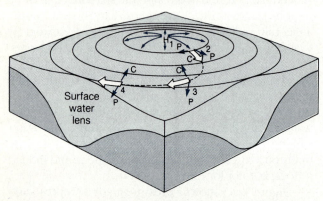

*Figure 12.15*  Pileups of sea surface water are acted on by gravity in the form of a pressure gradient force (*P*) directed away from the point of highest pressure (*H*). As water is driven outward by this force (*broad arrows*), it is deflected by the Coriolis effect (*C*), which increases with velocity until it is equal to the pressure gradient force and oppositely directed, as at point 4. This establishes geostrophic flow, which continues at a constant rate and direction, parallel to the isobars (lines of equal pressure).

## The General Circulation of the Oceans

The great oceanic gyres move vast quantities of water. The Gulf Stream, the western limb of the north Atlantic gyre, transports up to 90 million m³/sec of water (about 5000 times the mean discharge of the Mississippi River). This prodigious flow of the gyres tends to pile water up against the east coasts of Asia, Australia, the Americas, and Africa. To balance this input, there is a corresponding output in the form of *equatorial countercurrents* that return the accumulated water eastward between the westward-flowing low-latitude sections of the oceanic gyres, (i.e., between the north and south equatorial currents; see fig. 12.11).

South of the great gyres in the Southern Hemisphere, the pattern of oceanic circulation is quite simple. The eastward-flowing southern margins of the gyres (called the *west wind drift*) blend with the *Antarctic circumpolar current*. Both of these currents are driven by prevailing westerly winds at these latitudes (see fig. 14.6). This simple pattern is not repeated in the Northern Hemisphere, however, because of the presence of continental landmasses. From the Arctic Ocean, cold currents, such as the West Greenland current between Canada and Greenland, descend into lower latitudes through straits between landmasses. Other cold currents flow toward the equator from higher latitudes along west coasts. These are the eastern limbs of the oceanic gyres bringing cold water southward while their western limbs bring warm, equatorial waters to higher latitudes. Without this energy-redistributing action of the gyres, lower latitudes would be much hotter and higher latitudes much colder than they actually are.

## Local Currents

Where ocean waters interact with shorelines, local currents are often set up. Figure 12.7 shows how waves approaching a coastline at an oblique angle create a **longshore current** that flows parallel with the shoreline within the surf zone and away from the advancing waves. As the waves rush in, water piles up against the shore, and some of this piled up water escapes laterally to form the longshore current.

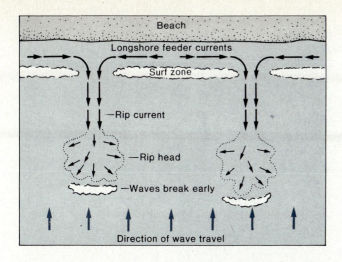

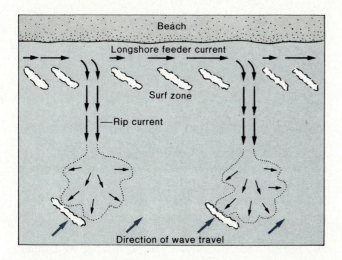

**Figure 12.16** Rip currents form where longshore currents converge or are deflected seaward by obstacles, or where low wave height presents an easy escape route for water piled onshore by advancing waves (note gaps in surf zone). Early-breaking waves around rip heads are clues to the presence of rip currents.

Longshore currents do not go on indefinitely, however. At irregular intervals along the shore, they turn seaward and escape as **rip currents**. These are strong, narrow flows that can reach a velocity of 1 m/sec, posing a severe hazard for unwary bathers (fig. 12.16). Rip currents often form along sheltered bays or coves where the breakers are lowest. In adjacent headlands, wave refraction generates higher breakers that create pileups of water along the shore. Water escapes laterally from these pileups and migrates toward adjacent bays as longshore currents. Two such flows then combine from opposite directions to create a rip current that can extend from 50 to 750 m seaward, where it spreads and dissipates in a *riphead*.

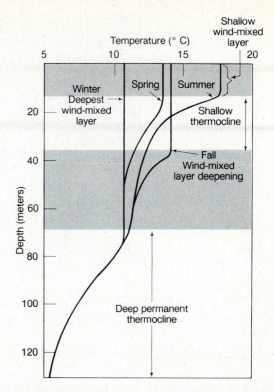

**Figure 12.17** A seasonal temperature gradient develops in the upper layers of the mid- to high-latitude ocean due to the warming effect of warm spring rains and solar heating in summer. The gradient is destroyed during the fall overturn, which is driven by storm winds and the sinking of cold, dense water.

## Gradients in the Sea

Although the world ocean is uniform in respect to volume, major ion ratios, and acidity, there are other ways in which it does vary. In Chapter 11, I note how the concentrations and proportions of dissolved gases can vary from place to place. Here, I consider variations in heat, salinity, and density.

## Thermoclines: Variations in Heat

Seasonal fluctuations in solar heating occur throughout the oceans due mainly to the tilt of Earth's rotational axis (see figs. 13.13 and 13.14). These seasonal variations are relatively slight in equatorial waters, and they increase toward the poles. Figure 12.17 shows the annual cycle of temperature variation with depth and season in near-surface ocean waters in the middle latitudes. Notice that such seasonal variation is limited to the upper 70 m or so of the ocean, the so-called *mixed layer*.

In winter, this surface layer is thoroughly mixed by strong, cold storm winds. In higher latitudes, the freezing of seawater also promotes mixing because it removes water, but not salt, from the ocean, producing

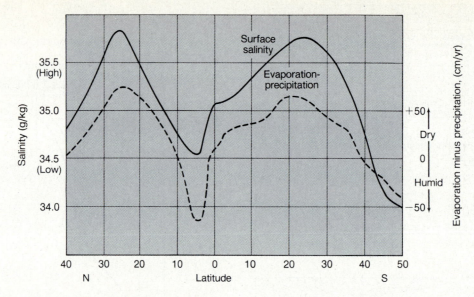

**Figure 12.18** The salinity of sea surface water is closely correlated with net evaporation (evaporation minus precipitation).

Copyright © 1960 by Arthur N. Strahler. Used by permission.

dense water that tends to sink to deeper levels. The result of such mixing is an *isothermal* temperature condition, or one in which the temperature is uniform from the surface to the base of the mixed layer.

In spring, warm rains heat the mixed layer, typically stepwise, each new rain adding a new installment of heat. This puts an end to mixing because it creates a layer of warm, low-density water over colder, denser water beneath. This condition is as inherently stable as an unshaken bottle of Italian salad dressing, in which the oil forms a layer above the vinegar. If nothing happened to mix the dressing, or the ocean, it would stay that way indefinitely. Maximum stability occurs in summer, when solar radiation raises the surface temperature several degrees above the temperature at depth. This creates a **seasonal thermocline**—a zone of rapidly declining temperature beneath a shallow, wind-mixed surface layer.

In fall, cold storm winds again mechanically mix the surface layer and cool the surface water, making it more dense and more prone to sink. This **fall overturn** destroys the seasonal thermocline and reestablishes the isothermal condition of winter.

Thermoclines can exist for other reasons, as when warm currents like the Gulf Stream flow into cooler waters, or cold currents, such as the Labrador Current, flow into warmer waters. Similarly, local thermoclines exist where cold river waters discharge into the ocean from glaciated highlands, as in Norway and the Pacific Northwest.

## Variations in Salinity and Density

In addition to thermoclines, there are also systematic variations in the salinity and density of seawater. Some of these variations are due to the same seasonal effects that produce thermoclines, but other processes are important as well. Where *net evaporation* (evaporation minus precipitation; fig. 12.18) is high, as it is between 20° and 30° latitude, surface salinity can rise well above 35.5‰. On the other hand, in equatorial waters where precipitation far exceeds evaporation, salinity can drop well below 35.0‰. Salinity is also low in latitudes above 30° where evaporation is low due to reduced solar heating, and where the polar jet stream is close enough to bring precipitation from cyclonic storms (see chapter 15). Other factors affecting salinity include the discharge of river water into the sea, the melting and freezing of ice, and submarine volcanism.

The density of seawater is also affected by all of these same factors, but much less than by changes in temperature.

## Water Masses

Large bodies of ocean water tend to assume and maintain characteristic values of salinity and temperature; hence, it has been possible to identify several major **water masses** within the world ocean. These water masses are created in certain geographic regions, from which some of them subsequently migrate into other regions. As they migrate, they retain enough of their original characteristics that they can still be recognized.

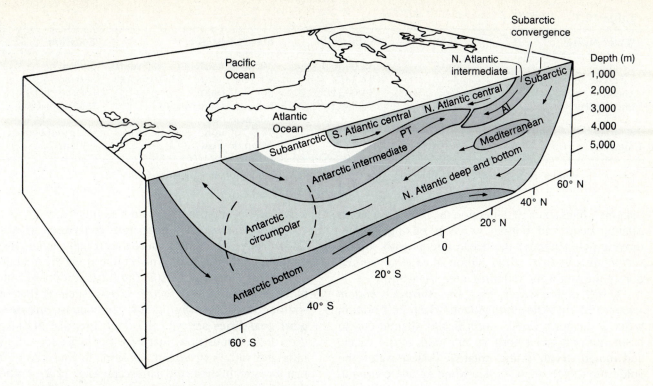

**Figure 12.19** Circulation patterns in the Atlantic Ocean correspond to the ocean's major water masses.

Figure 12.19 shows a vertical cross section through the Atlantic Ocean from north (*right*) to south (*left*). The different water masses are labelled, and their motions are indicated by arrows. Most of these water masses are also found in the Pacific and Indian Oceans.

At the ocean surface, extending from about 30° south to about 40° north, lies a broad but shallow water mass consisting of two parts: the *North Atlantic* and *South Atlantic central water masses.* These correspond to the north and south Atlantic gyres, and they are separated by the westward-flowing Atlantic equatorial countercurrent. They lie above colder, less saline water masses from which they are separated by the **permanent thermocline** (PT), a zone of rapid downward temperature decrease between about 300 and 900 m depth. Their salinity and temperature are high (table 12.1) because of high insolation and correspondingly high evaporation.

To the north of the central water masses lies the *subarctic water mass,* consisting of cold water brought southward from the Arctic Ocean by cold currents. To the south of the central water masses lies the *subantarctic water mass,* consisting of the west wind drift. The subarctic and subantarctic water masses have low to moderate temperature due to a lower input of solar heat, and they have the lowest salinity of all (table 12.1) due to their location beneath the polar jet streams, which carry cyclonic storm systems that generate large amounts of precipitation.

The *Antarctic circumpolar water mass,* consisting of the Antarctic circumpolar current, rings the globe in an unbroken belt. Being south of the main storm belt of the polar jet stream, it receives less precipitation than the adjacent subantarctic water mass, because of which it has a somewhat higher salinity, but its temperature is lower. It is unusual for a surface water mass in that it extends to a depth of about 3500 m.

Beneath the subarctic and subantarctic water masses, and actually derived from them, are the *Arctic* (AI in fig. 12.19) and *Antarctic intermediate water masses.* (Note the close similarity in salinity and temperature ranges of the second and fourth rows of table 12.1.) Summer evaporation and winter cooling make the surface water more dense, allowing it to sink to a maximum depth of about 1500 m, whereupon it flows toward the equator beneath the permanent thermocline.

The most extensive of all the water masses in the Atlantic Ocean is the *North Atlantic deep and bottom water mass.* Like the Arctic intermediate water mass, this water is derived from the subarctic water mass in the north Atlantic, but in this case the subarctic water mass is mixed with central water from the Gulf Stream that has a high salinity due to evaporation. When this saline water cools in winter, it becomes even denser than intermediate water, sinks to great depths, and flows southward beneath the Arctic intermediate water mass as far as 60° south latitude. This water is rich in oxygen due to the photosynthetic activity of marine algae (chiefly

**Table 12.1**

| Water Masses | Temp. °C | Salinity ‰ | Place of Origin | Location |
|---|---|---|---|---|
| Central | 6–19 | 34.0–36.5 | 30°–50° north, 30°–40° south | Surface |
| Subarctic and subantarctic | 3–10 | 33.5–34.7 | 40°–60° north, 30°–50° south | Surface |
| Antarctic circumpolar | 0–2 | 34.6–34.7 | 50°–70° south | Surface |
| Intermediate | 3–10 | 33.8–34.9 | Beneath subarctic and subantarctic | Intermediate |
| N. Atlantic deep and bottom | 2–4 | 34.8–35.1 | 50°–70° north, Atlantic Ocean | Deep and bottom |
| Antarctic | −0.4 | 34.7 | Antarctic coast | Bottom |

From H. U. Sverdrup, M. W. Johnson, and R. H. Fleming, *The Oceans,* © 1942. Prentice-Hall, Inc., Englewood Cliffs, NJ.

diatoms). In chapter 10, I mention the role of the North Atlantic deep and bottom water in mixing carbon-dioxide-rich waters of the ocean depths with surface waters, thus increasing the carbon dioxide concentration of the atmosphere during interglacial ages.

A still denser water mass, the *Antarctic bottom water,* underlies the north Atlantic deep and bottom water as far north as 30° north in the Atlantic Ocean basin, but only as far north as 20° south in the Pacific and Indian Ocean basins. Antarctic bottom water, the coldest in the ocean, is formed when Antarctic circumpolar water on the fringe of Antarctica freezes, leaving the remaining water highly saline. This dense water then sinks and mixes with about an equal proportion of Antarctic circumpolar water.

## Coastal Sculpture: Interactions of Land and Sea

About 14,000 years ago, the last great ice sheets began to melt, and sea level gradually rose by about 100 m over the next 9000 years. Since then, wave erosion and deposition have been tirelessly at work, reshaping Earth's coastlines into those configurations that provide for the most even and efficient dispersal of incident wave energy. The task is a formidable one, and in spite of a punishing, 24-hour work day, it is far from completed. In this section, I discuss the various forms of coastal sculpture that have resulted from this ceaseless endeavor and the processes that have produced them.

## Coastal Types

Figure 12.20 shows six common examples of the many types of coasts that border the world ocean. Variety in coastlines is due mainly to the interaction of two factors: (1) whether the coast is rising or sinking with respect to sea level and (2) whether the geological processes responsible for shaping the coastlines are dominantly marine or dominantly nonmarine. *Coastal emergence* results if (1) the continental margin is tectonically uplifted; (2) sea level falls, as when seawater is converted to glacier ice; or (3) the ocean floors sink due to a decrease in seafloor spreading rates. *Coastal submergence,* on the other hand, results if any of these conditions are reversed. **Emergent coastlines** are typically relatively straight because they usually either raise continental shelves as they have done along the Atlantic coast of the Florida peninsula (fig. 12.20*e*), or raise fault blocks, on tectonically active coasts, such as that of southern California (fig. 12.20*f*). In contrast, **submergent coastlines** are typically quite irregular because they drown landscapes that have been dissected into hills and valleys by stream erosion. In the case of a mature, stream-dissected landscape (fig. 12.20*a*), the drowning of valley mouths produces a rugged *ria coast* with deep bays, intervening headlands, and offshore islands. Such coasts are common in New England north of Cape Cod and adjacent Canada. *Fjord coasts* (fig. 12.20*b*) are similar, except that the coastline, which follows the walls of glacially scoured valleys, is even more irregular. Such coasts are common where glaciation occurs (or has occurred recently) at low altitudes near the coast, as in Norway, Scotland, Iceland, Greenland, Labrador, Alaska, western Canada, southern Chile, and New Zealand. The drowning of old age coastal landscapes produces *estuarine coasts* (fig. 12.20*c*).

Various things can happen to modify these fundamental coastal types. Where a sediment-laden stream feeds a submergent coastline, a delta often forms (fig. 12.20*d*). A classic example of this is the vast Mississippi delta complex on the coastline of Louisiana. Glacial deposits characterize some coasts, such as those of New York and southern New England, where Pleistocene ice sheets deposited glacial gravels that now underlie Long Island, Cape Cod, and the islands of Martha's Vineyard and Nantucket. Volcanism has produced or modified some coasts, as in Hawaii, Indonesia, and Japan. In California and Alaska, faulting has produced many straight, steep coastlines (fig. 12.20*f*), and some coasts have even been built by the vigorous growth of organisms, such as mangrove trees, seagrasses, and corals.

Marine erosion can extensively modify the shapes of submergent coasts, producing *cliffed headlands* and intervening bays (see fig. 12.30). Alternatively, marine deposition can prevail, resulting in a coast bordered by barrier islands, as in North Carolina and Texas (fig. 12.20*e*).

a. Ria coast

b. Fjord coast

Fjord

c. Estuarine coast

Estuary

d. Deltaic coast

e. Barrier island coast

Barrier island

Lagoon

Inlet

f. Raised platform coast

**Figure 12.20**  Sea level rise following the melting of ice age glaciers has submerged most coastlines, creating a variety of coastal types. Moderately embayed ria coasts (*a*) have formed on stream-sculpted landscapes. Deeply embayed fjord coasts (*b*) have formed where glaciers have deepened coastal valleys. Where rivers run clear (*c*), their drowned estuaries remain open, but deltas have filled those of muddy rivers (*d*). In places, uplift has outstripped sea level rise, exposing the continental shelf (*e*). Raised wave-cut platforms on a rugged coast (*f*) are a sure sign of an emergent coastline.

**Figure 12.24** A gently sloping sand beach along the Pacific Palisades, Los Angeles County, California.
Photo by J. T. McGill, U.S. Geological Survey

rivers, longshore currents, and the erosion of sea cliffs (a relatively minor source); storage in such shoreline features as deltas, beaches, spits, bars, barrier islands, lagoons, and estuaries (see fig. 12.30); and outputs of clastic sediment to longshore currents, backshore dunes, and deep water. If the input of sediment increases, storage also increases until output rises to equal input again. Similarly, decreased input results in corresponding decreases in storage and output.

### Sand from Rock and Shell

The clastic materials of beaches are derived from a variety of sources. *Quartz sand beaches* (fig. 12.24) are supplied mainly by rivers. They have gentle slopes, typically between 5° and 8°, because they are usually composed of the most readily erodible sediment (medium sand). *Shingle beaches* (fig. 12.25) are composed of sediment of pebble size (2–64 mm) or cobble size (64–256 mm), derived largely from the erosion of local headlands by waves. They are narrower and steeper than sand beaches because they are composed of less erodible material; hence, they require the "gravity assist" provided by a steeper slope. *Shell beaches* form in areas where silicate sand supply from rivers is negligible but where calcium carbonate secreting organisms live abundantly in the shallow, nearshore waters. This skeletal carbonate material is often comminuted to sand size (¹⁄₁₆–2 mm) as in the striking, white coral sand beaches of many tropical islands. On volcanic islands, where the supply of quartz or carbonate sand is minimal or absent,

**Figure 12.25** A steeply sloping shingle beach lies adjacent to a rocky promontory on the coast of Maine.

dark beaches of basalt sand can form. These beaches are usually not as extensive as quartz sand beaches because of the rapid weathering of basaltic minerals.

### Beach Face and Berm: The Structure of Beaches

The presence of a beach on a shoreline indicates that waves and longshore currents have not been active enough to remove all clastic sediment inputs to the littoral cell. Figure 12.26*a* shows the features of a typical beach, which extends from the *breaker bar* in the surf zone shoreward to the landward limit of the swash.

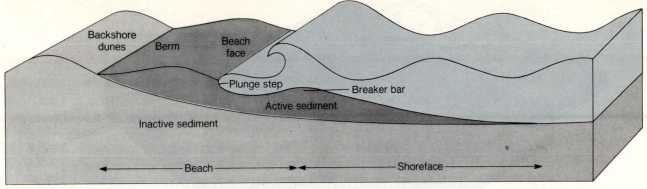

a. Summer

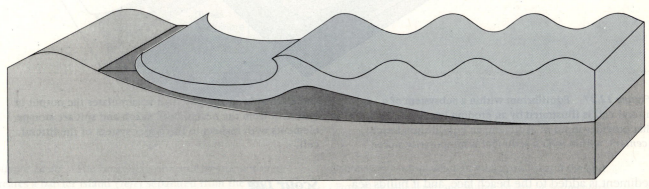

b. Winter

**Figure 12.26** The zone of active sand input, output, and storage, comprising the beach and the shoreface, extends from the backshore dunes to the outer limit of shoaling wave interactions with the seafloor. In summer (*a*), long, deep swells scour the shoreface, driving sediment onshore to rebuild the beach face. In winter (*b*), short, shallow storm waves erode the beach face and carry sediment seaward to rebuild the shoreface.

where windblown *backshore dunes* begin. Seaward of the beach is the *shoreface,* which extends from the breaker bar to the outer limit of wave-stirred bottom sediment, where the bases of shoaling waves begin to interact with the seafloor.

The **beach face** is a gently to steeply sloping surface generated and maintained by swash and backwash. On the landward edge of the beach face, the **berm** slopes gently toward the backshore dunes. The berm is rarely inundated. If it is overtopped by high waves, the spillover of sediment-laden swash simply builds the berm higher by depositing sand on it until a new equilibrium is reached and spillover can no longer occur.

Beneath the breaker zone, a *breaker bar* forms parallel to the shoreline. On its seaward side, this bar is built by shoaling waves. On its landward side, the bar is built by rip currents and the return of the backwash. Shoreward of the breaker bar is the **surf zone,** in which water travels on and offshore as shallow-water waves, rip currents, and backwash.

Both the beach and the shoreface are equilibrium forms, representing the storage factor in a littoral cell. Input to the cell occurs mainly as **littoral drift,** or clastic sediment moving parallel to the shoreline. That portion of littoral drift transported by swash and backwash is called *beach drift,* that portion transported by longshore currents is called *longshore drift* (see fig. 12.7). On the beach face, sediment is first washed upward at an oblique angle to the shoreline by the swash from obliquely approaching waves and then carried perpendicularly back down with the backwash. Longshore currents carry sediment (mainly sand) in suspension in the turbulent water of the surf zone. If the beach is in equilibrium, neither erosion nor deposition takes place as a result of littoral drift. Whatever sediment enters the system simply passes across the stable beach face or is exchanged, grain by grain, for sediment already stored in the beach. If the rate of sediment input by longshore currents increases, however, deposition is favored, more

Several interesting proposals have been made in recent years for harnessing the abundant energy that flows through the ocean system annually and turning it to human use, a logical extension of what humanity sees as its mandate to subdue and conquer the planet. These proposals focus on five possible oceanic energy sources. Together with the total power in watts within each source, these are *thermal systems* (oceanic thermoclines) $10^{13}$ watts; *evaporative systems* (restricted ocean basins) $10^9$–$10^{11}$ watts; *wave systems,* $10^{10}$ watts; *tidal systems,* $10^9$ watts; and *current systems,* $10^8$ watts. (One *watt,* or 1 $kgm^2/sec^3$, is the power required to raise one kilogram through a distance of 102 cm in one second at Earth's surface.)

Because the present total power demand of human civilization is $10^{13}$ watts, it might seem that the ocean offers a promising reserve of potential power sources. Unfortunately, however, the tapping of these sources presents a host of formidable problems. Most of the devices that have been proposed for tapping oceanic power are crude and inefficient. Furthermore, in most cases, the source is so widely dispersed that the prospect of exploiting significant amounts of it is dim indeed. Figure 12.32 shows a curious scheme for suspending enormous, tethered, buoyant turbines in a powerful current,

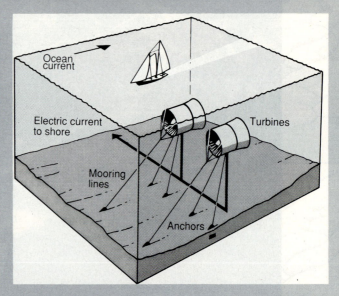

**Figure 12.32** This imaginative scheme for harnessing the power of ocean currents uses buoyant turbines tethered to the seafloor.

From D. A. Ross, *Introduction to Oceanography,* 3d ed. Copyright © Prentice-Hall, Inc., Englewood Cliffs, NJ.

such as the Gulf Stream. The idea might well have merit, although the technical problems of adequate material strength, protection of the machinery from fouling by marine life, and protection of marine life from the machinery, would be formidable.

Thermoclines might be tapped with a device called an OTEC (Ocean Thermal Energy Conversion), an enormous, vertical pipe suspended from the surface to a depth of about 300 m. Warm water, drawn into the upper end of the system, would be used to evaporate liquid ammonia, which would be used to drive turbines. Cold water, pumped from the lower end of the pipe, would then condense the ammonia, whereupon the cycle would repeat. There are several problems with this scheme. A thermal gradient of at least 20° C/1000 m is required, which restricts OTECs to the tropics. To minimize power

---

low **neap tides** result when these forces conflict at quarter moons. Slack water separates incoming **flood** and outgoing **ebb** tides.

**Parallels** are east-west latitude lines; **meridians** are north-south longitude lines. **Newton's first law of motion** states that objects persist in uniform, rectilinear motion, or at rest, unless acted on by an external force. **Newton's second law of motion** states that the application of a force to a mass produces a proportional acceleration ($F = ma$). The **Coriolis effect** makes currents turn right in the Northern Hemisphere and left in the Southern Hemisphere. The Coriolis effect is not a force, but an artifact of the interaction of two frames of reference: space and Earth's rotating surface. The **Ekman spiral** results from action of the Coriolis effect on the upper ± 100 m of the ocean. This layer moves at

about 90° to the wind, right in the Northern Hemisphere and left in the Southern Hemisphere. This process of **Ekman transport** moves water away from west coasts, causing upwelling. In the centers of **gyres,** Ekman transport piles up water, creating a **pressure gradient** and a **geostrophic flow** parallel to isobars, which helps to drive gyres. *Equatorial countercurrents* move water eastward, compensating for westward water transport in gyres. The *west wind drift* (poleward portions of southern gyres) blends with the wind-driven *Antarctic circumpolar current.* Water escaping from waves advancing onshore generates **longshore currents,** which return seaward here and there as **rip currents.**

The sea surface has constant temperature, salinity, and density in winter, but its density is reduced by spring rains and summer warmth, resulting in a **seasonal**

transmission losses, they would have to be located within about 160 km of land. At best, OTECs are only about 2% efficient. Corrosion and fouling would be persistent problems, as would storm damage. Finally, a large number of OTECs working together would lower sea surface temperatures in their vicinity. Inasmuch as sea surface temperature is a major factor in world climate patterns, this might have serious consequences for the future, possibly even to the extent of triggering a new ice age.

Evaporative systems harness currents flowing into a restricted sea, such as the Mediterranean or the Red Sea. In both these water bodies, high evaporation rates produce relatively saline seawater that sinks and flows out at the bottom of the sea's outlet to the main ocean while less saline water flows into the sea through the outlet as an overlying countercurrent. These flows could be used as they are, or dams could be built at the outlets and closed until evaporation caused the water level in the sea to fall by several meters, whereupon the dams would be opened and the inflow would be used to drive turbines. Problems with this latter scheme include severe disruption of marine ecology (and with it the fishing industry) and the lowering of water tables as streams cut downward in response to the lowering of the sea level.

Schemes to harness wave power would utilize wave motion to turn turbines, or to raise water levels in standpipes to create a hydrostatic head. The main problem with such devices, aside from the ever-present ones of chemical corrosion and biological fouling, is that any such device can only intercept a minute proportion of the total length of a given wave front.

Tidal power is already being harnessed by two plants, one in northwest Russia and the other in France. The French plant, which cost about $100 million, is a 750 m long dam across the Rance River estuary that can store up to 184 million m³ of water and produce about 240 megawatts (240 million watts) of power, a fairly modest output. Tides in the Rance estuary can reach up to 13.5 m. A minimum of 5 m is required for tidal power plants to function, and there are not many areas where such tidal amplitudes are available. One such area is the Bay of Fundy, Nova Scotia, but here there has been strong opposition from environmental groups, and with good reason, as the damming of tidal estuaries creates severe impacts on estuarine ecology.

Aside from the scheme illustrated in figure 12.32 for harnessing the power of ocean currents, another novel idea has been proposed in which huge parachutes would be attached to a

cable strung across the current. The parachutes would drag the cable forward to its maximum position, then collapse and allow the cable to relax before commencing the cycle again. The motion of the cable would then be converted to electricity.

No doubt some of these schemes—even the more bizarre ones—will be at least tried in the near future. Whether their contribution to our long-term power needs will be truly significant remains unclear. The one main advantage of ocean power sources is that they are all renewable, unlike most of our present sources of power (oil, gas, coal, and uranium). When these more limited resources have run out, ocean power might well gain ascendancy, together with solar, hydroelectric, and wind power. Meanwhile, there is the parallel problem of the development of oceanic mineral resources, which is fully as problematical as the development of oceanic power sources. It seems that because both these potential technologies are in such speculative states, it might make sense to have them join forces and to use only ocean power for the development of ocean resources. At the same time, we should remain alert to any possible ill effects that our exploitations might have on Earth's lifeblood.

thermocline. During **fall overturn,** cold winds raise density, and water sinks and mixes, restoring winter conditions. **Water masses** retain their salinity and temperature characteristics as they move. Warm, saline *central water* overlies the **permanent thermocline** between 40° north and 30° south. Cool, dilute *subarctic* and *subantarctic water* lie at the surface at higher latitudes up to 60°. *Intermediate water* underlies these surface masses. Cold *Antarctic circumpolar water* extends north beneath intermediate water. Cold *North Atlantic deep and bottom water* forms when saline central water cools in winter, mixes with subarctic water, and sinks to great depths. Cold *Antarctic bottom water* forms when ice formation raises salinity of Antarctic circumpolar water, allowing it to sink to great depths.

**Emergent coastlines** formed on continental shelves are straight and of low relief; those formed on tectonically active shorelines are straight and steep. **Submergent coastlines,** formed on drowned land-scapes, are irregular and include stream-eroded *ria coasts* with bays and cliffed headlands, *fjord* (glaciated) *coasts,* and some *organic coasts.* Coasts can be modified by *deltas, glacial drift, lava, faults,* and *barrier islands.* Waves erode **wave-cut notches** in headlands, which collapse to form **wave-cut cliffs.** The **wave-cut platform** caused by retreat of the notch eventually limits retreat by dissipating wave energy. **Sea caves,** *sea arches,* and *sea stacks* are carved by wave action on weak zones in headlands.

with input and output of **littoral drift** (comprising *beach drift* and *longshore drift*) and storage in a *shore-*

face and a *beach* (*sand, shingle,* or *shell*), consisting of a *breaker bar,* **beach face,** swash-built **berm,** and *backshore dunes.* The **surf zone** lies shoreward of the breaker bar. Short, steep, shallow, winter waves erode the beach, forming a high *winter berm.* Long, deep, summer swells sweep sand onshore, rebuilding the beach and forming a low *summer berm.* **Scour lag** keeps sediment onshore because grains are hard to re-suspend after settling. **Spits** form when beaches develop on *spit platforms* built away from headlands by longshore drift. Quiet **lagoons** form on landward sides of spits. Storms produce *washover fans* across spits or breach them and form **tidal inlets** with *flood deltas* and sometimes *ebb deltas.* Severed portions of spits become **barrier islands.** A **tombolo** is a necklike spit that forms in the sheltered zone between the mainland and a nearby island. Newly-formed submergent shorelines tend to accumulate sediment in sheltered bays and to be eroded on exposed headlands until an **equilibrium shoreline** is established.

Potential oceanic energy sources include *thermal, evaporative, wave, tidal,* and *current* systems, all of which are highly inefficient and likely to damage or be damaged by the surrounding environment.

## Key Terms

| | |
|---|---|
| water wave | gyre |
| wavelength | geostrophic flow |
| wave period | longshore current |
| wave base | rip current |
| deep-water waves | seasonal thermocline |
| shoaling waves | fall overturn |
| shallow-water waves | water masses |
| wave refraction | permanent thermocline |
| swash | emergent coast |
| backwash | submergent coast |
| tides | wave-cut notch |
| high tide | wave-cut cliff |
| low tide | wave-cut platform |
| semidiurnal tide | sea cave |
| mixed tide | littoral cell |
| diurnal tide | littoral drift |
| spring tide | beach face |
| neap tide | berm |
| flood tide | surf zone |
| ebb tide | scour lag |
| parallel | spit |
| meridian | lagoon |
| Newton's first and second | tidal inlet |
| laws of motion | barrier island |
| Coriolis effect | tombolo |
| Ekman spiral | equilibrium shoreline |
| Ekman transport | |

## Questions for Review

1. Explain wave refraction in terms of water depth, wavelength, wave base, and wave energy concentration.
2. Explain wave dispersion in terms of wavelength and velocity.
3. Explain the differences among deep-water, shoaling, and shallow-water waves in terms of wave base, water motion, and interaction with the bottom.
4. What is the lunitidal interval, and why does it occur?
5. If Earth did not rotate, what would the tidal period be?
6. Why is the Coriolis effect not a force?
7. How do Ekman transport and geostrophic flow augment the circulation of the oceanic gyres?
8. Describe and explain the cyclic, seasonal changes in temperature, salinity, and density of the upper 70 m of the oceans in the temperate zone.
9. Describe the locations and origins of the various water masses in the Atlantic Ocean.
10. How are sea cliffs eroded? What factor ultimately limits that process?
11. What are the various parts of a beach? How are they formed?
12. Describe and explain the cyclic, seasonal alternation of erosion and deposition on a beach.
13. Explain scour lag. In what way is it significant to equilibrium in a littoral cell?
14. Describe the process of spit formation.
15. Describe the various changes undergone by a submergent coast during the development of an equilibrium shoreline.

# Part Four

# Earth's Atmospheric Systems

Photo: A cyclonic storm system advances over the British Isles and Scandinavia in this false-color satellite image.

# 13

# The Dynamic Sky: Energy and the Atmosphere

## Outline

*G*o forth under the open sky,

*and list to Nature's teachings.*

*William Cullen Bryant*

***Thanatopsis***

Photo: Wyoming sunset paints the sky
with a burst of red and yellow.

## Probing the Sky

James Hutton was not the only Scotsman of his time to make pioneering contributions to Earth science. His contemporary, Alexander Wilson, performed the first documented experiment in the **sounding,** or probing and measuring, of the atmosphere. For this, he employed a series of connected kites, to which were attached several thermometers, each of which bore a small parachute. As this instrument cluster rose to a height of about half a mile, the thermometers were released at regular intervals by burning fuses. Wilson's assistant then gathered the thermometers and recorded their readings before they could readjust to the ground-level temperature. Three years later in America, Benjamin Franklin followed this crude experiment with another that was not only crude, but foolhardy. Fortunately, the lightning bolt that struck Franklin's famous kite illuminated the key that dangled from the kite wire but stopped short of illuminating the intrepid Mr. Franklin.

Since these early experiments, more sophisticated means of sounding the atmosphere have become available. These include *balloons, radio transmissions, aircraft,* and *rockets.* Following the invention of the hot air balloon by the French papermakers Josef and Jacques Montgolfier in 1783, several manned flights were made by scientists in both hot air and hydrogen balloons. Flights over southern England and the English Channel by François Blanchard and John Jeffries in 1784 and 1785 yielded some observations on pressure, temperature, and humidity. Two famous French physicists, J. L. Gay-Lussac and Jean Biot (for whom biotite was named), ascended from Paris to 4000 m above sea level in August 1804. In the following month, Gay-Lussac made a solo ascent to 7000 m. During these ascents, Biot and Gay-Lussac found that both the composition of the atmosphere and the geomagnetic field intensity remained constant except for a slight decrease in water vapor. Between 1862 and 1901, several British, French, and German balloonists ascended to altitudes between 8800 and 10,500 m and found that our atmosphere's ability to sustain life gives out within this range because of decreasing atmospheric pressure. Most of these men suffered unconsciousness, and a few of them died before they could descend to safe elevations again.

Fortunately, late in this same time interval, self-recording weather instruments became available, and it was no longer necessary to risk human lives in the scientific exploration of the atmosphere. Just before 1900, unmanned balloons ascended to 15,000 and 22,000 m in France and Germany, respectively. In some ways, these experiments were as crude as Wilson's. The balloons were only partly filled with hydrogen, which expanded as they rose into regions of lower pressure. Eventually, they would burst, and their instrument packages would parachute to Earth. Of course it was

**Figure 13.1**   A weather balloon with a radiosonde attached (*box*). When the balloon bursts, the radiosonde parachutes safely to the ground.

hard to predict where the instruments would land, so many flights were in vain, or at best, their data were collected days or weeks after the ascent.

This problem was solved gradually in the early 1900s with the development of radio and aircraft technology. The earliest experiments with radio were designed to measure wind velocities and directions high in the atmosphere by tracking radio signals from drifting balloons. Chief pioneer of this effort was Harry Diamond of the United States National Bureau of Standards. In 1927, French meteorologists P. Idrac and R. Bureau launched a sounding balloon equipped with instruments whose measurements were transmitted to Earth by radio signal. The principle involved is still in use in the modern, sophisticated *radiosonde,* a small, compact instrument box with a parachute suspended beneath a weather balloon (fig. 13.1). This equipment routinely measures temperature, pressure, and humidity up to an altitude of 16 km. Devices for measuring ionizing particles and radiant energy are often included as well. Finally, wind speed and direction can be determined by tracking the motions of sounding balloons.

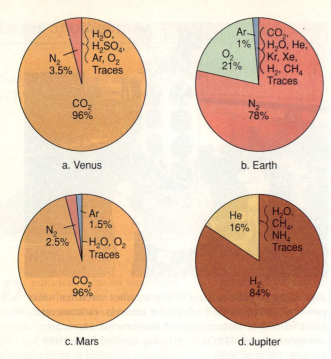

**Figure 13.4** Pie diagrams reveal a striking similarity in composition between the atmospheres of Venus (*a*) and Mars (*c*), but not their striking differences in temperature and pressure. Earth's atmosphere (*b*) shares with them a volcanic origin, but its chemistry has been profoundly altered by life processes. Jupiter's radically different atmosphere, representative of the giant planets, is shown here for contrast (*d*).

# Earth's Strange, Corrosive Atmosphere

Because we were born with it, we tend to regard Earth's atmosphere, laden as it is with corrosive oxygen gas, as "normal." A glance at figure 13.4, however, should raise some doubts about the validity of this provincial attitude. The atmospheres of Earth's two closest neighbors, Venus and Mars, are not only greatly different in composition from our own, but the chemical composition of each is virtually identical to the other, although the atmosphere of Venus contains traces of sulfuric acid and is about 13,000 times denser and almost 520° C hotter than that of Mars. It is unlikely that such remarkable chemical similarity is accidental. Rather, it suggests that a common process generated the atmospheres of both planets. There are three possible reasons for this similarity:

1. Atmospheric gases were inherited from the original dust cloud from which the Solar System formed.
2. Such gases were added by comets, which collided with the planets after they were formed.
3. After the planets were formed, atmospheric gases were *outgassed,* or "exhaled," from their crusts through volcanoes.

The possibility of inheritance must be ruled out because the chemical compositions of the atmospheres of Venus and Mars are radically different from that of the Sun's atmosphere. A low ratio of neon to the other noble gases (argon, krypton, xenon, and radon) in Earth's atmosphere supports this conclusion because this ratio is much higher in volatile-rich meteorites, comets, and the Sun, all of which are thought to have compositions similar to that of the primordial dust cloud. This deficiency also casts doubt on the otherwise reasonable hypothesis that volatile-rich comets in eccentric solar orbits crashed into the inner planets, thereby donating their volatiles to those planets to form their atmospheres. We are therefore left with outgassing by volcanoes as the only reasonable alternative.

What kind of an atmosphere would be produced by volcanoes? The exhalations of the Hawaiian volcanoes, which are thought to emanate from the deeper mantle via mantle plumes, consist of almost 70% water vapor. After this, in order of decreasing percentages, come carbon dioxide, nitrogen, sulfur oxides and hydrogen sulfide, carbon monoxide, hydrogen, and chlorine. Presumably, the composition of a volcanically produced atmosphere should closely resemble this list. If we ignore water, the atmospheres of Venus and Mars fit this model rather well. Earth's atmosphere, with its water content, fits the model even better if we allow for the conversion of carbon dioxide to oxygen by photosynthesis, for the storing of carbon dioxide in protoplasm and carbonate rocks, and for the regeneration of nitrogen by bacteria. There remains, however, the problem of why there is so much water on Earth, so little on Mars, and virtually none on Venus. I address that question in a following section.

The atmospheres of Jupiter (fig. 13.4*d*) and the other giant planets Saturn, Uranus, and Neptune, contrast drastically with those of the inner planets. Hydrogen and helium dominate, and there are various amounts of methane, ammonia, and water. It seems probable that these planets still retain their original atmospheres because they are too far away from the Sun for the young solar wind to have been able to blow them away (see chapter 1).

The modifications to Earth's atmosphere caused by life processes are highly unstable. Oxygen is continually reacting with a variety of materials at, above, and even below Earth's surface. If it were not constantly renewed by photosynthesis, all the oxygen in our atmosphere would vanish within a few thousand years because of the readiness with which it reacts with such substances as methane, ammonia, hydrogen sulfide, organic materials, and iron-bearing minerals exposed during weathering. Similarly, if denitrifying bacteria did not continually renew atmospheric nitrogen, it would soon be bound up in organic compounds and dissolved nitrate ions. Finally, both photosynthesis and carbonate secretion continually remove carbon dioxide from our atmosphere. Consequently, practically all the carbon dioxide that would otherwise be in the atmosphere is now locked up in organic material and carbonate rocks.

## Shells of Gas

If you have ever tried to remove one book from the bottom of a tall stack of books, you will appreciate that in any pile of things within a gravity field, whatever is near the bottom is under higher pressure from the weight of the things above it than are the things near the top. This principle applies equally to the rocks of Earth's interior, to the water in the oceans, and to the gases of the atmosphere. This pressure increase with depth is relatively slow in Earth's lithosphere and hydrosphere because the molecules of rocks and water are in mutual contact. In the atmosphere, however, the molecules of gases are separated by empty space except during collisions; hence, pressure and density increase much more rapidly with depth as increasing pressure from the weight of overlying molecules forces the molecules near the bottom of the atmosphere to crowd closer together. This trend is shown in figure 13.5 by the curve of **atmospheric pressure** at the left. (Note that all scales in this figure are logarithmic, a practical necessity in order to display all its information. On an arithmetic scale the pressure curve would slant strongly to the left at the base and become more vertical with increasing altitude.) At an altitude of about 5.5 km, the pressure has dropped to a value of 0.5 times the surface pressure. In other words, one half of the atmosphere's total mass (about $5.1 \times 10^{21}$ g) lies below 5.5 km. The other half extends from this altitude to the outer limit of the atmosphere at about 10,000 km, where atoms are about as concentrated as they are in interplanetary space.

## The Homosphere: A Uniform Base

Earth's materials are arranged with the densest substances (iron and nickel) in the core, substances of intermediate density (magnesium and iron silicates) in the surrounding mantle, and low density substances (aluminum silicates, quartz, carbonates, and water) in the outer crust and oceans. This same kind of *density stratification* continues into the atmosphere, which is subdivided into several concentric shells of gas, each with a different composition (see right side of fig. 13.5). At the base of this stack of gas shells lies the *homosphere,* a homogeneous layer having the composition shown in figure 13.4*b*. The homosphere extends upward to about 80 km, where it gives way to the *heterosphere,* which extends to the outer limits of the atmosphere.

### The Troposphere: The Weather Zone

Although the homosphere is indeed homogeneous in chemical composition, it is far from homogeneous in physical structure. It is, in fact, subdivided into three layers, each with its own, distinctive characteristics. The basal layer, the **troposphere,** extends upward to about 8 km in the polar regions and to about 18 km above the equator. As the air temperature curve in figure 13.5 shows, the temperature of this basal layer decreases steadily upward (apparent curvature is due to the logarithmic scale). The rate of this decrease, called the **environmental lapse rate,** averages 6.5° C/1000 m. At the **tropopause,** the upper "surface" of the troposphere, the lapse rate decreases, eventually reversing in the overlying stratosphere.

The reason for the lapse rate is simple: **Insolation,** or solar energy striking Earth's surface, is converted to thermal energy, which heats the lowermost meter or so of the atmosphere by *conduction,* or transfer of heat by direct contact (see fig. 13.18). With increasing altitude above Earth's solar-heated surface, the air temperature drops. This temperature drop with altitude would be much faster than it actually is if it were not for the fact that the surface-heated air expands, and becomes less dense than the surrounding air. Because of this, it rises, carrying its heat content upward, a process called **thermal convection** (see fig. 13.18).

As this air rises into regions of lower pressure, it expands. This results in cooling because there are fewer heat-releasing collisions among molecules in a given time interval than there are in a more densely packed gas. Eventually, the rising, cooling air reaches an altitude at which its temperature and density become equal to the temperature and density of the surrounding air. At this altitude, the air stops rising.

Within the troposphere, thermal convection keeps the atmosphere in a constant state of overturn (troposphere means "turning sphere" in Greek). This thermally driven churning action is the force behind both the hydrologic cycle of evaporation and precipitation (see chapter 1) and the general circulation of the atmosphere (see chapter 14). All of Earth's weather occurs within this thin, convective, basal layer, as does virtually all of its cloud cover.

### The Stratosphere: Earth's Weather Seal

Above the tropopause, the upward cooling trend of the troposphere gives way to an upward warming trend. (The temperature curve in fig. 13.5 is for the *equatorial* tropopause; over the poles, the curve reverses at a lower altitude.) The warming trend continues up to about 48 km, where it again reverses. Its effect is to place warmer, lighter, air above the cooler, denser air of the upper troposphere. This condition, known as a **temperature inversion,** is inherently stable, because air does not rise if it is denser than the air that lies above it. Because of its stability, there is virtually no vertical overturn (convection) of air within this atmospheric shell. Instead, the air forms horizontal layers and moves *within* these layers rather than across them. Because of its stratified nature, the shell is known as the **stratosphere.** Because the underlying tropopause halts the rise of moist, convecting air, the stratosphere is an exceedingly dry region, although trace amounts of water sometimes produce faint, pearly *nacreous clouds* at about 30 km altitude.

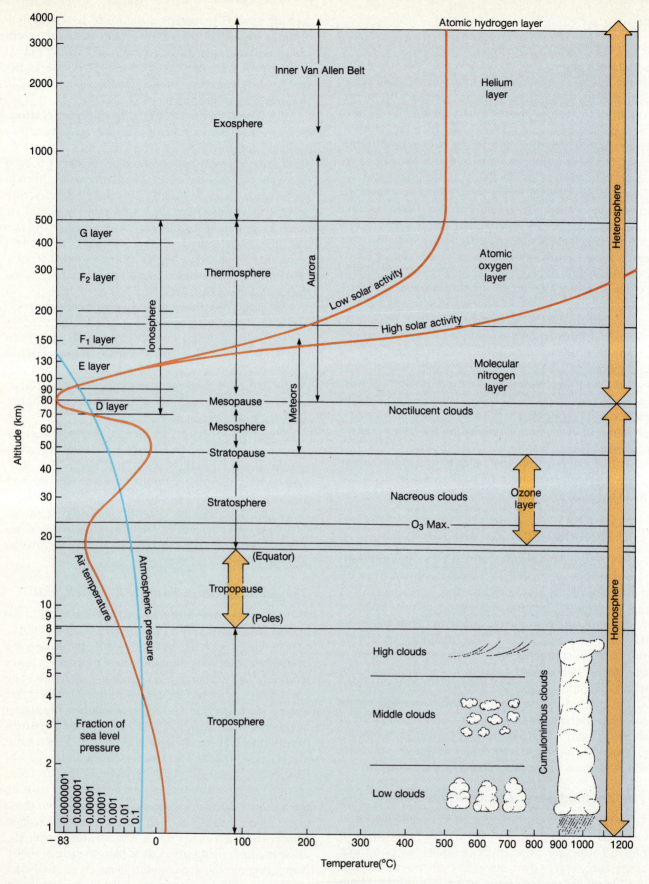

**Figure 13.5** A vertical profile through Earth's atmosphere. Its main divisions are given at the far right, and left of center. Curves of temperature and pressure are at the left. Compositional layers and various atmospheric phenomena, including cloud types, ozone, meteors, auroras, ionospheric layers, and a Van Allen belt, are also shown. All scales are logarithmic.

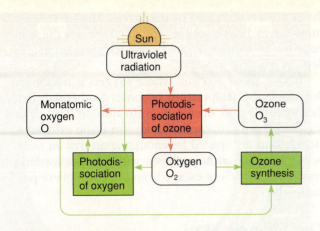

**Figure 13.6** The ozone cycle operates within the stratosphere, protecting Earth's surface against lethal UV radiation and the loss of surface water. In the cycle of ozone synthesis (*green arrows*) and destruction (*red arrows*), both processes result from the action of UV radiation.

The temperature inversion responsible for the stratosphere's great stability is caused by a phenomenon discussed briefly in chapter 1, the absorption of solar ultraviolet (UV) radiation within the ozone layer. In the following section, I examine this phenomenon in greater depth.

## The Creation and Destruction of Ozone

Figure 13.6 illustrates the process of ozone production, and one of the many processes that destroy ozone. High-energy solar UV rays in the band from 0.2 to 0.3 $\mu$m strike oxygen molecules, which consist of two atoms each ($O_2$), and split them into individual atoms of *monatomic oxygen* (O) by a process called *photodissociation* (see "photodissociation of oxygen" box at the lower *left* of fig. 13.6). Monatomic oxygen is exceedingly reactive, and in the subsequent process of "ozone synthesis" (*lower right* of fig. 13.6), each of these atoms combines with a normal oxygen molecule to form *triatomic* molecules of ozone ($O_3$).

The ozone molecule, however, is no less immune to attack by UV radiation than is the oxygen molecule. When struck, it splits into oxygen and monatomic oxygen (see "photodissociation of ozone" box in fig. 13.6). Here, then, we have a closed cycle, because as soon as monatomic oxygen is formed, it again combines with normal oxygen to form ozone. The heat generated during the splitting of oxygen and ozone molecules raises the temperature of the surrounding air, thereby generating the thermal inversion of the stratosphere, a feature lacking in the atmospheres of the other inner planets.

In spite of its great influence on terrestrial life and climate, ozone is present in only minute amounts in the ozone layer. Even at about 25 km altitude, where it is most abundant, ozone concentrations are only about 12 ppm (parts per million), but even this small amount generates enough heat to cause a reversal of the lapse rate. The temperature is highest at the top of the stratosphere because the absorption of UV radiation is strongest there, even though maximum concentration of ozone is about 10 km lower.

## Ozone's Vital Functions

It should be clear by now why the ozone layer shields Earth's surface against UV radiation. Nearly 100% of that radiation is intercepted and consumed in the splitting of either oxygen or ozone molecules. The small amount of UV radiation that penetrates the ozone shield causes severe sunburn, skin cancer, and genetic mutations in organisms. There is no doubt that except for those marine organisms that are shielded by a sufficient depth of seawater, all life on Earth would quickly be destroyed if all, or even most, of the solar UV radiation that reaches the top of the atmosphere were allowed to penetrate to Earth's surface.

Beyond shielding living organisms from UV radiation, the ozone layer also protects Earth's water, which is as prone to photodissociation by solar UV as oxygen. There is one important difference, however. In the case of water, photodissociation is forever! Hydrogen is too light to remain at Earth's surface, so when it is split from an $H_2O$ molecule, it rapidly rises to the top of the atmosphere, never to return.

The ozone layer guards against this unwelcome possibility in two ways. The first is simply its shielding action. The second is that the highly stable inversion of the stratosphere places an effective "lid" on tropospheric convective circulation at about 10 km altitude. If this barrier were not in place, convective circulation would be able to carry water vapor to much higher altitudes where UV radiation is stronger than at the surface. If it were not for the presence of the stratosphere and the ozone layer that maintains it, the convective overturn of the atmosphere would extend to much higher altitudes, and Earth's weather patterns would be much different from what they actually are.

Clearly, Earth's water supply survived the Azoic and Cryptozoic Eons despite an apparently oxygen-free atmosphere, but in the Azoic, solar radiation was some 30% less intense than it is today. Furthermore, it is possible that even trace amounts of photosynthetic oxygen could have produced a tenuous, but sufficient ozone shield at higher altitudes than the modern shield. Could Venus and Mars at one time also have had abundant water, which they eventually lost because, lacking photosynthetic life-forms, they never acquired oxygen-bearing atmospheres from which protective ozone layers could form? In view of what is at stake, it would be worthwhile to begin seeking answers to these questions.

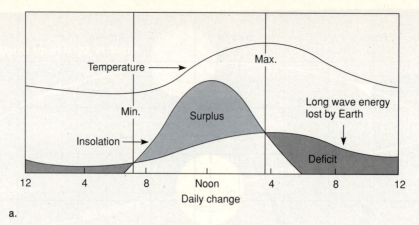

a.

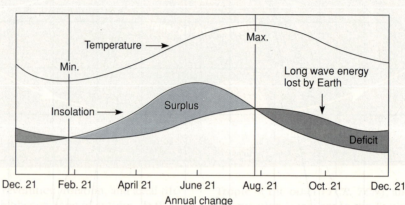

b.

**Figure 13.16** Daily (*a*) and annual (*b*) heat cycles for a hypothetical mid-latitude locality showing heat surpluses (*dark blue*) and deficits (*grey*). In the daily cycle, these fluctuations are caused by Earth's rotation. In the annual cycle, they are caused by seasonal shifts in the subsolar point and the circle of illumination due to the inclination of Earth's rotational axis.

Paradoxically, the peaks of these insolation curves increase with latitude despite the *decrease* in the insolation angle with latitude. This is because day length increases with latitude. If it were not for this effect, the polar regions would be much colder and less habitable than they actually are.

The dotted *net radiation* curves at the bottom of figure 13.15 show what happens to this insolation after it has interacted with the atmosphere and with Earth's surface. These curves are for stations close to the equator (Manaos), the Tropic of Cancer, (Aswan), and the Arctic circle (Yakutsk). They represent insolation augmented by counter radiation from the atmosphere and reduced by albedo and radiation from the ground surface. In general, these net radiation curves for Earth's surface resemble those for insolation at the top of the atmosphere, but their values are much lower, mainly because of ground radiation losses. They are also less regular, due largely to seasonal variations in cloud cover. Also, the net radiation curves have much lower amplitudes than the insolation curves (i.e., there is less vertical separation between peaks and troughs), indicating

that temperature is more evenly distributed over Earth's surface than we would expect it to be from the insolation curves alone. The reason for this lies in the phenomenon of planetary heat transfer, a topic addressed in the following section.

Figure 13.16 shows how energy input (insolation), storage (temperature), and output (long wave energy lost by Earth) vary at a single, midlatitude station over the course of a single day and a single year. In the daily cycle, insolation begins with sunrise occuring shortly after 6:00 A.M. Before sunrise, the temperature had been dropping steadily because of nighttime cooling due to long wave radiation from Earth's surface. Even after sunrise, the temperature continues to fall until this energy output is balanced by the input of solar energy, as indicated by the crossing of the two radiation curves shortly after 7:00 A.M. At this point, the nighttime energy deficit is replaced by a daytime energy surplus. In terms of systems, input is now exceeding output, and therefore temperature (storage) begins to increase. That increase continues as long as a surplus exists. Even though

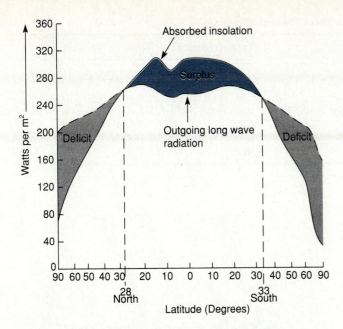

*Figure 13.17* Satellite measurements of absorbed short wave insolation (*solid curve*) and outgoing long wave radiation (*dashed curve*) reveal that between 28° north and 33° south, Earth receives more energy than it radiates. At higher latitudes, Earth radiates more energy than it receives. These imbalances are corrected by continual, poleward flows of energy as sensible and latent heat.

maximum insolation occurs at noon, temperature continues to rise until shortly after 3:00 P.M., when the surplus gives way again to a deficit.

Notice that in the afternoon the output curve crosses the input curve at a higher level than in the morning. This is because the Sun's increased input has increased Earth's radiant energy output. From now until shortly after sunrise on the following day, a deficit again prevails and temperature falls.

The annual energy cycle resembles the daily cycle, except that the input curve never falls to zero, and the amplitudes of both input and output curves are less than those of the daily cycle, although the amplitude of the temperature curve is often greater. Otherwise, the relationships among the three components of input, storage, and output are identical.

## Thermal Gradients and Heat Transfer

Figure 13.17, based on satellite data collected between June 1974 and February 1978, illustrates Earth's radiant energy balance. The solid curve, representing absorbed insolation, is high in the middle and low on the ends, indicating that energy absorption is high in the tropics and low near the poles. The dashed curve, representing outgoing long wave radiation, is flatter, indicating that Earth radiates energy much more evenly than it receives

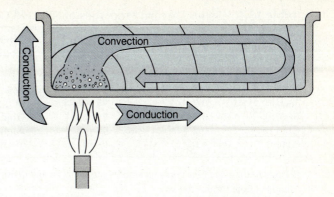

*Figure 13.18* The heat of a flame is transferred through a liquid-filled, metal basin by two mechanisms. Conduction, in both media, involves the transfer of molecular motion down a thermal gradient (*concentric isotherms*) by direct contact among molecules. Convection, a far more efficient mechanism of heat transfer, involves the bulk motion of warm, low-density fluid away from the heat source (*broad arrow*).

it. Between 28° north and 33° south, energy input exceeds output, yielding a surplus. At higher latitudes, the situation is reversed, yielding a deficit. Significantly, the surplus and the deficit are equal (the surplus appears smaller than the deficit, but it is distributed over a much larger area). This distribution of equal energy surplus and deficits implies three things:

1. Earth's surface is neither warming nor cooling over time.
2. Nonsolar contributions to Earth's energy budget (e.g., geothermal heat) are relatively insignificant (otherwise the surplus would exceed the deficit).
3. Energy must be flowing "down" an energy gradient from the tropical region of surplus to the higher latitude regions of deficit.

How this energy transfer is accomplished is suggested by figure 13.18, in which a flame is set beneath one end of a liquid-filled, metal basin. The flame's heat sets up *thermal gradients* in both the metal and the liquid, shown (for the liquid) by contours of equal temperature called *isotherms*. Molecular motion is most vigorous above the flame and is transferred across the isotherms toward lower temperatures by **conduction,** or the intermolecular transfer of heat. Thermal convection, represented by the broad arrow in figure 13.18, is also at work here, and it is a faster, more efficient mechanism of heat transfer than conduction. At the vessel's far end, its walls and the water surface force the cooling flow to sink and return to the warm end. This *convective circulation* stirs the fluid, breaking down the thermal gradient and homogenizing the liquid with respect to temperature and density.

**Table 13.1**  Some Thermal Properties of Selected Common Natural Materials

| Material | Specific Heat | Albedo (percent) | Thermal Conductivity | Transparency (sec/cm° C) | Relative Heating Rate (Water = 1.0) |
|---|---|---|---|---|---|
| Air (sea level) | 0.17 | 6 | Very low | High | 5600 |
| Soil | 0.2 | 5–20 | Low to high | Low | 3.5–4.4 |
| Snow | 0.2–0.5 | 75–95 | Medium (packed) | Low | 1.1–5.6 |
| Basalt | 0.2 | 7 | High | Low | 1.7–2.1 |
| Granite | 0.2 | 35–45 | High | Low | 1.1–1.4 |
| Water | 1.00 | 6–10 | Medium | Med | 1.0 |

Relative heating rate based only on albedo, specific heat, and density.
Compiled from various sources.

Convective transfer of both warm ocean water and warm, humid air is the main process whereby solar heat is redistributed from equatorial regions to higher latitudes. Conduction, on the other hand, is of little importance in planetary heat transfer because Earth materials are very poor conductors of heat.

As we have seen, another mechanism for redistributing heat that works hand-in-hand with thermal convection is latent heat transfer (see fig. 13.12). In latent heat transfer, winds carry water vapor from warmer regions to cooler regions, where the vapor condenses to cloud droplets or snow crystals, releasing the latent heat that was stored in the water vapor when it evaporated from the sea surface. At the same time, the airmass carrying this water vapor also transfers heat poleward by convection.

## Heat Reservoirs: Land, Sea, and Air

Water has a much greater capacity for absorbing and storing heat than other Earth materials have. Because heat is simply an expression of the vigor with which atoms vibrate, it is independent of the size and weight of the type of atom involved. A given mass of water contains more atoms than an equal mass of most other substances (recall that hydrogen is the lightest atom); hence, it takes more heat to produce a given amount of vibration in water than is required for the same mass of most other substances. As I note in chapter 12, water's specific heat of 1.00 is among the highest in Nature. Pure hydrogen has the highest specific heat of any substance (2.42) because it packs so many more atoms into a given mass than do most other elements.

Differences in specific heats (table 13.1) influence the rates at which the Sun's rays heat the different materials of Earth's surface, but there are many more factors involved than specific heat. As figure 13.19 indicates, albedo is also significant because it determines the percentage of insolation that is absorbed rather than reflected. In general, the lighter, smoother, and less transparent the material, the higher its albedo. The heating rates (relative to water) of the various materials in figure 13.19 reflect the combined effects of albedo, specific heat, and density. (Higher density results in slower heating rates because it concentrates atoms.) Air

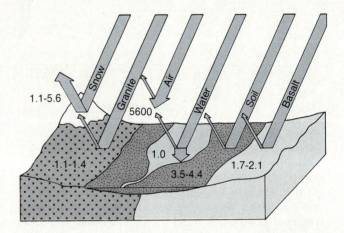

**Figure 13.19**  The rate at which insolation heats a land or water surface depends primarily on the albedo, specific heat, and density of the surface. Effects due to emissivity, thermal conductivity, and transparency, though important, are ignored here.

heats more than 1000 times faster than soil, which has a similar albedo and specific heat, but a much greater density. The heating rate of basalt is about half that of soil, and here, the difference is due mainly to the greater density of basalt. A further source of variability in soil is water, which can fill pore spaces and increase both the density and the specific heat of soil, both of which decrease the heating rate. Variability in the heating rate of basalt is due to vesicles (air bubbles), common in the tops of lava flows. Granite is comparable to basalt in its specific heat and its density. Its lower heating rate is due to its significantly higher albedo (table 13.1).

The large range of heating rates for snow results from its tendency to become compacted with age. As with soil and basalt, air space in snow greatly reduces density, and this increases the heating rate. Compaction squeezes out air space and replaces it with ice, which has a higher density and a higher specific heat (0.5 cal/cm³/°C), so it has a lower heating rate. Because of its low density, snow can be melted readily by sunlight if any dark colored material is sprinkled on it. The dark material greatly lowers the snow's albedo, allowing more energy to enter the snow beneath it. This is why snow in contact with a dark object, such as a tree trunk, an asphalt pavement, or road dirt, melts more quickly than adjacent snow (fig. 13.20).

**Figure 13.20** The absorption of solar energy by the dark trunks and foliage of these spruce trees has warmed the snow cover around their bases, hastening its melting.

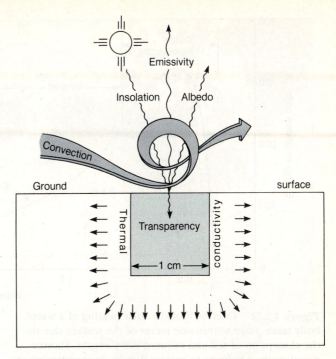

**Figure 13.21** Factors other than specific heat and density that affect the heating time of a cubic centimeter of material at Earth's surface.

Water has the lowest heating rate of all common natural materials. This is due to both its unusually high specific heat and to its unusually low albedo. Because of this remarkable heat-absorbing property, water is a great moderator of environmental temperature fluctuations, a property I consider in more detail in a following section.

The heating rates in figure 13.19 assume that all nonreflected radiation is absorbed instantaneously within the top 1 cm of the surface with no loss due to wind or air convection, ground conduction, or long wave radiation. In reality, all these factors are significant, however, and they should be taken into account (fig. 13.21). Ground conduction is measured as **thermal conductivity** (column 4, table 13.1), the amount of heat, in calories, that can pass through a 1 cm thickness of a given material in 1 sec when there is a temperature gradient of 1° C between the two ends of the centimeter. Thermal conductivity is roughly proportional to density, another reason why the surfaces of dense materials heat slowly. This makes sense when you consider that thermal vibrations are more readily transferred from one atom to another in a denser material than in a less dense material. Materials with low thermal conductivities heat quickly at the surface because heat does not drain away rapidly through the material. On the other hand, materials that have high thermal conductivities heat slowly at the surface but more rapidly in the interior.

The thermal conductivity of air is extremely low, a property that makes air an excellent insulator if it can be prevented from moving and transferring heat by convection. One familiar use of air as an insulator is in sealed, double-pane storm windows. Another is in down- or fiber-filled sleeping bags, in which "dead" air is trapped among the fibers of the filling.

The thermal conductivity of fresh snow is low because of its high air content. As snow becomes compacted and loses some of this air, its thermal conductivity

rises. The same is true of soil, which has a correspondingly wide range of thermal conductivities. Most thermally conductive of all are rocks, and water is intermediate in conductivity.

Another complicating factor in the heating of surfaces is *transparency* (column 5, table 13.1). Electromagnetic energy can not penetrate the surfaces of opaque materials, such as basalt and granite; hence, its transformation to heat energy must occur at the surface of such materials. On the other hand, electromagnetic energy penetrates readily beneath the surfaces of such transparent materials as air and water, so it can be transformed to heat throughout considerable depths within such materials. Figure 13.22 shows the changes that take place in the spectrum of solar energy as it penetrates to various depths within a body of water. The long, infrared "tail" of the energy spectrum is almost completely absorbed within the upper 10 cm, leaving 57% of the original energy to penetrate to greater depths. At 1 m depth, all but 39% of the solar beam has been absorbed, and at 10 m, only 22% still remains. With increasing penetration, increasingly shorter wavelengths are absorbed. At 10 m and below, the unabsorbed light is decidedly bluish, and its reflection back to the surface is what gives most water bodies their blue color. At a depth of 100 m, only 4% of the original energy remains unabsorbed.

efficient "heat sponge." Because insolation can penetrate so deeply into clear water, it heats not only the surface layer, but a considerable thickness of water below

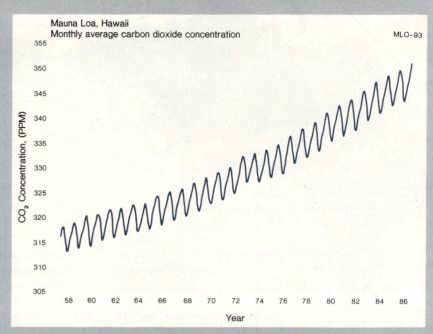

Because the greenhouse gas carbon dioxide plays such an important role in regulating the rate of escape of long wave radiation from Earth's surface, climatologists have been keeping a weather eye on the trends shown in figure 13.25. This graph shows monthly mean concentrations of atmospheric carbon dioxide in parts per million (ppm) measured at the Mauna Loa Observatory in Hawaii. Two trends are evident. The first is an annual oscillation in carbon dioxide concentration with an amplitude of about 6 ppm. The troughs of the oscillation result from increased photosynthesis in spring and summer. The peaks in fall and winter occur as respiration and decay in extratropical regions overshadow photosynthesis in the tropics.

The second trend is the overall rise in the curve from 1958 to 1987. This is only the most recent portion of this trend. Prior to about 1850, carbon dioxide had a mean concentration of about 280 ppm. Since that time, the concentration has steadily risen largely due to the burning of fossil fuels. Since about 1974, the rise has been noticeably steeper, and many scientists think that this is, in itself, cause for concern. There are several factors that could be contributing to this increase in the upward trend. First, in recent years many developing nations have been desperately trying to put more land into production to reduce their enormous foreign debts. This has led to the wholesale destruction of tropical rainforests, which constitute one of the two major "sinks" for carbon dioxide.

**Figure 13.25** Mean monthly values of atmospheric carbon dioxide concentration at Mauna Loa, Hawaii, reveal two trends, one normal and the other disconcerting. The annual fluctuation is due to seasonal variation in the rate of photosynthesis; the general rise through time is due to the burning of fossil fuel and, most recently, to the destruction of rainforests.

The only other important sink is the oceans (weathering is a pathway to the oceanic sink, and not a sink in itself), which appear to have absorbed about half of the carbon dioxide that human activities have introduced into the atmosphere.

Other contributions to the accelerating accumulation of greenhouse gases in our atmosphere illustrate the complexity of the problem. In addition to carbon dioxide, there are several other common greenhouse gases, including methane, nitrogen oxides, and chlorofluorocarbons (CFCs). Methane, which is about 20 times more effective as a greenhouse gas than carbon dioxide, is a product of the bacterial decomposition of organic matter, mainly in animal intestines and refuse dumps. A proliferation of feedlots and landfills in recent years has greatly increased the rate of methane

---

regions and increases toward higher latitudes. The increase is slight in the Southern Hemisphere, which is dominated by ocean water, but dramatic in the Northern Hemisphere, which is dominated by landmasses. The annual temperature range is greatest in continental interiors, and it diminishes toward continental margins. For a given latitude, the range is greater in desert areas (e.g., north Africa and Australia) than it is in heavily vegetated regions (e.g., central Africa and northern South America).

The poleward increase in annual temperature range occurs because seasonal changes in insolation are more pronounced at higher latitudes (see fig. 13.15). Annual temperature range increases more slowly in the Southern Hemisphere than in the Northern Hemisphere because the oceanic "heat sponge" absorbs and releases heat more slowly, and holds more heat in storage, than do land surfaces. Oceans are therefore far more effective than land surfaces in moderating temperature extremes. The increase in annual temperature

generation, especially in the developed nations. Nitrogen oxides, unavoidably produced in all high-temperature combustion processes, are linked mainly to automobile emissions. The only effective means of reducing nitrogen oxide production is to reduce combustion temperature. This, however, has the effect of increasing the output of unburned, carbon-rich particulate matter, which contributes to the generation of smog. CFCs, also much more potent greenhouse gases than carbon dioxide, are used in refrigeration and air conditioning units, and as inflatants in styrofoam. When any of these products are discarded, as is happening to an increasing extent, CFCs escape to the atmosphere, where they cannot only increase the greenhouse effect, but also contribute to the destruction of stratospheric ozone, the consequences of which are discussed in this chapter.

What does the increase in greenhouse gases portend for the future of Earth's environment? In theory, the implications are quite clear. An increased greenhouse effect translates into a global warming trend. In the 1980s, a statistically improbable number of years with above normal mean annual global temperature lends strength to this conclusion. Because climate and weather are so complex, however, it is difficult to prove a definite link between human-caused increases in the concentrations of atmospheric greenhouse gases and global warming.

That difficulty has been underscored by a recent analysis of satellite data that shows wild swings in global temperature over the decade of the 1980s but no evidence of an overall warming trend. In the face of such equivocal evidence, it is understandable that policymakers remain reluctant to impose regulations on the production of greenhouse gases. Constrained by corporate interests on one side and relative public indifference on the other, they see little sense in pushing for draconian restrictions on human practices that so far have produced no clear evidence of the global warming that should, in theory, be happening for reasons that have been detailed in this chapter.

Is there something wrong with the theory? Have we overlooked something? The answers to these questions are almost certainly no and almost certainly yes, respectively. It should be clear by now that the Earth system is replete with steady states, such as the salinity, acidity, and ionic composition of the ocean. Subsystems that are characterized by such steady states are typically cybernetic, and they have a high degree of *persistence,* meaning that they have an ability to resist change, just as a human body has an ability to resist heat, cold, and disease. Because of that resistance, such subsystems can continue to function in their accustomed manner even when they are subjected to harmful stresses. Their resistances are not limitless, however, and if they are ultimately exceeded by the harmful stresses, the systems break down, typically abruptly and without warning. Equilibrium systems, on the other hand, respond to stresses *immediately,* and because of this, it is easy to tell when a given stress is causing an undesirable effect.

In the case of global warming, we might not have the luxury of such telltale warning signs, at least not of the sort we have been looking for. In consideration of the many mechanisms available for redistributing heat over Earth's surface (see chapter 15), global warming is likely to be the very *last* symptom we see, and the most dire, as it will signify the ultimate failure of those heat-transfer mechanisms. One symptom that should be quite clear right now, however, is the fact that those mechanisms are working overtime. That evidence might not be especially convincing to corporate executives, whose management policies are determined by current market trends and current competition. It also might not be especially convincing to the public, whose main concern (justifiably!) is financial security within a global economy that has been overstressed by decades of end-use military expenditures and financial mismanagement. As the decade of the 1990s wears on, however, the U.S. Global Change Research Program, a satellite-based, interagency data acquisition and analysis effort, should provide much new evidence on the extent to which our planet is coping with the many human influences that *should be* producing global warming. The big question, of course, is: will we find the answers we are looking for in time to adjust our socioeconomic policies before those influences stress Earth's climate system beyond the limits of its ability to persist in its present state?

range toward continental interiors is a reflection of this difference because the farther a land area is from the coast, the less moisture is likely to be brought to it from the sea by storm winds. This effect, known as **continentality,** is less noticeable in humid regions than in arid regions because of the moderating effect of moisture in the atmosphere and in vegetative cover.

## Summary

Routine **sounding** of Earth's atmosphere is done with balloons equipped with *radiosondes.* Rockets, aircraft, and satellites with television and *radiometers* are also used. *Crustal outgassing* probably produced our atmosphere. Life processes maintain its chemically unstable composition. **Atmospheric pressure** decreases rapidly with altitude. Half our atmosphere's mass lies below 5.5 km altitude. The compositionally homogeneous *homosphere* comprises (1) the turbulent, cloudy **troposphere** (0–18 km), in which temperature decreases upward at the **environmental lapse rate** to the **tropopause**; (2) the tranquil, virtually cloudless, layered **stratosphere** (8–48 km), in which ozone creation and

# 14

# The Sphere of Aeolus: Winds and Deserts

*H*oist up saile while gale doth last;

*Tide and wind stay no man's pleasure.*

*Robert Southwell*

**St. Peter's Complaint**

Photo: Wind turbine generators rise above
the sands of a California desert.

# Barometric Pressure: The Force behind the Wind

We seldom think of air as having weight. Nevertheless, the atmosphere exerts more than 10 metric tons of pressure per square meter of Earth's surface! Because this great pressure is the same inside our bodies as it is outside, we are unable to sense it unless the outside pressure changes, as it does during the approach of a severe storm or when we suddenly change elevation, as in an elevator or an aircraft. Atmospheric pressure is of great consequence to Earth's weather, because differences in pressure are the driving mechanism behind the phenomenon of **wind,** or the horizontal movement of air. I discuss some of the more important characteristics of this phenomenon later in this chapter, but we first need to consider just what atmospheric pressure is and how we measure it.

Let us suppose that Earth has been stripped of all its atmosphere except for one thin, vertical column, the lower end of which is placed on one pan of a laboratory balance. The column is 1 cm square and extends upward into interplanetary space. We must further assume that this column maintains all the characteristics it would have had if it were still surrounded by the rest of the atmosphere. Given these necessary (if impossible) assumptions, we can now find the mass of the column by placing weights on the other pan of the balance until the scale reads zero. Using this method, we find that we need 1033 g of weights to match the mass of a 1 cm² column of atmosphere under average conditions at sea level.

In order to find the *weight* of the column, we must first understand that a weight is a force ($F$). Accordingly, by Newton's second law, $F = ma$, we must multiply this mass ($m$) by an acceleration ($a$). The appropriate acceleration to use is the rate at which Earth's gravity accelerates freely falling objects at sea level (980.7 cm/sec²). Doing this, we find that the column's weight is 1,013,000 *dynes* (dyn). One dyne is a force sufficient to accelerate a mass of 1 g at a rate of 1 cm/sec². The final step in our investigation is to find the *pressure,* or weight per unit area of the column on the ground surface. Because the area on which the column rests is 1 cm², we need only divide by 1 ("weight per unit area" means divide the weight by the area). The pressure of the column is therefore 1,013,000 dyn/cm², a quantity that has been called one **standard atmosphere.**

Another, nearly equivalent measure of pressure is the *bar,* which is equal to 1,000,000 dyn/cm². In practice, the **millibar** (1 mb = $10^{-3}$ bar) is a more convenient unit; hence, most meteorologists use it in preference to the bar. *One standard atmosphere is equal to 1013 mb.* Note, however, that this is only an *average* value! As I explain in a following section, various things can happen to this average value to raise it as high as 1084 mb or to depress it as low as 870 mb.

Figure 14.1*a* shows the earliest method of measuring atmospheric pressure. In 1643, an Italian physicist, Evangelista Torricelli (Tore-ee-*chel*-lee) inverted a long glass tube, closed at one end and filled with mercury, in a mercury-filled dish. Torricelli found that the fluid level in the tube dropped below the closed end until it stood at a height of 760 mm above the surface of the mercury in the dish. It was found subsequently that aside from the presence of a trace of mercury vapor, the space above the mercury column in such a tube is a total vacuum.

The reason for this is that the mercury column in the tube is held up by the pressure of the atmosphere on the mercury in the dish. In other words, the pressure exerted on the mercury by the atmosphere is exactly the same as the pressure exerted on it by the mercury column. If the atmosphere were suddenly to vanish, the mercury column would collapse into the dish. (Like the mercury column, Earth's atmosphere also has a nearly total vacuum at its top.) As atmospheric pressure increases, the mercury level in the tube rises slightly, and as it decreases, the mercury level falls. Torricelli's instrument came to be known as a **barometer** (from the Greek *baros,* "weight"), and it is still the standard instrument for measuring atmospheric pressure.

Because it is a rather heavy and cumbersome device, the mercury barometer has been replaced, in most applications, with the *aneroid barometer,* the heart of which is a small, corrugated, air-tight box (fig. 14.1*b*). The box contains a partial vacuum, and as it expands and contracts with variations in atmospheric pressure, these changes are measured on a dial by a needle linked to the box through a lever system.

## Isobars and Pressure Gradients

Atmospheric pressure decreases rapidly upward as the concentration of air molecules decreases with altitude (see fig. 13.5). This pressure decrease is shown in figure 14.2 as a stack of sloping planes, each of which represents a surface of equal pressure. These *isobaric surfaces* are spaced farther apart on the right, where they lie above warmer ground, than they are on the left, where they lie above cooler ground. Because of this, the atmospheric pressure decreases upward faster on the left than it does on the right.

Let us now consider a surface of constant elevation (horizontal plane in fig. 14.2) passing through these isobaric surfaces. The intersections between this horizontal surface and the isobaric surfaces, shown as dark blue lines, are contours of equal pressure, or **isobars.** It is clear, from the values of these isobars, that the horizontal surface intersects higher pressures on the right than on the left. In other words, there is a pressure gradient at this level, along which pressure decreases from right to left. The gradient is indicated by the longer straight arrow running toward the left, perpendicularly

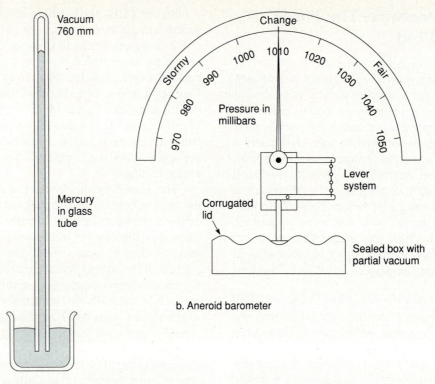

**Figure 14.1** Two common methods for measuring atmospheric pressure. In a mercury barometer (*a*), the height of the mercury column can be used to measure atmospheric pressure because the column's height is proportional to the pressure at its base, which is equal to the pressure of the surrounding air. The aneroid barometer (*b*) is less accurate, but more convenient than the mercury barometer. Fluctuations in atmospheric pressure cause the sealed, corrugated box to expand and contract. These small motions are amplified by a lever system to produce large motions in the indicator needle.

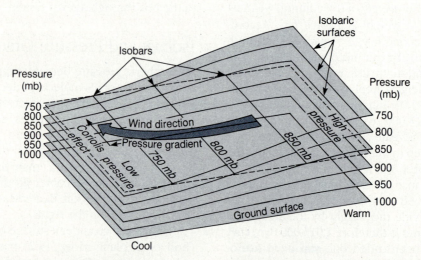

**Figure 14.2** Isobaric surfaces (*sloping planes*) portray the upward decrease in atmospheric pressure above a given region. This decrease is more rapid over cool surfaces (*left*) than over warm surfaces (*right*). Isobars (*dark blue lines*) on a horizontal plane portray a gradient from high pressure at right to low pressure at left (*long straight arrow*). Air attempts to flow perpendicularly to the isobars from the high pressure to the low (*broad arrow*), but is deflected by the Coriolis effect (*short straight arrow*).

to the isobars. Wherever such a pressure gradient exists, air responds to it by flowing "down" the gradient away from regions of high pressure toward regions of low pressure, thereby generating the phenomenon we call wind (broad arrow). Another, perhaps easier, way of looking at it is that *wind always blows in the direction in which the isobaric surfaces are sloping*. Note, however, that the Coriolis effect (shorter straight arrow in fig. 14.2) exacts its due and deflects this pressure gradient wind toward the right (left in the Southern Hemisphere), just as it deflects pressure gradient currents in the ocean (see fig. 12.15). Such deflections have important consequences for atmospheric circulation, but before examining these we need first to consider why pressure gradients exist in the atmosphere.

## Convective Circulation

If insolation were uniform over the globe, Earth's surface would have a uniform temperature, and the overlying atmosphere would be evenly heated. There would be no differences in density at a given altitude under these conditions because atmospheric density is dependent on temperature. In figure 14.3*a*, we have such a situation, a real possibility for only a very limited area of Earth's surface. The isobaric surfaces here are strictly horizontal.

In figure 14.3*b,* insolation heats the ground surface in the center of the land area, and the corners are cooled, perhaps by cloud cover. Because the air molecules above the center are now moving faster, they push each other farther apart, which results in expansion and a decrease in air density. The air molecules above the corners, on the other hand, are slowing down and crowding closer together, which results in an increase in density. Expansion of the air above the center causes an uparching of the isobaric surfaces there, while at the corners, contraction causes a downwarping of the surfaces. Because of this, a pressure gradient develops from high pressure at the center to low pressure at the corners. In response to this gradient, upper air begins to flow away from the center toward the corners. This results in **divergence,** or an outward flow of wind, which is strongest at high altitudes where the gradients are steepest.

Before these divergent winds develop (fig. 14.3*a*), the atmospheric pressure at ground level is uniform over the entire region because the same weight of air lies over each point. As air flows away from the center, however, the weight of the central air column is reduced, resulting in reduced pressure at the surface (fig. 14.3*c*). At the same time, air flowing toward the corners increases their weight, thus increasing the surface pressure beneath them. This raises the lower isobaric surfaces at the corners and sets up a second pressure gradient at ground level with an orientation that is the reverse of the one aloft. This results in convergence, or an inward flow of wind toward the center.

The return flow of these low-level, convergent winds sets up a bagel-shaped **convective circulation** that persists as long as the center continues to be heated more than the corners. If that difference were to be eliminated, the opposing wind systems would only continue long enough to reestablish the conditions in figure 14.3*a.*

## Hadley Cells

The sequence of events illustrated in figure 14.3 represents a local effect, but it can also be applied globally, as in figure 14.4, which portrays a cross section of Earth and its atmosphere. In this model, the equator corresponds to the center of the region in figure 14.3, and the poles correspond to the corners. In the interest of simplicity, we shall assume that the planet is uniformly covered by water and does not rotate. In figure 14.4*a,* the planet is experiencing even heating from all directions. Its surface is therefore at a uniform temperature, and consequently, the isobaric surfaces are everywhere parallel to it. Now, in figure 14.4*b,* imagine that a "ring of fire" has developed above Earth's equator. The planet is therefore heated most at the equator and least at the poles, creating a temperature gradient, which, in turn, generates a pressure gradient from the equator toward the poles. High temperatures and pressures arch the isobaric surfaces above the equator, whereas lower temperatures and pressures depress the isobaric surfaces above the polar regions.

In figure 14.4*c,* high-level, divergent winds flowing poleward away from the high pressure aloft transfer atmospheric mass toward the poles, creating a surface low pressure area at the equator and surface highs at the poles. This produces two equatorward pressure gradients at the surface to balance the two poleward pressure gradients aloft. In figure 14.4*d,* a pair of large-scale convection cells has developed in response to these opposing gradients.

Such a large-scale convective circulation does in fact operate in Earth's atmosphere, but this so-called **Hadley cell** circulation (named after George Hadley, the eighteenth-century English meteorologist who first proposed the concept) is actually restricted to the belts between the equator and about 30° north and south latitude. Let us see, now, why this is so.

## The Geostrophic Wind: The Coriolis Force Revisited

In figure 12.15, we saw that water flows directly away from a region of high pressure, moving perpendicularly to the isobars along the pressure gradient. We also saw that the Coriolis effect, acting at a right angle to the direction of motion, progressively deflects the water's path toward the right in the Northern Hemisphere and toward

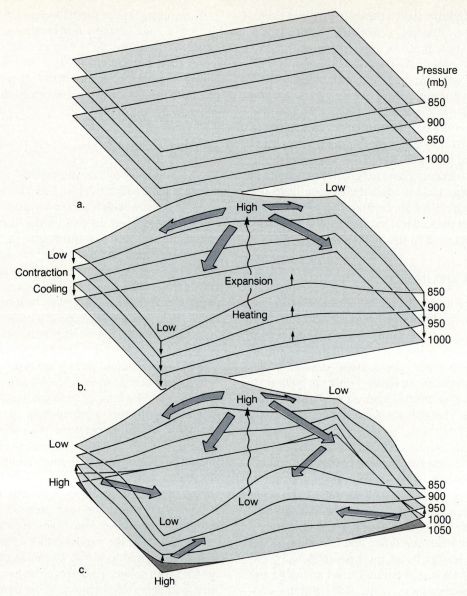

Pressure
(mb)

850
900
950
1000

a.

Low

High

Low
Contraction
Cooling

Expansion

Heating

850
900
950
1000

Low

b.

Low

High

Low

High

Low

850
900
950
1000
1050

Low

Low

c.

High

**Figure 14.3** The development of convective circulation. Planes are isobaric surfaces, broad arrows are surface and high altitude winds, and wavy lines are convective heat rise.

the left in the Southern Hemisphere, the effect increasing with the flow velocity. As you might expect, the Coriolis effect acts the same way on wind as it does on marine currents and with the same results. As in figure 12.15, the wind's velocity and direction continue to change only until the Coriolis effect becomes equal and opposite to the pressure gradient force, whereupon the wind direction is parallel with the isobars. An air flow of this type is known as a **geostrophic wind.** In the Northern Hemisphere, *geostrophic winds always travel with higher pressure to the right and lower pressure to the left.* In the Southern Hemisphere, these relations are reversed.

## The General Circulation of the Atmosphere

When we add the Coriolis effect to the convective circulation shown in figure 14.4, we find that a rather different picture emerges, as in figure 14.5. This new model assumes a rotating Earth covered uniformly with ocean water and therefore free of the disturbing influences of continental landmasses. It is evident in figure 14.5 that air rising by convection at the equator gradually turns eastward until it becomes geostrophic between 20° and 30° north and south latitude. The relatively stable wind bands resulting from this geostrophic flow have been called **subtropical jet streams,** and they mark the poleward limits of the Hadley cells. They circle the

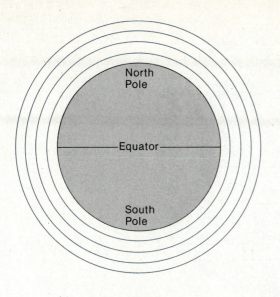

a. Uniformly heated globe

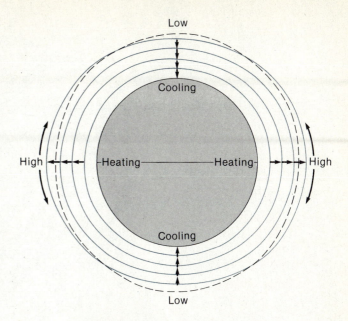

b. Heating at equator; cooling at poles. Poleward pressure gradient forms aloft.

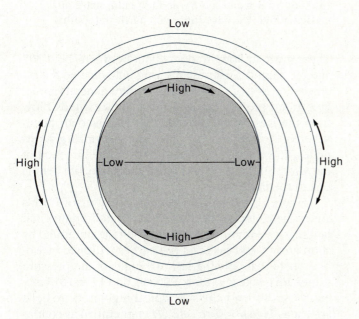

c. Equatorward pressure gradient forms at surface.

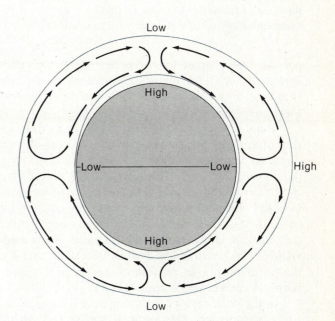

d. Hadley cell circulation forms.

**Figure 14.4** Convective circulation on a nonrotating planet covered uniformly by water. Here, diagrams *a, b,* and *c* correspond to diagrams *a, b,* and *c* in figure 14.3. In this model, the convective circulation takes the form of two great, bowl-shaped "Hadley cells," one in each hemisphere (*d*).

globe at about 30° north and south latitude, and their axes of maximum wind velocity (up to 385 km/hr) lie at an altitude of about 13 to 14 km.

Because the air flow converges at these latitudes, it becomes locally denser and sinks, creating high pressure at the surface. In these **subtropical high pressure belts,** air that had risen from the equatorial region descends to the surface again. On its upward journey,

that air had lost much of its moisture by precipitation and much of its heat by radiation. As it descends into the subtropical high pressure belts, it becomes warmer because of more frequent molecular collisions, and it also becomes drier because a given volume of warm air can absorb more water vapor than the same volume of

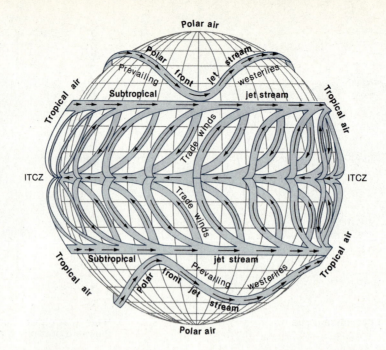

**Figure 14.5** Atmospheric circulation on a hypothetical, rotating, water-covered globe would be dominated by three major features: (1) a pair of Hadley cells maintaining convective circulation at low latitudes (0°–30° north and south latitude); (2) dynamically meandering polar front jet streams at middle latitudes (30°–60° north and south); and (3) cold, stable air masses at high latitudes (60°–90° north and south).

cool air (see chapter 15). Accordingly, the high pressure belts are characteristically extremely dry and cloudless, and most of the world's deserts lie beneath them (see figs. 14.15 and 1.20).

About one-third of the air that falls to Earth in the subtropical high pressure belts continues in a poleward direction, leaving the remaining two-thirds to return to the equator (fig. 14.5). This return flow is also affected by the Coriolis effect, which causes it to veer westward in both hemispheres as it approaches the equator. Because these surface winds blow from an easterly direction, they are known as the *tropical easterlies,* or **trade winds.** The equatorial belt where the trade winds of the two hemispheres converge is known as the **intertropical convergence zone (ITCZ).**

In the polar regions, outgoing long wave radiation exceeds insolation. This results in surface cooling and downwarping of the isobaric surfaces there. This in turn creates low pressure aloft over the poles, toward which a convergent air flow develops. High surface pressure results from this convergence aloft, and this induces a *divergent* air flow away from the poles at the surface (see fig. 14.4c). At the **polar front** this divergent flow meets tropical air carried northward by the Hadley circulation, and here powerful temperature and pressure gradients arise. These set up correspondingly powerful winds that are quickly turned eastward by the Coriolis effect to become the geostrophic **polar jet streams.**

These great air flows wander about between about 30° and 60° latitude, exerting a strong influence on the average direction of surface winds in their general vicinity (see frontispiece). Because most surface winds blow from the west within this latitude belt in both hemispheres, these winds are called the **prevailing westerlies.** In spite of the name, however, the direction of the winds in these belts is much more variable than is that of the trade winds.

Figure 14.6 shows the *actual* circulation of Earth's atmosphere, including the effects of the continental landmasses. Prevailing wind directions are indicated by short arrows, and sea level pressure is indicated by isobars (note that only the last two digits of the pressure values are given; e.g., 96 = 996 mb, and 17 = 1017 mb, etc.). On this map, the subtropical high pressure belts appear as *high pressure cells* (*H*) rather than as continuous belts. The reason for this is that the strong heating of land areas in summer produces surface *low pressure cells* (*L*) that interrupt the high pressure belts. This effect is most noticeable in the Northern Hemisphere in July (fig. 14.6b). Low pressure cells are also present over ocean areas in the winter hemispheres (especially the northern, fig. 14.6a). I discuss the reason why these features develop in chapter 15.

The ITCZ, shown by a dot-dash line in figure 14.6, coincides roughly with the equator, but it normally lies a few degrees north of the equator in July and (in part) a few degrees south of the equator in January.

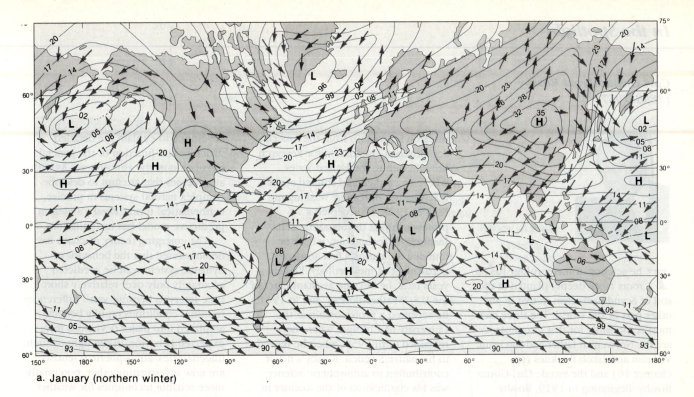

a. January (northern winter)

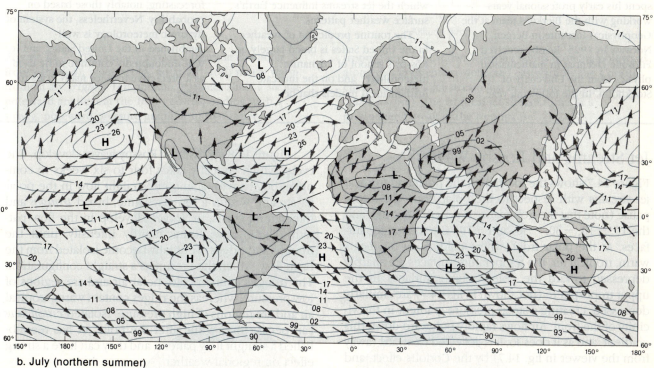

b. July (northern summer)

**Figure 14.6** Earth's actual atmospheric circulation is modified from that in figure 14.5 by the presence of land masses, particularly in the Northern Hemisphere. Mean January conditions (northern winter) are shown in *a*, mean July conditions (northern summer) in *b*. Contour lines are isobars, or lines of equal sea level atmospheric pressure, labelled in millibars (only the last two digits are given). Also shown are cells of high (*H*) and low (*L*) pressure and surface winds (*short arrows*).

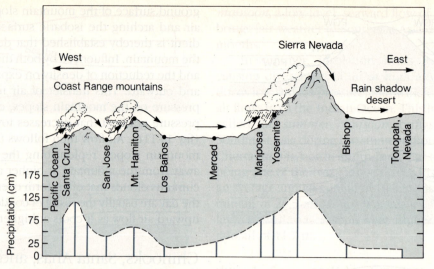

**Figure 14.13** Bar graphs of precipitation (*bottom*) across central California show the rain shadow effects of mountain barriers on moisture-laden air passing over them. West slopes receive abundant precipitation, but east slopes get little or none.

When *dry* air rises over a mountain barrier, it cools by expansion at a rate of 10° C/km of altitude gain (see chapter 15). If the original air temperature is 10° C, the journey to the crest of a 3 km high mountain range would lower its temperature to −20° C. As the air descends the leeward side of the mountain range, it warms again by compression at the same rate, until it once more has a temperature of 10° C at the base of the range. If this descending air replaces cold, cP air in the valley (as it often does in the Front Range Urban Corridor east of the Colorado Rockies), a sharp temperature rise accompanies its arrival.

If the air is *moist,* it will be considerably warmer and drier when it arrives at the leeward base of the range than it was at the windward base. Figure 14.12*a* shows what happens to a parcel of moist air as it flows up and over a 3 km mountain barrier. In this case, latent heat is released into the rising air as water vapor condenses and clouds form on the windward side of the range. Depending on the amount of condensation that occurs, which in turn depends on the moisture content and temperature of the air, this heat input can reduce the rate at which the rising air cools to between 3° and 9° C/km. If condensation begins at an altitude of 1 km, the air cools 10° C in the first kilometer, but in the next 2 km, as clouds form, it only cools 12° C (assuming an intermediate rate of 6° C/km). Under these conditions, air that has a temperature of 10° C at the base of the mountain range will cool to only −12° C at the crest of the range instead of −20° C, as in the preceding example. On the leeward side of the range, however, there is no release or absorption of latent heat, so the air warms

at the dry rate of 10° C/km. Consequently, this air remains 8° C warmer than the air in the preceding example all the way to the valley bottom. When it arrives there, it has a temperature of a balmy 18° C.

Typically, these warm, dry chinook winds produce **rain shadows** on the leeward sides of mountain ranges. Figure 14.13 shows the effect of the Coast Range and Sierra Nevada mountains on precipitation from the prevailing westerlies. Notice that precipitation amounts are highest on west-facing slopes, and lowest on the east-facing slopes downwind of them. The presence of these north-south mountain barriers is in large measure responsible for the general dryness of the western United States. Evidence from fossil plants indicates that the flora of this region changed drastically from a warm, humid, subtropical type to a cool, semiarid, temperate type as the Coast Ranges, the Sierra Nevada, and the western states in general were uplifted during the Pliocene Epoch (5.3 to 1.6 million years ago).

**Santa Ana winds,** named for Santa Ana Canyon in southern California, are hot, dry winds that occur when a high pressure cell lies over the western United States, as in figure 14.14. Air flowing away from this cell is already quite dry, because it has descended on the northern edge of a Hadley cell. As it traverses the southwest desert, it becomes even drier, and decidedly hotter (note spot temperature readings in fig. 14.14). Descending toward sea level in southern California, it warms still more. As the Santa Ana wind passes over the land, it dries out the chaparral vegetation (consisting of oil-rich evergreen shrubs), rendering it extremely flammable. Consequently, Santa Ana winds create extreme fire hazards in southern California.

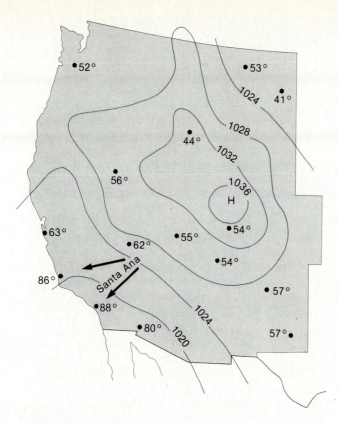

**Figure 14.14** Warm, dry, descending winds circulating clockwise around a high pressure cell situated over the western United States become still warmer and drier as they drop toward sea level through canyons in the Coast Range (*arrows*). These Santa Ana winds bring extreme fire hazards to southern California.

## Deserts

The same conditions that give rise to Santa Ana winds are responsible for desert conditions on many of those portions of Earth's landmasses that lie between about 20° and 40° north and south latitude. The descent of warm, dry air on the poleward sides of Hadley cells is a principal cause of the desert belts shown in figure 14.15. There are, however, several other arid regions shown in that figure that are not related to the Hadley circulation, although the aridity of these regions is also attributable in some degree to the presence of high pressure.

The Arctic and the Antarctic, despite their heavy snow and ice cover, are regions of extremely low precipitation (see fig. 15.1). This is because of the presence of permanent, surface high pressure cells over the polar regions. Their low prevailing temperatures result in extremely low rates of evaporation, however, and because of this, the ground actually remains moist throughout the year. This effect is greatly enhanced by the presence of permafrost throughout large areas of the polar regions. In summer, only a shallow surface layer of ground thaws above the permafrost. The resulting

meltwater is prevented from percolating downward by the permafrost layer, and this renders the soil much wetter than it would otherwise be. Because of this ponding of water at the ground surface, polar soils support a humid type of vegetation. The stunted growth of this *tundra* vegetation is due not to a lack of ground moisture, but to high winds that dry out and kill any plant parts that protrude above the snow surface in winter when the ground moisture is frozen and therefore unavailable to the plants.

The rain shadow effect is another major inducer of desert conditions in regions downwind of mountain barriers that lie transverse to the prevailing wind. The world's principal rain shadow deserts are in the western United States, central Asia, and southeastern Argentina, east of the Andes Range (see fig. 15.1).

Midlatitude **west coast deserts** form as a result of the interaction of three factors (fig. 14.16). First, subtropical high pressure cells are normally present off such coasts (see fig. 14.6). Warming, descending air on the east limbs of these cells favors arid conditions.

Second, air flow in the east limbs of the high pressure cells is directed toward the equator. This equatorward flow generates a seaward Ekman transport at the ocean surface, which causes upwelling of cold, deep ocean water along the coast (see fig. 12.14). The cold sea surface cools the overlying air, creating a persistent temperature inversion that prevents the air from rising by thermal convection during the day. This, in turn, prevents condensation, cloud formation, and precipitation.

The third factor is the presence of a coastal, north-south mountain barrier, which prevents the prevailing westerly winds from dissipating the inversion. Thus, fog and smog can be trapped over the coast for extended periods of time, as it is with distressing frequency in Los Angeles, California. Fogs are caused by the condensation of moisture in the surface air as it comes in contact with the cold, upwelling ocean water. Notable locations for coastal desert conditions, aside from southern California, include northern Chile, southwest Africa, and Northwest Africa.

## Erosion and Deposition by Wind

Wind was at one time thought to be a relatively effective agent of erosion. Quantitative studies have shown, however, that in comparison with running water and flowing ice, the effects of wind are relatively insignificant. This is true even in semiarid and arid regions, where infrequent cloudburst floods move much more sediment in a given time interval than is moved by winds during the same interval.

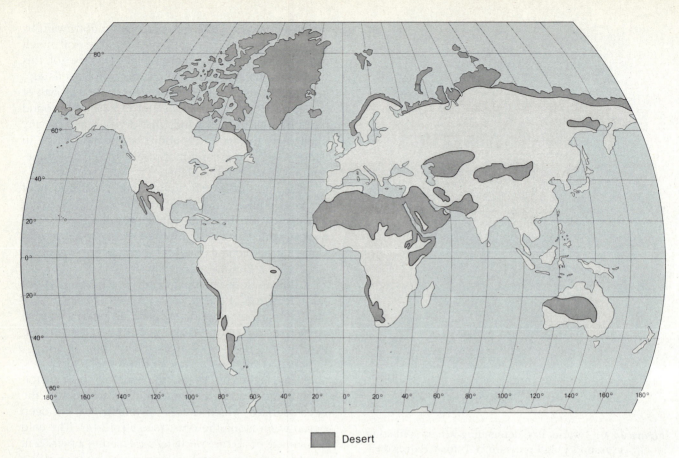

Desert

**Figure 14.15** Earth's deserts have various origins. Polar regions are dominated by high pressure. The deserts of Australia, North and South Africa, Arabia, Pakistan, and the southwestern United States lie in subtropical high pressure belts. All but Pakistan's are also influenced by west coast effects. These alone create the desert of coastal Peru and Chile. Rain shadow effects are responsible for the deserts of central Asia and southern Argentina and, in part, for the arid regions of the southwestern United States.

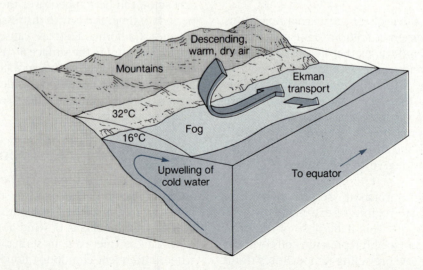

**Figure 14.16** The extreme dryness of a west coast desert results from a combination of warm, dry air descending on the east side of a subtropical high pressure cell and cold, upwelling ocean water. This cold water chills the overlying air, producing fog and creating a temperature inversion that prevents the air from rising convectively. Where high mountain barriers prevent the eastward dissipation of the inversion, severe smog problems can arise from pollutants that mingle with the fog.

*Figure 14.17* Water and wind leave different erosional imprints on horizontally bedded, poorly consolidated, clastic sediments of the Great Plains. The dendritic pattern on the left is characteristic of stream erosion. The pitted pattern on the right, in which stratification planes have the appearance of contour lines, is characteristic of wind erosion.

Reprinted with permission of Macmillan Publishing Company from *The Earth's Dynamic Systems,* 4e by W. K. Hamblin

The effects of both water and wind erosion in a semiarid environment are shown in figure 14.17. Here, wind action has produced an array of small **deflation basins** (from the Latin *deflare,* "to blow down") within poorly-consolidated, flat-lying, clastic sediments of the Great Plains. Such sediments are usually lightly cemented by calcium carbonate precipitated from evaporating groundwater. During times of heavy rainfall and high water tables, enough of this carbonate cement can be dissolved to free the clastic grains. The smaller grains are then swept up and away by high winds, leaving depressions in the bedrock. These depressions range in diameter from a few meters to several kilometers, and they can be up to 50 m or more in depth. Nonetheless, even in the strongly wind-impacted area of figure 14.17, the greater efficacy of water erosion is evident in the "badland" topography that has developed on the left side of the photograph. During heavy rains, this intermittent drainage system eats its way progressively into the wind-deflated upland to the right by the process of headward erosion (see chapter 9).

In arid environments, such as the one shown in figure 14.18, one often finds extensive flats extending outward from mountain fronts. These flats are the sur-

*Figure 14.18* Desert pavement, a veneer of pebbles, formed on the surface of this flat basin in Death Valley, California, as high winds gradually blew away the finer clastic material. An ancient trail, worn by Native American traders, winds across the pavement.

Photo by R. J. Ross, U.S. Geological Survey

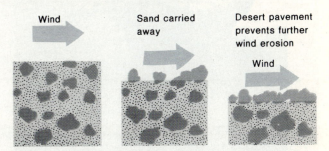

*Figure 14.19* The creation of a lag concentrate. Wind can erode sand-sized particles, but not larger clasts, which become concentrated in an interlocking surface layer as their finer sand matrix is gradually blown away.

faces of alluvial fans (see fig. 9.14). They are usually thinly veneered with a **desert pavement,** consisting of pebbles that range from a few millimeters to a few centimeters in diameter. This pebbly veneer results from deflation by strong winds that winnow out the finer surface material and leave the coarser material behind as a *lag concentrate* (fig. 14.19).

Many of these pebbles in desert pavements have two or three flat, smooth facets. Figure 14.20 shows how such surfaces develop as a result of centuries of sandblasting by surface winds. Sandblasting is a fairly effective erosional agent within the first few centimeters above the ground surface. After sandblasting has produced a well-developed facet on one side, a pebble can be rotated by some disturbance, such as the passage of a large animal, or alternatively, the prevailing wind direction can shift. A new facet then forms on whatever side of

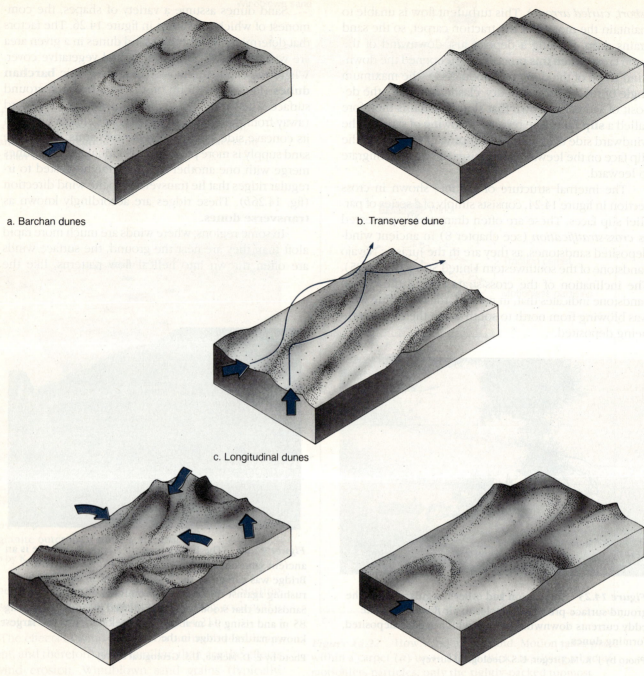

a. Barchan dunes

b. Transverse dune

c. Longitudinal dunes

d. Star dunes

e. Parabolic dunes

**Figure 14.26** Sand dunes assume a variety of forms. Barchans (*a*) form where steady winds move relatively small supplies of sand. The "horns" of a barchan always point downwind. Where sand supply is larger, barchans coalesce to form transverse dunes (*b*). Longitudinal dunes

form where strong winds aloft produce helical air flow with convergence near the ground (*c*). Variable wind direction produces relatively immobile star dunes (*d*). Parabolic dunes (*e*) result where blowouts occur in unconsolidated sand covered by vegetation.

stripes on a barber pole (fig. 14.26*c*). Sand in such regions is swept into more or less smooth, longitudinal ridges by the converging air flows between adjacent helical cells. These ridges are known as **longitudinal dunes,** or seifs (*seefs;* Arabic for "swords"), because they run parallel with the prevailing wind direction.

Where wind direction is erratic and variable, sand is piled up in massive, multi-ridged **star dunes** (fig. 14.26*d*). Star dunes are often of mountainous proportions, and unlike other types of dune, they are normally immobile. In Arabia and the Sahara, star dunes are large and persistent enough that they are often named and used as landmarks.

Although it is clear that most of Earth's semiarid and desert regions have been created by natural, climatic processes and by changes in the distribution of those processes over time, it is also clear that a large and increasing percentage of these regions has been created by human activity. Figure 14.27 shows the regions of the planet that are subject to moderate, high, and very high levels of **desertification,** or conversion to wastelands through human agency, as determined by the U.S. Council on Environmental Quality (1978). Altogether, these moderate-to-high-risk areas total about 48 million km², or roughly 36% of the habitable area of the world. Of this total, an estimated 9 million km², or about 7%, has already been lost to desertification, and an additional 13 million km², or about 10%, has been so severely affected as to be beyond reasonable hope of recovery.

Desertification takes a number of different forms, including soil erosion (see fig. 8.23), the salinization of soils, the lowering of water tables, overgrazing, and deforestation. All of these conditions relate to poor land use practices, and all are potentially correctable but only if corrective measures are taken before the land is destroyed or so severely weakened that its recovery is unlikely. In the "Food for Thought" section of chapter 9, I describe two apparent incidents of poor land use that occurred sequentially within the same drainage system in the semiarid American southwest, separated by an interval of some 600 years. This sequence seems, of itself, to offer some hope, but to most human societies, waiting several hundred years for a ruined ecosystem to recover its productive capacity is a less than satisfactory approach to land use management.

Other examples abound. Recently, the Sahel region on the southern border of the Sahara desert has experienced severe desertification as a result of the combined effects of overgrazing and drought. In the United States, the experience of the Dust Bowl in the 1930s left many hopeful Americans sadder but wiser to the hard reality that farming techniques suited to humid climates do not work in the semiarid prairies of the midcontinent. Following several years of drought, severe winds stripped the topsoil from a 100 million acre region comprising parts of the states of Colorado, New Mexico, Kansas, Oklahoma, and Texas. Fortunately, the lesson was not learned in vain, and as a result of that calamity, the U.S. Soil Conservation Service was organized in order to develop policies of land use management designed to prevent such disasters in the future.

An earlier and less well-managed incident of desertification in the United States resulted from the introduction of cattle into the semiarid grasslands of the southwest in the second half of the nineteenth century. This led to a severe and widespread episode of erosion by arroyo cutting in that region. The effects of that episode still dominate the landscape, where sagebrush has largely replaced formerly lush grasslands because of the general lowering of water tables that accompanied the arroyo cutting.

Although the lowering of water tables as a result of erosion is an indirect effect of human land use, water tables have also been, and are still being, lowered directly. A recent study by the U.S. Geological Survey (1978) has revealed that in two-thirds of the 18 major drainage basins in the United States (see fig. 9.7), groundwater is being withdrawn at rates that exceed the average annual recharge rate. Nationwide, withdrawals exceed recharge by 25%. One particularly glaring example of the effects of such excessive use of groundwater is the city of Tucson,

Arizona. Until this century, the Santa Cruz and Rillito Rivers flowed through lush cottonwood forests in the broad Tucson basin. Because of the pumping of million-year-old groundwater from beneath the basin to supply the burgeoning city and the rapid lowering of the water table that has resulted, the rivers have dried up completely, the forests have long since died, and great cracks have begun to form on the margins of the Tucson basin due to ground subsidence. Despite these patently obvious evidences of desertification, people still continue to settle in Tucson at an ever-increasing rate. Many similar examples can be seen in other parts of the western United States.

Obviously, such trends cannot be sustained for long, and most of the schemes that have been proposed to deal with the problem are not very realistic. The Central Arizona Project was devised to divert water from the already overcommitted flow of the Colorado River and reroute it to Tucson and Phoenix. Recently, the North American Water and Power Alliance has proposed a large-scale diversion project that would bring water southward from western Canada to serve a much broader region of the American southwest. Not surprisingly, this scheme has met with robust opposition from Canadians and Americans alike, and it seems unlikely that it will ever happen, despite the vigorous lobbying efforts of those who would profit from the venture. From both functional and aesthetic standpoints, this is probably just as well. Inasmuch as the forms of fluvial landscapes are "living" responses to "living" streams, then to the extent that we control and even eliminate the flow of streams, we find ourselves living in "dead" landscapes.

A more fruitful approach to the problem of overuse of our water resources would be to make more

---

The influence of vegetation on dune formation is seen in figure 14.26*e*, in which a series of *blowouts* has formed in an area where dune sand has been stabilized by a vegetative cover. Natural or human-made breaches in this cover have allowed the prevailing wind to excavate the underlying sand. As this sand is blown down- wind, it forms a ridge which advances over and kills the vegetation on the leeward edge of the blowout. This ridge, called a **parabolic dune,** resembles a barchan, but its arms point into the wind, and its slip face is on the dune's convex side.

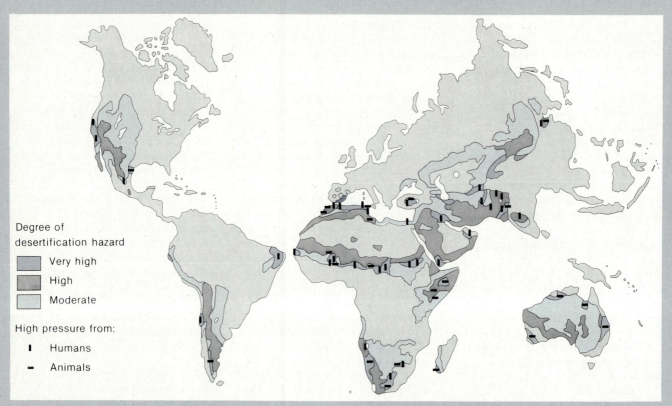

**Degree of desertification hazard**

Very high

High

Moderate

**High pressure from:**

I Humans

- Animals

**Figure 14.27** Most areas at risk from desertification, as determined by the U.S. Council on Environmental Quality, lie near Earth's natural desert regions (cf. fig. 14.15).

Source: Council on Environmental Quality, Annual Report 1978. Washington, DC: U.S. Government Printing Office.

efficient use of our water as it is presently distributed, particularly in the practice of agriculture, which currently consumes about 80% of it. Changes in diet would be of great value, as it takes anywhere between 40 and 167 times more water to produce a ton of beef than to produce a ton of wheat. Programs to reduce population through family planning supported by tough but realistic tax incentives would be of even greater value. Such approaches stress *prevention,* the most reasonable strategy, as in most cases of desertification, *cure* is generally too costly or too time-consuming to be practical.

Can human beings ever manage to live harmoniously and nondestructively on Earth? This question is being raised more and more frequently, and with good reason. The objective of Earth-appropriate living has certainly not been very well met by the tradition of human civilization that arose about

11,000 years ago in the semiarid "fertile crescent" between the Mediterranean Sea and the Persian Gulf. That tradition developed around the domestication of sheep, goats, and cattle and the farming of wheat and barley. The pattern has been one of increasing exploitation and manipulation of Nature through the use of increasingly complex technology. It has also been a pattern of increasing concentration of wealth, civil power, and religious authority in the hands of a few favored and aggressive individuals who have tended to place the quality of their own lives ahead of concern for the quality of life enjoyed by the rest of Earth's inhabitants, human and otherwise.

As deserts spread and Earth's skies and waters turn foul, it is becoming increasingly evident even to the power-hungry, that a high quality of life can not be the exclusive privilege of a select few. Unless the quality of *all* life

on Earth is assured, even that of the select few must inevitably decline. Perhaps, as the economically privileged class comes to accept this self-evident truth, a more responsible approach to the natural environment will emerge. For better or for worse, if such change is to occur, it will probably have to come from the economically privileged, as the less privileged majority, traditionally, has been too concerned with fulfilling its various economic obligations to that class to be able to direct much of its energies toward the kinds of societal reforms that will be required to insure the continued health of the biosphere. Unfortunately, however, all too often the key to economic privilege has been the exploitation and manipulation of Nature. The crisis facing our way of life, then, appears to be the challenge of finding keys to financial success that are less damaging to Earth's vital systems.

# Summary

**Wind** is air movement along a horizontal pressure gradient from high to low pressure. The mass of an atmospheric column of 1 cm² basal area at sea level is 1033 g. Its *weight* is 1,013,000 *dynes*. Its sea level *pressure* is 1,013,000 dyn/cm², or one **standard atmosphere.** One standard atmosphere equals 1013 **millibars.** In a Torricelli **barometer,** it can support a column of mercury 760 mm high.

If Earth's surface were uniform in substance and temperature, *isobaric surfaces* would be parallel to it. **Isobars** are horizontal lines on an isobaric surface. Solar heating of Earth's surface expands overlying air, raising isobaric surfaces. Cooling of Earth's surface contracts overlying air, lowering isobaric surfaces. Air flows in the direction in which the isobaric surfaces slope. Air flow into an air column adds weight, increasing surface pressure. Air flow out of an air column removes weight, reducing surface pressure. Surface solar heating expands air, creating **divergence** aloft. This reduces surface pressure and initiates **convergence** at the surface, completing the **convective circulation.**

**Hadley cells** are large-scale convection systems that carry air aloft from the equator to about 30° north and south latitude, and return it to the equator at Earth's surface. Their poleward limits are marked by **subtropical jet streams,** where poleward air flow turns eastward and becomes a **geostrophic wind.** Below the subtropical jet streams, **subtropical high pressure belts** persist, and dry, descending air produces desert conditions. Summer heating of continents creates *low pressure cells* that interupt the belts, breaking them into *high pressure cells.* About two-thirds of the descending air returns to the equator as the **trade winds,** which merge in the **intertropical convergence zone (ITCZ).** Low polar temperatures result in low pressure aloft, hence convergence aloft, hence high surface pressure.

The **polar jet streams** are geostrophic flows at the tropopause above the **polar front,** which separates cold polar air from warm tropical air. The polar jet alternates between tranquil, straight flow and turbulent, meandering flow with high amplitude **Rossby waves,** which have high-pressure *ridges* and low-pressure *troughs.* The mid-latitude **prevailing westerlies** reflect the mean direction of the polar jet stream.

**Airmasses** retain their temperature and moisture characteristics as they move. **Weather** is the local, day-to-day state of the troposphere; **climate** is the aggregate of local weather conditions over many years. Cold, dry, stable *continental polar* (cP) air develops over high latitude land surfaces. *Polar outbreaks* of cP air bring bitter weather in winter and cool, dry weather in summer.

Cool, moist, unstable *maritime polar* (mP) air is cP air warmed and moistened by ocean water. Warm, moist, unstable *maritime tropical* (mT) air originates over oceans in the subtropical high pressure belts. *Continental tropical* (cT) air forms over land surfaces in subtropical high pressure belts and is hot, dry, and stable at night, and unstable during the day.

A **sea breeze** is the convergent, surface part of a convective circulation resulting from daytime heating of coastal land. Nighttime cooling of the land surface reverses the circulation, resulting in a divergent **land breeze.** Winter and summer **monsoons** are persistent land and sea breezes resulting from winter cooling and summer heating of the land. Cold, dense air, produced when nighttime long wave radiation loss cools upland surfaces, flows down slopes as **katabatic wind,** creating stable temperature inversions in valley bottoms and often trapping pollutants. Daytime solar heating of south-facing uplands creates convective circulation with air diverging aloft and low level **anabatic winds** flowing toward the upland. Dry **chinook winds** warm at 10° C/km as they descend mountain slopes. If precipitation has occurred during its windward ascent, the air is warmer as it descends than it would be otherwise. Its dryness normally produces a **rain shadow. Santa Ana winds** are southwestward flows of hot, dry air from a high pressure cell over the western United States.

Descending air produces *deserts* within subtropical highs, polar regions, and rain shadows. In **west coast deserts,** upwelling cold water chills overlying air, producing temperature inversions that prevent convective circulation. Descending air is also present, as are mountain barriers, which prevent eastward escape of air, favoring pollution. Wind action erodes **deflation basins** in fine sediment and erodes finer material from coarse alluvial fan sediment, leaving a **desert pavement** of pebbles, some sandblasted. Windblown sand grains move by **saltation** ($\pm 75\%$) and *surface creep* ($\pm 25\%$).

Surface obstacles create **wind shadows,** in which **sand dunes** form. A dune migrates downwind as sand eroded from its windward slope avalanches down the **slip face.** The maximum angle of repose for dry sand is 34°. Crescentic **barchan dunes** form where sand supply is limited. Where sand is more plentiful, barchans coalesce into **transverse dunes.** Where high winds aloft produce helical flow cells in surface winds, **longitudinal dunes** develop. Large, immobile **star dunes** form where wind direction is erratic. *Blowouts* in sand fields stabilized by vegetation produce **parabolic dunes.** Human populations that fail to restrict their numbers to levels that local environments (or the global environment) can support are likely to experience **desertification.**

## Key Terms

wind
standard atmosphere
millibar
barometer
isobar
divergence
convergence
convective circulation
Hadley cell
geostrophic wind
subtropical jet stream
subtropical high pressure
   belts
trade winds
intertropical convergence
   zone (ITCZ)
polar front
polar jet stream
prevailing westerlies
Rossby wave
air mass
weather

climate
sea breeze
land breeze
monsoon
katabatic wind
anabatic wind
chinook wind
rain shadow
Santa Ana wind
west coast desert
deflation basin
desert pavement
saltation
wind shadow
slip face
sand dune
barchan dune
transverse dune
longitudinal dune
star dune
parabolic dune
desertification

## Questions for Review

1. Why is it incorrect to say that an average atmospheric column of 1 cm² basal area weighs 1033 g at sea level?

2. In order to convert standard atmospheres to millibars, what conversion factor would you use?

3. What must first happen to a horizontal isobaric surface before a wind can develop? In what direction will the wind move?

4. Explain convective circulation in terms of surface heating, isobaric surfaces, pressure gradients, and divergent and convergent air flows.

5. Explain the mechanisms that generate the subtropical jet stream and the polar jet stream.

6. Describe the general circulation of the atmosphere in terms of Hadley cells, trade winds, the intertropical convergence zone, subtropical and polar jet streams, and prevailing westerlies.

7. Where does each of the major air masses originate? How does each affect the weather of the United States?

8. Explain the conditions that give rise to summer and winter monsoons in terms of surface heating, isobaric surfaces, pressure gradients, and convergent and divergent air flows.

9. Explain the origins of katabatic and anabatic winds in terms of surface heating and cooling, isobaric surfaces, pressure gradients, and air flows.

10. Describe the conditions that give rise to chinook and Santa Ana winds.

11. Describe the conditions that give rise to subtropical deserts, west coast deserts, and high-latitude deserts.

12. In what ways do the effects of deflation differ in well-sorted, fine sediment and in poorly sorted gravelly sand?

13. Describe the processes of dune formation and advance.

14. Under what conditions do the five fundamental types of sand dune originate?

15. Suggest some avenues to financial success that might have little adverse effect on Earth's vital systems.

# The Whims of Thor: Storms and Rain

## Outline

*T*he sky is changed,—and such a change! O night

*And storm and darkness! ye are wondrous strong,*

*Yet lovely in your strength . . .*

*Lord Byron*

**Childe Harold's Pilgrimage**

Photo: A lightning storm over the San Pedro Valley, southeast Arizona.

## The Rai

Rain is an
nomena th
tainly exis
possibly (t
galaxy. The
mean annu
Notice that
three broad
roughly co
approximat
tudes. The
from the co
tropical air
tion of the
belts, on th
mechanism
ciated with
nomenon l
the phenor
that contro

## Water i

Earth's atm
gases:

1. Nobl
   becau
   and l
2. Chen
   are c
   contr
   oxyge
   sulfic
3. Chen
   are v
   physi
   (e.g.,
   dioxi
   conce

All these g
sure of the
centration.

In chap
in the atm
condensati
equilibriun
occurring a
overlying a
saturated w
vapor in th
pressure.

**Figure 15.8**   A bizarre denizen of the virtually rainless desert of coastal Angola and Namibia, *Welwitschia mirabilis* is a close relative of the conifers but has more evolved reproductive characteristics. Photosynthesis occurs in two large (2 m long), straplike, wind-shredded leaves. Starch and water are stored in the deep, carrotlike taproot, which bears a crown of small cones.

**Table 15.1**   Cloud Groups and Types

| Groups | Types | Altitude range (m) |
| --- | --- | --- |
| 1. Low clouds | | 0–2000 |
| | Stratus | |
| | Stratocumulus | |
| | Nimbostratus | |
| 2. Middle clouds | | 2000–4000 (poles) |
| | Altostratus | 2000–8000 (equator) |
| | Altocumulus | |
| 3. High clouds | | 3000–8000 (poles) |
| | Cirrus | 6000–18,000 (equator) |
| | Cirrostratus | |
| | Cirrocumulus | |
| 4. Clouds with vertical development | | |
| | Cumulus | Mainly below 2000 |
| | Cumulonimbus | Up to tropopause |

The various types in the classification are shown in the photographs of figure 15.9, and their distribution in the troposphere is shown in figure 15.10.

The primary division of Howard's classification is by altitude of the cloud base. **Low clouds,** typically uniform gray or variably gray-toned, occur from near the ground surface to an altitude of 2000 m. **Middle clouds,** typically uniform gray or gray and white, occur from 2000 to 4000 m above the poles and from 2000 to 8000 m above the equator. **High clouds,** typically thin and white, occur from 3000 to 8000 m above the poles and from 6000 to 18,000 m above the equator. A fourth group includes **clouds with vertical development.** These clouds can extend from a few hundred meters above the ground to 18,000 m altitude.

### Low Cloud Types

**Stratus** is a low, uniformly gray cloud that typically covers the entire sky. Stratus is indistinguishable from radiation fog that has lifted during the day, except that radiation fog is usually stationary, whereas stratus is normally wind-driven. When patches of blue sky are visible among rounded, variably gray-toned masses of stratus, the name changes to **stratocumulus.** When rain or snow are falling from stratus, it becomes **nimbostratus.** Ragged, shreddy gray clouds, called *stratus fractus,* or "scud," often race along with the wind below the main mass of nimbostratus.

### Middle Cloud Types

**Altostratus** closely resembles stratus but is higher and of a lighter shade of gray, and it does not produce rain (although it often precedes rain from lower nimbostratus clouds). The Sun or Moon can also be seen through thinner sections of an altostratus cloud layer, but this is seldom true of stratus. **Altocumulus** resembles stratocumulus but is higher, and the individual cloud masses are usually gray rimmed with white.

**Figure 15.9** Some important cloud types. (*a*) stratus, (*b*) stratocumulus, (*c*) nimbostratus, (*d*) altostratus, (*e*) altocumulus, (*f*) cirrus, (*g*) cirrostratus, (*h*) cirrocumulus, (*i*) cumulus, (*j*) cumulonimbus.

Photos *C, E, H,* and *J* by NOAA

## High Cloud Types

**Cirrus,** like all other high cloud types, is composed of ice crystals, and it appears pure white. Cirrus clouds are usually wispy and plumelike, often resembling the tail of a frisking horse, hence the common name, "mares' tails." Cirrus and other high cloud types form on the leading edges of cyclonic storms that might or might not bring precipitation to the observer's location, depending on the path followed by the storm center and various other factors. **Cirrocumulus,** often called "mackerel sky," resembles altocumulus, but the individual cloud masses appear much smaller and are pure white. Mackerel sky has traditionally been considered a more reliable indicator of impending precipitation

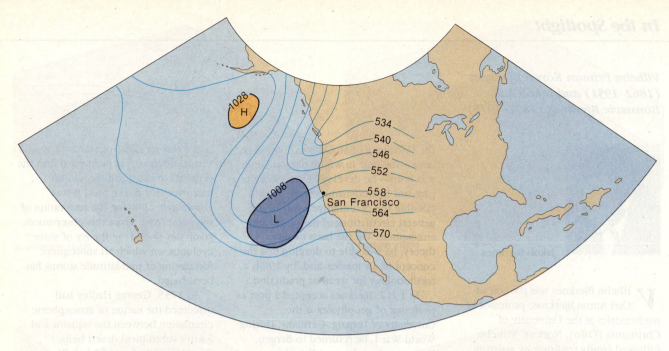

Figure 15.23    A weather map for 4:00 A.M., 26 March 1979 shows the relationship between convergence and divergence in upper air flow (here expressed as contours of altitude on the 500 mb isobaric surface in tens of meters) and ground level highs and lows (expressed as closed isobars, in millibars). A surface high in the northeast Pacific Ocean is associated with upper air convergence, and a surface low to the south with divergence.

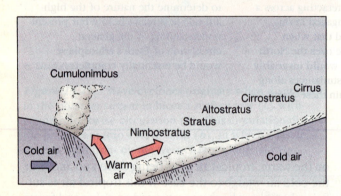

Figure 15.24    The abrupt uplift of air by the steep face of a cold front (*left*) favors the formation of cumulonimbus clouds. The gradual uplift of air by the gentler slope of a warm front (*right*) favors the formation of stratiform cloud types as warm air rides up and over cold air to the east. Slopes are greatly exaggerated here.

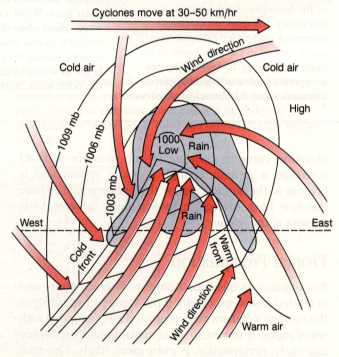

Figure 15.25    The essential features of a mid-latitude wave cyclone.

marked rise in temperature. Notice that the south-westerly wind direction persists until the passage of the cold front, whereupon it undergoes a second abrupt shift to northwesterly. As the cold front passes, the temperature drops sharply, and the barometric pressure rises abruptly.

## Tornadoes, Whirlwinds, and Waterspouts

A notorious atmospheric hazard that sometimes accompanies frontal storm systems is the **tornado** (fig. 15.26), a narrowly cylindrical downward extension of a cumulonimbus cloud within which intensely rotating winds surround a column of low-pressure air. The low pressure draws air inward into an ever-tightening funnel, in which it is accelerated to high speeds by the need to conserve angular momentum. As the rotating air expands in the low pressure and cools to its dew point, condensation takes place, whereupon the form of the tornado becomes visible. If the tornado touches the ground, it picks up large quantities of dust and debris, which further add to its visibility.

Tornadoes range in diameter from a few meters to as much as 1600 m in diameter, but most fall in the range between 100 and 600 m. Few measurements of wind speeds within tornadoes exist because most of the few anemometers (wind gages) that have been hit directly by a tornado have been destroyed by the event. Such data as are available indicate that speeds of 200 km/hr are typical, but that the most intense winds can top 400 km/hr. The pressure in the interior of a tornado can be more than 100 mb lower than the pressure of the surrounding air.

Tornadoes spend variable amounts of time on the ground. Some fail to touch ground altogether, whereas others have remained on the ground for over 7 hours while cutting paths of devastation nearly 500 km long. Because of their destructive potential and their unpredictability, much study has been devoted to the behavior of tornadoes. This effort has yet to yield a clear picture of how they form, but two factors seem to be fundamental: severe thunderstorms and unstable air.

Typically, tornadoes form in the wedge of warm air east of a cyclonic cold front where warm, moist air advected from the south has formed an inversion layer at about 2 km altitude. At the same time, cold, dry air is being advected from the southwest at about 3 km altitude. This advection of cold air aloft creates a high lapse rate there, which promotes instability, but the inversion layer beneath prevents surface-heated air from rising above about 2 km. Eventually, however, a few extremely powerful updrafts do succeed in breaking through this thermal lid, whereupon they build, almost explosively,

**Figure 15.26**    Tornadoes, spawned by severe thunderstorms, are swirling, destructive cylinders of cloud droplets, dust, and debris surrounding columns of extremely low pressure.
**Photo by NOAA**

into severe or supercell thunderclouds. Strong downdrafts often distend the bases of such clouds downward into pouchlike forms known as *cumulus mammatus.* The appearance of this cloud form is just cause for alarm, and if any of these "pouches" are rotating, it is highly advisable to take cover as they can quickly develop into tornadoes.

**Whirlwinds,** or dust devils (figure 15.27), are small vortexes that bear a superficial resemblance to tornadoes but are quite different in origin. They are associated neither with severe thunderstorms nor with cyclonic, frontal weather, but with clear, hot, summer days when strong surface heating creates a lapse rate higher than 34° C/km. Beyond this critical rate, air becomes denser upward and overturns spontaneously. As the air plummets Earthward, it is spun by the Coriolis effect into a narrow vortex. Whirlwinds are rarely more than a few meters in diameter, they seldom persist for more than a few minutes, and their wind speeds rarely exceed 140 km/hr. Such speeds can be damaging, but most whirlwinds generate much less powerful winds.

**Waterspouts** (fig. 15.28), an occasional phenomenon of warm, tropical, shallow seas, bear a closer resemblance to tornadoes than to whirlwinds, and they

# Food for Thought

In recent years, both scientists and citizens have expressed growing concern about human interference with the weather. The problems are varied and complex, and they raise many thorny issues that do not have easy solutions. The owner of a ski resort in the American southwest recently brought suit against a government agency for seeding clouds upwind of his resort, thereby allegedly robbing the clouds of moisture that would otherwise have fallen as snow on the mountains in which the resort is located. The legal action is not as absurd as it sounds. The resort owner was highlighting a significant principle: we tend to choose our places of residence and work, and our business enterprises, with an eye for the weather, and we do so with the expectation that the prevailing local weather patterns will remain constant. Some occupations, such as agriculture and resort area management, are more weather-dependent than others, such as arms manufacturing, but there are few places in the world, if any, where no one cares about possible alterations to the weather.

Just as it is becoming increasingly difficult to identify the villain in modern movies and television, it is also becoming increasingly difficult to identify environmental villains. A cloud seeding operation can benefit one commercial concern while spelling disaster for another. Probably no one would dispute that hurricane abatement techniques are good, in principle. Some experiments have shown that seeding the clouds a short distance from the eye of a hurricane starves the more energetic, inner regions, thereby downgrading the storm, but what if such modifications should cause the hurricane to shift course and make landfall on a coastline that might otherwise have been unaffected? Is the effort worth the risk, or would it be better to leave well enough alone and be adequately prepared to deal with the consequences of a landfall if one should occur?

Looking at the situation in reverse, we might also question the wisdom of using climate as a basis for selecting our places to live and work. In the Sahel, a semiarid region south of the Sahara desert, for example, famine has claimed hundreds of thousands of lives because of the "failure" of expected rains. In this notorious incident of desertification, pastoral people had migrated northward into the northern Sahel in the early and middle 1960s in response to favorable summer rains. In 1968, however, these rains "failed," and they continued to fail for several succeeding years. This led to overgrazing of the range and eventually to mass starvation of both grazing animals and the humans who depended on them. In the mid-1960s a major climate shift occurred from tranquil flow in the polar jet streams to meandering flow with high-amplitude Rossby waves. This shift, which probably occurs on a 50 to 100 year cycle, is likely responsible for the drought in the Sahel and for similar droughts that have occurred since our once "predictable" weather turned "crazy" in the late 1960s. It is questionable whether these unfortunate herds people would have thought twice about migrating closer to the barren Sahara if they had understood the concept of such shifts in jet stream flow. Human history is littered with the wrecks of "good ideas" undertaken for the sake of short-term advantage, and it seems likely that it will continue to be so in the future.

In addition to the climatic changes that we create willfully, and to those natural changes that we choose to ignore, there are other changes that we create inadvertently and with which we then have to live. Modern city dwellers are all too familiar with the "brown clouds" of pollution that blanket urban areas during temperature inversions and the "no swimming" signs posted beside polluted waters. Far more alarming than these, however, is the more recently recognized threat of **acid rain.**

When fossil fuels are burned in industrial processes, electricity-generating plants, and automobiles, the main products of combustion are carbon dioxide and water. Some fossil fuels (notably coal), however, contain variable amounts of sulfur, and when this is burned, it yields *sulfur dioxide.* In addition, all high-temperature combustion processes combine atmospheric nitrogen and oxygen to form *nitrogen oxides.* The automobile is a prime offender in this regard. Both sulfur dioxide and nitrogen oxides are exceedingly hygroscopic, because of which they act as condensation nuclei in clouds. Raindrops formed on such nuclei are highly acidic because the oxides of sulfur and nitrogen form sulfuric and nitric acids when they dissolve in water. Normal rain is slightly acid—about 5.6 on the pH scale, on which 7.0 is neutral—due to dissolved atmospheric carbon dioxide (see fig. 1.22), but the pH of rain acidified by human activities is always less than 5.

In the northeastern United States, where the problem is most severe, pH values of acid rain as low as 2.1 (equivalent to undiluted lemon juice) have been recorded. When this rain falls, it has a number of disastrous effects. It leaches nutrients from soil and foliage, kills fish and soil organisms (notably nitrogen-fixing bacteria), and releases metallic toxins from the soil. In heavily affected areas, such as Sudbury, Ontario and Ducktown, Tennessee, airborne acid emissions can obliterate all vegetation from the land, and subsequent erosion can strip the soil down to bare bedrock. In New England, many ponds, lakes, and streams have become sterile, and sugar maples, for whose colorful fall foliage the region is justly famous, have been suffering from an affliction that has been euphemistically termed "maple decline."

Figure 15.33 shows the distribution of wet sulfate deposition in North America as of 1983. In the most strongly affected areas, the equivalent of over two bathtubs full of sulfuric acid is being dumped on each square kilometer of land every year. As sulfur-rich coal begins to replace scarce oil as an energy source, the problem can only be expected to worsen in the future unless stringent measures are taken to prevent it.

Acid rain actually poses a far more potent threat to life on Earth than nuclear holocaust because it can irreversibly disrupt and destroy all living things in a given ecosystem from the soil bacteria on up to the trees, and because it is happening *right now.* Furthermore, the destruction is a direct result of everyday activities that are essential to the functioning of our technoindustrial civilization. This calls into question the appropriateness and long-term viability of our fossil fuel dependent life-style. In the words of Pogo, "we have met the enemy, and he is us." In order to control the deadly threat of acid rain, it will be necessary to establish and observe strict limits on all industrial *and personal* activities that depend heavily on the combustion of fossil fuels. The choice is a hard one, but the alternative is unthinkable.

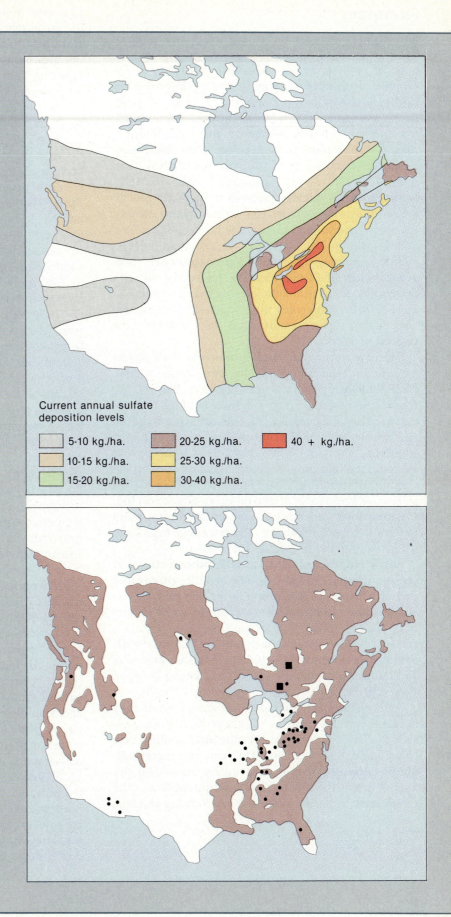

Current annual sulfate deposition levels

- 5-10 kg./ha.
- 10-15 kg./ha.
- 15-20 kg./ha.
- 20-25 kg./ha.
- 25-30 kg./ha.
- 30-40 kg./ha.
- 40 + kg./ha.

*Figure 15.33* The deposition of wet sulfate in Canada and the United States (*upper map*) and areas most sensitive to acid rain because they lack natural buffers (*lower map*). Black squares are sources emitting over 500,000 metric tons sulfate per year. Dots are sources emitting 100,000 to 500,000 metric tons per year. Smoke emissions from electricity-generating power plants account for 65% of the total. Industry and transportation account for the rest.

# Summary

Annual precipitation exceeds 100 cm in an equatorial belt of convective thunderstorms and two mid-latitude belts of cyclonic storms. **Absolute humidity** is the number of grams of water in 1 m³ of air. **Relative humidity** is the ratio of the absolute humidity of an air parcel at a given temperature to the absolute humidity it would have if it were saturated at that temperature. It rises as temperature drops until the **dew point temperature** (temperature of saturation) is reached. Condensation occurs on suspended dust particles (**condensation nuclei**). Unclean air increases precipitation downwind of cities (the **urban island effect**). Raindrop *terminal velocities* range from 1 to 10 m/sec. Raindrops grow as electrically charged cloud droplets collide randomly (the **collision-coalescence process**). Below 0° C, nucleation of ice occurs on such *freezing nuclei* as clay particles. Ice crystals grow by acquiring water molecules from water droplets, which have higher vapor pressures than ice (Wegener's **ice crystal process**).

**Sleet** is frozen rain. **Hail** forms when snow pellets are lofted repeatedly by updrafts. **Dew** forms on cold, clear nights with high absolute humidity. Below 0° C, **frost** forms instead of dew. *Dry* **haze** is a high concentration of mineral particles ≤0.1 μm diameter in the air. Higher relative humidity converts dry haze to *wet haze,* which becomes *fog* when *visibility* is <1 km. Dry air aloft and a slight ground breeze increase cooling, favoring **radiation fog** (*valley fog* in river valleys). Moist air flow over cold water or land produces **advection fog.** **Upslope fogs** form where moist air flows from warm lowlands to cool highlands. **Evaporation fogs** form where warm water evaporates into cold air.

**Low clouds** include **stratus, nimbostratus,** and **stratocumulus. Middle clouds** include **altostratus** and **altocumulus. High clouds** include **cirrus, cirrocumulus,** and **cirrostratus.** *Halos* (rings) and *sundogs* (paired spots) form around the Sun when ice crystals in cirrostratus refract sunlight. **Clouds with vertical development** include fair weather **cumulus** and **cumulonimbus,** the latter often having an *anvil head* of ice crystals.

The **environmental lapse rate** varies from strongly negative (inversion) to 34° C/km or higher but averages about 6.5° C/km. The **dry adiabatic cooling rate** of unsaturated air is about 10° C/km. The **moist adiabatic cooling rate** is close to the dry rate in cold, dry air, but it is as low as 3° C/km in warm, moist air. The **cloud base** marks the altitude at which rising air cools to its dew point temperature. In **absolutely stable** air, the lapse rate is lower than the moist rate. In **conditionally unstable** air, the lapse rate lies between the dry and moist rates. In **absolutely unstable** air, the lapse rate is higher than the dry rate. Advection of cold air aloft and solar heating of surface air favor instability.

In the *cumulus phase* of a thunderstorm, "bubbles" of air rise in thermal cells. In the *mature stage,* raindrops are large enough to be pulled down against updrafts by gravity. Downdrafts in thunderstorms are cold because of evaporation of water into descending air. The system of updraft plus downdraft is a *thunderstorm cell.* In the *dissipating stage,* downdrafts suppress updrafts, rain stops, and the lower part of the cloud evaporates. Severe thunderstorms result when high winds aloft tilt the cell so that precipitation does not fall through and suppress the updraft. Complexes of inclined storm cells in extremely moist, unstable air produce *supercell thunderstorms.*

In **wave cyclone theory,** divergent air flow in the polar jet stream induces surface low pressure at the polar front. Counterclockwise air flow (Northern Hemisphere) into this low pressure warps the polar front into a wave form whose apex points poleward. The west margin of the wave is a **cold front.** The east margin is a **warm front.** Cold fronts eventually overtake warm fronts, yielding **occluded fronts.** High-pressure **anticyclones** typically chase cyclones eastward. Air flowing out from an anticyclone circulates clockwise (Northern Hemisphere). **Frontal wedging** at a warm front produces a sequence of stratiform cloud types from cirrus to nimbostratus. As a warm front passes, pressure drops, temperature rises, and wind shifts from the southeast to the southwest. As a cold front passes, pressure increases sharply, temperature drops abruptly, and wind shifts from the southwest to the northwest.

**Tornadoes** have diameters of 100–600 m, wind speeds of ±200 km/hr, and internal pressures up to 100 mb below the pressure of surrounding air. **Whirlwinds** occur on clear, hot, summer days when strong surface heating produces a lapse rate high enough to create an upward increase in air density. **Waterspouts** are small, marine tornadoes composed of cloud droplets. **Hurricanes** form within and travel west with the trade winds and require sea surface temperature ≥26° C, unstable air, no *trade wind inversion* aloft, and surface air convergence, as in an *easterly wave.* Wind speed in *tropical disturbances* is <37 km/hr; in **tropical depressions** 37–63 km/hr; in **tropical storms** 63–120 km/hr; and in hurricanes >120 km/hr. Clear weather prevails in the hurricane's **eye,** where warm air descends. *Storm surges* are rises in sea level due to wind-driven water and low pressure within the storm. **Acid rain** poses an immediate threat to all life.

## Key Terms

absolute humidity
relative humidity
dew point temperature
condensation nucleus
hygroscopic
urban island effect
collision-coalescence
    process
ice crystal process
sleet
ice storm
hail
dew
frost
haze
radiation fog
advection fog
upslope fog
evaporation fog
low clouds
middle clouds
high clouds
clouds with vertical
    development
stratus clouds
stratocumulus clouds
nimbostratus clouds
altostratus clouds
altocumulus clouds
cirrus clouds

cirrocumulus clouds
cirrostratus clouds
cumulus clouds
cumulonimbus clouds
adiabatic rate
dry adiabatic cooling
    rate
moist adiabatic cooling
    rate
cloud base
absolute stability
conditional instability
absolute instability
thunderstorm cell
Bjerknes
wave cyclone
cold front
warm front
occluded front
anticyclone
frontal wedging
tornado
whirlwind
waterspout
hurricane
eye
tropical depression
tropical storm
acid rain

## Questions for Review

1. If the relative humidity of an airmass is 50% at 10° C, what is its absolute humidity in $g/m^3$? What would its relative humidity be at 5° C? What is its dew point temperature? (Hint: see fig. 15.2.)
2. Compare and contrast the collision-coalescence and ice crystal processes.
3. Describe the conditions that give rise to radiation fog, advection fog, upslope fog, and evaporation fog.
4. Locate each of the cloud types of table 15.1 in its customary environment within a wave cyclone system.
5. What are the environmental prerequisites for absolute stability, conditional instability, and absolute instability?
6. Describe the cumulus, mature, and dissipating stages of a cumulonimbus cloud in terms of the environmental lapse rate, the dry and moist adiabatic rates, updrafts, downdrafts, rainfall, and latent heat.
7. Describe the evolution of a typical wave cyclone in terms of air masses, fronts (polar, cold, and warm), convergence, divergence, pressure, and wind direction.
8. What are the environmental conditions favoring the development of tornadoes, whirlwinds, and waterspouts?
9. Describe the dynamics of a hurricane in terms of sea surface temperature, lapse rate, high level trade wind inversion, convergence, latent heat, and pressure.
10. What are the causes and the dangers of acid rain?

# Part Five

# Beyond Earth

Photo: The Pleiades star cluster in Taurus.

# 16

# The Celestial Sphere: A Growing Awareness

## Outline

*The space between heaven and Earth is like a bellows:*

*Hollow, but not in vain,*

*When it moves it yields more than its measure.*

*Words in excess are worthless.*

*Therefore, hold to the core.*

*Lao Tzu*

**Tao-Te Ching, 5**

Photo: The constellations Auriga (*left*) and Taurus (*right*) gleam above a moonlit waterfall.

## From Geocentrism to Heliocentrism

Today, most people accept the central position of the Sun in the Solar System as an article of faith. Prior to the late seventeenth century, however, a different belief prevailed, and by the test of everyday experience it was a more reasonable one. After all, what would lead you to conclude that Earth revolves around the Sun instead of the other way around? Our perceptions tell us that the Sun "rises" in the east and "sets" in the west, and the same is true of the Moon, stars, and planets. Those of us who pay closer attention to celestial phenomena might notice, however, that whereas the stars appear fixed in their transit of the heavens, the positions of the Sun, the Moon, and the five visible planets are variable with respect to the stars and to each other.

All of these heavenly bodies move within a narrow belt, called the **zodiac** (see fig. 16.12 and appendix V). This belt, 16° of arc wide, is centered about the *plane of the ecliptic,* that is, the plane of Earth's orbit around the Sun, (see chapter 13), which is tilted at an angle of about 23½° to Earth's equatorial plane. Within the belt of the zodiac, 12 star groups, or **constellations,** are fairly evenly spaced, one constellation for each 30° sector of the belt. These star groups provide an unchanging backdrop for the wanderings of Sun, Moon, and planets. The name *planet,* in fact, comes from the Greek word *planetes* (plah-*nay*-tace), which means "wanderer."

## The Vault of Heaven: Early Perceptions of the Universe

Ancient astronomers, convinced that these solar, lunar, and planetary wanderings had important consequences for the welfare of humankind, watched them as closely as their crude technology would permit. Many impressive stone structures, including the great stone circle of Stonehenge in southern England, the Mayan and Aztec temples in Middle America, and the Babylonian temples in Mesopotamia, were designed to be used at least in part for the routine observation of the heavens.

All of these early societies recognized the solstice phenomenon (from the Latin *solstitium,* a standing of the Sun), the disturbing tendency of the Sun to shift ever lower toward the horizon with each daily passage across the sky until the day of the solstice when the trend would be reversed and the Sun's path would again climb gradually higher in the sky. In some societies, such as the Aztec's, a fear developed that the trend might not be reversed without some form of propitiation. Among the Aztecs, that fear mounted to the point of paranoia and led to the practice of human sacrifice, thought to be necessary to ensure the annual return of the Sun. Cortez the conqueror, indignant at such barbarity, proceeded to replace the Aztec version of human sacrifice with a

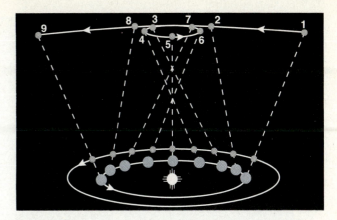

**Figure 16.1** Differing orbital velocities account for the illusion of retrograde motion in one planet as viewed from another against the backdrop of the fixed stars. Here, as Earth passes Mars on the inside, the red planet appears to slow (1–4), reverse (4–6), and then reaccelerate into direct motion (6–9).

more comprehensive version of his own, but he did so with a view to retrieving something more golden than the Sun.

### Geocentrism: The Most Logical Explanation

As viewed from Earth, the Sun's position shifts gradually eastward relative to the fixed stars. If the Sun could actually be seen against its starry backdrop, it would appear to shift eastward about 1° of arc with respect to the stars in each 24 hr period. The Moon, which *can* be seen against the stars, also shifts its position eastward, but at a much faster rate: about 13° in 24 hr.

The motions of the Moon and the Sun are regular, but the same is not true of the motions of the planets. As seen against the background of the fixed stars, the planets appear to follow looping paths, at some times seeming to move, as do the Sun and the Moon, with an eastward, or *direct,* motion with respect to the stars and at other times seeming to move with a westward, or *retrograde,* motion. The illusion of retrograde planetary motion arises because Earth's orbital velocity differs from those of the other planets. Figure 16.1 shows that as Earth passes it on the inside, Mars appears first to slow down (points 1–4), then to reverse (points 4–6), and finally to accelerate again into a direct path (points 6–9).

Ancient astronomers, who had no reason to suspect that the stars and planets were anything other than lights in the sky, explained retrograde motion in a way that granted Earth a central position in the Universe. They assumed that the planets move around Earth in circular orbits, called **deferents,** within the zodiac but that each planet also describes a small circular path called an **epicycle** around a point on the deferent much as the Moon follows its slightly elliptical path around Earth as the

**Figure 16.2** In an Earth-centered Solar System, the retrograde motions of the planets can only be explained if the planets describe small circles, or epicycles, as they move in their circular, clockwise orbits, or deferents. The motions of Venus and Mercury, which lie between Earth and Sun, were thought to be locked into that of the Sun.

latter revolves in its orbit. This **geocentric,** or Earth-centered, scheme (fig. 16.2), was perfected around 140 A.D. by the Greek-Egyptian Ptolemy (*Toll*-uh-mee, or Claudius Ptolemaeus). In Ptolemy's geocentric system the Moon and the Sun also have circular deferents, without epicycles, and the fixed stars rotate westward on a crystal sphere, the *Vault of Heaven,* at a greater distance and at a slightly greater angular velocity than the Moon and the Sun.

To these early astronomers, the heavens gave no hint of the enormous distances that separate us from even the nearest of the heavenly bodies, so they had no way of knowing that these celestial lights were actually great spheres anywhere from 0.27 to 250,000 times the size of Earth. For that matter, few of them accepted the fact that Earth is a sphere, nor had they any awareness of how big it actually is.

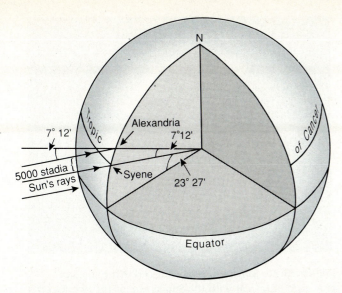

*Figure 16.3*  By simultaneously observing the angle between a plumb bob and the Sun's parallel rays in Syene and Alexandria, Eratosthenes found that the angle between the plumb bobs was 7°12'. He then calculated Earth's circumference by multiplying the distance between the two stations (5000 stadia, or 925 km) by 360° and dividing by 7°12'. The result, 46,250 km, is reasonably close to the actual figure of 40,009 km.

## Heliocentrism: *The Dark Horse*

One early exception to this general unawareness was another Greek-Egyptian astronomer named Eratosthenes (Air-ra-*toss*-thuh-neez), who conducted a clever experiment in about 235 B.C. that showed Earth to be a sphere with a radius only 14% larger than the one accepted today (fig. 16.3). At noon on the June solstice, Eratosthenes and an assistant simultaneously measured the angle between the Sun's rays and a plumb bob in two different locations on the same meridian in Egypt: Alexandria and Syene (now Aswan). In Syene, which lies on the Tropic of Cancer, the angle was 0°. In Alexandria, 925 km to the north, the angle was 7°12'.

The reasoning behind Eratosthenes's experiment is as follows: If the Sun's rays are parallel and if Earth is a sphere, then along any meridian there must be only one subsolar point, such as Syene, at which the direction of the Sun's rays would coincide with that of a plumb bob seeking Earth's center, and two points at 90° from that subsolar point at which the Sun's rays would make a 90° angle with a plumb bob. Therefore, any angle between the Sun's rays and a plumb bob measured at some point on the meridian must represent the difference in latitude between that point and the subsolar point (fig. 16.3). The ratio between that angle (in this case 7°12') and 360° should then be equal to the ratio between the distance separating the two points (in this case 925 km) and Earth's circumference.

The idea of a spherical Earth was not original with Eratosthenes. Thales (*Thay*-leez) of Miletus had expressed that concept about 600 B.C., and Plato (427–347 B.C.) had cited the appearance of Earth's circular shadow on the Moon during eclipses as evidence for its roundness. Plato had also maintained that Earth was not at the center of the Universe. In this, he agreed with Pythagoras, who had stated in the fifth century B.C. that the Sun stands at the center. In a world without benefit of either the scientific method or sophisticated instrumentation, however, the concepts of a round Earth and a Sun-centered, or **heliocentric,** Universe remained minority opinions, and the geocentric Universe of Ptolemy became generally accepted. (Even today, there persists a Flat Earth Society, whose motto is: "Anyone can look out the window and see the world is flat.") The writings of Ptolemy were translated into Arabic in 827 A.D., and in this form they were introduced into western Europe as the *Almagest* (*Ahl*-mah-jest; Arabic for "the greatest"). The Catholic Church, responsible for virtually all scholarship in Europe during the Middle Ages, accepted and sanctioned the pronouncements of the Almagest because they agreed with orthodox interpretations of the Bible. Even in modern speech and writing, we still find vestiges of the strict ecclesiastical filter through which all knowledge was required to pass in the Middle Ages. For example, the common practice of rendering the names Earth, Sun, and Moon, as "the earth," "the sun," and "the moon," are examples of a church effort to eliminate the worship of Nature.

## Copernicus to Newton: The 200-Year Awakening

It was not until astronomers began studying the heavens with improved instruments—including mechanical clocks for measuring elapsed time—that serious discrepancies between the Ptolemaic model and reality began to appear. Among the first to notice such discrepancies was the Polish astronomer Mikoíaj **Kopernik** (Kaw-*purr*-neek, 1473–1543, better known by his Latinized name, Nicholas Copernicus). Although he did not reject the Ptolemaic system altogether, Kopernik modified it by placing the Sun at the center because he felt that such an arrangement would be more consistent with observed facts.

One major advantage of a heliocentric system over a geocentric one is that it eliminates the necessity for epicycles, although Kopernik himself still continued to make use of epicycles in some cases. Kopernik's ideas were not published until shortly before his death, but even so, they were denounced as heretical by Martin Luther and other prominent churchmen. In 1600, the great Italian philosopher Giordomo Bruno was burned at the stake in Rome for refusing to recant his advocacy of the Copernican system (see chapter 19). In the end,

Kopernik himself became a churchman, but others continued to probe the heavens with increasingly sophisticated equipment, and the results of these investigations were anything but auspicious for the geocentric tradition. One of these researchers was the Danish nobleman-astronomer Tycho **Brahe** (*Tee-ko Bra-eh*, 1546–1601), who established an observatory near København (Copenhagen) under the auspices of the Danish king. Brahe did not accept the Copernican system, but he amassed a great quantity of data on the motions of the heavenly bodies which would later be used by his brilliant student, Johann **Kepler** (1571–1630), in support of the heliocentric theory. Brahe moved to Praha (Prague), Bohemia, late in life and there met Kepler, who used both Brahe's data and his own to derive the following three fundamental laws of planetary motion:

1. The path of a planet in orbit around the Sun is an ellipse with the Sun at one focus.
2. A radius connecting the Sun and a planet sweeps over equal areas in equal times.
3. The square of a planet's orbital period is proportional to the cube of its distance from the Sun.

Now let us examine each of these laws in somewhat greater detail.

Early astronomers believed that the planets moved in circular orbits because they regarded the circle as a natural expression of perfection. Kepler originally shared that belief but was ultimately forced to abandon it when careful observation showed that planetary orbits are actually not perfectly circular, but very slightly elliptical. An *ellipse* (fig. 16.4) differs from a circle in that it is more or less flattened and has two *foci* in place of a single center. (None of the actual planetary orbits of the Solar System is as strongly elliptical as the one illustrated in fig. 16.4.) The sum of the lengths of two radii drawn from any point on an ellipse, such as $P_1$ or $P_2$ in figure 16.4, to the two foci is always the same. Because of this, if you were to attach to the two foci the ends of a length of string just long enough to stretch to $P_1$ or $P_2$ when taut, you would be able to trace out the ellipse by moving a pencil point at the limit of the taut string. The positions of perihelion and aphelion (see chapter 13) occur at either end of the ellipse's long axis, which passes through the two foci.

Kepler's second law followed from the observation that a planet moves faster in its orbit in the vicinity of perihelion than it does in the vicinity of aphelion. In this way, it compensates for shorter distances with higher velocities. Kepler, who had a passion for discovering mathematical order in Nature, toyed with this situation until it yielded the precise, mathematical relationship stated in his second law. In figure 16.4, the two shaded

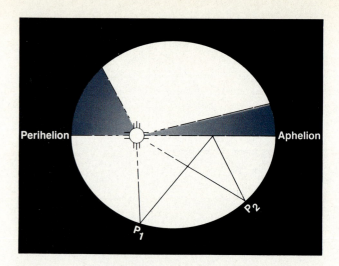

**Figure 16.4** Johann Kepler deduced that planetary orbits are elliptical with the Sun at one of the two foci of the ellipse. He also found that the radius between the planet and the Sun sweeps over equal areas in equal time, as indicated here by the two shaded areas. This requires a higher orbital velocity at perihelion than at aphelion. The sum of any two radii drawn from the two foci to the same point, such as $P_1$ or $P_2$, on an ellipse is always the same.

areas represent sectors of the ellipse traversed by the orbiting planet during two time intervals of equal length. As Kepler's second law predicts, the areas are also equal.

The third law, simply stated, is

$$P^2 = D^3,$$

where $P$ is the planet's period in Earth years (i.e., the time it takes to make one revolution around the Sun), and $D$ is the planet's distance from the Sun in **astronomical units** (AU). One AU is Earth's mean distance from the Sun, approximately 150 million km. Although the actual value of an AU was not known in Kepler's time, the *relative* distances of the planets from the Sun could be found from angular relationships.

While Kepler was discovering his laws of planetary motion in Bohemia, his contemporary, **Galileo** Galilei (1564–1642), was making equally myth-shattering discoveries in Italy. In 1609, Galileo turned his first crude telescope toward the heavens and opened a new age of understanding by extending human vision far beyond the limits imposed on it by Nature. Among other things, Galileo discovered that (1) the Moon's surface is rough, (2) the planets appear as round disks, (3) Jupiter has moons (he saw the largest four), (4) the Sun has spots which reveal a rotational period of a little less than a month (unusually slow for a star), and (5) Venus displays phases identical to those of the Moon. Especially significant was his observation that the disk of Venus becomes much smaller in its full phase than in its new phase. In the geocentric, Ptolemaic system, such a pronounced diminution in size would be impossible. In the heliocentric, Copernican system, however, it would be perfectly logical because that system allows the distance between Earth and Venus to vary. Spurred by these exciting findings, Galileo took a bold step in publishing a book expounding the Copernican system, but he was cautious enough to write it in the form of a dialogue between a proponent of heliocentrism and a proponent of geocentrism. Nonetheless, the Catholic church was quick to discern a clear bias in favor of heliocentrism, and Galileo was promptly forced to recant his views before the Inquisition and to spend the rest of his life under house arrest.

Galileo is considered to be the first scientist to use experimentation in his investigations of the natural world. His famous experiments in which he dropped a variety of different objects from the leaning tower of Pisa demonstrated that Aristotle (384–322 B.C.) had been wrong in asserting that heavier objects fall faster than lighter ones. Galileo correctly attributed the small variations in falling times he observed to different amounts of air resistance encountered by the various falling objects. He observed that the distance covered by a falling object varies as the square of the time, and from this he drew the fundamental inference that the acceleration of falling objects is constant.

Sir Isaac **Newton** (1642–1727) was born in England in the year of Galileo's death. Building on the discoveries of Kepler and Galileo, Newton worked out the three fundamental laws that govern the motions of both Earthly and heavenly bodies. In order to explain those laws, he also invented a powerful new branch of mathematics, the *calculus*. Newton's three laws of motion, which form the basis for classical mechanics, are:

1. An object must remain either at rest or in uniform motion in a straight line unless it is affected by an external force.
2. The application of an external force to an object produces a proportional acceleration of the object. In symbols,

$$F = ma,$$

where $F$ is force, $m$ is mass, and $a$ is acceleration.

3. A force applied by one object to a second object is simultaneously opposed by an equal and oppositely directed force applied by the second object to the first.

In terms of everyday experience, one could restate Newton's laws as follows:

1. It takes a lot of energy to get a freight train moving, but once it is in motion, it is hard to stop and hard to turn.
2. The more you push on something, the faster it goes (and the heavier it is, the harder you have to push in order to get the same result).

3. The harder you push on something, the more it seems to push back at you.

If Newton had been living in a space module, deriving his three laws would have been a much simpler task than it actually was. Living as he did on the surface of a planet, where the motion of all objects is influenced by gravity and friction against other substances, the task was a formidable one. Newton rose to this added challenge by developing the *law of universal gravitation,* which states that the attractive force between two bodies is directly proportional to the product of their masses and inversely proportional to the square of the distance separating their centers. In symbols:

$$F_g = Gm_1m_2/r^2,$$

where $m_1$ and $m_2$ are the two masses, $r$ is the distance between their centers, and $G$ is a constant.

Using this law in conjunction with his first and second laws of motion, Newton deduced that the orbital motions of the planets result from two opposing tendencies: that of the planets to maintain constant velocities in straight-line paths (first law), and the *centripetal* (center-seeking) force of gravity tending to draw the planets inward toward the Sun. The observed orbital motion of a given planet is simply a compromise between these two tendencies.

The significance of Newton's discoveries is that they explain such a wide variety of natural phenomena. Not only did they unravel the mysteries of planetary motion, but they also explained tides, rivers, glaciers, rainfall, the trajectories of cannon balls, and to a large extent the circulation of the atmosphere and oceans, among many other things. With the publication in 1687 of Newton's *Mathematical Principles of Natural Philosophy,* most of the mystery that had previously enshrouded our perception and understanding of the natural world suddenly fell away. The human world was quick to recognize and appreciate this, and Newton won international fame and the first knighthood ever accorded a scientist by the British Crown.

This fame and honor would have been well enough deserved on the basis of the laws of motion and gravitation alone, but Newton went on to discover the principles of optics, on which all subsequent investigations of the heavens have been based. Let us now turn to some of the fruits of Newton's optical theories.

## Observing the Heavens

Techniques for perceiving what is beyond the reach of normal human vision have been best developed in those sciences that deal with the exceedingly small and with the exceedingly large and distant. Magnifying lenses have been in use since antiquity, but it was not until about 1600 A.D. that the first *compound microscope* (i.e., having more than one lens) was invented, either by the

**Sir Isaac Newton (1642–1727)**

Isaac Newton, Sr., a farmer in Woolsthorpe, Linconshire, England, died in October 1642, a few months before the birth of the son who would forever change our perceptions of the physical world. When young Isaac was three, his widowed mother remarried and gave him to his grandmother, who raised him until the age of 14. When he was 12 years old, Isaac, shy and diffident, was sent to the Grantham grammar school, where he failed to distinguish himself until a triumphant scuffle with a schoolmate bolstered his self-esteem enough that he became the school's best student. In 1656, his mother was widowed for the second time, and she called him home to work on the farm. By then, however, Isaac's fancy was more taken with numbers than with cows. An uncle who was a fellow of Trinity College at Cambridge recognized his talent for natural philosophy and persuaded his

mother to send him back to school. In 1661, he enrolled at Trinity, where he studied with and befriended Isaac Barrow, a well-known mathematician.

Newton received a B.A. degree from Trinity in 1665. In the fall of that year, the Black Plague struck England, and Cambridge closed for almost two years. Newton returned to Woolsthorpe, where instead of tending the farm, he invented the calculus and the binomial theorem. With the aid of a glass prism and a falling apple, he also discovered the fundamental principles of optics and celestial mechanics.

Newton returned to Trinity in 1667 and was made a fellow of the college. He made his discoveries known to Barrow, who was impressed enough to yield his professorship to his young protégé when he retired to study theology in 1669. In the preceding year, Newton invented the reflecting telescope, on the strength of which he was elected to fellowship in the Royal Society in 1672. Thereafter, he became involved in a lengthy series of disagreements and debates with various colleagues, most notably with Robert Hooke in regard to optics and with the German Baron Gottfried Wilhelm von Leibnitz (*Lipe*-nitz), who had independently developed the calculus.

On the insistence of his astronomer friend Edmund Halley, Newton reluctantly published his ideas on planetary motions in 1687 in his *Principles,* the work which won him international acclaim. Almost as successful was his *Opticks,* published in 1704.

Newton's childhood deprivations and his bitter conflicts with colleagues contributed to lifelong feelings of paranoia and insecurity, and to a prolonged nervous collapse in 1692–1694. Perhaps it was these difficulties that ulimately led him away from his spectacularly successful search for truth into a more closely circumscribed existence as president of the Royal Society (1703–1727), and as warden (1696–1699) and master (1699–1727) of the British mint, a post from which he condemned many a convicted counterfeiter to the gallows. At the same time, and largely in secret, he delved into studies of theology and alchemy, efforts that have left virtually no legacy. Nonetheless, many would still agree with the tribute accorded him by the French mathematician Joseph Lagrange (La-*grawnzh*), who declared that Newton was "the greatest genius that ever existed."

Dutchman Cornelis Drebbel (see chapter 11) or by his countrymen Hans and Zacharias Janssen. In 1608, another Dutchman, Hans Lippershey, invented the first telescope. In the following year, Galileo produced the first of his own telescopes, modelled after Lippershey's design.

## Refracting Telescopes

The telescopes built and used by Galileo are of a type known as **refracting telescopes** (fig. 16.5). These employ two lenses: an *objective* at the far end of the telescope tube and an *eyepiece* at the near end. The function of the objective lens is to concentrate the light rays it receives into a small area. It accomplishes this by **refraction,** the tendency of light to bend as it crosses a boundary between two transparent media of different

densities. It does this because light travels more slowly in a more dense medium than in a less dense one. The principle is analogous to the bending of earthquake shock waves discussed in chapter 6 (but *not* to the refraction of water waves, discussed in chapter 12!).

In figure 16.6, a set of parallel light rays from a distant source approaches the outer surface of the objective lens from the left along a path that is parallel to the lens axis. (Light is actually continuous and is represented in fig. 16.6 as separate lines only for the sake of discussion.) As the light rays enter the glass of the lens, they slow down. The direction of the ray that is travelling along the lens axis is unaffected by this slowing, but all other rays are bent inward to an increasing degree with increasing distance from the axis. This bending can be understood if an individual ray is thought of not as a line but as a band. Because of the curvature of the lens

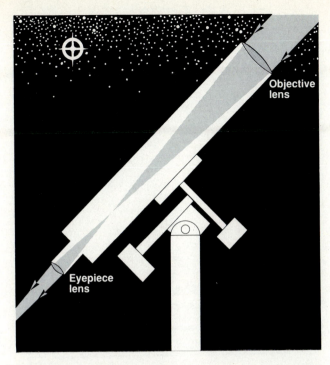

**Figure 16.5** Refracting telescopes combine an objective lens on the outer end of a telescope tube and an eyepiece lens on the inner end.

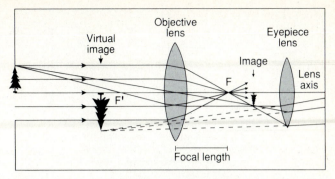

Focal length

**Figure 16.6** In refracting telescopes, light rays that travel parallel with the lens axis are refracted inward to meet at a focus (*F*). Rays from objects in the field of view (such as the tree, *far left*) are refracted through the lens to form an inverted image on the far side of the focus. As they pass through the eyepiece lens, rays from this image are refracted in such a way that an enlarged virtual image appears to form in the vertical plane in which those rays appear to converge (*dashed lines*).

surface, the light at the edge of the band closest to the lens axis strikes that surface slightly earlier than the light on the other side of the band. Consequently, the inner edge of the band is slowed slightly before the outer edge. This results in an inward bending of the ray.

On emerging again on the other side of the lens, the light rays once again accelerate to their former velocity. As before, this does not affect the direction of the central ray, but all other rays are bent inward to an increasing degree with increasing distance from the lens axis. If the lens has been properly ground, the rays all converge at a single point, the **focus** (*F*). The distance of the focus from the center of the objective lens is called the **focal length** of the lens.

In figure 16.6, the tree on the left represents the object that is emitting the light rays. Note that the image of this object does not form at the focus. Three representative rays are shown emanating from the tip of the tree. Because the upper ray is parallel with the lens axis, it must bend in such a way as to pass through the focus. The ray that passes through the center of the lens is undeflected because here the two lens surfaces are parallel. Consequently, the slight upward refraction the ray experiences on entering the lens is exactly compensated by an equivalent downward refraction on leaving it. The lower ray passes through the focus *F'* on the left side of the lens. This ray is refracted in such a way as to intersect the other two rays at the same point. Similar

sets of rays from other parts of the tree intersect within a plane directly above that point, thereby reconstructing an inverted image of the tree.

This image is often photographed directly, but if the telescope is to be used visually, an eyepiece is placed in the lower end of the telescope tube (fig. 16.6 *right*). The eyepiece refracts the rays emanating from the image in such a way as to create a *virtual image* (*dashed arrow*) in a vertical plane within which the projections of the refracted rays passing through the eyepiece intersect (*dashed lines*). The image is called virtual because if a screen were placed at the point where it appears to form, no actual image of the object would be seen.

Refracting telescopes with lenses as much as 102 cm in diameter have been built, but such large lenses have an inherent problem that is hard to correct. When a ray of white light passes through a prism (such as the edge of a lens) it does not emerge as the same, single ray of white light that it was before it entered the prism, but it is spread out into a rainbowlike *spectrum* (see fig. 18.5). As I note in chapter 18, this behavior of light is extremely useful in some astronomical studies. When it comes to obtaining a crisp image of a celestial object, however, it is a nuisance because light rays passing through the objective do not meet at a precise focus, but within a range of foci, closer to the lens for violet light and farther away from it for red light. The problem has been called *chromatic aberration,* and one solution to it is to make the objective lens as thin as possible, a practice that also increases the focal length. One refracting telescope, built by the Dutch astronomer Christiaan Huygens (*High*-guns, 1629–1695), measured 64 m in length! Another solution used more commonly today is to add correcting lenses to the objective to reduce chromatic aberration.

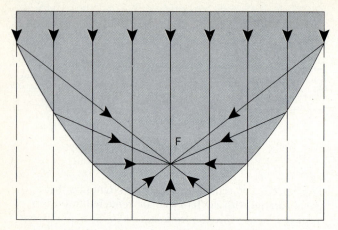

**Figure 16.7** A parabola is the locus of points equidistant from a point (*F*) and a line (*bottom*). Light rays travelling parallel with the axis of a parabolic mirror are reflected toward the focus.

## Reflecting Telescopes

A simpler solution to the problem of chromatic aberration came from an expectable source, Sir Isaac Newton. From his studies of analytic geometry, Newton knew that a *parabola,* the locus of points equidistant from a focal point and a line, reflects incoming rays to its focal point if those rays are parallel to its axis (fig. 16.7). Similarly, a *paraboloid,* formed by rotating a parabola about its axis, also reflects such axis-parallel rays to its focal point. Consequently, if a mirror were made in the shape of a shallow paraboloid, it should concentrate light rays just as a lens does but without the problem of chromatic aberration.

Newton's **reflecting telescope** used such a mirror mounted at the base of a large tube (see fig. 16.9*b*). In his design, light was not allowed to reach the focal point within the telescope but was reflected out of the tube by a second, flat mirror mounted at a 45° angle to the telescope axis. Contrary to what you might expect, this mirror did not block the central part of the image, but it simply dimmed the entire image slightly. Since Newton's time, some very large reflecting telescopes have been built. The world's largest is located in the Caucasus Mountains of the Soviet Union, and its mirror measures 600 cm in diameter. The second largest is the 508 cm Hale telescope on Mount Palomar near Pasadena, California (fig. 16.8).

In place of the Newtonian type of focus, the Hale telescope utilizes three different types of focussing methods: the *prime focus* method (fig. 16.9*a*), in which the observer sits in a cage mounted within the telescope tube at the focus; the *Cassegrain focus* method (fig. 16.9*c*), which reflects the light back down through a hole in the center of the mirror; and the *Coudé focus* method (fig. 16.9*d*), which is like the Cassegrain but employs a third mirror to reflect the light out the side

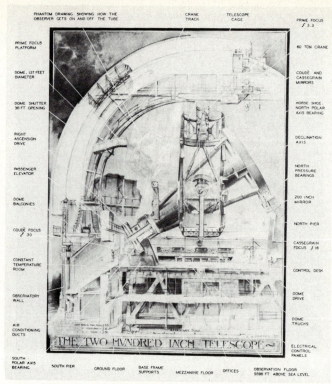

**Figure 16.8** A detailed drawing of the 508 cm Hale reflecting telescope on Mount Palomar, California.

**Photo by Hale Observatories**

of the tube and through the mountings to an observation room. In addition to this variety of focussing methods, all modern research telescopes also have computer-driven tracking systems that allow them to be trained on distant objects for many hours at a time.

## Extending the Range of Perception

From Galileo's time to the mid-nineteenth century, the primary objective of telescope design was to improve the visual image. In 1850, however, a new approach was tried at the Harvard College Observatory. There, human observers were replaced by photographic plates. This approach proved so successful that it has become routine in observational astronomy. Photography has two principal advantages over the human eye: (1) through prolonged exposure, photographic emulsions can record objects that are too faint to be detected by even the keenest human eye; and (2) photographs provide a permanent record of astronomical observations that can be referred to at any time in the future.

At about the same time, the **spectrograph**—a device for dispersing narrow strips of light into their spectral colors and photographing them—was invented (see fig. 18.5). Spectrographic analysis of the light from different stars showed characteristic differences that have enabled astronomers to determine a remarkable amount of information about heavenly objects that could not

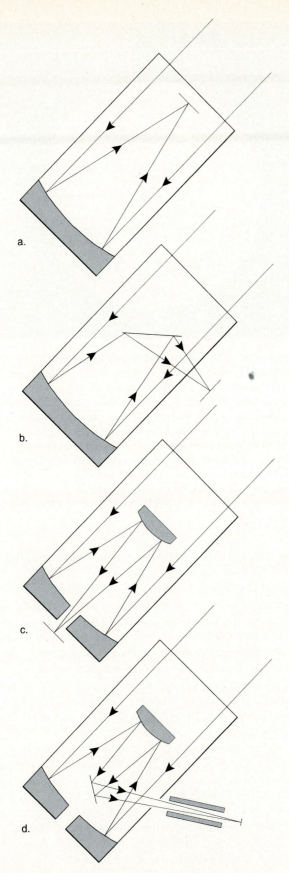

**Figure 16.9** **A variety of reflecting telescopes with different types of focus.** *a.* Prime; *b.* Newtonian; *c.* Cassegrainian; *d.* Coudé.

otherwise have been obtained. Among the things revealed by spectrographic astronomy are the temperatures, velocities, and chemical compositions of stars, nebulas, and galaxies (see chapter 18).

Another development of the mid-nineteenth century was the work of James Clerk (pronounced Clark) Maxwell (1831–1879), a Scottish physicist who demonstrated that visible light is only a small part of a vast spectrum of electromagnetic energy ranging from gamma rays at the short-wave end to radio waves at the long-wave end (see fig. 3.1). Our own star, the Sun, emits most of its spectral energy in the visible range, but there are many phenomena in the Universe that radiate energy at wavelengths other than those of visible light. Nevertheless, it was not until the 1920s and 1930s that the first studies of radio signals from space were made by Karl Jansky of the Bell Telephone Laboratories. The practical objective of these studies was to find the source of unexplained static in transoceanic radiotelephone communications. With the help of a rotating, 18 m radio antenna of his own devising, Jansky found that most of the static was due to thunderstorm activity, but that a small, steady, irreducible residue came from extraterrestrial sources.

Since World War II, the science of radio astronomy has become a respectable subdiscipline of observational astronomy. The principal cosmic source of radio signals is clouds of hydrogen gas, but there are many other sources, ranging from sunspots to exploding galaxies (see fig. 18.17). Radio astronomy has yielded much information about the center of our own galaxy, about which little was previously known due to light-obscuring clouds of cosmic dust.

The standard tool for observing such radio disturbances is the **radio telescope,** which resembles the reflecting telescope in that it has a paraboloid receiving surface. That surface, however, is not aluminized glass but wire mesh, and there is no telescope tube. Most radio telescopes are movable and can be aimed at specific targets in the heavens, but the world's largest is a stationary, 300 m paraboloid dish set into a natural, bowl-shaped depression in karst topography near Arecibo, Puerto Rico (fig. 16.10). Near Soccoro, New Mexico, a group of 27 radio telescopes forms a "superscope" called the Very Large Array (fig. 16.11). Each telescope in the array can be oriented separately and moved laterally along a Y-shaped track with a 21 km radius. This gives the effect of a single radio telescope of gigantic dimensions, an advantage in the precise location of radio sources outside the Milky Way galaxy.

## Measuring the Heavens

Probably the most useful aspect of the legacy left to modern astronomy by the earliest astronomers is the system whereby we locate objects in the heavens. Figure

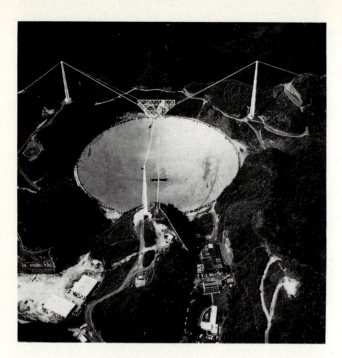

*Figure 16.10*  The 304.8 m diameter dish of the Arecibo radio telescope, operated by Cornell University's National Astronomy and Ionosphere Center is set into a natural hollow in a karst topographic region of Puerto Rico.

Photo by NASA

*Figure 16.11*  The National Radio Astronomy Observatory's Very Large Array (VLA) consists of 27 mobile radio telescopes, each 25 m in diameter, set on tracks on a high desert plateau to the west of Socorro, New Mexico.

Photo by NASA

16.12 shows the essential features of this ancient design, known as the **equatorial system** of celestial coordinates. This system retains the ancients' incorrect view of the heavens as a "celestial sphere" with Earth at its center purely for the purpose of locating the positions of objects in the sky. This is done by extending Earth's rotational axis and its equatorial plane (shaded surface

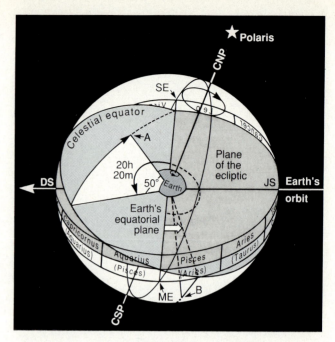

*Figure 16.12*  The equatorial system of celestial coordinates locates heavenly objects (*A*, *B*) by declination north or south of the celestial equator and right ascension eastward from the March equinox (*ME*). Constellations of the zodiac are shown in their present positions above the plane of the ecliptic and as they were about 2200 years ago below, the change being due to precession of the equinoxes. Broad arrow represents the rising of the Sun above the celestial equator at the March equinox. *CNP* = celestial North Pole, *CSP* = celestial South Pole, *SE* = September equinox, *JS* = June solstice, *DS* = December solstice.

in fig. 16.12) until they intersect this imaginary sphere at the *celestial north* and *south poles* (CNP, CSP) and the *celestial equator,* respectively. The significance of this arrangement is that as Earth rotates eastward on its axis (note circular arrow on Earth in fig. 16.12) the celestial sphere *appears* to be rotating westward (note circular arrow on celestial sphere). This apparent rotation is best illustrated in time exposures of the stars made with a camera pointed at Polaris, the North Star, which is located approximately in line with the axis of the celestial sphere (fig. 16.13). In such exposures, the stars make circular streaks concentric on Polaris.

To the celestial sphere we now add the plane of the ecliptic and the 16° wide band of the zodiac, centered on that plane. Because of Earth's axial tilt, the celestial equator is inclined at an angle of 23°27' to the plane of the ecliptic. The two points where the celestial equator intersects that plane are called the *equinoxes* (Latin for "equal night." See chapter 13). At the *March equinox* (ME), the Sun appears to rise above the celestial equator (broad arrow in fig. 16.12); and at the *September equinox* (SE), the Sun appears to sink beneath it.

**Figure 16.13** Earth's rotation is evident in this time exposure of the region around Polaris, the North Star. As the planet turns, the stars appear to move in concentric circular arcs around Polaris.

Photo by Hansen Planetarium

This geometry provides us with a coordinate system within which any object in the heavens can be precisely located. The celestial equator serves as a reference line from which the location of a celestial object is measured in degrees of north or south **declination.** A second reference line is provided by the meridian, or north-south line, that passes through the March equinox. The location of a celestial object is described in terms of its **right ascension,** or its bearing to the east of the March equinox, expressed as hours, minutes, and seconds. For example, a star at point *A* in figure 16.12 would have a north declination of 50° and a right ascension of 20 hours, 20 minutes. A star at point *B* would have a south declination of 20° and a right ascension of 1 hr 20 min. The star charts of appendix V show the locations of the major stars and constellations according to this location scheme.

A related but simpler problem is the location of a celestial object with respect to the position of an observer at a particular point on Earth. In this case, the object's position is described by another method, called the **horizon system** of celestial coordinates. In this system, an object's position (point *P* in fig. 16.14) is given in terms of its **altitude** in degrees above the observer's horizon and its **azimuth** in degrees east of the *celestial meridian.* The latter is a north-south longitude line on the celestial sphere that passes through the *zenith,* the point that lies directly above the observer. The point on the celestial meridian opposite the zenith is called the *nadir.* Horizon system coordinates have relevance only to the particular location on Earth's surface where they are made, and at the time and date when

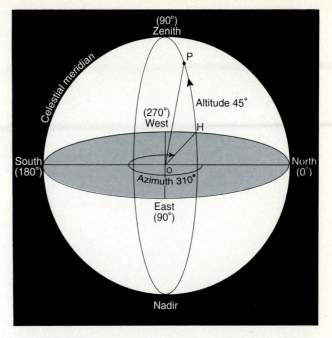

**Figure 16.14** The horizon system of celestial coordinates locates celestial objects relative to the observer's local position (*O*). Here, the point *P* is located at an altitude of 45° above the point *H* on the observer's horizon, which has an azimuth of 310° east of north. The zenith lies directly above, and the nadir, directly below, the observer. The celestial meridian is a north-south line on the celestial sphere that passes through the zenith and the nadir.

they are made. They are used mainly to advise astronomers or navigators where to look in the sky at a particular time and date to observe some particular heavenly object or phenomenon. Equatorial system coordinates of stars, on the other hand, are permanent and fixed on the celestial sphere.

## The Motions of Earth and Moon

In chapters 12 and 13, I discuss the motions of the Earth-Moon system in terms of their effects on the tides and seasons (see figs. 12.8, 12.10, and 13.13). Figure 16.15 shows the geometric relations among Earth, Moon, and Sun in somewhat greater detail.

One of the great regularities about the Earth-Moon system—and indeed of the entire Solar System—is that with few exceptions, everything turns counterclockwise, as viewed from above the North Pole. Sun, Earth, and Moon all rotate counterclockwise, and both Earth and Moon follow counterclockwise orbits, the Moon completing one slow rotation for each revolution, always keeping the same face turned toward Earth as it does so.

Earth's orbit defines the plane of the ecliptic, the fundamental reference plane in astronomy. Along that orbit, Earth approaches the Sun most closely at perihelion, on 3 January (fig. 16.15 *right*), and is most distant

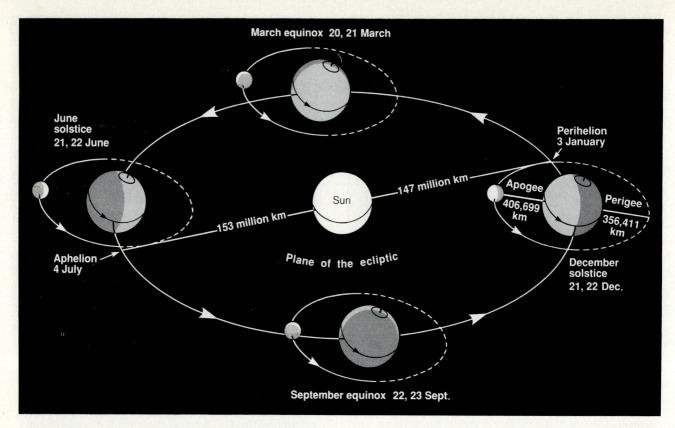

*Figure 16.15* The motions of Earth and Moon with respect to the Sun form the basis for our perception of time.

at aphelion, on 4 July (fig. 16.15 *left*). The perihelion and aphelion points are not fixed on Earth's orbit, however, and they are also not fixed with respect to the solstices and equinoxes. Because of gravitational tugs on Earth by the other planets, these points shift gradually around the orbit, completing one counterclockwise revolution every 108,000 years. At the same time, the solstices and equinoxes shift *clockwise* along Earth's orbit by about 50 seconds of arc per year, a phenomenon called **precession of the equinoxes.** That phenomenon is caused by the Moon and Sun, which exert gravitational pulls on Earth's equatorial bulge, causing the rotational axis to execute a slow, clockwise, conical revolution that requires 25,800 years to complete (fig. 16.16*a*). This revolution is modified by a slight wobble in the rotation axis, due to varying amounts of gravitational pull exerted on Earth's equatorial bulge by the Moon and Sun as they constantly change positions relative to Earth. Because of precession, the position of the North Star is filled by five different stars in turn. At present, the North Star is Polaris, but 5000 years ago, it was Alpha Draconis, and in another 12,000 years, it will be Vega (fig. 16.16*a*).

Another effect of the precession of the equinoxes is evident in figure 16.12. Notice that the names of the constellations in the upper half of the band of the zodiac are shifted by 30° to the east of their counterparts in parentheses in the lower half of the band. The names in parentheses represent the positions of these constellations 2200 years ago. Since then, precession of the equinoxes has shifted them into the positions indicated in the upper half of the band.

The Moon revolves in an orbit that is almost four times as eccentric as Earth's. At **perigee** (Greek for "near Earth"), it is 356,411 km from Earth, and at **apogee** ("far from Earth"), it is 406,699 km distant (fig. 16.15, *right*). The plane of the Moon's orbit does not quite coincide with the plane of the ecliptic but is inclined to it at an angle of 5°09'. In figure 16.15, the Moon's orbit is shown with a solid line where it lies above the plane of the ecliptic and with a dashed line where it lies below it. The gravitational pull of the Sun on the Moon causes the Moon's orbital plane to rotate clockwise, making one complete revolution every 18.6 years. Because of this, the Moon's declination on the solstices can be as much as 28°36' or as little as 18°18' above or below the celestial equator, depending on whether the inclination of the Moon's orbit is added to or subtracted from the inclination of Earth's equatorial plane. In figure 16.15, the declination is shown at its minimum value. If the Moon's equatorial plane were inclined the other way (as it is in fig. 16.16*a*), the declination would be at its maximum.

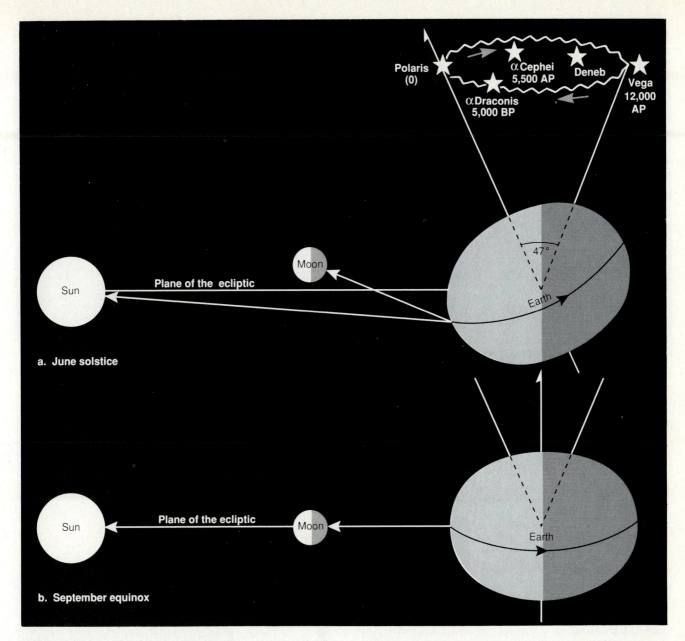

**Polaris (0)** α **Cephei 5,500 AP** **Deneb** **Vega 12,000 AP** α **Draconis 5,000 BP**

47°

Earth

**Plane of the ecliptic**

Sun

Moon

**a. June solstice**

**Plane of the ecliptic**

Sun

Moon

Earth

**b. September equinox**

***Figure 16.16*** Gravitational attractions of the Sun and Moon act on Earth's equatorial bulge (here greatly exaggerated), causing the planet's rotational axis to describe a conical path of precession with an internal angle twice that of its axial inclination. Under the conditions shown in *a* (June solstice with the Moon at maximum declination), the effect is at its highest intensity. In *b* (September equinox with the Moon on the ecliptic), the effect is zero. Variations in the effect cause a slight wobble in the precessional motion.

At the right of figure 16.15, the Moon is aligned between the Sun and Earth (the fact that the alignment is occurring at the December solstice is a coincidence). This condition is known as **conjunction** (see also fig. 16.17*a*). The only other possible alignment, called **opposition,** occurs when Earth is located between the Moon and the Sun, as at the June solstice in figure 16.15 (left) and in figure 16.17*c*. Halfway between conjunction and opposition, the three bodies form a right angle, as they happen to do at the March and September equinoxes in figure 16.15. This condition is known as **quadrature** (see also fig. 16.17*b* and *d*).

Notice the appearance of the Moon's disk in relation to conjunction, quadrature, and opposition in figure 16.17. These are called the **phases of the Moon.** When the Moon lies between Earth and the Sun at conjunction (fig. 16.17*a*), its illuminated side faces away from us, so we see at most a thin crescent, called a *new Moon.* At this phase, the Moon and the Sun rise and set together. As the Moon moves into quadrature (fig. 16.17*b*), the right half of its disk appears illuminated. This phase is called the *first quarter Moon* because the Moon is a quarter of the way through its orbital cycle, and it rises and sets about six hours later than the Sun. At opposition, when Earth lies between the Moon and the Sun

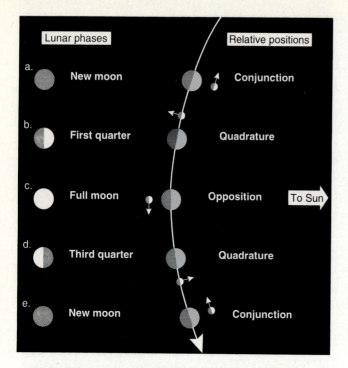

**Figure 16.17** The phases of the Moon in relation to conjunction (*a, e*), quadrature (*b, d*), and opposition (*c*).

(fig. 16.17*c*), we see the full, illuminated disk, which we designate the *full Moon.* Now the Moon rises as the Sun sets, and vice versa. At the following quadrature (fig. 16.17*d*), the left half of the Moon's disk appears illuminated, and we call this phase the *third quarter Moon.* Here, the Moon rises and sets about six hours earlier than the Sun.

A **solar eclipse** can occur during conjunction if the three bodies are all *perfectly* aligned so that the Moon's shadow can fall on Earth's surface (fig. 16.18). Similarly, a **lunar eclipse** can occur during opposition under the same limiting conditions (fig. 16.19). In both types of eclipse, an outward-widening cone of partial shadow, called a **penumbra** (Latin for "almost a shadow"), encloses an outward-narrowing cone of total shadow called an **umbra** (figs. 16.20 and 16.21). These two zones result from the crossing of light rays emanating from the Sun, of which only those from the Sun's rim are shown in the two figures.

Lunar eclipses are both commoner and more visible worldwide than solar eclipses because Earth's shadow is much larger in relation to the Moon than the Moon's shadow is in relation to Earth. Because the Moon is so much smaller than Earth, its shadow only makes a small dot on Earth's surface, if it reaches Earth at all. At perihelion (fig. 16.21*b*), the umbra cone is shorter than at aphelion (fig. 16.21*a*). If apogee happens to coincide with perihelion, as in figure 16.20*b,* the umbra does not reach Earth. When perigee and aphelion happen to coincide, however, as in figure 16.21*a,* the umbra cone

**Figure 16.18** A total solar eclipse photographed on 26 February 1979.

Photo by Hansen Planetarium

**Figure 16.19** A lunar eclipse.

Photo by W. A. Banaszewski

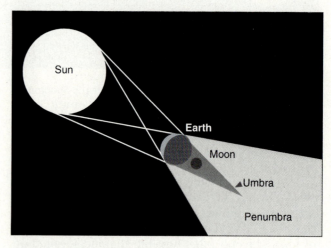

**Figure 16.20** Earth's shadow is broad enough to immerse the Moon if it is aligned with Earth and Sun during opposition.

does reach Earth and can produce a dot of total darkness with a maximum diameter of 270 km. Total solar eclipses can only be viewed within the narrow path that the dot follows as it crosses the terrestrial disk.

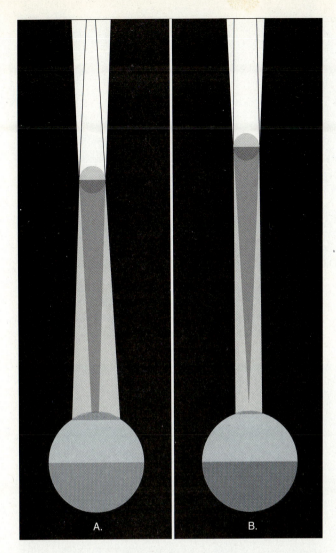

*Figure 16.21* The Moon's shadow is never large enough to immerse Earth totally. At aphelion and perigee (*a*: Earth farthest from Sun, Moon closest to Earth), conditions are most favorable for total solar eclipses, which occur along narrow paths, but at perihelion and apogee (*b*: Earth closest to Sun, Moon farthest from Earth), the umbra cone is too short to reach Earth.

## The Celestial Basis of Time

The various motions of Earth, Moon, and Sun form the basis of our measurement of the passage of **time,** the fourth dimension of existence that produces changes in systems. All such changes require *duration* in which to happen, and by measuring the durations of certain changes, we mark the progress of time. To be useful, those durations must have a regular periodicity, or *recurrence interval* (see chapter 19). As Earth-dwellers, we have understandably settled on the durations of three natural rhythms of Earth, Moon, and Sun: the day, the month, and the year, as the most convenient standards for the measurement of time.

The **mean solar day** is a rhythm whose recurrence interval is the average time (the rhythm varies slightly) that it takes for Earth to make one complete rotation about its axis with respect to the Sun (fig. 16.22; recall that rotation means turning on an axis). For convenience, we have further subdivided this basic time unit into 24 *mean solar hours,* each consisting of 60 *mean solar minutes,* each consisting in turn of 60 *mean solar seconds.* There is nothing sacred about these subdivisions beyond the fact that both 24 and 60 are evenly divisible by the numbers 2, 3, 4, 6, and 12.

Careful studies have revealed that the rate of Earth's rotation varies slightly due mainly to the effects of tidal friction, intermittent core/mantle interactions, and storm winds. Because of these factors, and other less influential ones, the length of a solar day varies by about 50 seconds throughout the year. This variation is taken into account by averaging day length to yield the standard value of the mean solar day. The hour, then, is defined as $\frac{1}{24}$, the minute as $\frac{1}{1440}$, and the second as $\frac{1}{86,400}$ of a mean solar day. Because of its inherent variability, the second was redefined in 1967 as 9,192,631,770 cycles of cesium-133 radiation, a nonvarying and easily observable rhythm.

The **synodic month** is a rhythm whose recurrence interval is the time it takes for the Moon to make one complete revolution around Earth with respect to the Sun (recall that revolution means following an orbit). Synodic means "coming together," which is what the Moon and Sun appear to do once every 29.53 mean solar days, on average (fig. 16.23). The 7-day week is based on the changing phases of the Moon, each of which requires one-fourth of a synodic month, or 7.38 days, to complete.

Finally, the recurrence interval of the **tropical year** is the time required for Earth to make one complete revolution in its orbit with respect to the Sun (see fig. 16.15). Its average value is 365.242 mean solar days.

## Solar and Sidereal Time

At first, it might seem that a day would be simply the time required for Earth to make one complete rotation on its axis, but the situation is not quite that simple. A glance at figure 16.22 shows why. Because Earth is constantly moving *past* the Sun as it revolves in its orbit, it must rotate not 360°, but 360.9856° before it can complete one full turn *with respect to the Sun.* There are, in other words, two slightly different kinds of day: the **sidereal day,** or the time required for Earth to make one complete revolution with respect to the fixed stars (23 hr, 56 min, 4 sec), and the *solar day,* or the time required for one complete revolution of Earth with respect to the Sun (24 hr).

Just as there is a sidereal day corresponding with the solar day, there is also a sidereal month and a sidereal year corresponding with the synodic month and

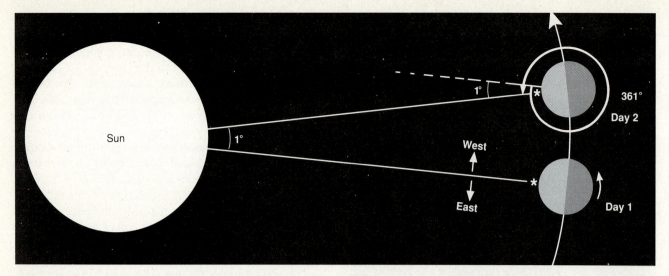

**Figure 16.22** One mean solar day is the time required for Earth to make one complete rotation with respect to the Sun. This is about 361°, and not the expected 360°, because of Earth's orbital motion. A 360° turn would bring a given point on Earth around as far as the dashed line. An additional turn of about 1° (actually 0.9856°) is required to bring the point directly in line with the Sun. Angles are greatly exaggerated here for clarity.

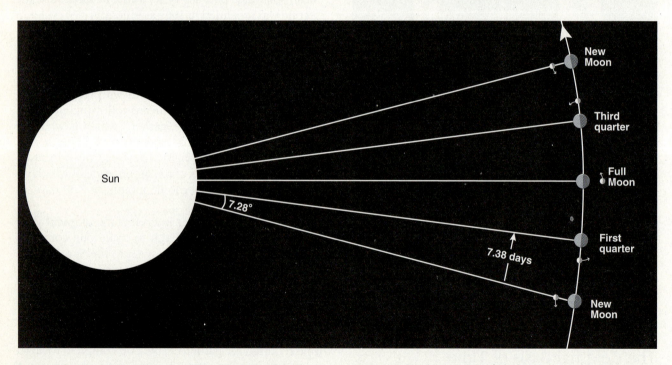

**Figure 16.23** One synodic month is the elapsed time between one new Moon and the next. At the new Moon (*bottom*), Earth, Moon, and Sun are all aligned; 29.53 days later (*top*), they are again aligned. The intervening phases of the Moon are separated from one another by one-quarter of this interval: 7.38 days, or approximately one week, during which time Earth moves through an angle of 7.28° along its orbit. Angles are not exaggerated.

the tropical year, respectively. The **sidereal month,** which averages 27.32 mean solar days, is the time required for the Moon to make one complete orbital revolution *with respect to the fixed stars,* in other words, to return to the same celestial coordinates. The reason for this difference of a little more than two days between the sidereal and synodic months is shown in figure 16.24.

Starting with its new phase at conjunction on day 0, the Moon executes a single revolution around Earth with respect to a distant star in 27.32 days (point *B* in fig. 16.24). In this same time interval, Earth has revolved through an angle of about 27°, so the Moon must continue to revolve through this same angle before it is once again in conjunction with the Sun (point *C* in fig. 16.24). This requires an additional 2.21 days.

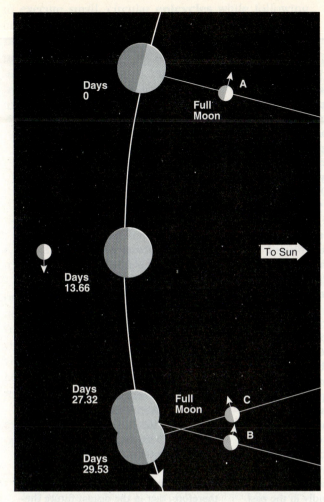

**Figure 16.24** The difference between a sidereal month and a synodic month. The interval between two successive conjunctions of the Moon with a given star (*A–B*) is 27.32 mean solar days, or one sidereal month. The interval between two successive conjunctions of the Moon with the Sun, however, is 29.53 mean solar days, or one synodic month (*A–C*). All angular relationships and relative sizes of Earth and Moon are to scale. Note that the Earth-Moon barycenter, not Earth's center, lies on the orbit.

You can easily verify all this by observing the Moon's motions and timing them with a wristwatch. You will find that, on average, the Moon rises 48 min 46 sec later each night than it did the night before, corresponding to an eastward shift with respect to the Sun of 12.19° in one mean solar day (15° of arc correspond to one hour; 360° to one day). Dividing 360° by 12.19° gives 29.53 days, the mean length of the synodic month. You will also find that in one mean solar day the Moon rises 52 min 42 sec later than a given star each night, resulting in an eastward shift with respect to the fixed stars of 13.18° in one mean solar day. Dividing 360° by this amount gives 27.32, the mean length of the sidereal month.

Earth completes one full revolution from one March equinox to the next in one mean *tropical year* of 365.242 days (the value varies slightly due to perturbations of

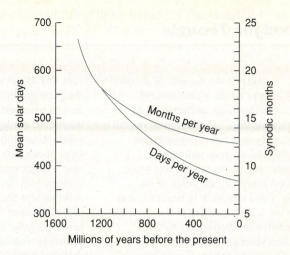

**Figure 16.25** The number of mean solar days per year (*lower curve*) and the number of synodic months per year (*upper curve*) have both decreased at decreasing rates through geologic time (note different scales).

Earth's orbit by the other planets). With respect to the fixed stars, however, Earth completes one full revolution in one **sidereal year** of 365.256 days. Because of the precession of the equinoxes, the sidereal year is 0.014 solar day longer than the tropical year.

## Calendars

It would be most convenient for us if one Earth revolution corresponded to exactly 360 mean solar days, if one synodic month were exactly 30 mean solar days long, and if the phases of the Moon lasted exactly seven days (six would be even better). Unfortunately, however, the celestial motions of Earth, Moon, and Sun were not established with human needs in view. To make matters still worse, it is now clear that all these rhythms are gradually slowing down, mainly because of gravitational forces (fig. 16.25). Nonetheless, imprecise as they are, these three natural periodicities of the Solar System are still the best means we have to measure the frequencies and the rates at which all other phenomena occur.

As we have seen, both the tropical and sidereal years have a small fraction of a day over 365, and neither is evenly divisible by the 29.53 days in a synodic month. Because of this, our calendar months vary arbitrarily in duration from 28 to 31 days. This awkward situation was addressed in 46 B.C. by Julius Caesar, whose *Julian calendar* added a day to every fourth year, thereby instituting the familiar tradition of **leap years.** In assuming a year of 365.250 days' duration, the Julian calendar generated an annual excess of 11 min 30 sec. By the sixteenth century, this excess built up to about a 10 day lead. In 1582, the astronomer Clavius persuaded Pope Gregory to reform the Julian calendar. Accordingly, the

## Key Terms

zodiac  
constellation  
deferent  
epicycle  
geocentrism  
heliocentrism  
Kopernik  
Brahe  
Kepler  
astronomical unit  
Galileo  
Newton  
refracting telescope  
refraction  
focus  
focal length  
reflecting telescope  
spectrograph  
radio telescope  
equatorial system  
declination  
right ascension  
horizon system  

altitude  
azimuth  
precession of the  
    equinoxes  
perigee  
apogee  
conjunction  
opposition  
quadrature  
phases of the Moon  
solar eclipse  
lunar eclipse  
penumbra  
umbra  
time  
mean solar day  
synodic month  
tropical year  
sidereal day  
sidereal month  
sidereal year  
leap year  
Gregorian calendar  

## Questions for Review

1. If you could view the motions of the planets from the positions of Earth, Moon, and Sun, from which vantage would they appear most complex, least complex, and intermediate in complexity?

2. In the light of Kepler's second law, should day length be longer or shorter at perihelion than at aphelion? Why?

3. Compare and contrast the forms and functions of refracting and reflecting telescopes.

4. Describe three examples from Nature that illustrate each of Newton's three laws of motion.

5. Explain why the same celestial object will have different coordinates in the equatorial and horizon systems of celestial coordinates.

6. At what season does the Moon appear highest in the sky? Under what conditions does it reach its highest altitude? How often does it do so?

7. What lunar phase occurs, and what kind of eclipse can occur, at conjunction, at opposition, and at quadrature?

8. Which occur more frequently, solar eclipses or lunar eclipses? Why? When it occurs, which is more likely to be seen? Why?

9. Explain the differences between a sidereal day and a solar day, a sidereal month and a synodic month, and a sidereal year and a tropical year.

10. Describe and explain the long-term trends in the rotation rates of Earth and Moon.

# Earth's Context: The Solar System

# *17*

## *Outline*

*That very law which moulds a tear*

*And bids it trickle from its source,—*

*That law preserves the Earth a sphere,*

*And guides the planets in their course.*

*Samuel Rogers*

***On a Tear***

Photo: A full-disk view of Saturn and four of its larger satellites from the space probe *Voyager 2.*

## Our Average Star: The Sun

Early astronomers had little reason to suppose that there was anything in common between the Sun and the stars. The Sun is apparently a much larger and much brighter object, and whereas other stars appear to maintain fixed positions in the heavens relative to one another, the Sun does not. To date, no one has ever actually seen another star in the way that we can see the Sun, because even the nearest star, Alpha Centauri, is much too distant for even the largest reflecting telescopes to be able to resolve the features of its disk. What we see is the light from the stars, hopelessly distorted by its passage through Earth's shimmering atmosphere. Proof that the Sun is a star came with the development of spectroscopy, which is discussed in chapter 18. For now, I ask you to accept that identity without proof while I use the Sun as a shining example of an average star, and one that is conveniently close and observable.

### Vital Statistics of the Sun

The Sun is by far the largest and brightest object in the Solar System. Its 1.991 million billion billion billion $(1.991 \times 10^{30})$ g of mass compose 99.9% of the total mass of the Solar System, excluding comets. The Sun has a density of 1.41 g/cm³, 41% denser than water and about one-fourth the density of Earth. The Sun's average diameter is 1,391,900 km, 109 times that of Earth, and 10 times that of Jupiter, the largest planet. Unlike most of the planets, which have been flattened to some degree at their poles due to rotation, the Sun is almost perfectly spherical, yet it does rotate slowly, as attested by the passage of sunspots across the Sun's face. The Sun's rotation is unique, however, in that it is most rapid at the equator and diminishes toward the poles. The rotation period is approximately 25 days at the equator, increasing to about 35 days in the polar regions. The reasons for this uneven rotation are not clear.

Studies of the electromagnetic energy emitted by the Sun have revealed much information about our local star. Because most of the Sun's energy is radiated in the visible part of the spectrum (see fig. 13.11), we know that the temperature of the Sun's surface is about 5500° C, but this seemingly extreme temperature is a deep freeze in comparison with the temperatures of the Sun's interior. From theoretical considerations, astrophysicists have been able to place the Sun's core temperature at a blistering 14,000,000° C.

The presence of 63 chemical elements has been revealed by spectrographic analysis of the Sun's atmosphere. Because even the infernal temperatures of the Sun's core are still much too low to have created any of these elements except helium, we know that the Sun is made of recycled materials. The primordial nebula from which the Sun and planets were formed some 5 billion years ago must have been derived from the explosions of far hotter, far more massive stars. Most of the Sun's high temperature elements, however, are present only in trace amounts. The bulk of the Sun's composition is hydrogen (81.76% by volume) and helium (18.17% by volume). Nevertheless, this residue of heavier elements (0.07% by volume) constitutes a mass of more than 50 times that of all the planets combined.

Finally, the Sun is a prodigious producer of radiant energy, generating about $3.84 \times 10^{33}$ ergs per second, or about five million times the amount of energy consumed by the United States in a year (1985). Let us turn our attention now to the source of this abundant energy.

### Nature's Nuclear Reactor: The Sun's Core

Before the discovery of radioactivity by A. H. Becquerel in 1896, it was thought that the source of the Sun's heat was the conversion of potential energy to heat energy by slow gravitational collapse. The German physicist Hermann von Helmholtz (1821–1894), calculated that a radial contraction of 85 m/yr would be sufficient to account for the Sun's observed energy output. Although we now know that gravitational contraction is not necessary to explain the Sun's present energy output, it was almost certainly the cause of the Sun's initial heating while it was still forming within the primordial solar nebula.

As the Sun's core temperature rose during gravitational collapse, the mean velocity of protons (hydrogen nuclei) in the core increased until, above about 4 million°C, their mutually repulsive positive charges could no longer shield them against mutual collisions. At this point, protons began to combine with one another in a process called the **proton-proton chain,** the simplest example of a **thermonuclear reaction** (fig. 17.1). Inputs to this equilibrium system are high energy protons ($_1^1$H) and electrons ($e^-$), and outputs are protons, normal helium ($_2^4$He), photons of short wave electromagnetic energy called **gamma rays** ($\gamma$), and massless, uncharged neutrinos ($\nu$). The chain reaction begins when two protons collide, yielding a neutrino, a nucleus of the heavy hydrogen isotope *deuterium* ($_1^2$H), and a positively charged electron called a *positron* ($e^+$). The positron soon collides with an electron, annihilating both particles. The neutrino, having neither charge nor mass, flies away from the reaction into outer space totally unaffected by the crush of surrounding matter through which it must pass. The deuterium nucleus soon collides with a third proton, yielding a light helium isotope called helium-3 ($_2$He³) and a gamma ray. The process ends when two helium-3 nuclei collide to form a normal helium nucleus and two protons.

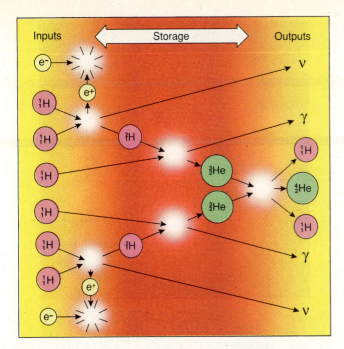

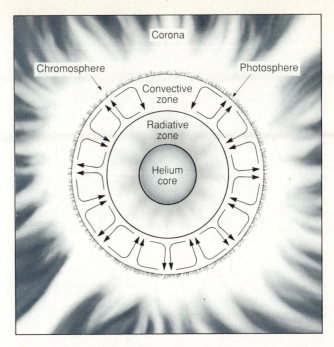

**Figure 17.1** In this diagram of the proton-proton chain, inputs are shown at left, outputs at right, and intermediate products in storage in the center. Six protons ($^1_1H$) combine with two electrons ($e^-$) to form a helium ($^4_2He$) nucleus and two protons, releasing two massless, chargeless neutrinos ($\nu$), and two gamma rays ($\gamma$), the source of the Sun's energy.

The release of gamma rays during this process is the source of most of the Sun's energy. Another chain reaction not described here, the *carbon-nitrogen chain,* also occurs in the Sun's core at higher temperatures. This reaction also converts protons into helium and gamma rays, but it uses carbon atoms as proton carriers. Both these chain reactions involve the conversion of mass into energy according to a simple but consequential equation developed by a Swiss patent officer named Albert Einstein (1879–1955). That equation is

$$E = mc^2,$$

where $E$ is energy produced in Joules (1J $= 10^7$ ergs), $m$ is mass in kg, and $c$ is the speed of light in m/sec. Because $c$ is an extremely large number to begin with, when squared it makes $E$ truly gigantic. In the proton-proton chain, out of each 140 kg of protons consumed in the making of helium nuclei, 1 kg is converted to gamma ray energy. If we substitute this value in the equation, we find that the energy released is (1 kg) $\times$ ($3 \times 10^8$ m/sec)$^2 = 9 \times 10^{16}$ J, a quantity of energy sufficient to satisfy all of humanity's present energy needs for about 2.7 hr.

It might come as a surprise to learn that the initial step of proton collision in the proton-proton chain is a highly improbable event, occurring on the average only once every 7000 million years or so *for a given proton.* When you consider that there are about $8 \times 10^{54}$ protons in the Sun, however, this becomes somewhat less

**Figure 17.2** The structure of the Sun and its atmosphere.

surprising. Dividing the latter number by the former yields about $1.14 \times 10^{45}$ proton collisions per year, or about $3.6 \times 10^{37}$ collisions per second! If a given proton were to expect a collision more often than once every 7000 million years, the Sun would quickly explode.

The thermonuclear reactions in the Sun's core not only serve as energy sources, but they also prevent the Sun from collapsing into gravitational oblivion. It is only the ceaseless outpouring of energy from the Sun's core that supports the enormous mass of the overlying gases against the pull of gravity. As its hydrogen fuel is used up, however, and a core of nonreactive helium "ash" grows in its place, the Sun will eventually shrink to a white-hot, Mercury-sized sphere of inconceivable density after passing through two brief episodes of enormous expansion (see chapter 18). Fortunately, this is not scheduled to happen for at least another 5000 million years.

Overlying the Sun's core is a thick "mantle" of dense, hot, nonreacting gases that makes up the bulk of the Sun (fig. 17.2). The lower, thicker part of this mantle—the **radiative zone**—is thought to consist of relatively rigid matter. In the upper, somewhat thinner part—the **convective zone**—convection is thought to occur. A dense plasma of dismembered atoms within the radiative zone absorbs and contains much of the gamma radiation generated in the outer core, reradiating energy of longer wavelengths to the convective zone. Here, in less densely packed plasma, convection probably joins this reradiation in transferring energy outward toward the Sun's surface.

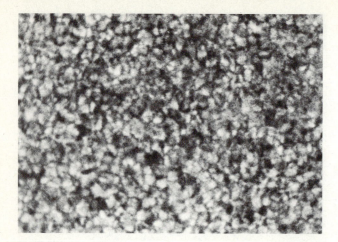

**Figure 17.3**  An image of the Sun's granular photosphere. Hot, rising gas brightens the centers of the closely-packed granules, while cooler, descending gas darkens their edges.

Photo by Hale Observatories

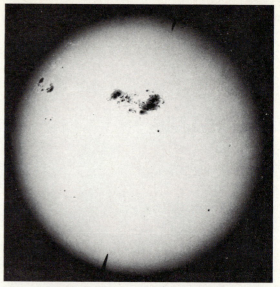

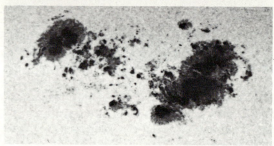

**Figure 17.5**  Sunspots are powerful magnetic storms on the Sun's surface that typically form in pairs of opposite polarity. The dark, cooler, central umbra is surrounded by the lighter, hotter penumbra. This general view (*top*) and close-up (*bottom*), show an unusually large group, photographed April 1947.

Photos by Hale Observatories

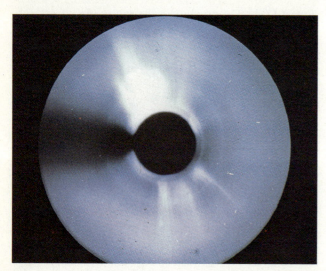

**Figure 17.4**  The solar corona as viewed in ultraviolet light by Skylab's Solar Telescope in 1973. The bright region in the upper left of the corona consists of helium ionized by the eruption of a solar prominence.

Photo by NASA

Above the upper mantle lies the **photosphere** (the "light sphere"), the surface layer of the Sun, equivalent in position to Earth's lithosphere. Here the equivalence ends, however, for the photosphere is a boiling, convective layer of rarefied gases some 300 km thick with a density of only about 0.01 Earth atmospheres. The most characteristic and abundant feature of the photosphere is the **granule,** a small convection cell up to a few hundred km in diameter (fig. 17.3). Granules are delineated by cooler, darker gas on their descending margins. They lie closely packed like the seeds in a sunflower, and they change shape constantly and rapidly. Often, their hotter centers eject fountains of glowing gas into the overlying **chromosphere** (the "color sphere"), the lower layer of the Sun's atmosphere, imparting to it a reddish color. Beyond the chromosphere, beginning at an altitude of about 6500 km above the photosphere, lies the Sun's outer atmosphere, a broad but tenuous zone of glowing, rarefied gas called the **corona** (fig. 17.4), whose outer limits extend as far as the orbit of Mercury.

**Sunspots** (figs. 17.5 and 17.7) are immense magnetic storms that normally exhibit a cool, dark center, called the *umbra,* surrounded by a brighter, mottled halo called the *penumbra* (not to be confused with the shadows of Earth and Moon). Often, large, loop-shaped **solar prominences** are seen arching up to 800,000 km above sunspots, following magnetic lines of force where they emerge from the Sun (fig. 17.6). Sometimes, enormous explosions, known as **solar flares,** erupt from the

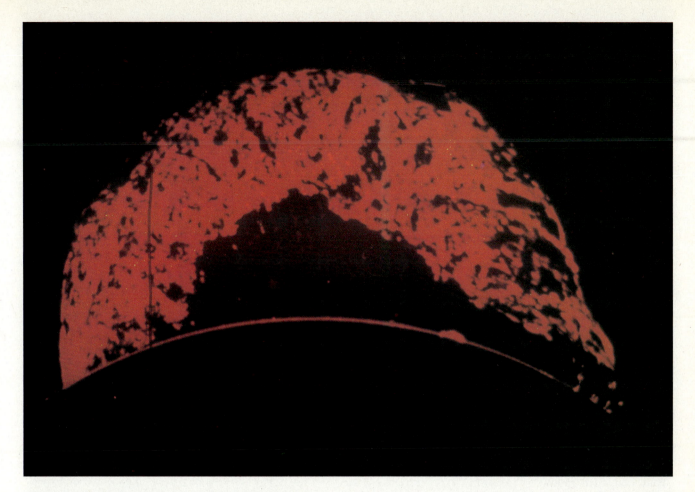

*Figure 17.6*    Great solar prominences can soar up to
800,000 km above the surface of the Sun, usually from
the vicinity of sunspots. Rich in heavier elements, these
brilliant scarlet plumes must originate from the deeper
layers of the Sun's atmosphere.

Photo by NOAA

vicinity of sunspots and send out showers of ionized
particles that result in auroras and magnetic storms on
Earth (see chapter 13).

It has long been noticed that there is a roughly 11-
year cycle in sunspot activity. At the start of each cycle,
sunspots form in pairs with opposite magnetic polarities in the middle latitudes of both hemispheres. Individual sunspot pairs tend to persist for hours to months.
During the course of the 11-year cycle, the two opposed
regions of sunspot activity migrate slowly eastward and
toward the solar equator (fig. 17.7). This cycle is thought
to result from the Sun's peculiar pattern of rotation,
which wraps the lines of its magnetic force field around
the equator like a tightening watch spring. When this
process brings the lines close enough together that their
mutual repulsion overcomes the force of the Sun's
gravity, they erupt into sunspots, disrupting the field. The
field then reforms, and the cycle repeats.

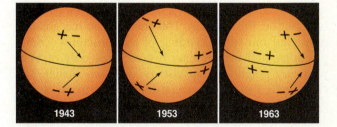

*Figure 17.7*    In a 20-year cycle from 1943 to 1963,
three sets of sunspots—four in each set—formed at high
latitudes and migrated eastward and toward the Sun's
equator. Note that sunspots form in pairs with opposite
polarities and that corresponding pairs form in the
opposite hemisphere with their individual polarities
reversed.

**Table 17.1**  Characteristics of the Sun, Planets, Moons, and Asteroids

| Body | Mean Orbital Radius (10^6 km) | Eccentricity of Orbit[a] | Inclination of Orbit[b] (degrees) | Inclination of Equator[c] (degrees) | Mean Diameter (km) | Mass (Earth = 1) | Density (g/cm³) | Sidereal Period of Revolution[d] |
|---|---|---|---|---|---|---|---|---|
| Sun | — | — | — | — | 1,391,900 | 332,999.00 | 1.41 | — |
| Mercury | 57.95 | 0.21 | 7.0 | 0.0 | 4866 | 0.05 | 5.43 | 88.0 |
| Venus | 108.11 | 0.01 | 3.4 | 177.3[e] | 12,106 | 0.82 | 5.26 | 224.7 |
| Earth | 149.57 | 0.02 | 0.0 | 23.5 | 12,743 | 1.00 | 5.52 | 365.3 |
| Moon | 0.38 | 0.05 | 18.3–28.6 | 6.7 | 3477 | 0.01 | 3.34 | 27.3 |
| Mars | 227.84 | 0.09 | 1.9 | 25.2 | 6760 | 0.11 | 3.91 | 687.0 |
| Asteroids[h] | 396.62 | 0.06–0.83 | Varies | — | 20 | 0.00 | 0.7–8.1 | 986.4 |
| Jupiter | 778.14 | 0.05 | 1.3 | 3.1 | 139,516 | 317.93 | 1.34 | 4332.4 |
| Io | 0.35 | 0.00 | 0.0 | 0.0 | 3638 | 0.01 | 3.53 | 1.8 |
| Europa | 0.60 | 0.00 | 0.5 | 0.0 | 3126 | 0.01 | 3.03 | 3.6 |
| Ganymede | 1.00 | 0.00 | 0.2 | 0.0 | 5276 | 0.01 | 1.93 | 7.2 |
| Callisto | 1.81 | 0.01 | 0.2 | 0.0 | 4848 | 0.01 | 1.79 | 16.7 |
| Saturn | 1427.0 | 0.05 | 2.5 | 26.7 | 58,219 | 95.07 | 0.69 | 10,759.3 |
| Mimas | 0.18 | 0.02 | 1.5 | 0.0 | 390 | 0.00 | 1.2 | 0.9 |
| Enceladus | 0.24 | 0.01 | 0.0 | 0.0 | 510 | 0.00 | 1.2? | 1.4 |
| Tethys | 0.30 | 0.00 | 1.9 | 0.0 | 1050 | 0.00 | 1.2 | 1.9 |
| Dione | 0.38 | 0.00 | 0.0 | 0.0 | 1120 | 0.00 | 1.5 | 2.7 |
| Rhea | 0.53 | 0.00 | 0.4 | 0.0 | 1530 | 0.00 | 1.1? | 4.5 |
| Titan | 1.23 | 0.03 | 0.3 | 0.0 | 5150 | 0.02 | 1.9 | 16.0 |
| Iapetus | 3.57 | 0.03 | 14.7 | 0.0 | 1440 | 0.00 | 1.1 | 79.3 |
| Uranus | 2870.3 | 0.05 | 0.8 | 97.9[e] | 46,940 | 14.52 | 1.27 | 30,684.5 |
| Miranda | 0.13 | 0.01 | 3.4 | 0.0 | 480 | 0.00 | 1.3 | 1.4 |
| Ariel | 0.19 | 0.00 | 0.3 | 0.0 | 1170 | 0.00 | 1.6 | 2.5 |
| Umbriel | 0.27 | 0.00 | 0.4 | 0.0 | 1190 | 0.00 | 1.5 | 4.1 |
| Titania | 0.44 | 0.00 | 0.1 | 0.0 | 1590 | 0.00 | 1.7 | 8.7 |
| Oberon | 0.59 | 0.00 | 0.1 | 0.0 | 1550 | 0.00 | 1.6 | 13.4 |
| Neptune | 4499.9 | 0.01 | 1.78 | 29.6 | 45,432 | 17.18 | 1.64 | 60,188.3 |
| Triton | 0.35 | 0.00 | 160.0[e] | 0.0 | 2720 | 0.00 | 2.03 | 5.9[e] |
| Pluto | 5909 | 0.25 | 17.2 | 1.22 | 2300 | 0.002 | 2.0? | 90,710.1 |
| Charon | 0.02 | ? | 94 | 0.0 | 1186 | 0.00 | 2.0? | 6.4 |

## The Solar Family: The Planets

Since the *Mariner II* space probe was launched on its exploratory journey to Venus on 27 August 1962, the science of *planetology* has amassed an enormous data base on the Solar System. This information has greatly expanded our understanding of the structures, compositions, and evolutionary histories not only of our planetary neighbors but also of our own planet. In chapter 16, I mention several regularities in the Solar System that were discovered by early astronomers, among them Kepler's laws and the eastward, counterclockwise sense in which most planets and moons both spin and revolve. These regularities, and many others, are boundary conditions which must be taken into account by any hypothesis that proposes to explain the origin and evolution of the Solar System. I consider that as-yet-unresolved problem at the end of this chapter, but first, let us embark on a brief Cook's tour of some salient features of the planets and their moons. During this tour, it will be helpful to refer to the planetary and lunar statistics listed in table 17.1, and to figures 17.8 and 17.9, which show the relative sizes of the bodies of the Solar System and the orbits of the planets around the Sun respectively.

## Mercury

The Sun's closest neighbor was first photographed by the *Mariner 10* space probe in March 1974 (fig. 17.10). Mercury (named for the Roman god of commerce) had long been a mystery among planets because its proximity to the Sun (never more than 28° distant) means that it can only be viewed for about an hour before sunrise and an hour after sunset through a great thickness of Earth's atmosphere, which severely distorts its image. It was no great surprise, however, when the *Mariner 10* photographs revealed a planetary surface that strongly resembles the Moon's. Mercury has a basaltic crust that has been liberally peppered with meteorite **impact craters.** In addition to the smaller craters, there is one enormous **multi-ringed basin,** an impact structure 1300 km in diameter called the *Caloris basin* (fig. 17.11). This structure was created by the impact of an asteroid-sized body early in the history of the Solar System. Following the impact, basalt erupted within the basin. The basalt was then radially compressed, possibly due to slow rebound of the crust after the impact. On the *antipode* (opposite point on the planet) of the Caloris basin, the crust has been thrown into chaotic, hilly terrain, probably by shock waves from the Caloris impact that were focussed on the far side of the planet. Our

**Table 17.1**  *continued*

| Body | Sidereal Period of Rotation[d] | Polar Flattening (%)[g] | Force of Gravity (Earth = 1) | Albedo (%) | Surface Temp. Range (°C) | Surface Pressure (atm) | Atmospheric Ingredients (%) | Number of Asteroidal Moons |
|---|---|---|---|---|---|---|---|---|
| Sun | 24.6 | 0.2 | 27.91 | — | 5500 | ? | $H_2(89)He(11)$ | — |
| Mercury | 58.7 | 2.9 | 0.36 | 8 | $\leq -170$ to 430 | 0.0 | — | 0 |
| Venus | 243.9[e] | 0.0 | 0.90 | 76 | 460 | $\geq 90.0$ | $CO_2(96)N_2(3.5)$ | 0 |
| Earth | 1.0[f] | 0.3 | 1.00 | 36 | $-60$ to 50 | 1.0 | $N_2(78)O_2(21)Ar(1)$ | 0 |
| Moon | 27.3 | 0.6 | 0.17 | 9 | $-170$ to 90 | 0.0 | | |
| Mars | 1.0 | 0.5 | 0.38 | 15 | $-100$ to 30 | 5–7 | $CO_2(96)N_2(2.5)Ar(1.5)$ | 2 |
| Asteroids[h] | 0.5 | 0.0 | 0.00 | var. | — | 0.0 | | |
| Jupiter | 0.4 | 6.6 | 2.65 | 54 | ? | $10^4$–$10^6$ | $H_2(90)He(10)$ | 13 |
| Io | 1.8 | 0.0 | 0.18 | 40 | $-213$ to $-153$ | $>0.0$ | $SO_2$, Na (traces) | — |
| Europa | 3.6 | 0.0 | 0.13 | 39 | ? | 0.0 | — | — |
| Ganymede | 7.2 | 0.0 | 0.08 | 20 | ? | 0.0 | — | — |
| Callisto | 16.7 | 0.0 | 0.07 | 3 | ? | 0.0 | — | — |
| Saturn | 0.4 | 10.3 | 1.14 | 57 | ? | $10^2$–$10^4$ | $H_2(94?)He(6?)$ | 10 |
| Mimas | 0.9 | 0.0 | 0.01 | 27 | ? | 0.0 | — | — |
| Enceladus | 1.4 | 0.0 | 0.01 | 67 | ? | 0.0 | — | — |
| Tethys | 1.9 | 0.0 | 0.02 | 59 | ? | 0.0 | — | — |
| Dione | 2.7 | 0.0 | 0.02 | 43 | ? | 0.0 | — | — |
| Rhea | 4.5 | 0.0 | 0.02 | 43 | ? | 0.0 | — | — |
| Titan | 16.0 | 0.0 | 0.14 | 4 | $-180$ | 1.5 | $N_2(88)Ar?(12)$ | — |
| Iapetus | 79.3 | 0.0 | 0.02 | 3–50 | ? | 0.0 | — | — |
| Uranus | 0.7[e] | 7.0 | 1.07 | 65 | | 2–3 | $H_2(88)He(12)CH_4$ | 10 |
| Miranda | 1.4 | 0.0 | 0.01? | 34 | $-193$ | 0.0 | — | — |
| Ariel | 2.5 | 0.0 | 0.03 | 40 | $-193$ | 0.0 | — | — |
| Umbriel | 4.1 | 0.0 | 0.03 | 19 | $-193$ | 0.0 | — | — |
| Titania | 8.7 | 0.0 | 0.04 | 28 | $-193$ | 0.0 | — | — |
| Oberon | 13.4 | 0.0 | 0.04 | 24 | $-193$ | 0.0 | — | — |
| Neptune | 0.7 | $\leq 8.0$ | 1.35 | 68 | ? | 5–10? | $CH_4$, $H_2S$, He(?) | 7 |
| Triton | 5.9 | 0.0 | 0.25? | ? | $-236$ | 0.00001 | $N_2$, $CH_4$(?) | — |
| Pluto | 6.4 | $\leq 15.6$ | 0.06? | 50 | $-230$ to $-215$? | 0.00001 | $CH_4$, CO(?)$N_2$(?) | — |
| Charon | 6.4 | 0.0 | 0.03? | <40 | $-230$ to $-215$? | 0.0 | — | — |

[a](Perihelion − aphelion)/2 × mean orbital radius

[b]To plane of ecliptic or to equator of planet

[c]To orbit

[d]Mean solar days

[e]Retrograde motion

[f]Accurate value 0.9973

[g]$100(e - p)/e$, where $e$ = equatorial radius and $p$ = polar radius

[h]Average

Sources: *CRC Handbook of Chemistry & Physics*, 62d ed., 1982; D.H. Menzel and J. M. Pasachoff, *A Field Guide to the Stars and Planets*, 2d ed., revised; Peterson Field Guide Series #15, Houghton Mifflin Co., 1983; and the Jet Propulsion Laboratory, Pasadena, CA.

***Figure 17.8***  True relative sizes of the Sun and the planets and their satellites (*white*).

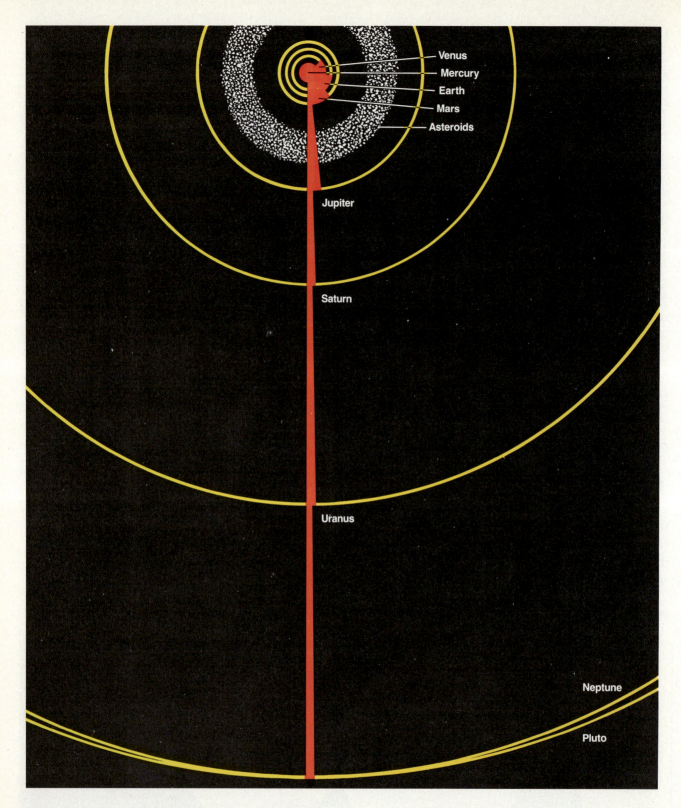

**Venus**

**Mercury**

**Earth**

**Mars**

**Asteroids**

Jupiter

Saturn

Uranus

Neptune

Pluto

*Figure 17.9*    Planetary orbits are nearly circular ellipses
that lie in, or very close to, the plane of Earth's orbit (the
ecliptic). Pluto could be an asteroid, or an escaped moon
of Neptune, hence its anomalous orbit. Colored orbital
sectors indicate the distance each planet travels during
one complete revolution of Mercury, the innermost
planet.

**Figure 17.10**   Mercury, as seen by the Mariner 10 spacecraft on 29 March 1974 from a distance of 234,000 km. The planet's basaltic surface is densely covered with impact craters. Lava flows within the older craters subdue their topographic relief. Bright rays and ejecta blankets splash outward from the younger craters.

Photo by NASA

**Figure 17.11**   Mercury's Caloris basin is a giant, multi-ringed impact structure partially filled with lava, subsequently wrinkled by isostatic uplift, and impacted by smaller, younger, deeper craters. Odin plains display several large, lava-filled, prebasin craters (*upper right*); smaller, more rugged postbasin craters; and ridges in the lava plain that could indicate shrinkage of the planet.

Photo by NASA

probings of the Solar System have revealed that the meteorite peppering event recorded on Mercury was system-wide, affecting planets and moons alike. Henceforth, I shall refer to this event as the *Great Bombardment.*

Mercury's high density (5.4 g/cm³) requires the presence beneath its cratered surface of an iron-nickel core that is considerably larger in relation to the planet than is Earth's core. Earth is actually denser than Mercury, but it is also about 20 times more massive, and is therefore more compressed by its own gravity. Spectrographic analysis suggests that Mercury's crust has been greatly depleted in iron relative to Earth. This in turn suggests that at some time Mercury was heated to a higher temperature than Earth, allowing a greater percentage of its iron to sink from its mantle to form its massive core. Great lobate scarps on Mercury's surface, indicating a general compression of the crust, could be evidence of the shrinking of this large core as it cooled. Another notable feature of the core is a magnetic field, only about 1% as strong as Earth's, but otherwise identical in form (see fig. 1.23). The presence of such a field implies some fluid motion within the core.

A final noteworthy aspect of Mercury is its period of revolution. The planet rotates exactly one and one half times during each revolution around the Sun so that on each return to its perihelion, Mercury presents an opposite face to the Sun. Such locking of a planet's rotational period into synchronism with its period of revolution is called **spin-orbit coupling.** Because of this phenomenon, two antipodal points on Mercury's surface always receive two and one half times more solar radiation than the points that lie 90° to the east and west. At these two subsolar points on Mercury, the Sun appears to stop, hover, and reverse slightly for about eight days at perihelion as the temperature rises to a blistering 425° C. Spin-orbit coupling is also characteristic of all natural moons (i.e., not captured asteroids) in the Solar System, including our own, all of which rotate just once during each revolution around their central planet.

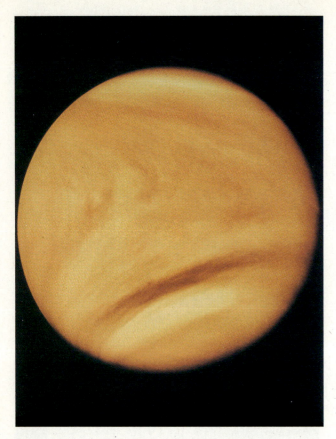

***Figure 17.12*** The atmosphere of Venus sweeps furiously westward in this ultraviolet-sensitive image, taken by the Pioneer-Venus Orbiter on 10 February 1979. The dense clouds are composed of droplets of sulfuric acid. The greenhouse effect of the carbon dioxide atmosphere keeps the planet's surface at a uniform, hellish 460° C. Photo by NASA

## Venus

If there is a misnomer in the Solar System, it is surely the name of the Roman goddess of love applied to the second planet. In its diameter and its density, Venus (fig. 17.12) closely resembles Earth, but in practically all other respects it is radically different, having more the aspects of a hellish inferno than of a planetary paradise. The surface of Venus, at a fairly constant 460° C, is hotter even than the subsolar points on Mercury. The surface pressure of its dense, carbon dioxide atmosphere is a crushing 90 bars or more, equivalent to a depth of 1 km in Earth's ocean. The thick clouds that hide the searing surface of Venus are composed of droplets of sulfuric acid, one of the most corrosive chemicals known.

Another peculiar attribute of this contrary planet is its slow, retrograde, westward rotation with a period of 243 Earth days. It has been suggested that its dense atmosphere, rotating westward about 60 times faster than Venus itself, has gradually reversed the planet's original eastward rotation by frictional drag. Another possibility

is that Earth's gravitational pull has locked Venus into its unusual rotational style. This is supported by the observation that Venus completes precisely four rotations between successive times of closest approach to Earth and therefore always presents the same face toward Earth at those times. A third possibility is that the planet has somehow been turned upside down, a far-fetched idea, but one that must be considered because the rotational axis of Uranus has clearly been turned through an angle of 98°. Venus lacks a magnetic field, probably because of its slow rotation rate rather than, as some have suggested, because it lacks a liquid iron core.

Aside from its density, the atmosphere of Venus is noteworthy in several other respects. In addition to being 96% carbon dioxide, it also contains 3.5% nitrogen and trace amounts of water, sulfuric acid, argon, and oxygen. All of these gases are characteristic of volcanic exhalations, which can be presumed to be their source. It is possible that Venus once had as much water as Earth, but life probably never developed on Venus; therefore, its atmosphere would not have contained enough oxygen to create an ozone layer or its byproducts, a tropopause and a stratosphere. In chapter 13, I consider the possibility that in the absence of a stratosphere, thermal convection might have lofted the planet's water to altitudes at which it was gradually lost to photodissociation. Figure 17.13 compares the vertical temperature profile of Earth's atmosphere, which has a low-level inversion due to ozone, with the atmospheric temperature profiles of Venus and Mars, which have no such inversion, and no water.

Venus is the Solar System's classic example of the greenhouse effect. Even though it is much closer to the Sun, the surface of Venus actually receives less solar radiation than Earth's surface receives due to the density and high albedo of its atmosphere. Nonetheless, because of its insulating blanket of carbon dioxide, a much greater quantity of solar heat is held in storage in the atmosphere of Venus than is held in Earth's atmosphere before being reradiated to space. (In terms of the bathtub model in chapter 2, the atmosphere of Venus is a "bigger bathtub" than Earth's atmosphere.) Under this insulating layer, Venus lacks a thermal gradient from equator to poles, and its infernal temperature of 460° C prevails uniformly over the planet's entire surface. Another contrast with Earth is that the rotation rate of Venus is too slow to produce a Coriolis effect. Hadley cell circulation on Venus (see chapter 15) therefore extends all the way from equator to poles rather than ending in geostrophic flow in the middle latitudes, as on Earth.

Information generated by the Soviet *Venera* landing probes, the United States *Pioneer Venus* orbiting probes, and the giant radio telescope at Arecibo, Puerto Rico, has greatly increased our knowledge of the veiled surface of Venus. The relief map of figure 17.14 shows the distribution of highlands and lowlands on the planetary surface, revealing a predominance of flat, featureless

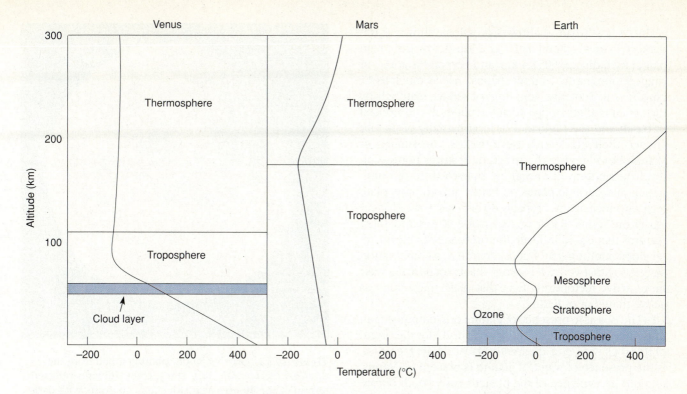

**Figure 17.13** In the tropospheres of Venus, Mars, and Earth, temperature drops with altitude, and convection occurs readily. Earth is unique in having an exceedingly low troposphere overlain by a stratosphere in which the absorption of solar energy by ozone creates a pronounced temperature inversion. As a result, water vapor (*blue*) seldom rises to altitudes at which it can be photodissociated.

From "The Atmosphere of Venus" by G. Schubert and C. Covey. Copyright © 1981 by SCIENTIFIC AMERICAN, Inc. All rights reserved.

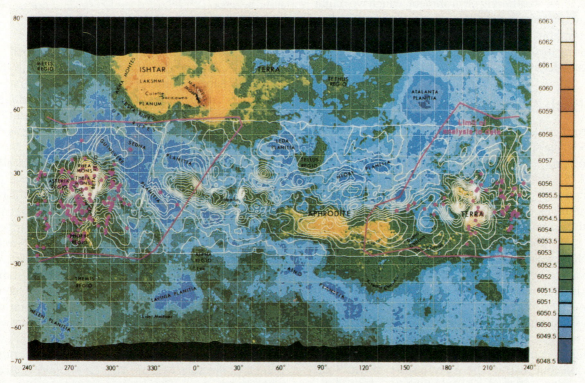

**Figure 17.14** The surface of Venus, revealed beneath its deadly clouds in this radar image by the *Pioneer Venus* orbiter, shows widespread lowlands (*blue and green*) interrupted by two large and several small highlands (*yellow to red and white*). Circular structures, such as Artemis Chasma (*lower right*) could be impact basins. Some highlands, such as Beta Regio (*left*) and the right end of Aphrodite Terra, are characterized by high gravity (*white contour lines*) and frequent lightning strikes (*pink dots*), both suggestive of volcanic activity.

Photo by NASA

plains. A few broad, shallow depressions, shown as the darkest shades, extend as far as 2 km below the mean planetary radius of 6051.4 km, and several "islands" of rugged topography (lighter shades) rise as much as 11 km above it. Tentative samplings of surface materials by the Soviet landers suggest that the plains are composed of basalt and the highlands at least partially of granite (hence their elevation). Nevertheless, the surface of Venus shows neither the structural features nor the bi-level distribution of elevations characteristic of continental and oceanic plates on Earth. Accordingly, plate tectonic mechanisms probably do not exist (or no longer exist) on Venus. Radar images made at Arecibo show features that could indicate the presence of volcanism, tensional faulting, and impact cratering, a combination suggesting that Venus has a crust that is (or perhaps was) more active tectonically than that of Mercury but less active than that of Earth.

The possible presence of granite on Venus raises an intriguing question because a prerequisite for the origin of this rock type, and apparently also for plate tectonics, is the presence of water. If granite is present on Venus, it could be evidence of the planet's early youth before the loss of its water.

Before the Soviet landers failed due to the high temperature, they transmitted images of the surface of Venus showing a bedrock surface littered with coarse "soil" and an abundance of large rock fragments. The presence of such clastic materials on the surface implies an active weathering environment, which we might well have expected, given the existence of so corrosive an atmosphere. Undoubtedly, the dominant form of weathering is chemical, although sandblasting is probably also effective.

## The Earth-Moon System

Because the features of Earth are the principal topic of this book, we shall skip that port of call on our Cook's tour, and proceed directly to Earth's satellite, the Moon (fig. 17.15). Because of its proximity and accessibility, the Moon has served, and will continue to serve, as a convenient laboratory for the testing of hypotheses about the origin and evolution of the Solar System. Most notably, the Moon is one of two extraterrestrial sources from which we have actually obtained rock samples, the other source being meteorites. Six of the Apollo missions carried human explorers and scientists to a variety of sites on the Moon. The investigations and samplings of the lunar surface conducted by these pioneering space voyagers have greatly expanded our understanding not only of the structure and composition of the Moon, but also of the early history of the Solar System, which still lies preserved in the features of the lunar surface.

**Figure 17.15** Earth's Moon, photographed from the *Apollo 11* spacecraft, July 1969. Light, ancient, cratered terrain of the Moon's near side (*left*) contrasts with dark maria basalts that fill enormous, multiringed basins, such as Mare Crisium (*above center*) and some older craters. Young craters, such as the two at the upper right and lower left, have bright rays and ejecta blankets. The Moon's far side (partly visible on right) lacks large, lava-filled, multiringed basins.

Photo by NASA

Probably the most remarkable of the Moon's features is its large size relative to its planet. With over a quarter of Earth's diameter (but only about a hundredth of its mass), it is regarded by many as Earth's twin rather than its satellite. In the Solar System, only distant Pluto and its moon Charon are as comparable in size as Earth and the Moon.

Like Mercury, the Moon is inactive geologically, as a result of which many ancient structures have been preserved on its surface. The Moon's oldest rocks are brecciated *anorthosite* (feldspar-rich gabbro) radiometrically dated at 4600 million years. The youngest of the Moon rocks yield a radiometric age of 3100 million years, attesting to the great stability of the lunar crust since that time. Samples of pulverized rock from the large crater Copernicus have been dated at between 900 and 800 million years old. Analysis of the Moon rocks in relation to their ages and the structures from which they came has yielded a tentative chronology of events that we may regard as typical of at least the terrestrial planets of the Solar System. That chronology is given in table 17.2.

**Table 17.2**   Lunar Chronology

| Date (10⁶ yr ago) | Time Interval | Events | Rock Units |
|---|---|---|---|
| <900 | Copernican | Rare impacts | Bright-rayed craters (Tycho, Aristarchus, Copernicus) |
| 3300–900 | Eratosthenian | Rare impacts; Late basaltic eruptions | Craters with rays no longer clearly visible Fresh-surfaced lavas |
| 3900–3300 | Imbrian | Rare impacts Voluminous basalt flows | Craters with no sign of rays Maria basalts |
| 4600–3900 | Pre-Imbrian | Frequent, large impacts | Multi-ringed impact basins and craters; lunar breccia |
| >4600 | Unnamed | Solidification of the Moon | Anorthosite crust |

The early formation of an anorthosite crust implies that at least the outer portions of the Moon were molten in Early Pre-Imbrian time (see table 17.2), allowing relatively light calcic plagioclase crystals to float to the surface to form the anorthosite. The early anorthosite crust was subsequently smashed during the Great Bombardment by large objects that impacted the Moon in Late Pre-Imbrian time. These impacts have been radiometrically dated to between 4500 and 3800 million years ago. The result of this rain of cosmic rubbish was the development of great, multi-ringed basins, equivalent to the Caloris basin on Mercury, but subsequently filled with vast, dark basalt plains. Galileo called these plains **maria** (*mah*-ree-ah; Latin for "seas") because he thought they might be water-filled. Each of these huge impact structures consists of a series of concentric rings of upthrust, shattered rock, but because of the filling of the lunar maria with later deposits, only the outermost, and highest, ring of each basin is visible. Enormous circular structures in the eroded bedrock of Earth's Archaean shield areas suggest that Earth was as liberally impacted by flying debris as was Mercury and the Moon, but erosion, deposition, and plate tectonics have erased most of the evidence.

As on Mercury, smaller impact structures also abound on the Moon. A compact cluster of craters of various sizes is shown in figure 17.16. Notice that the smaller craters have simple bowl shapes, whereas the larger ones have central peaks caused by the rebounding of the crater floor following impact. The larger the crater, the larger is the central peak area. Beyond a certain diameter, the central peaks expand into a central ring, within which a second central peak, or a third ring, can form if the structure is large enough. Thus, there is a genetic relation between simple impact craters and multi-ringed basins. A final note about craters:

**Figure 17.16**   Lunar impact craters of a variety of sizes show definite trends with increasing diameter. In larger craters, rims are more broken and rugged, floors are flatter and shallower (from isostatic rebound), and central peaks appear. In the largest craters, these peaks are often expanded into central rings.

**Photo by NASA**

the number of them in a given area gives a rough, relative indication of the age of a planetary surface. Thus, the higher crater density of the light-colored highlands in figure 17.15 shows the highlands to be older than the smooth, relatively crater-free basalt plains of the lunar maria.

During Imbrian time, following the formation of the multi-ringed basins, great volumes of basalt welled up from the Moon's interior and filled the basins to their outer rings and formed the maria. These great eruptions

*Figure 17.17* Mars, as seen by the *Viking I* space probe on 17 June 1976 from a distance of 560,000 km. The darker Southern Hemisphere consists of cratered basaltic highlands, whereas the lighter Northern Hemisphere (*upper right*) consists of lowlands whose craters largely have been obscured by blowing dust. The four dark spots are great basaltic shield volcanoes that rise above the dusty plains. At the bottom of the image, a thin layer of carbon dioxide frost fills the impact basin of Argyre Planitia. Photo by NASA

lasted from 3900 million to 3100 million years ago. Since the eruption of the maria basalts, little has happened to modify the face of the Moon. Numerous small impact craters and a few large ones, such as Copernicus and Tycho, have been formed. The larger craters have covered older features nearby with **ejecta blankets** of crushed and brecciated rock and with bright **rays** that splash out radially from the craters in all directions (see fig. 17.15). The only other agent (aside from a few human footprints and the tracks of a "Moon buggy") that has affected the Moon's surface since Imbrian time is an incessant sandblasting by micrometeorites, which has gradually dimmed the sharp outlines of the older craters.

## Mars

When the *Mariner IV* space probe radioed the first images of the Martian surface back to Earth in 1965, it shattered a long-held romantic tradition that the "Red Planet" was a garden oasis akin to Earth and habitable by humans. With stark impartiality, the scanning electron beam traced out the unmistakable image of a landscape almost as pockmarked with impact craters as the Moon's. Since those first disappointing glimpses, the *Mariner 9* and *Viking* missions to Mars (named for the Roman god of war) have shown our outer planetary neighbor to be a geologically complex system with some great surprises and some valuable lessons that have even enriched the fabric of terrestrial geology (fig. 17.17).

In addition to impact craters, the later orbiters sent back pictures of ice caps, giant volcanoes, great tensional rifts, enormous canyons, dry channels indicative of phenomenal floods, dune fields, and peculiar hummocky terrain suggestive of the collapse of water-saturated, unconsolidated surface material. The geologic maps of figure 17.18 show the distribution of these various structures.

The diversity of these seemingly unrelated phenomena created much confusion at first, but this soon gave way to an intelligible picture in which the artists are the agents of volcanism, gravity, and wind, and the media are water, basalt, and dust. The picture begins with the "undivided cratered terrain" that covers most of the Southern Hemisphere of Mars. That this is a geologically old surface is evident from its high crater density, but this density is still only about a quarter of that found on the lunar highlands. High-resolution images of the surface of the Martian cratered terrain suggest that basalt flows have flooded many of the larger craters nearly to their rims, and in so doing probably buried most of the smaller craters.

Much of this terrain has also been lightly dissected by a network of sinuous, dendritic channels similar to youthful stream systems on Earth. In many places, these channels are interrupted by craters, indicating that they must be ancient features, perhaps as much as 4000 million years old, assuming that the bombardment of Mars was contemporaneous with that of the Moon. Because the channels are short, steep-walled, and the drainage networks far less elaborate than their terrestrial counterparts, it is suspected that they might only have been active for a very short time and that they could represent the channels of lahars or debris flows generated by the sudden volcanic heating of ice-saturated ground.

Within the southern cratered terrain are two enormous impact structures—Argyre and Hellas—probably equivalent to the Caloris basin of Mercury and the Imbrium basin on the Moon. As with the lunar maria, these basins appear to have been filled with basalt flows, which have suffered only minor impacts since the time of their

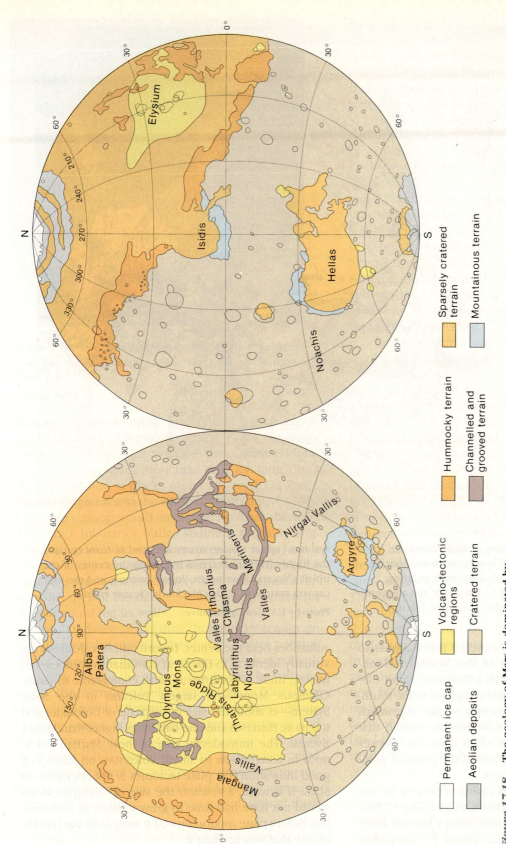

*Figure 17.18*  The geology of Mars is dominated by seven major types of terrain: (1) old, moderately cratered plains (*red*); (2) young, sparsely cratered plains and lava-filled impact basins (*gold*); (3) broad, volcano-tectonic domes (*yellow*) surmounted by huge volcanoes; (4) channelled and grooved terrain (*brown*); (5) hummocky terrain (*orange*); (6) aeolian deposits beneath polar ice (*grey*); and (7) mountainous terrain (*blue*).

From "Mars" by J. B. Pollack. Copyright © 1975 by SCIENTIFIC AMERICAN, Inc. All rights reserved.

Permanent ice cap

Volcano-tectonic regions

Aeolian deposits

Cratered terrain

Hummocky terrain

Channelled and grooved terrain

Sparsely cratered terrain

Mountainous terrain

**Figure 17.21** Most of the geological processes that have shaped the Martian surface appear in this view of the region between Lunae Planum and Chryse Planitia. Moderately cratered lava plains (*left* and *right*) flank a gigantic, braided flood channel, which has divided around the ejecta blanket of a large impact crater. Collapse structures in the lava plains (*lower left*) suggest the melting of permafrost-saturated dust underlying the lava. Light dust streaks trending north from small craters in the channel bed indicate the direction of the prevailing southerly wind.

**Photos by NASA**

## Asteroids and Meteorites

The space between the terrestrial and the great planets (see fig. 17.9) is occupied by a belt of orbiting cosmic rubbish once thought to be the remnants of a shattered planet, but now considered to be the remains of several small, moonlike bodies that shattered during mutual collisions. The largest of these remnants is the asteroid Ceres, which has a diameter of 700 km. Nine others with diameters exceeding 100 km have been found, but most are much smaller than 1 km. Together, they are estimated to amount to no more than 1/500th the mass of Earth, hardly sufficient to make up a respectable planet. It could be that the bulk of the debris from the asteroid belt has been lost to space or swallowed by other planets and the Sun. In addition, most of the Solar System's smaller moons are undoubtedly of asteroidal origin, and some consider Pluto to be an asteroid. If so, the total mass would be much greater.

Some planetary geologists argue that the asteroid belt is the source of the bodies that created the impact craters and multi-ringed basins of the Great Bombardment from Mercury to the moons of Uranus. Meteorites are suspected to be the meagre residue of this ancient episode of interplanetary peppering. There are two serious drawbacks to this hypothesis, however. The first is the statistical improbability that such a peppering could have been so efficient as to have scored so many direct hits on such small and widely spaced targets, and in so doing, to have depleted the supply of asteroidal rubbish to such an extent in so short a time. The second is the high percentage of meteorites with young radiometric ages. Certain of these younger meteorites give strong evidence of having originated on Mars, and some give evidence of a violent shock dated to 180 million years ago. Could this evidence relate to the possible major impact structure in the Northern Hemisphere of Mars mentioned in a preceding section? Might the Solar System's present stock of meteoroids have originated in the collision of a large asteroid, moon, or small planet with Mars at that time? Many respected scientific hypotheses, including the currently popular concept that the extinction of the dinosaurs was caused by an asteroid impact on Earth at the end of the Mesozoic Era, have been erected on the basis of this kind of circumstantial evidence. Until such intriguing possibilities have been exhaustively researched and all alternative explanations systematically disproven, however, they should be regarded as speculative.

Analyses of meteorites support the view that asteroids are fragments of small moonlike bodies with iron cores. About 4.5% of the meteorites that fall to Earth are *iron meteorites* composed of a coarsely crystalline iron-nickel alloy that must have cooled extremely slowly under high pressure (fig. 17.22*a*). These would correspond to a lunar core. Most of the remainder are *stony meteorites,* comprising *chondrites* (about 84.5%) and

a.  b.  c.

**Figure 17.22** Meteorites: *a,* a polished and etched section of the Casas Grandes iron meteorite showing the triangular Widmanstätten structure, a crystalline intergrowth of two different nickel-iron alloys; *b,* a polished section of a chondrite meteorite showing flecks of nickel-iron alloy within a matrix of silicate minerals; *c,*

the Allende carbonaceous chondrite meteorite, which fell in northern Mexico in 1969. One of the most studied meteorites, Allende contains the oldest known matter on Earth, with a radiometric age of 4560 million years.

Photos by Geological Survey of Canada

*achondrites* (about 9.5%). Achondrites consist of pyroxene and plagioclase with some olivine and ironnickel, and they resemble the mafic plutonic rocks of Earth's crust. Chondrites (fig. 17.22*b*), analogous to the peridotites of Earth's mantle, resemble achondrites in composition, but their minerals occur in rounded grains. About 1.5% of meteorites are *stony iron meteorites,* which share the compositions of stones and irons in roughly equal proportions. Stony irons could represent mixed materials of a mantle-core boundary.

A fraction of a percent of meteorites is of an even stranger type called *carbonaceous chondrites* (fig. 17.22*c*). These consist of serpentine or chlorite, olivine, hematite, clays, carbonates, sulfates, calcite, and organic compounds. The presence of hematite and sulfate ion indicates that they are highly oxidized. It is also noteworthy that many of these minerals were clearly deposited from a liquid water medium. Might all these curious facts suggest that carbonaceous condrites are surface deposits from a former planet that had developed life and an oxygen-rich atmosphere? The thought is appealing, but evidence from isotopic ratios suggests that these materials are the oldest in the solar system, and that they probably resulted from the early accretionary heating of moon-sized bodies of mixed rock and ice.

## Jupiter

The close encounters of the *Voyager 1* and *Voyager 2* probes with Jupiter (named for the chief god of the Romans) and its moons on 5 March and 9 July 1979 added a wealth of crisp, photographic data and instrumental measurements to our store of knowledge about the Solar System's largest planet (fig. 17.23). Its great mass, 318 times that of Earth, is still less than a thousandth that of the entire Solar System, yet Jupiter possesses almost 60% of the Solar System's angular

**Figure 17.23** This photomontage of the Jovian System shows Jupiter and its four natural, Galilean moons. Cloud bands parallel Jupiter's equator, each flowing in the opposite direction to the adjacent one. Io (*left*) is closest to Jupiter, followed by Europa (*center*), Ganymede (*lower left*), and Callisto (*lower right*).

Photo by NASA

momentum, a fact that cosmologists have been hard put to explain. With its four large, *Galilean* moons, Io, Europa, Ganymede, and Callisto (fig. 17.24), the Jovian system is a miniature replica of the larger Solar System, an analogy that is all the more compelling when you consider that if Jupiter had been about ten times more massive, it would have been a star.

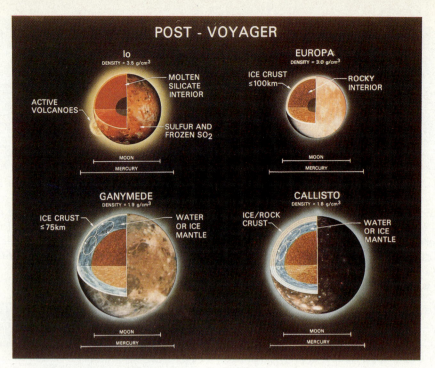

POST - VOYAGER

**Figure 17.24** Jupiter's four natural moons exhibit marked differences. Io is volcanically active and covered with sulfur and frozen sulfur dioxide. Europa is covered by a thick, global ice sheet that is active enough to have erased all but a few impact craters. Ganymede and Callisto are similarly ice-shrouded but are less active; hence, they retain more craters.

Photo by Jet Propulsion Lab

Spectrographic analysis of Jupiter's atmosphere has revealed that hydrogen and helium dominate the surface layer, and that methane, ammonia, carbon monoxide, and a few other gases are also present below the surface. These gases have all been found within the *Great Red Spot,* an enormous, permanent, cyclonic storm longer than two Earth diameters (fig. 17.25).

Below its surface, Jupiter is thought to be composed mainly of hydrogen and helium in about the same 9:1 ratio in which they are present in the Sun. Given this assumption and the planet's density (1.33), the internal structure of Jupiter can be approximately determined. Beneath a gaseous envelope a few hundred kilometers thick, the atmospheric gases become compressed into a liquid layer that extends downward for about 20% to 25% of the way from the surface to the center. Here, the pressure becomes great enough to compress the atoms into a metallic structure, which prevails down to a fairly large core composed of a probably molten mixture of silicate minerals, water, and iron. Jupiter's powerful magnetic field indicates that fluid dynamo motions are active either in the metallic gas layer or in a molten iron core.

Because of Jupiter's short rotation period of 9.84 hours, the planet is somewhat flattened at the poles, and its equatorial radius is about 7% longer than its polar radius. This rapid rotation also affects the dynamics of Jupiter's atmosphere, producing the striking cloud bands that run parallel with the equator. Studies of the motions of spots embedded within these bands have shown that alternate bands travel in opposite directions at speeds of up to 150 m/sec. The bands are brightly colored red, white, brown, and blue in order of decreasing altitude, and they are interrupted by huge, stationary or semistationary cyclonic storms, of which the Great Red Spot is the principal example. To date, no completely satisfactory hypothesis has emerged to explain the circulation of Jupiter's atmosphere.

Images transmitted by the *Voyager* probes have revealed that Jupiter is surrounded by a ring similar to, but less elaborate than, the rings of Saturn. Jupiter's ring consists of a bright outer section about 6000 km wide and less than 30 km thick and a diffuse inner disk that could extend inward to the top of the planet's atmosphere. Although the composition of the ring is still unknown, it probably consists of stony particles no larger than a few centimeters in diameter.

Fully as exciting as the *Voyager* images of Jupiter itself are the images of its four Galilean satellites (fig. 17.24). These small-planet-sized bodies exhibit several interesting trends with proximity to Jupiter. Callisto, the outermost moon, is about the size of the planet Mercury, and it has a similar appearance, as its surface is densely mottled with both craters and multiringed basins. In contrast to Mercury, however, its density is only 1.24 g/cm³ because it is thickly shrouded in ice. The icy crust of this outer moon was heavily impacted during the Great Bombardment and has suffered little

**Figure 17.25** Jupiter's Great Red Spot is an enormous counterclockwise cyclone that could swallow Earth with room to spare. The cloud band to the north of the spot moves westward as the one to the south moves eastward.

Photo by NASA/JPL

**Figure 17.26** Seven natural moons are shown in this unnatural photomontage of the Saturnian System. Mimas (*center,* against Saturn's disk), bears a huge impact crater and orbits closest to the planet. Ice-encrusted Enceladus (*left foreground*) is next in sequence, followed by wrinkled Tethys (*lower right*). The "washed out" craters of Dione (*upper left*) suggest a crust of low rigidity. Enigmatic white wisps cross the face of Rhea (*right of Dione*). Saturn's largest moon, Titan (*upper left*), shrouded in a smoggy nitrogen atmosphere, orbits farthest from the planet except for Iapetus (*right of center*), one face of which is light, the other dark.

Photo by NASA

alteration since. Ganymede, next closer to Jupiter, is slightly larger and denser (1.93 g/cm³) than Callisto, and its icy crust shows evidence of recent geologic activity in the form of peculiar "grooved terrain," which has obliterated many of the moon's ancient craters.

Europa is a smaller moon, whose surface, like that of Ganymede and Callisto, is also covered with a thick mantle of ice. Europa's mantle is featureless and smooth, however, except for a network of intersecting fractures filled with recrystallized ice. Only three impact craters have been tentatively identified on Europa's face, indicating that this moon has been far more tectonically active than its outer neighbors. That activity results from powerful gravitational stresses imposed on the nearer moons by Jupiter. Europa's density (3.03 g/cm³) is markedly higher than that of the outer moons, presumably because much of its ice was driven from it by intense heat from Jupiter during its early youth.

Io, the innermost moon, is a few kilometers smaller than our own Moon and is the most geologically active body in the Solar System. Constantly flexed by Jupiter's gravity as it follows its eccentric orbit, Io experiences heat-generating tidal distortions that result in extremely active volcanism. Io's bizarre colors give it an appearance that has been deservedly likened to a pizza pie, complete with yellow pie crust, red-orange tomato, white cheese, and black olives. The colors are due to elemental sulfur, which appears to be the principal (if not the only) constituent of the bright orange lavas. Io, with a density of 3.53 g/cm³, has neither ice nor water, but not surprisingly, it has a tenuous atmosphere of sulfur dioxide, which is probably present in large quantities as a liquid in underground "aquifers." The Ionian volcanoes appear to erupt continuously. Seven of the eight volcanoes observed in *Voyager 1* images were still erupting four months later when the *Voyager 2* images were made. Because of Io's extreme geologic activity, not a single impact crater has been found on its surface.

The effectiveness of tidal forces in stimulating tectonic activity on Io suggests that the tidal stresses exerted on Earth by the Moon might also be a significant contributing factor to Earth's tectonic activity. Perhaps, if it were not for Earth's large satellite, our planet would be considerably less tectonically active than it actually is. Conversely, if the Moon had ever had a significant content of water, it might also have developed a low velocity zone and a plate tectonic system.

In addition to the four Galilean moons, Jupiter has 10 other satellites, mostly in strongly inclined, and in some cases retrograde, orbits. All are less than 100 km in diameter and are of nonspherical shapes. All these features suggest that these small moons, like those of Mars, are captured asteroids.

## Saturn

*Voyager 1* encountered Saturn (named for the Roman god of agriculture) and its moons in November 1980 (fig. 17.26) and found a system much like that of Jupiter but with important differences. Like Jupiter, Saturn has a hydrogen/helium atmosphere with colored cloud bands paralleling the equator and flowing in alternately

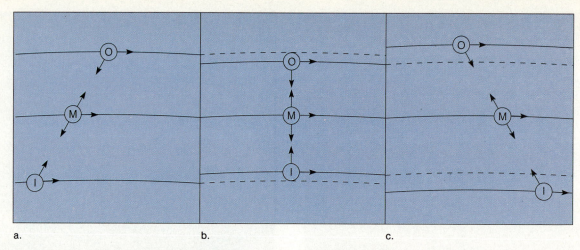

a.                                        b.                                        c.

*Figure 17.27*    The shepherding effect guards against collisions among objects in Saturn's rings. In *a,* gravitational attractions accelerate the inner object (*I*),

raising its orbit, and retard the outer (*O*), lowering it. The middle object (*M*) is unaffected. In *c,* attractions retard *I,* lowering its orbit, and accelerate *O,* raising it.

opposite directions, but these bands attain wind speeds of nearly 500 m/sec, more than three times the speed of their Jovian counterparts. Saturn's rotational period is about the same as Jupiter's (10.2 days), but because of its lower density (0.69 g/cm³), it is more flattened at the poles than Jupiter. Saturn has a magnetic field, but it is weaker than Jupiter's, indicating less activity in its core.

Saturn's greatest glory is the system of rings that surrounds the planet in its equatorial plane. In all, there are seven distinct rings, of which only two, the *A* and *B,* are plainly visible in figure 17.26 (the *A* ring is outside the *B* ring, and has two parts). The *A, B,* and *C* rings have a combined width of 275,000 km but a thickness of less than 1 km. The *Voyager* images show the rings to be much more complex than was anticipated. They are thought to consist of chunks of rock and ice ranging between 10 cm and 10 m in diameter. This material is mustered into narrow subrings that appear as finely spaced as the grooves of a phonograph record in the *Voyager* images. Radial spokes and bands form within the rings but soon become distorted and obliterated because the inner parts of the rings orbit at higher velocities than the outer parts.

Motion within the rings appears to be closely controlled by a sort of natural selection process. Departures from uniform, circular, orbital motion result in destructive collisions; hence, those particles that do not deviate from such motion are the ones that survive. Furthermore, adjacent particles have a sort of "shepherding" effect on one another, which tends to maintain spacing (fig. 17.27). Thus, a state of dynamic equilibrium prevails in the rings in which particles are constantly shifting outward and inward within the plane of the rings but seldom colliding.

Saturn's seven natural moons are an odd lot. Titan, with a diameter of 5150 km, and a density of 1.9 is, after Jupiter's Ganymede, the second largest moon in the Solar System. Titan is noteworthy in having a nitrogen atmosphere 1.5 times as dense as Earth's. Under a uniform surface temperature of −179° C, methane is thought to behave in much the same way on Titan as water behaves on Earth. Thus, there could be methane clouds, methane rain, and probably a universal methane ocean. Ultraviolet radiation creates a smog of hydrogen cyanide (HCN) and a variety of hydrocarbons in the atmosphere. The larger smog particles probably coagulate and slowly drift down to Titan's surface to accumulate as sediment at the bottom of the methane ocean.

Saturn's six other natural moons range between 500 and 1500 km in diameter, and their densities range between 1.1 and 1.5, indicating a composition of about half rock, half ice. One moon, Iapetus, appears to have had eruptions of a pitchlike, possibly organic material on its surface. All are cratered, but as with Jupiter's Galilean moons, crater density appears lower on the inner moons than on the outer, notably so on Enceladus, which shows signs of geological activity.

In addition to its seven natural satellites, Saturn has 10 smaller, irregular asteroidal moons. One of these, Phoebe, is Saturn's outermost moon, and it lies in a strongly inclined, retrograde orbit.

## Uranus, Neptune, and Pluto

The last two planetary visits of *Voyager 2* were to Uranus (*You-ray*-nus or *You*-ra-nus, father of Kronos (time) in Greek mythology) in January 1986 (fig. 17.28) and to Neptune (named for the Roman god of the sea) in August 1989. Spectrographic analyses indicate that the atmospheres of both Uranus and Neptune are similar to

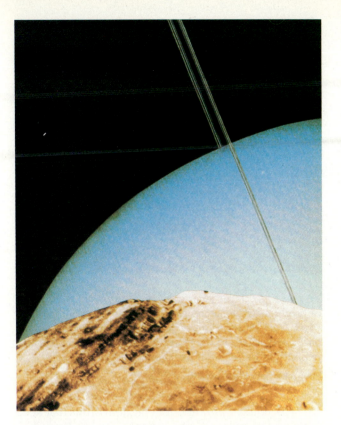

**Figure 17.28** This photomontage views Uranus and its thin rings across the upper limb of its geologically active closest satellite, Miranda.

Photo by NASA/JPL

those of Saturn, Jupiter, and the Sun. They are composed primarily of hydrogen and some helium, but with larger proportions of methane, ammonia, and water. This poses a problem because it implies that the mean molecular weight of the material in the primordial dust cloud from which the Solar System formed increased radially away from the center, a highly unlikely circumstance.

The plane that includes Uranus's equator, its nine diffuse rings, and the orbits of its five satellites, is inclined at an angle of 98° to the plane of the ecliptic. The reason for this inclination is not known, but a collision with a former moon is suspected. Uranus's rings were discovered accidentally when they symmetrically dimmed the light of a star as it passed behind the planet. The density of Uranus and that of all five of its natural moons is about 1.6. The moons are relatively small, with diameters between 480 and 1590 km, and they appear to consist of a mixture of rock and ice. Miranda, the smallest and closest moon, is tectonically active and displays oddly bent and truncated structures on its surface (fig. 17.28), suggesting that the satellite might have been fragmented and reassembled. The other four natural moons, like those of Saturn and Jupiter, show less tectonic activity and higher crater density with increasing distance from the planet.

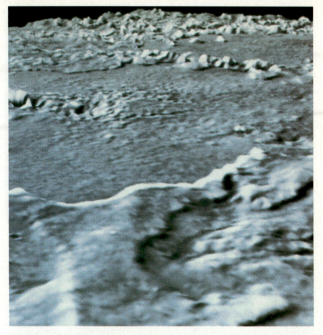

a.

b.

**Figure 17.29** Coated with reflective ices of hydrogen oxide and methane, Triton's silvery surface (*a*) gleams in brilliant contrast to the more somber, sapphire jewel of Neptune, (*b*) whose thick, methane-rich atmosphere glows blue because it absorbs the longer, reddish wavelengths of the Sun's rays.

Photos by NASA

Neptune (fig. 17.29) is slightly smaller than Uranus, and its atmosphere is a beautiful sapphire blue due to a high percentage of methane, which absorbs red light. Like Uranus, it has a magnetic field, which is tipped at a large angle to its rotational axis. Neptune has three

**Figure 17.30** An artist's conception of the mystery "planet" Pluto and its relatively large moon, Charon. The dim and distant Sun gleams faintly in the background.

Photo by NASA

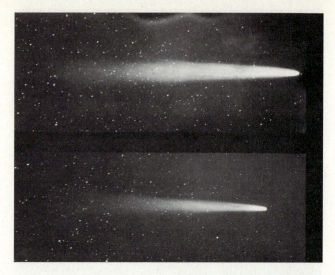

**Figure 17.31** The solar wind blows a spectacular plume of bright methane, ammonia, carbon dioxide, and water vapor from Halley's comet during its perihelion passage on 12 May 1910. In the lower photo taken three days after perihelion, Halley's tail has shortened by about 25%.

Photos by Hale Observatories

faint rings, the outermost of which has three thickened segments on one side. Neptune's large natural satellite, Triton (son of Neptune in Greek mythology), is noteworthy in several respects. Inexplicably, it circles its planet in a tilted, retrograde orbit. Its surface is covered with highly reflective ices—probably of methane and hydrogen oxide—and it shares with Earth and Jupiter's closest moon, Io, the distinction of active volcanic eruptions. On Earth, the erupted material is silicate lava, water, and carbon dioxide, and on Io it is sulfur and sulfur dioxide. On Triton, however, it appears to be black organic compounds and liquid nitrogen.

Pluto (named for the Roman god of the underworld), with a diameter of only 2300 km, is smaller than Earth's Moon. Its orbit carries it within the orbit of Neptune at perihelion, because of which the planet and its large moon, Charon, are suspected of being escaped Neptunian moons or possibly even large asteroids (fig. 17.30).

## Comets

The far fringes of the Solar System are populated by a little-understood class of objects called **comets** (from Latin, *stellae comatae,* "hairy stars"). These are large, cosmic "snowballs" with three main components: (1) microscopic dust particles composed of iron and magnesium silicates; (2) *ices* composed mainly of hydrogen oxide with lesser amounts of carbon monoxide, carbon dioxide, methane, ammonia, and hydrogen cyanide; and (3) polymerized organic compounds composed of carbon, hydrogen, oxygen, and nitrogen. This mixture probably closely reflects the original composition of the solar nebula.

Most comets whiz about in random orbits that keep them far from the orderly, ecliptic system of the Sun, planets, and moons, but a few have been trapped into highly eccentric solar orbits. One example is Halley's Comet (*Hal*-eez, after Sir Edmund Halley, 1656–1742, a British astronomer), which returns approximately every 77 years (fig. 17.31). Its most recent perihelion was in April 1986, and in the preceding month, it was closely scrutinized by spacecraft from a variety of nations. The European space module *Giotto* (*Jot*-toe) photographed the nucleus, which is potato-shaped, about 16 km long, and has a mass of about 100 billion tons and a density of 0.1 to 0.3 g/cm³.

As a comet approaches the Sun, it is struck by the solar wind, which heats the nucleus and causes it to eject jets of gas and dust, forming a bright halo, or **coma,** around the nucleus. Pressure from the solar wind causes the gases and dust to stream out from the comet in the form of a long *tail,* which always extends radially away from the Sun (the tail does *not* "follow" the comet!) and can be up to 80 million km in length. From *Giotto's* observations, it has been estimated that on each orbital pass, Halley's comet loses about 100 million tons of dust and gas. At this rate, it should last for at least another 77,000 years.

Many hypotheses have been advanced to explain the origin of the Solar System. These are too numerous and for the most part too elaborate to permit a detailed accounting here. In general, they fall into three broad categories:

1. The *nebular hypotheses* of Descartes, Kant, and Laplace (eighteenth century) and of von Weizsäcker, Kuiper, and Urey (twentieth century), in which the planets condense from a flattened, rotating cloud of dust and gas called the *solar nebula* (fig. 17.32).

2. The *close encounter hypotheses* of Buffon (eighteenth century), Chamberlin and Moulton (twentieth century), and Jeans and Jeffreys (twentieth century), in which the planets are gravitationally drawn, or even gouged, from the Sun by a passing comet or star (fig. 17.33).

3. The *exploding companion hypothesis* of Lyttleton (twentieth century), in which the planets are the condensed remnants of a companion star to the Sun which exploded (fig. 17.34).

Close encounter hypotheses have been discounted largely on the basis of the extreme improbability of such an event. Exploding companion hypotheses have been discounted partly because the velocity of the outrushing material would have exceeded the escape velocity required to free them of the Sun's gravitational field and partly because observation of stellar explosions has revealed the presence of residual masses, called pulsars (see chapter 18), that emit phenomenal amounts of radiant energy. No such object exists in the vicinity of the Solar System.

Current cosmological thinking favors one or another version of the nebular hypothesis, but each encounters problems that have generated more arm waving than convincing explanation. In particular, the models fail to account adequately for the transfer of angular momentum from the Sun, which now has only

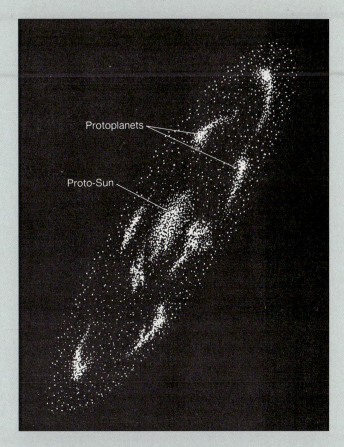

**Figure 17.32** Nebular hypotheses propose that the Solar System evolved from a rapidly spinning disk of cosmic dust. Collapse of the disk would logically produce a central body, the Sun, but mechanisms for accreting the planets are problematic.

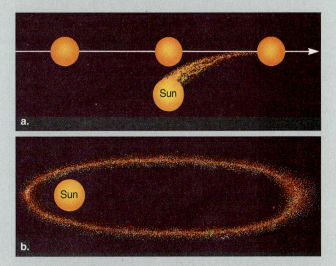

**Figure 17.33** The origin of the Solar System from materials torn from the Sun by a passing star presents the problems of extreme improbability and high orbital eccentricity of the extracted material about the Sun.

*continued*

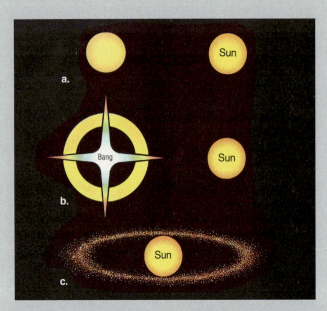

***Figure 17.34*** The exploding companion hypothesis for the origin of the Solar System fails to explain

2.74% of it, to the planets, which have the rest. The mechanisms invoked—magnetic fields and the solar wind—have recently been found to be unviable on theoretical grounds. Equally problematical is the lack of a clear-cut mechanism to space the planets with the exponential regularity they display. Finally, the increase in mean molecular weight of planetary atmospheres from Jupiter to Neptune is hard to explain with the currently favored models, as is the Great Bombardment.

One possible way out of this dilemma lies in an expansion of a hypothesis first proposed by George Darwin (son of Charles) in 1879 and recently revived by D. U. Wise and J. A. O'Keefe. Darwin (see *In the Spotlight*) proposed that as Earth gained new matter by accretion from the solar nebula, its rotation rate would have

the absence of a neutron star, an apparently universal residual feature of such explosions.

increased to the point at which it would have become flattened into a cushion shape (fig. 17.35*a*). When Earth's polar diameter shortened to 7/12 of its equatorial diameter, it would have developed a bulge on one side (fig. 17.35*b*). If the diameter across that bulge exceeded 23/8 of Earth's polar diameter, the bulge would then have separated by fission from Earth to form the Moon (fig. 17.35*c* and *d* ). Because fission depends only on spin rate and density, Darwin could calculate that a body of Earth's density would have to spin at a rate of at least 9.06 revolutions per day in order that such fission might occur. This rate corresponds to a year of 3308 days, a reasonable possibility in the light of the curves in figure 16.25 and their supporting evidence from the geologic record. Furthermore, the geochemistry, crustal structure, and residual

magnetism of the Moon are all consistent with its having been extracted from Earth.

How, then, does this bear on the origin of the Solar System? If we assume that the initial collapse of the solar nebula (probably initiated by the explosion of a nearby star) produced a rapidly spinning, solid body in its center, which would later become the Sun, then this same fission process that could have created the Moon should also have affected that body. Furthermore, on the reasonable assumption that this "proto-Sun" had a density of about 1.6 g/cm³, equivalent to that of Uranus, a spin rate of at least 4.87 revolutions per day would have sufficed to initiate fission (only about half that required for Earth to spin off the Moon). Inasmuch as the collapse of the solar nebula should have produced spin rates far greater than this, it seems logical that the proto-Sun must have sequentially spun off *several* "protoplanets" as the nebula collapsed around it (fig. 17.36).

Each new fission would have reduced the spin rate of the proto-Sun (fig. 17.36*b* and *d* ), but ongoing accretion from the collapsing nebula would have increased it again until another fission occurred (fig. 17.36*c*). Furthermore, the reacceleration of the proto-Sun's spin rate would also have induced increases in the velocities of the orbiting protoplanets, which would then have moved into increasingly distant orbits (fig. 17.36*c*). This mechanism would thus explain the problematical distribution of angular momentum in the Solar System. The exponential spacing of the planets would also be explained primarily by an increasing rate of nebular collapse, which would have shortened the interval between fissions. The increasing mean molecular weight of planetary atmospheres with distance from the Sun would be a logical result of the sequence of protoplanetary spin-offs. The early ones would have scavenged the heavier gases from the collapsing nebula, leaving lighter gases to accrete on the later ones. Recent observations of so-called "Herbig-Haro objects"—young stars that fling out gouts of matter—and of other young stars surrounded by nebular disks that appear to be coring themselves out from the inside, lend strength to this scenario.

While all this would have been going on, the solar nebula would also have been collapsing on the protoplanets, increasing *their* spin rates and causing them to spin out moons in the same fashion in which the proto-Sun was spinning out protoplanets (fig. 17.36c and *d* ). Because the moons would have been too small to undergo much accretion, the process would have stopped there. Debris from each spin-off would accumulate between planet and moon, and would eventually fall into one or the other, producing the impact craters and multi-ringed basins of the Great Bombardment (fig. 17.36b and *d* ). Because the moons would have undergone negligible accretion, their spin rates would not have increased, and they would have remained spin-orbit coupled to their planets, as they are today.

It is possible that the Earth-Moon system was spun out as a double planet, separating as it left the proto-Sun rather than resulting from a secondary fission process, as with the moons of Jupiter. This would help account for the Moon's relatively large size as well as the problematical distribution of angular momentum in the Earth-Moon system.

By the time the proto-Sun began to spin off the inner planets, it would have become quite hot from gravitational contraction. This would have vaporized its more volatile constituents; hence, the planets from

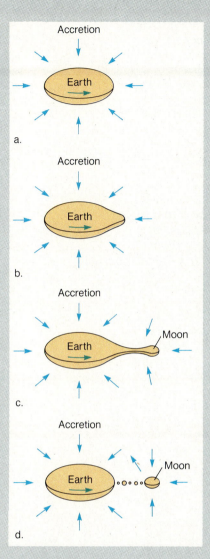

**Figure 17.35** Sir George Darwin's hypothesis for the origin of the Moon depends on the concept of the conservation of angular momentum. As nebular material accretes on it, Earth's spin rate increases, causing the planet to flatten (*a*) until a bulge forms at the equator (*b*). If the spin rate exceeds 9.06 revolutions per day, the bulge breaks away to form the Moon (*c, d*).

*continued*

Mars through Mercury would have been made of increasingly denser material. Shortly after the spinning off of Mercury, the temperature of the proto-Sun's core would have risen high enough to initiate thermonuclear reactions, which would have converted the solid, proto-solar body into a full-fledged star and thus put an end to protoplanet formation. It also would have blown away the volatile atmospheres of the inner planets.

An interesting corollary of this hypothesis is the inference that the Sun failed to divide into a binary or multiple star system (as many other stars have done *after* they began thermonuclear burning) because the process of planet formation robbed the Sun of the necessary angular momentum for fission after it became a star. This could be further evidence that "planet fission" is a universal characteristic of "dirty" nebulae, such as the one that produced the Solar System.

The model just presented, however reasonable it might seem in the light of the evidence and boundary conditions I have outlined here, is, of course, yet another speculative idea. Before it could be elevated to the status of a *theory*, it would have to be subjected to rigorous testing to determine whether it is entirely consistent with all the constraints and boundary conditions already established with respect to the Solar System. In this case, the burden of proof would be particularly heavy, as the model departs radically from currently accepted views. It is just this quality, however— the reluctance to accept new ideas unless they are supported by demonstrated consistency with established knowledge—that makes the scientific method such a powerful tool in the investigation of truth.

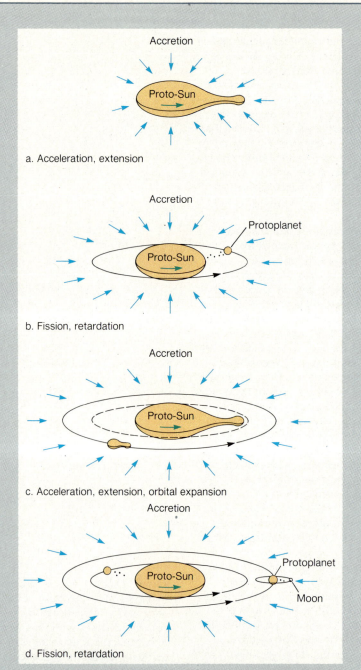

**Figure 17.36** A possible mechanism for the origin of the Solar System by Darwinian fission. In *a,* nebular accretion accelerates the spin rate of the proto-Sun. In *b,* fission occurs, creating a protoplanet and reducing the proto-Sun's spin rate. Continued accretion increases the spin of the proto-Sun, which expands the radius of the protoplanet, and it also increases the spin of the protoplanet (*c*). Both bodies subsequently undergo fission, yielding a second protoplanet from the proto-Sun and a moon from the first protoplanet (*d*). Repetition of this cycle generates the Solar System.

### Sir George Howard Darwin (1845–1912)

Occasionally, as in the case of the Bach dynasty of German musicians, the child of a famous parent will choose to enter the parent's profession despite the risk of being overshadowed by the parent's achievements. In the case of the 20 children of Johann Sebastian Bach, four of those children became distinguished composers in their own right, perhaps because along with their father's talent they also inherited his sunny and optimistic disposition. A similar dynasty was set in motion in 1731 with the birth of Erasmus Darwin, an English physician, poet, and amateur naturalist. Sir Charles Robert Darwin, the grandson of Erasmus, was the author of the theory of evolution. He was also the author of two daughters and five sons, among them the botanist Sir Francis Darwin and the geologist-astronomer Sir George Howard Darwin. At Trinity College, Cambridge, George Darwin distinguished himself, graduating in 1868 with high honors in astronomy and second place in mathematics. He was then made a fellow of the college, and in 1883 was appointed Plumian professor of astronomy and experimental philosophy.

Inspired by the theoretical work of Newton, Kelvin, and Laplace on the effects of lunar and solar gravity on the tides, Darwin further elaborated and refined these pioneering studies. In 1884, he published a work that laid the foundations of our modern understanding of tidal phenomena, influenced as they are by landmasses and friction against the irregular ocean floor. Following this contribution, Darwin became increasingly interested in the effects of tidal friction on the rotational periods and angular momentum of the Earth-Moon system. In a work titled *The Tides and Kindred Phenomena in the Solar System,* published in 1898, he evaluated the probable past and future states of the Earth-Moon system on the basis of mathematical theory. Looking ahead, he envisioned a time far in the future when the Moon will have moved far enough away from Earth to increase the latter's day length to 1320 hours, whereupon it will become spin-orbit coupled to the Moon, thereby preventing any further increase in the distance between the two bodies. Looking backward, he envisioned a time in the distant past when Earth and Moon would have been nearly in contact, whereupon Earth's day length would have been about four hours.

This led Darwin to consider the possibility that the Moon might have been derived from Earth during this early phase of rapid rotation. Applying sophisticated theory of the behavior of rotating bodies developed by him and his French colleague J. H. Poincaré (Pwon-car-*ray*), he found that the hypothesis was, indeed, theoretically possible. Unwittingly, however, he set a trap for his hypothesis that would be as damaging as Alfred Wegener's Polarfluchtkraft was to the hypothesis of continental drift. Darwin proposed the apparently reasonable idea that the Pacific Ocean basin represents the scar left by the avulsion of the Moon, which would have carried away a granite cover that he suggested had once overspread the Pacific basin. Unfortunately, the discovery that there is no granite on the Moon discredited that idea, and the development of plate tectonic theory discredited it further, as plate tectonics was a sufficient mechanism to explain the origin of the Pacific basin. In addition, even though the necessary rotation rate for Earth-Moon fission might have been attained, the present distribution of angular momentum in the Earth-Moon system presents difficulties for the concept. These difficulties are not severe enough to invalidate it, however, and it is still considered to be a contender for the Moon's origin.

In any case, Darwin, unlike Wegener, moved on to other things before his death, applying the same theory of rotational fission to double and multiple star systems. In this case, Darwin's fission model remains unchallenged, and following his death it was further elaborated by the British astronomer Sir James Jeans.

In recognition of his achievements, Darwin was elected president of the Royal Astronomical Society in 1899, and in 1905 he was knighted by King Edward VII.

# 18

# Beyond the Solar System:
# Our Cosmic Neighbors

*A broad and ample road, whose dust is gold,*

*And pavement stars, —as stars to thee appear*

*Seen in the galaxy, that milky way*

*Which nightly as a circling zone thou seest*

*Powder'd with stars.*

*John Milton*

***Paradise Lost***

Photo: The spiral galaxy NGC 6946 in
Cepheus.

# Cosmic Distance

The expression "fixed stars," which I use repeatedly in chapter 16, is a convenient misconception rooted in ancient astronomical tradition. To Earth-dwellers, the stars appear fixed in position relative to one another as they revolve endlessly around Earth on a "celestial sphere" of finite radius. To all appearances, all the stars are the same distance from Earth, and they are simply one-dimensional points of light, some brighter than others. To a seasoned star-traveller, however, it would be quite clear that (1) the stars are all moving at enormous velocities, (2) they are at such great distances from Earth that their motions are not apparent except when measured over long periods of time, and (3) they are similar to our Sun in general appearance, although some are over 100 times smaller, and others over 2000 times larger, than the Sun. Furthermore, stars come in a variety of colors and temperatures: some yellow, like our own, some bluer and hotter than the Sun, and others redder and cooler. Finally, stars have a tendency to cluster into groups that range in size from closely associated double or triple star systems through globular clusters of stars numbering in the tens of thousands to galaxies with stars numbering in the tens to hundreds of billions. To the unaided eye, all these diverse objects, including those clusters and galaxies that are visible without a telescope, all look much the same.

It was not until optical and radio telescopes and spectrographs became available that astronomers could begin to differentiate among these various objects, and it was not until they had developed reliable means for measuring cosmic distances that all this wealth of stellar information could be mustered into a meaningful picture. Measuring the distances to extraterrestrial objects from an Earthbound stance is no simple challenge, and some ingenious methods have been devised to meet that challenge. In the following sections, I examine the most important of these methods.

## Parallax

Before the acceptance of the heliocentric theory, astronomers had available to them a fairly sizable cosmic yardstick in the form of Earth itself. In figure 18.1, two observatories are located at points 1 and 2, and the distance between them is known. If a nearby celestial object, O, is sighted simultaneously from both observatories, there will be a slight angular difference in the object's celestial coordinates, as measured from the two stations. This is the **parallax angle, $\pi$**, of the object, and it can be used to calculate the distance, D, to the celestial object if the length of the *baseline, B,* between the two observatories is known. If the baseline is drawn so that it forms the short leg of a right triangle, as in figure 18.1, then the distance can be easily solved by

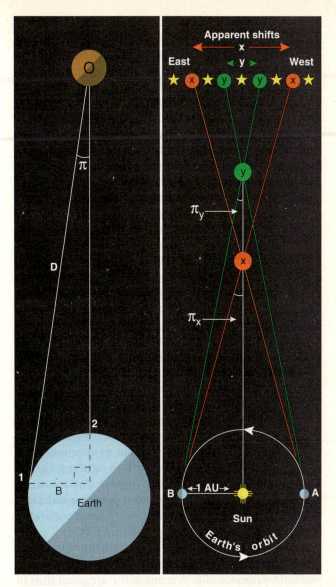

***Figure 18.1 (left)*** One way to determine the distance of a celestial object, O, from Earth is to observe it simultaneously from two different stations and then calculate the parallax angle, $\pi$, between the object's celestial coordinates as noted at each station. Here, the sightings are made when the object is directly above station 2. The baseline, B, is drawn perpendicular to the line connecting the object and station 2. The distance, D, is then equal to $B/\sin\pi$.

***Figure 18.2 (right)*** When Earth's orbital radius is used as a baseline, the distances to some of the nearest stars can be measured. A parallax angle can be measured from a star's apparent shift against the backdrop of more distant stars over a six month period. Note that as Earth moves from point A to point B, star X, being nearer than star Y, exhibits a greater parallactic shift. The parallax angle, $\pi$, for a baseline of one astronomical unit (AU) is one-half the apparent shift.

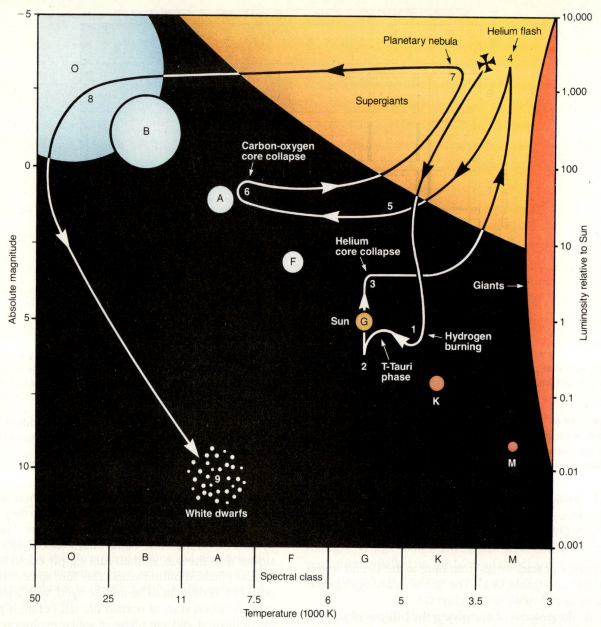

**Figure 18.9**   In the Hertzsprung-Russell diagram, spectral classes and temperatures (*horizontal axis*) are plotted against luminosity and absolute magnitude (*vertical axis*). Most stars fall along the main sequence from blue giants (*upper left*) to red dwarfs (*lower right*). Red to white giants and supergiants lie above the main sequence on the right. White dwarfs lie below the central portion of the main sequence. The solid line traces the evolutionary path of the Sun. In 3 to 5 billion years, the Sun will expand twice into a red giant, then shrink to a white dwarf. Relative sizes of stars are to scale.

## Abnormal Stars

Most stars fall comfortably within one or another of the spectral classes, but there are some that do not. In the 1900s, Ejnar Hertzsprung and Henry Russell demonstrated this fact when they independently produced plots of absolute magnitude against spectral class (fig. 18.9). These plots have come to be known as **Hertzsprung-Russell diagrams,** or simply **H-R diagrams,** and they have proved extremely useful to astrophysicists. Notice that most stars form an array that makes a practically straight line sloping down toward the lower right corner of the diagram in figure 18.9, confirming the direct relationship between temperature and brightness. This primary band has been called the **main sequence.** The Sun's position in the main sequence corresponds to that of the yellow star of class G. In the upper right corner of the H-R diagram, however, there is a group of stars that are too big, too bright, and too cool for their spectral classes. In the lower part of figure 18.9, below the main sequence, lies a second group of "abnormal" stars that are too small, too dim, and too hot for their spectral classes. The first of these star groups includes the **giants** and **supergiants.** As an example of a red supergiant, Antares (An-*tare*-eez) in the constellation Scorpio (see the star charts of appendix V) has only half the surface temperature of the Sun but 30 times the Sun's mass, 480 times its diameter, 3500 times its brightness, and only 270 billionths of its density. The second cluster includes the **white dwarfs,** of which the companion star to Sirius in Canis Major is an example. Its mass is about the same as the Sun's, and its surface temperature is, at 7500 K, about 1700 K hotter than the Sun's, but it has only one-thirtieth the Sun's diameter, one-tenth its brightness, and an incredible 24,000 times its density. The atoms within this star must be so tightly compressed that their electron shells have been crushed, allowing the nuclei to pack more closely together.

In addition to the supergiants, giants, and dwarfs, there are some other peculiarities within the stellar population. These fall under the general category of **variable stars,** so called because the light received from them varies cyclically in intensity. This variability can have either an internal or an external cause. In the latter category are the **eclipsing binaries,** double stars whose orbital plane is so oriented that the two members of the pair eclipse one another fully or partially with respect to Earth (fig. 18.10). Notice that the brightness of the pair is diminished far more when the smaller, dimmer star is in conjunction (point *A*) than it is when it is in opposition (point *C*). The most notable eclipsing binary is Algol in Perseus. Its Arabic name (Al Ghul, "The Demon") reflects the star pair's mysterious habit of "winking" down to about one-third of its normal brightness as the darker companion eclipses the brighter for about 4.5 hours during its 68.8 hour orbital period.

In general, neither star of an eclipsing binary system is abnormal, but the **pulsating variable** stars are definitely so. Among these are the RR-Lyrae variables, small, dim stars that pulsate with periods of less than one day, and the **Cepheid variables,** yellow supergiants with periods ranging from 1 to 50 days. In addition to the RR-Lyrae stars and the Cepheids, there are also brilliant, variable blue giants with fast pulse rates ranging from one hour to one day, and smaller, dimmer, longer-period variables with pulse rates between 50 and 1000 days. We have seen that the longer the period of a pulsating Cepheid, the greater is its brightness. These stars change diameter by as much as 30% with each pulse. As they collapse their brightness increases, and as they expand it decreases again. This happens because during collapse core temperature rises sharply, which stimulates more energetic nuclear reactions, which, in turn, expand the star again. *T-Tauri variables* are a group of irregularly variable stars, surrounded by gas clouds, that are

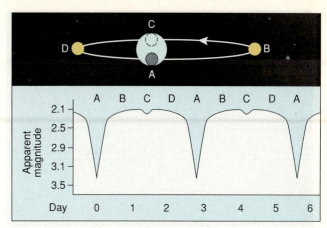

**Figure 18.10** The light curve (*bottom*) of the eclipsing binary Algol shows a strong dip as the dimmer star transits the disk of the brighter (point *a*), and a less pronounced dip as it passes into opposition (point *c*).

thought to be very young and therefore still in the process of working toward an equilibrium between gravitational contraction and thermonuclear expansion.

More dramatic than the pulsating variables are the *exploding variables.* These spectacularly unstable stars fall into two categories: the *novae* (*no*-vee, singular: *nova*; Latin for "new") and the *supernovae*. The novae were so called by early astronomers because of their tendency to appear suddenly in the night sky in places where no star had ever been seen before. Novae increase suddenly in brightness on recurrent schedules, most of them dropping about 7 to 10 magnitudes once every million years or so. Supernovae, as their name suggests, are far more spectacular, typically undergoing a single, nonrecurrent, catastrophic explosion with a decrease of at least 14 magnitudes at the end of the star's lifetime.

**Novae** are relatively common events, a given galaxy experiencing on the order of 10 per year. They occur in double star systems in which one star of the pair is a bloated giant and the other an aging dwarf. Gas from the giant is sucked in by the dwarf's gravitational field and accumulates until it becomes deep enough and hot enough to initiate nuclear fusion. This evidently occurs with such suddenness as to blow the entire accumulation clear of the star's surface. Shells of rapidly expanding gas have been detected around most novae.

Supernovae, on the other hand, are quite rare, occurring only about once every 30 years in a given galaxy. These dramatic events destroy the stars that experience them, and the debris from the explosion flies outward in a chaotic splash of furiously excited gas (fig. 18.11). Because these great, cosmic eruptions are involved in stellar evolution, I discuss them in the following section.

**Figure 18.17**  M82, an irregular galaxy in Ursa Major, is emitting over 10 times the energy emitted by the Milky Way. It appears to be undergoing a catastrophic explosion.

Photo by U.S. Naval Observatory

**Figure 18.18**  M51, the Whirlpool Galaxy in Canes Venatici, is a typical spiral with a small, irregular companion galaxy, NGC5195, attached to the end of one of its spiral arms.

Photo by U.S. Naval Observatory

**Figure 18.19**  NGC147, an elliptical galaxy in Casseopeia, photographed in red light.

Photo by Hale Observatories

**Figure 18.20**  M83, a barred spiral galaxy in Eridanus.

© Science VU-NOAA/Visuals Unlimited

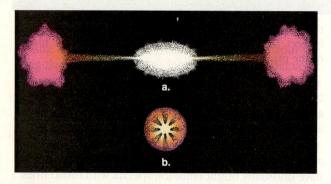

**Figure 18.21**  Radio galaxies (*a*) and quasars (*b*), are probably identical phenomena viewed from different angles.

radio power than normal galaxies. Visual images of such objects have revealed that many are experiencing cataclysmic explosions, which in some cases are obviously tearing the galaxy to pieces. Typically, radio galaxies exhibit a dumbbell shape, in which the ends of the dumbbell consist of clouds of charged particles ejected from the galactic center at enormous velocities by oppositely directed jets (fig. 18.21*a*). One hypothesis attributes these energetic jets to the destruction of stars as they become crowded together in old, elliptical galaxies. This would be expected to result in supermassive black holes that would attract and engulf nearby stars at a rapid rate.

Closely related to the radio galaxies, and possibly identical to them are the **quasars** (quasi-stellar radio sources), mentioned in chapter 16. Quasars appear starlike in telescopic view, but many of them exhibit large

# Food for Thought

How did the Universe originate? Inasmuch as the human species has existed for only about 200 millionths of the probable age of the Universe, the question might appear to be the height of impertinence. As it is in our nature to question all things, however, it was inevitable that this ultimate question should be among those asked. Unfortunately, the origin of the Universe, like that of the Solar System, is not subject to experimental verification. Any answers we come up with must stand or fall purely on the basis of the boundary conditions that must have constrained the event.

The most significant of these boundary conditions is **Hubble's law,** the observation that the Universe is everywhere expanding at a uniform rate of 75 km/sec/megaparsec (1 megaparsec = $10^6$ parsecs). Hubble proposed this law following his observation that the amount of red shift exhibited by celestial objects increases regularly in all directions with distance from Earth. If Earth were stationary, this observation would imply that all celestial objects are moving radially away from Earth at velocities that increase with their distance. This is possible, but exceedingly unlikely. A more probable explanation is that the Universe is expanding uniformly; hence, all celestial objects are steadily increasing their distances from one another, much as the raisins in a loaf of raisin bread spread apart from one another as the dough rises. Inasmuch as these separations are cumulative, adding to one another with distance, the effect is that more distant objects are receding from Earth much faster than are nearer objects, but this translates into relative, and not absolute, velocities. The most distant objects we can "see" with radio telescopes appear to be fleeing from us at nearly nine-tenths the speed of light, but that velocity is only relative to our own, unknown velocity. As we have not yet identified any absolute, universal point of reference, we have no way of establishing the "true" velocity of any object in the heavens.

Three hypotheses of the origin of the Universe are currently in vogue. The **big bang hypothesis** proposes that some 13 billion years ago all the matter in the Universe was concentrated within a supermassive point source. That source was so hot that it existed in the form of a single, undifferentiated type of particle that contained, symmetrically distributed within itself, the six *fundamental properties* of charge, mass-energy, gravitational force, electrostatic force, and the strong and weak nuclear forces. (How the point source got there in the first place is not explained.) Suddenly, this primordial point source exploded, dispersing itself into space. In the process, the temperature of the original, uniform particles decreased until they resolved themselves into derivative particles of lower symmetry. These particles were the quarks, the *leptons* (including electrons, positrons, and neutrinos), and the photons that constitute the structure of ordinary matter and energy. According to the big bang hypothesis, the uniform expansion we observe today is simply a continuation of the dispersal that began with the original explosion.

The existence of gravity implies that this great, centrifugal outburst should have experienced a steady braking effect over time, so that the expansion should be somewhat slower now than it was at first. This is confirmed by the observation that the apparent radial velocities of more distant objects are greater than are those of closer objects. Because we see those more distant objects as they were millions to billions of years ago, their apparent velocities should indeed appear to be higher, and those of nearby objects should appear to be lower, if gravitation has been effective in braking the expansion of the Universe.

Just how effective this braking will ultimately prove to have been is the point of contention between the big bang and the **pulsating Universe hypotheses.** Whereas the former postulates a single expansion that continues forever, the latter favors endless cycles of expansion and contraction in which each big bang is both followed and preceded by a "big crunch."

A third, more peaceful prescription for the origin of the Universe supposes that all matter was not created in one grand burst, but is continually being created as the Universe expands. This **continuous creation,** or steady state, hypothesis requires that, on average, 45 atoms of hydrogen per liter of space be created every 100 years to account for the observed expansion. Because the ratio between numbers of galaxies and the energy they emit over time is only about half of that required by this hypothesis, it has largely been abandoned.

Many more small-scale questions need to be answered before this biggest of all questions can be approached with confidence. Among these are the following:

1. What is the ultimate fate of the neutrinos that are released in thermonuclear processes?
2. What happens to the matter that is sucked down into black holes? Are there, as some have argued, "white holes" that spew out matter as black holes consume it?
3. Does the Universe have zero curvature like a flat sheet, negative curvature like a saddle, or positive curvature like a beach ball (fig. 18.22)?

Limited evidence indicates that the answer to the last question might be that the Universe is positively curved, i.e., parallel lines in space might ultimately converge. If this is true, the Universe is finite in size; therefore, infinite expansion would not be possible. Hence, the pulsating Universe model could be the only admissable one. This conclusion carries with it the satisfying implication that, as Hutton declared, there truly is "no vestige of a beginning; no prospect of an end."

*continued*

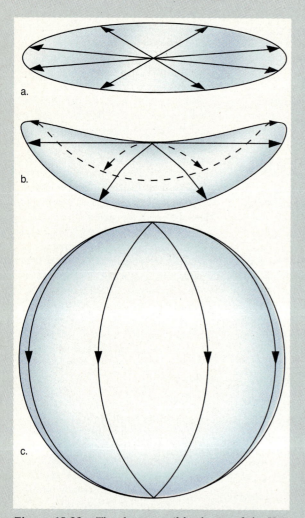

**Figure 18.22** The three possible shapes of the Universe. In a noncurved Universe (*a*) and a negatively curved Universe (*b*), light rays diverge, never to meet again. In a positively curved Universe (*c*), however, light rays ultimately converge.

red shifts, indicating that they must be as much as 15 billion light-years away. Hence, they are the most distant, and therefore the oldest, visible objects in the Universe (fig. 18.21*b*). To be as visible as they are at such distances, they must emit energy at rates that are hundreds of times greater than those of normal galaxies, although other evidence suggests that quasars might be much smaller than normal galaxies. Long, straight jets of glowing material have been seen erupting from some quasars. Recently, Peter Barthel, a Dutch astronomer, has proposed that quasars are simply radio galaxies viewed along the axis of the jets that form the dumbbell. Thus oriented, a radio galaxy would function as a beacon of phenomenal brightness, beaming forth unimaginable quantities of light from the destruction of stars as they swirl into the central black hole.

## Summary

The **parallax angle** of a nearby celestial object can be used to determine its distance from Earth. One **parsec** (30.8 trillion km or 3.26 **light-years**) is the distance at which one AU subtends 1″ of arc. Parallactic distances of more than 25 parsecs are unreliable. *Dynamical parallax* can be used to measure distance to stars in **binary systems** (about 25% of all stars).

A star of **apparent magnitude** $X$ is 2.512 times brighter than a star of apparent magnitude $X + 1$. A star's **absolute magnitude** is its apparent magnitude at a distance of 10 parsecs. The brighter a *Cepheid variable,* the longer is its period. From this and its apparent magnitude, the distance to the Cepheid can be calculated, up to about 3 million light-years. Hubble and Humason used apparent magnitudes of blue giant stars and of galaxies to estimate distances between 3 million and 1 billion light-years. There are 100 to 200 billion stars in the Milky Way and about as many galaxies in the Universe.

Wavelength of peak "black body" radiation varies inversely as the Kelvin temperature. White light passing through a slit is **dispersed** by a prism, forming a **continuous light spectrum.** Violet light is refracted more than red light. Gases in the path of a light beam absorb characteristic wavelengths, shown as dark lines in **absorption spectra.** A glowing gas produces an *emission spectrum* with bright lines corresponding to the dark lines in its absorption spectrum. Absorption spectra of approaching objects are shifted by the **Doppler effect** toward the blue end. Absorption spectra of receding objects are shifted toward the red end. The **red shift** for most galaxies is directly proportional to distance, indicating a uniformly expanding Universe. Objects beyond the **light horizon,** about 13 billion light-years away, recede faster than light; hence, they are invisible to us. Images of objects are as old as their distance in light-years; hence, the Universe is at least 13 billion years old. The **time horizon** is probably not much farther out than this.

Trends in stars from O to M in the **spectral classes** O, B, A, F, G, K, and M include lower temperature, blue to yellow to red color, and increasingly heavy elements and molecules. Hotter stars are larger, more massive, more rapidly burning, shorter lived, and less abundant than cooler stars. The middle-aged Sun has 3 to 5 billion years of normal life left. A star's **color index** is its apparent magnitude on a blue-sensitive meter minus its apparent magnitude on a yellow-sensitive meter. O and B stars have negative color indices. Stars in classes A through M have positive indices.

On an **H-R (Hertzsprung-Russell) diagram,** normal stars plot along the **main sequence,** a line showing decrease in magnitude with increasing temperature. **Giants** and **supergiants** plot above and to the right of the main sequence. **White dwarfs** plot below the main sequence.

**Variable stars** include **eclipsing binaries** and **pulsating variables,** the latter comprising blue giants with pulse rates of 1 hour to 1 day, small RR-Lyrae variables with pulse rates of less than a day, yellow supergiant **Cepheid variables** with pulse rates of 1 to 50 days, and longer-period variables with pulse rates of 50 to 1000 days. *T-Tauri variables* are young stars with irregular pulses, surrounded by gas clouds. *Exploding variables* include recurrent **novae** and nonrecurring **supernovae. Population I,** characteristic of galactic spiral arms, is of mixed age, and includes young, hot, blue giant stars and abundant gas. **Population II,** characteristic of the galactic hub and **globular clusters** surrounding the Milky Way, is older and contains abundant red giants but no blue giants and little gas.

Stars condense from **nebulas,** contracting and heating until thermonuclear ignition occurs. After a star has consumed about 10% of its hydrogen, its helium core collapses, and it expands into a *red giant.* When it has consumed about 40% of its hydrogen, the core collapses again, producing a *helium flash,* after which the star dims and shrinks as its core heats to the ignition point of helium. Helium burning produces a core of carbon and oxygen ash, which collapses, again expanding the star into a red giant and ultimately expelling up to 20% of the star's mass as a **planetary nebula.** The star then slowly cools and shrinks to a white dwarf. In a star more than 4 times as massive as the Sun, core collapse results in a supernova, in which heavy elements are created, and in a dense, rapidly spinning **neutron star,** or **pulsar,** which emits strong radio beacons. If the residue exceeds 10 solar masses, it collapses further to a **black hole,** from which not even light can escape.

The Milky Way galaxy is about 80,000 light-years across, about 15,000 light-years thick at the center, rotates fastest near its center, and is surrounded by a halo of stars and globular clusters in random orbits. The Sun is located about three-fourths of the way from hub to outer edge. The Sun revolves around the galaxy in about 250 million years. Doppler shifts show that other galaxies also rotate. About 10% of normal galaxies are **irregular galaxies,** about 30% **spiral galaxies,** and about 60% **elliptical galaxies.** Abnormal galaxies include **barred spirals,** constituting one-third of all spirals, and *eruptive galaxies,* including **radio galaxies** and **quasars,** both of which contain massive black holes and are probably identical but seen from different angles.

By **Hubble's Law,** the Universe is expanding uniformly at a rate of 75 km/sec/megaparsec. Of the **big bang, pulsating Universe,** and **continuous creation** hypotheses, the second appears to be the most probable and the third the least probable.

## Key Terms

parallax angle
light-year
parsec
binary system
apparent magnitude
absolute magnitude
continuous light
   spectrum
dispersion
absorption spectrum
Doppler effect
red shift
light horizon
time horizon
spectral class
color index
Hertzsprung-Russell
   diagram
main sequence
giant
supergiant
white dwarf
variable stars
eclipsing binary

pulsating variable
Cepheid variable
nova
population I star
globular cluster
population II star
nebula
planetary nebula
supernova
neutron star
pulsar
black hole
irregular galaxy
spiral galaxy
barred spiral galaxy
elliptical galaxy
radio galaxy
quasar
Hubble's law
big bang hypothesis
pulsating Universe
   hypothesis
continuous creation
   hypothesis

# Questions for Review

1. What is the parallax angle of a celestial object having a distance from Earth of 1 parsec?
2. How many times brighter is Algenib than Formalhaut?
3. Explain how Cepheid and RR-Lyrae variables can be used to measure distance.
4. What trend is shown by the red shift with increasing distance from Earth? What does this imply about the state of the Universe?
5. What three main trends are evident in the progression from class O stars to class M stars?
6. In what ways are giants, supergiants, and dwarfs abnormal with respect to the main sequence stars on the H-R diagram?
7. What is the relation between the period of a pulsating variable and its brightness?
8. Explain the differences in the processes involved in nova and supernova events.
9. Describe the differences between populations I and II in terms of characteristic star type, age, nonstellar material, and location within galaxies.
10. Describe the life histories of stars that are of (1) less than 4 solar masses, (2) between 4 and 10 solar masses, and (3) over 10 solar masses.
11. Compare and contrast the structures of the four galactic types.
12. Describe radio galaxies and quasars. State the probable reason for their apparent differences.
13. Describe the three principal hypotheses for the origin of the Universe. Which is the least probable and which the most probable? Why?

# Part Six

# Earth and Humanity

Photo: The Kenai Range viewed from across Kacheniak Bay, Alaska; one of the few places on Earth where Nature still dwarfs the human presence.

# 19 Care and Maintenance

## Outline

*I* met a traveler from an antique land

Who said: 'Two vast and trunkless legs of stone

Stand in the desert. Near them, on the sand,

Half sunk, a shattered visage lies, whose frown

And wrinkled lip, and sneer of cold command,

Tell that its sculptor well those passions read

Which yet survive, stamped on these lifeless things,

The hand that mocked them and the heart that fed.

And on the pedestal these words appear—

"My name is Ozymandias, king of kings:

Look on my works, ye Mighty, and despair!"

Nothing beside remains. Round the decay

Of that colossal wreck, boundless and bare,

The lone and level sands stretch far away.'

*Percy Bysshe Shelley*

***Ozymandias***

*B*ú dàu dzáu yi.

*(Whatever deviates from the Path soon vanishes.)*

*Lao Tzu*

***Tao, Te Ching***

Photo: What price progress? Ironically, the pursuit of the American dream, and its equivalent in other industrialized nations, threatens to destroy the very thing that those who pursue it are seeking to improve: the quality of life.

## How to Use a Living Planet

On purchasing a new automobile, garden tractor, snowblower, personal computer, or any other piece of sophisticated equipment, we are accustomed to receiving with the purchase an owner's manual that describes the equipment and its functions, and tells us what we should take pains to do (and not do) if we wish to get the best use and the longest possible service life out of the equipment. Unfortunately, no such owner's manual was handed to the human species when it decided to take possession of planet Earth. This chapter was written with that pressing need in mind.

## Safety

As with any complex apparatus, there are many hazards involved in the use of a planet, and the user should be well informed about them in order to avoid serious injury or death. The following discussion covers the situations that most frequently lead to personal loss as a result of imprudent exposure to the many environmental hazards of planet Earth. It also includes preventive measures, sources of information, and a list of the most reliable warning signs that signal the impending occurrence of a hazardous event.

**Frequency** (how often) and **intensity** (how much) are the two most significant factors in assessing the potential risk associated with a given hazard. Areas subject to a high frequency of hazardous events should probably not be developed for any use regardless of the intensity of such events. Areas in which hazardous events are infrequent and of low intensity are suitable for many uses, although the possibility of occasional damage to permanent structures or of personal injury should be kept in mind, especially when considering insurance coverage. Areas in which hazardous events are infrequent but of high intensity should be restricted to such low-risk development as parks and open space, roads and trails, and possibly to the storage of nonhazardous materials. *In no case* should an area be used for any purpose, such as dam construction, nuclear power plant siting, hazardous materials storage or disposal, or high-density recreation, that would significantly increase the damage potential of a natural hazard.

## Endogenetic Hazards

An **endogenetic hazard** is one that results from processes originating at a depth below Earth's surface beneath which surficial processes are inactive. This category comprises earthquakes, volcanism, and natural radioactivity. With the exception of radioactivity, endogenetic hazards tend to be of high intensity and low frequency. This is an insidious combination because long periods of inactivity lead to complacency and a false sense of security, and the temptation is high to put land

**Table 19.1** Average Annual Losses from Selected Natural Hazards in the United States

| Hazard | Property Damage, $ million | Deaths |
|---|---|---|
| Earthquakes | 1420 | 12 |
| Volcanoes | 25 | (1–2) |
| Floods | 3800 | 200 |
| Mass wasting | 1000 | 25 |
| Tsunami | 26 | 5–7 |
| Subsidence | 200 | (1–5) |
| Swelling soil | 5600 | (1) |
| Hurricanes | 500 | 75 |
| Tornadoes | 75 | 125 |
| Lightning | 200 | >150 |
| Hail | 284 | 0 |
| Total | 13,130 | ±600 |

Sources: J. E. Costa and V. R. Baker, *Surficial Geology: Building with the Earth,* 1981, John Wiley & Sons, Inc., New York, NY, and various others.

subject to such hazards to permanent use. Unless the statistical risk of damage from exposure is low, however, it is unwise to locate any permanent structure, or to engage in any long-term activity, within a region that is known to be subject to significant endogenetic hazards.

### Earthquakes

Faulting and earthquake shock waves have dramatically destructive effects both on Earth materials and on human-made structures. In addition, they have been responsible for extensive losses of human life (table 19.1). Most such losses could be avoided if adequate precautions were taken to evaluate the potential risk and to minimize exposure to it. Following are the most significant and troublesome hazards associated with large earthquakes:

**1. Shaking.** The most directly damaging effect of an earthquake is the shaking of the ground surface by Love and Rayleigh waves (see chapter 6). Any potentially unstable structure, whether natural or human-made, can be dislodged or toppled, as was the upper deck of the Nimitz expressway in San Francisco during the Loma Prieta earthquake of October 1989, which had a Richter magnitude of 7.1 (fig. 19.1). During the same earthquake, San Francisco's waterfront district also suffered heavy damage, as it was built on artificially drained wetland soil. The principal effects of shaking on natural materials are discussed later in this section.

**2. Ground level changes.** Extensive areas of land or seafloor are sometimes elevated or depressed by as much as several meters during an earthquake. In the great Alaskan earthquake at Prince William Sound in March 1964, some parts of the affected region were permanently uplifted by more than 13 m, and others sank by as much as 3 m (fig. 19.2).

*Figure 19.1*    The upper deck of the Cypress Street viaduct on Interstate 880 in Oakland, California lies collapsed on the lower deck following the magnitude 7.1 Loma Prieta earthquake of 17 October 1989.

Photo by H. Wilshire, U.S. Geological Survey, courtesy of Peter L. Ward

**3.   *Liquefaction.***    In areas of loose or poorly consolidated sediments, especially where these are saturated with water, the violent shaking of the ground by an earthquake can so disrupt the microstructure of clay-rich materials that they assume the properties of a fluid. This process is called **liquefaction.** During the 1964 Alaskan earthquake, the 1 km² residential area of Turnagain Heights, built on montmorillonite clays, was destroyed when the clay was partially liquefied by the earthquake shocks (fig. 19.3).

**4.   *Tsunami.***    During an earthquake or volcanic eruption, the seafloor can be displaced suddenly or buried by submarine landslides. These disturbances can, in turn, displace large volumes of seawater, resulting in the generation of seismic sea waves, or **tsunami** (tsoo-*nah*-me; Japanese for "harbor wave"), that travel outward from the point of disturbance. Tsunami are fundamentally different from ordinary, wind-generated waves (see chapter 12). The wavelength of a tsunami can be anywhere from 120 to 720 km, and its velocity in deep water can be as high as 250 m/sec. The wavelengths of wind-generated waves, on the other hand, are less than 1 km, and their velocities are less than 50 m/sec.

Another important difference between seismic and wind-generated sea waves is that because of their great wavelengths, tsunami set the entire body of the sea in motion (recall that wave depth is one half the wavelength). In spite of this, tsunami are undetectable on

*Figure 19.2*    During the great Alaskan earthquake of 27 March 1964, a large block of the seafloor (*right*) in Prince William Sound was raised as much as 4.9 m, forming a shallow lagoon in which mud and sand were deposited.

Photo by U.S. Geological Survey

*Figure 19.3*    The Bootlegger Cove Clay, a deposit of Pleistocene marine silt, proved to be an inadequate foundation for the elegant homes of the Turnagain Heights development in Anchorage, Alaska, during the 1964 earthquake.

Photo by U.S. Geological Survey

the open ocean because their height is rarely more than 1 m. When these waves impinge on shallow, nearshore water, however, all the kinetic energy that was previously distributed throughout the full depth of the ocean becomes concentrated into a much shallower column of water. The result of this is a sudden, drastic amplification of the wave height to as much as 35 m.

Such monster waves can inflict great destruction on seacoast settlements. Figure 19.4 shows the chaos that resulted when a tsunami struck Hilo, Hawaii, in May 1960 after travelling across thousands of kilometers of ocean from an earthquake of magnitude 8.4 on the coast of Chile. Occasionally, an unusual geological event can produce truly colossal waves, such as the 410 m high wave that resulted when 30 million m³ of rock fell into

**Figure 19.4** An earthquake of magnitude 8.4 in Chile in May 1960 generated a seismic sea wave that created wholesale destruction in Hilo, Hawaii, 15 hours after it occurred.

Photo by U.S. Geological Survey

**Figure 19.5** During the 1959 earthquake at Hebgen Lake, 32 million m³ of bedrock slid from a mountainside, (*right*), burying a campground and damming the Madison River to form a lake 8 km long and over 30 m deep.

Photo by J. R. Stacy, U.S. Geological Survey

Lituya Bay, Alaska, during an earthquake in 1958. The dense forests that once lined the bay shore were totally obliterated by the wave.

**5. Landslides.** Although most landslides are not endogenetic, some of the worst are triggered by earthquakes in mountains or waterfront terrains. Figure 19.5 shows a massive landslide that dammed the Madison River in Montana when an adjacent mountainside collapsed during the 1959 Hebgen Lake earthquake, forming Earthquake Lake. In this case, the mountainside was composed of waterlogged sedimentary rock strata that dipped steeply toward the river. Hillslopes with this kind of geologic structure are inherently unstable. Wherever such unstable landforms can be identified by geological field studies in earthquake-prone

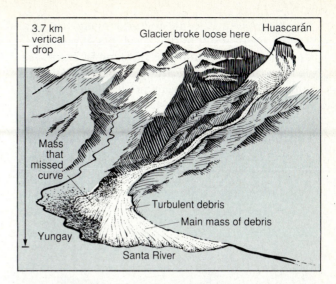

**Figure 19.6** During an earthquake in May 1970, a massive avalanche broke loose from near the summit of Nevado Huascarán, a 6663 m high Peruvian volcano. A volume of glacier ice, rock, and mud sufficient to cover a square kilometer to a depth of 50 m fell 3.7 km at speeds up to 435 km/hr. A portion of the debris flew over a ridge and descended in a thick blanket, burying the town of Yungay and an estimated 17,000 people. Geologic evidence indicates that similar events have occurred frequently here in the past.

areas, they should be avoided as sites for human land use. The 28 people who died in the Madison Canyon landslide might still be alive if a public campsite had not been located in the path of the slide.

Sometimes, whole towns are destroyed by earthquake-triggered landslides. Such was the case during the Peruvian earthquake of 31 May 1970. In that event, a portion of the north peak of the giant volcano Nevados Huascarán collapsed, sending an avalanche of muddy debris 14.5 km down the slopes of the Cordillera Blanca at speeds of 400 km/hr to bury completely the towns of Ranrahirca and Yungay and nearly 21,000 of their inhabitants (fig. 19.6). There were only about 300 survivors.

In the province of Kansu, northwest China, an earthquake in 1920 triggered massive landslides in thick deposits of windblown loess. These slides took a toll of 180,000 lives, making the event one of the greatest earthquake disasters in history.

**6. Hazards related to human activities.** Aside from natural earthquake-related hazards, there are others of human origin that can be even more costly of lives and property. In the San Francisco earthquake of 1906, many more people were killed by fire and the collapse of poorly constructed buildings and "gingerbread" cornices than by the shock of the earthquake itself (fig. 19.7). Modern building codes require that buildings be constructed to withstand severe earthquake shocks that can last for up to three minutes, as did the Alaskan earthquake of 1964. Furthermore, water systems in

Care and Maintenance    **499**

earthquake-prone cities are now so designed that breaks in water mains result in only local loss of pressure, so that fires that result from the breaking of electric and gas lines can still be effectively suppressed. Such precautions can help to alleviate the deadliness, if not the destructiveness, of future large quakes.

**Warning Signs.** The most reliable indicators of an impending earthquake, together with their approximate lead times (in parentheses) are

1. Ground level tilting and deformation (hours to years) (fig. 19.8)

*Figure 19.7* Ornate masonry structures such as the Hibernia Bank building, pictured here, fared worse than other types of structures during the great San Francisco earthquake of 1906.

Photo by W. C. Mendenhall, U.S. Geological Survey

2. Swarms of microearthquakes (hours to days)
3. Decrease in the velocity of P waves (due to a decrease in rock rigidity), followed by a return to normal velocity (months to years)
4. Stress-induced changes in electrical and magnetic properties of rocks (hours)
5. Prolonged seismic inactivity in a local area (a so-called *seismic gap*) within an otherwise active seismic belt (years)
6. Increased emission of radon due to the dilation of bedrock (days to months)
7. Abnormal behavior in various animals from fish (weeks) to horses (minutes)

**Precautionary Measures.** Avoid building any structure for human habitation within 30 m of the mapped trace of a fault known to have been active in historic time. Avoid building such structures on unconsolidated materials, such as drained swamp muck and quick clay, that have a high potential for shaking or liquefying during earthquakes. Avoid building beneath potentially unstable cliffs or slopes that could fall or slide during an earthquake. Use approved construction methods and materials when building in earthquake prone areas. Avoid the use of unreinforced masonry and nonstructural building ornaments.

**Sources of Information.** The seismic risk zone map of the United States in figure 19.9, compiled by the U.S. Coast and Geodetic Survey, divides the country into four categories of seismic risk, from no damage (0) to major

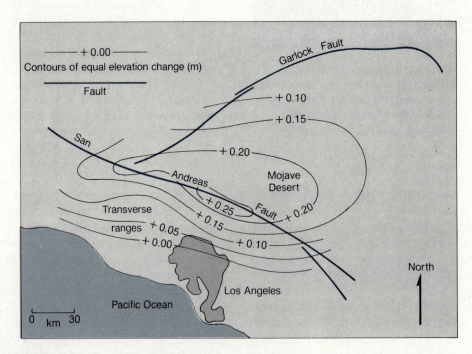

*Figure 19.8* Accumulating strain along the San Andreas fault has produced an uplift of as much as 0.25 m in the Palmdale area north of Los Angeles.

Source: U.S. Geological Survey.

damage (3). All the maximum risk zones are either broken by extensive systems of faults or are regions of rapid upwarping of the crust, as on the Idaho-Montana border and the coast of California. Seismic risk maps for smaller regions are available from state geological survey offices. Also available from most state surveys and the U.S. Geological Survey are bedrock geologic maps, which show the locations of faults. In California, fault zone maps for active fault systems are available from local government offices.

In 1948, the U.S. Coast and Geodetic Survey established a seismic sea wave warning system (SSWWS) to alert coastal Pacific settlements to the threat of tsunami following major earthquakes. Maps such as the one in figure 19.10, showing tsunami travel times to specified locations, can be obtained from the National Oceanic and Atmospheric Administration (NOAA) in Boulder, Colorado.

## Volcanism

Although volcanoes have been less destructive of human life and property than earthquakes, they still pose a formidable threat (see table 19.1). One reason they have been less destructive is that volcanoes, with their prominent, distinctive forms, are more obvious than faults. Another is that they normally give more warning of impending activity than earthquakes do. Nonetheless, some kinds of volcanic eruptions are potentially far more destructive than earthquakes. Within the past 2 million years, thick layers of volcanic ash have repeatedly covered large areas comprising all or parts of as many as 11 western states. The scale of the felsic caldera eruptions that caused them is beyond both our experience and our imagination, and the most recent occurred only 2400 years ago. There is little reason to suppose that similar eruptions will not occur in the future. The most likely

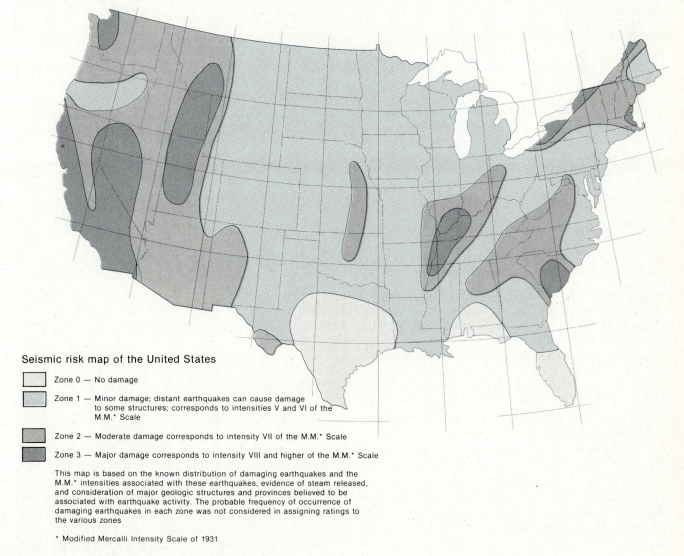

Seismic risk map of the United States

Zone 0 — No damage

Zone 1 — Minor damage; distant earthquakes can cause damage to some structures; corresponds to intensities V and VI of the M.M.* Scale

Zone 2 — Moderate damage corresponds to intensity VII of the M.M.* Scale

Zone 3 — Major damage corresponds to intensity VIII and higher of the M.M.* Scale

This map is based on the known distribution of damaging earthquakes and the M.M.* intensities associated with these earthquakes, evidence of steam released, and consideration of major geologic structures and provinces believed to be associated with earthquake activity. The probable frequency of occurrence of damaging earthquakes in each zone was not considered in assigning ratings to the various zones

* Modified Mercalli Intensity Scale of 1931

**Figure 19.9**    Seismic risks in the United States.

Source: U.S. Coast and Geodetic Survey, 1969.

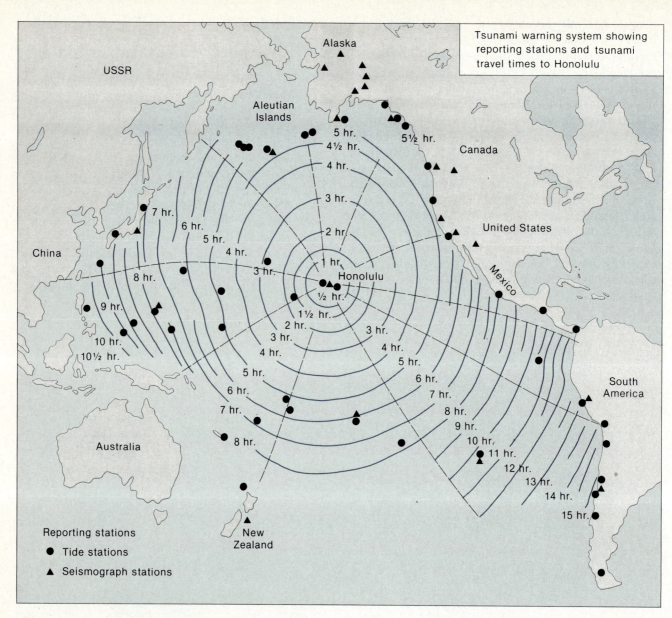

**Figure 19.10** Since 1948, a seismic sea wave warning system (SSWWS) has been in effect for the Pacific islands. This map shows travel times for tsunami to Honolulu from any location in the Pacific Ocean.

Source: National Oceanographic and Atmospheric Administration.

candidates are Yellowstone caldera in northwest Wyoming (fig. 19.11) and Long Valley caldera in east central California. The following list comprises the major hazards associated with volcanic eruptions:

**1. Tephra.** Bombs and lapilli can inflict severe damage and injury, and if hot they can also pose a fire hazard. Volcanic ash is abrasive and often toxic and corrosive. Animals eating ash-coated grass often die from digestive obstruction, poisoning, or simply from such severe tooth abrasion that they are unable to eat. Automobile engines are quickly destroyed if allowed to run in ash-filled air. In Pompeii, Italy, about 16,000 people died from suffocation as the city was inundated by ash from Mount Vesuvius during its great eruption in 79 A.D.

Finally, ash can collapse roofs. More than 100 people were killed during the 1906 eruption of Vesuvius when a church roof collapsed under the weight of accumulating ash.

**2. Ash flows.** Blasts of hot gas, dust, and ash from volcanic vents pose a severe hazard to nearby areas. The eruption of Mont Pelée on the morning of 8 May 1902 blew out the side of the mountain overlooking the city of Saint Pierre. Immediately thereafter, an ash flow roared south from the mountain at speeds of about 150 km/hr. The flow obliterated the city and virtually all of its 28,000 inhabitants (fig. 19.12). Ships at anchor in Saint Pierre harbor were set afire and capsized by the fiery blast. The sole survivor in the city was a prisoner who

**Figure 19.11** Yellowstone Lake partly fills a vast, Pleistocene caldera in northwestern Wyoming.

Photo by W. B. Hamilton, U.S. Geological Survey

**Figure 19.12** In this view from Morne d'Orange, little other than rubble remains of the city of St. Pierre following its engulfment by an ash flow from Mt. Pelée (*background*) in 1902.

Photo by I. C. Russell, U.S. Geological Survey

was locked in an underground dungeon. The prisoner, badly burned, was released and pardoned, whereupon he began a new career with a side show as the person who lived through the eruption of Mont Pelée. Ironically, the mountain had given ample warning of its eruption, and those warnings had been recognized by local scientists, but the governor chose to downplay the danger until after the coming election on the 10th of May. He even went so far as to station troops around the city to keep the frightened populace from fleeing. It was an election that nobody won.

**3. Toxic gas.** Volcanoes emit gases that either do not support life (hydrogen oxide, carbon dioxide, methane, hydrogen, nitrogen) or actively destroy it (carbon monoxide, hydrogen sulfide, hydrochloric and hydrofluoric acids, sulfur dioxide). Volcanic gases tend to flow down valleys and accumulate in hollows, where they can prove to be deadly even in the short time before

**Figure 19.13** A bold and successful attempt to prevent a lava flow from encroaching on Vestmannaeyjar harbor, Iceland, 4 May 1973. Seawater sprayed on the lava front checked its advance.

Photo by U.S. Geological Survey

atmospheric circulations can disperse them. In 1986, an eruption of carbon dioxide gas from Lake Nyos in Cameroon, Africa, killed 1700 humans and an undetermined number of other animals in a valley located below the lake.

**4. Lava flows.** Although they pose a relatively minor threat to human life, lava flows are highly destructive of property that lies in their paths. Flank eruptions of Kilauea volcano in Hawaii often descend on homes, roads, and cropland on the slopes below. Until 1973, no attempt was made to protect property from lava flows, but in that year, basaltic flows from the eruption of Eldfell volcano on the Icelandic island of Heimaey (*Hay*-ma-eh) were successfully deflected from the port town of Vestmannaeyjar (*Vest*-man-na-*eh*-yar) by directing jets of seawater against them with high-pressure hoses (fig. 19.13).

**5. Lahars.** Volcanic mudflows are the leading cause of destruction, injury, and death from volcanic phenomena (fig. 19.14). They are characteristic of composite volcanoes and result from the mixing of water with unconsolidated tephra on the cone. This water can come from many possible sources, including heavy rain, groundwater, a crater lake destroyed either by erosion or eruption, the rapid melting of glacier ice and snow on the cone by lava, and streams dammed by lava, tephra, or glacier ice. Lahars often occur during eruptions, but they can also be triggered by earthquakes, heavy rains, the melting of ice dams, the failure of dams formed by landslides, and other events.

**6. Tsunami.** Some of the world's greatest disasters have been caused by tsunami generated by volcanic explosions. The prosperous, powerful Minoan civilization on the Mediterranean island of Crete was fatally weakened by the tsunami that resulted from the phenomenal explosion of Santorini volcano about 3400

**Figure 19.14** The remains of the village of Armero, Columbia, after it was inundated by a lahar generated by the eruption of Nevado del Ruiz in November 1985.

years ago. When the Indonesian volcano Krakatao (Crack-ah-*tah*-oh; popularly, "Krakatoa") blew up in 1883, it generated a tsunami that reached a maximum height of 40 m and killed 36,000 persons in Indonesia before circling the globe.

**Warning Signs.**    The most reliable indicators of an impending eruption are

1. Rapid increase in frequency of small, shallow earthquakes as magma rises within the volcanic conduit
2. The occurrence of **harmonic tremor,** a continuous, low-intensity seismic signal given off by rising magma
3. Bulging of the volcanic cone (recorded by tiltmeters) as it fills with magma
4. Increases in emissions of hydrochloric and hydrofluoric acids and sulfur dioxide
5. Increased heat flow from the ground surrounding a volcano
6. Changes in local magnetic, gravitational, and electric fields
7. Abandonment of the area by animals

**Precautionary Measures.**    As with areas prone to earthquake hazards, it is often more practical to evacuate areas prone to volcanic hazards in case of an emergency rather than avoid them altogether. Fortunately, indicators of impending volcanic activity are usually much clearer than those of impending earthquake activity, and normally they allow adequate lead time for orderly evacuation.

**Sources of Information.**    Figure 19.15 is a map of volcanic hazard zones in the western United States prepared by the U.S. Geological Survey (modified to include the Yellowstone region, northwest Wyoming). In addition, many reports on volcanic hazard potential in specific areas are available from that agency and from some state surveys. The U.S. Geological Survey provides an efficient early warning service to alert residents of the hazard area through the various media.

### Natural Radioactivity

Until the mid-1980s, natural radioactivity was not considered a hazard. Recently, however, medical investigations of the incidence of lung cancer in miners exposed to radon gas have raised the question whether the seepage of radon into homes might pose a health risk. Although the link between radon and lung cancer is by no means conclusive, high radon concentrations have been found in many American homes, especially those that have been well-sealed against air flow for energy efficiency. Because radon is a decay product of uranium, it is often found in high concentrations in regions underlain by granite, shale, or phosphate rock, all of which are relatively rich in uranium. The gas can seep into houses through joints, cracks, and drain holes in foundations. A prudent minimum precaution in well-sealed houses in such regions is to test for radon with an inexpensive home test kit. If the test results indicate a radioactivity level higher than about 200 picocuries per liter, additional air flow can be provided through the basement. Alternatively, the basement can be sealed with mortar and urethane foam. Further information is available in free publications of the U.S. Environmental Protection Agency.

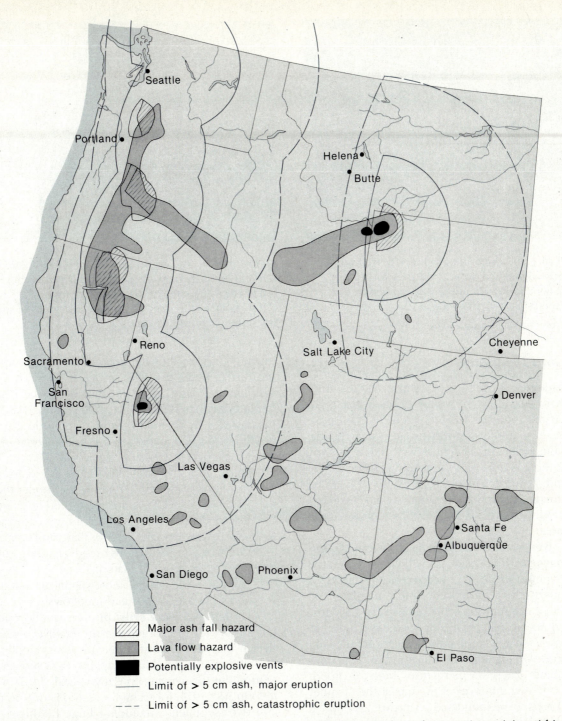

**Figure 19.15** Volcanic hazard zones of the western United States according to the U.S. Geological Survey. (modified to include the Yellowstone region). Potentially explosive vents have shown major activity within the past 2 million years.

Source: U.S. Geological Survey.

## Exogenetic Hazards

An **exogenetic hazard** is one that results from processes originating at Earth's surface. This category comprises flooding, many types of mass wasting, avalanches, ground subsidence and collapse, swelling and corrosive soils, high water tables, coastal erosion, tornadoes, lightning, and hail. Exogenetic hazards vary with respect to intensity and frequency, but in general, they are much more predictable than endogenetic hazards. As with the latter, it is unwise to use regions known to be subject to significant exogenetic hazards for any purposes that might put life or property at risk.

**Figure 19.16** The end of the road on U.S. Highway 34 was abruptly engineered by Nature during the 31 July 1976 flash flood in Big Thompson Canyon between Loveland and Estes Park, Colorado.

Photo courtesy of the Colorado Geological Survey

**Figure 19.17** The main street of the town of Fort Fairfield in northern Maine is only centimeters above the top of the south bank of the Aroostook River. Consequently, it frequently experiences floods in spring when the ice breaks up on the river.

## Floods and Mudflows

Stream systems can become flooded by runoff resulting from either snowmelt in spring or cloudbursts in summer. If the drainage basin is protected by dense vegetation, these events normally produce water floods, in which the sediment content is well below 40% by volume. If there is little or no vegetation and an appreciable depth of loose soil and rock overlying bedrock in the drainage basin, however, such events are more likely to produce mudflows, which consist of more than 40% sediment.

For a given stream system, there is a certain **bankfull discharge** of floodwater that will fill its channel to capacity. For most streams, the **recurrence interval** for the bankfull discharge is 1.5 years, meaning that a discharge of that magnitude occurs once every 1.5 years, on average. Accordingly, such a discharge is referred to as a *1.5-year flood.* Any flood flow in excess of the bankfull discharge will overflow the banks and spill out onto the floodplain. Therefore, structures built on floodplains at or below the level of the top of the channel bank will probably be inundated on an average of once every couple of years unless the channel is bordered by natural or artificial levees. For some types of structure, minor flooding on such a schedule is acceptable. Unfortunately, however, bank level localities are also subject to a range of less frequent, but more damaging floods. Of these, the **100-year flood,** which has a recurrence interval of 100 years, is of particular interest, as it is the largest flood that most large structures, such as buildings and bridges, are designed to withstand.

Even less frequent are truly catastrophic floods with recurrence intervals of 500, 1000, and even 5000 years or more. Such events can be phenomenally destructive (see table 19.1), and they can occur at any time. Unfor-

tunately, estimates of recurrence interval can only indicate the frequency with which a flood of a given magnitude should occur and not *when* the next one might occur. It is as likely that a 100-year flood will occur the day after another of the same magnitude as it is that it will not occur for another 200 years. *Three* 100-year floods occurred in Houston, Texas, in 1979!

That portion of a floodplain that is inundated by the 100-year flood is ineligible for coverage under the Federal Flood Insurance Program and is therefore regarded by most land use planners and zoning boards as suitable only for low-risk uses, such as parks and open space, and (with certain restrictions) roads, parking, and the storage of nonhazardous materials.

The following are the most significant hazards associated with large floods and mudflows:

**1. Bed and bank erosion.** Stream flow velocity increases with discharge, and the erosive power of a stream increases dramatically with flow velocity. Consequently, large floods are capable of severe erosion. During the Big Thompson Canyon flash flood in Colorado in July 1976, for example, millions of tons of soil and river sediment—including huge boulders—were flushed downriver along with trees, houses, cars, 139 human beings, countless other animals, and several kilometers of U.S. Highway 34 (fig. 19.16). Damage was assessed at $35 million. One analysis indicated that the Big Thompson disaster was a 5000-year flood. Normally, such severe erosion from flood events is limited to young, steep, headwater streams that have not yet become graded or have just barely become so.

**2. Inundation.** Overbank flooding on the floodplains of mature, graded streams is a perennial problem that results in major losses annually due to water damage to buildings, crops, livestock, stored goods, machinery, and automobiles (fig. 19.17).

*Figure 19.18*    A house and cars lie buried in a deposit of muddy, bouldery debris following a severe mudflow in La Crescenta, California, in 1978.

Photo by U.S. Geological Survey

**3.  *Deposition.***    When floodwaters have abated, there is normally a residue of sediment of variable coarseness and thickness. Where the deposits are coarse and thick, as in a mudflow, hazards to life and property can be severe (fig. 19.18). Where they are finer, they simply bury things and clog them up.

**4.  *Hazards resulting from human activities.***  Any modification of a drainage basin that reduces its ability to absorb and slow the runoff of rainwater or snowmelt increases the potential for flooding and erosion. Such activities include removal or alteration of the natural vegetation by logging (especially clearcutting), grazing, farming (especially corn), and road building; and the replacement of soil with impervious surfaces, such as pavement and roofs. Dams have often been built to control the cloudburst flooding of headwaters, but if their construction is faulty, they themselves can create a severe flood hazard.

**Warning Signs.**    The most reliable indicators of a potential flood hazard are

1.  Sudden warming in winter or spring when there is still a heavy snowpack in stream headwaters
2.  Intense rainfall over a short time period, especially after a prolonged period of less intense rainfall that has saturated the soil
3.  Extensive human alteration of land surfaces within the drainage basin
4.  Recent damming of a stream system by a landslide, forming a temporary lake
5.  Blockage of glacial drainage by ice, forming a temporary lake within a glacier
6.  The presence of floodplains and alluvial fans

Flood-prone areas can be recognized by any of the following characteristics:

1.  Alluvial fan and floodplain features, such as braided and meandering channels, point bars, oxbows, and natural levees
2.  Alluvial soils
3.  Leaning and flood-damaged trees
4.  Flood debris lodged in tree branches, bridge abutments, banks, etc.

**Precautionary Measures.**    Building within the limits of the 100-year flood is an invitation to disaster. As in the case of active faults, avoidance is by far the best alternative. Cyclists and motorists caught in mountain canyons at times of heavy rainfall should abandon their vehicles and climb straight up the canyon sideslope at the first sign of a wall of floodwater rushing downcanyon. Hikers should not enter drainage systems entrenched in desert bedrock when thunderclouds hang over the headwaters.

**Sources of Information.**    Under the National Flood Insurance Program, flood-prone communities are eligible for a 90% federal subsidy of the cost of flood insurance if they adopt and enforce a floodplain management plan under guidelines established by the Federal Insurance Administration. Central to the planning process is the flood insurance rate map (FIRM), on which the limits of the 100-year flood in a particular stream valley are delineated on the basis of data on floodplain features, soils, vegetation, high water marks and other types of historic flood information and hydrologic engineering studies (fig. 19.19). Such maps are available from the town office of any community that participates in the Program. They are also available from the Federal Emergency Management Agency in Baltimore, Maryland.

## Avalanches, Mass Wasting, and Potentially Unstable Slopes

When gravitational stress exceeds cohesive forces on natural or human-made slopes, material breaks loose from such slopes and creeps, flows, or falls downslope until the gravitational force transporting it is overcome by friction (see chapter 8). Materials so affected include snow, ice, bedrock, regolith, and incidental substances, including vegetation and human-made structures and artifacts. Falls of snow and ice are called **avalanches,** whereas falls of earth materials are called *landslides.* Losses from mass wasting are comparable to those from earthquakes (see table 19.1).

**Potentially unstable slopes** are stable under existing conditions, but would probably fail if those conditions were altered in certain ways. Accordingly, such slopes are especially problematical and require careful analysis before they are developed. Human activities that

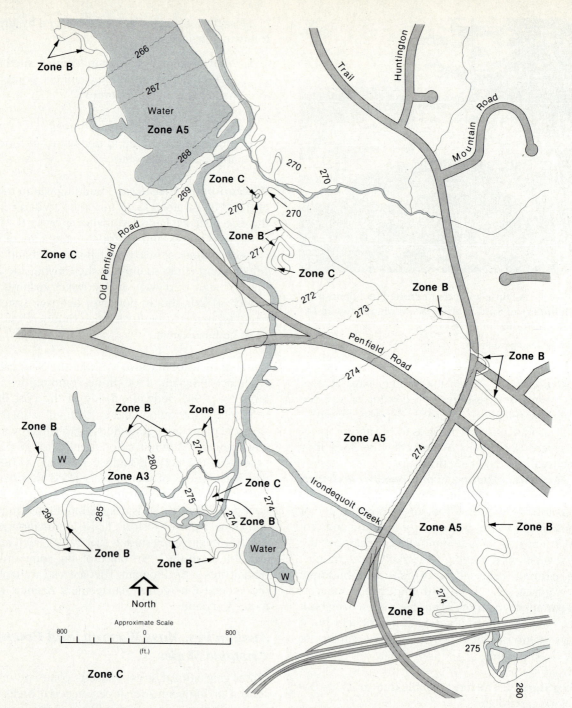

**Figure 19.19** A flood insurance rate map (FIRM) showing areas subject to inundation by a 100-year flood (zone *A*), by a 500-year flood (zones *A* and *B*), and areas of minimal flooding (zone *C*). Elevations of the flood water surface in feet are indicated by wavy lines.

Source: Federal Insurance Administration (Federal Emergency Management Agency).

increase the instability of slopes include the undercutting, loading, and wetting of slopes. Road and house construction and irrigation, including the watering of lawns, have all resulted in mass wasting that would not otherwise have occurred.

**Warning Signs.** The most reliable indicators of an avalanche hazard are

1. Vertical swaths bare of large trees on an otherwise forested, steep hillslope (fig. 19.20; small trees are often present, having grown since the last major avalanche)

**Figure 19.20** A series of narrow, treeless paths cutting down through this otherwise forested valley sidewall near Carbondale, Colorado, bespeaks frequent avalanches from a snow cornice that forms in winter on the cliff top. Note the troughlike hanging glacial valley in the foreground.

2. A large bowl above timberline, resembling a cirque, but steeper, that narrows downward into a chute (fig. 19.21)
3. Broken tree debris mixed with rooted trees and shrubs that have all been pushed over in the same direction (fig. 19.22; these can be pushed over uphill if the avalanche path crosses a valley bottom), or trees with no leaves or buds (these having been blown away by powder avalanches)
4. Rapid accumulation of a great depth of snow in a mountainous area
5. Fracture lines in high, steep snowfields
6. Mounds or blocks of snow at the base of a steep slope

Indicators of a mass wasting hazard include:

1. Accumulations of angular, freshly broken talus at the base of a rocky bluff or cliff (rockfall; see fig. 8.19)
2. Cracks and scarps in earth and pavements, trees tilted in all directions and hummocky ground (slumps and slides; fig. 19.23).

Indicators of potentially unstable slopes include:

1. Displacement of trees, fences, and other human-made structures
2. Cracks in the ground
3. Horizontal soil ripples on steep slopes (fig. 19.24)

**Precautionary Measures.** Avalanche and rockfall zones should not be developed under any circumstances. Construction sites on, or at the foot of, any slope should be carefully evaluated for stability before proceeding with development, with particular attention

**Figure 19.21** An avalanche bowl high above Maroon Creek near Aspen, Colorado, delivers snow to the valley floor in winter and rock debris in summer (note prominent alluvial fan).

**Figure 19.22** Aspen trees leaning uphill (*foreground*) were pushed over by a major avalanche from the mountainside on the opposite side of the valley (*background*).

Photo by Nicholas Lampiris

**Figure 19.23** A slump on the coast of the Bay of Fundy, Nova Scotia.

**Figure 19.24** Soil ripples on a steep mountainside in Rocky Mountain National Park indicate a potentially unstable slope.

Photo by W. B. Hamilton, U.S. Geological Survey

given to the underlying bedrock. Sites with slide potential should not be developed unless they can be adequately stabilized by such measures as adding fill to the base of the slope and installing drainage pipes to remove water.

**Sources of Information.** Both the U.S. Geological Survey and the various state geological surveys publish maps outlining areas of existing and potential mass wasting hazards.

### Other Exogenetic Hazards

Ground subsidence and collapse, swelling soils and rock, high water tables, and corrosive soils all pose additional exogenetic hazards that are less dramatic, perhaps, but that can be equally damaging or more so. State geologic surveys normally provide publications and maps addressing these problems where such hazards are significant.

**Figure 19.25** The collapse of this 97 m wide, 30 m deep sinkhole in Winter Park, Florida on 8–9 May 1981 caused economic losses in excess of $2 million.

Photo by A. S. Navoy, U.S. Geological Survey

*Sinkhole collapse* is possible in areas underlain by soluble bedrock, such as salt rock, gypsum rock, or by bedrock that is subject to rapid hydrolysis, such as limestone and dolomite. Sinkhole collapse is a major hazard in karst terrain (fig. 19.25). **Ground subsidence** can occur in unconsolidated aquifers that have been extensively drilled for groundwater or oil. Pumping of water has resulted in subsidence of almost 3 m in Houston, Texas, and production of oil from underground reservoirs in California has resulted in subsidence of almost 9 m. Losses can be heavy (see table 19.1).

Swelling soil is a major problem in Colorado, as much of the state is underlain by Cretaceous shales that swell when wet, causing foundations to warp and crack. Nationwide, a 1976 survey revealed that swelling soil has caused at least $5.6 billion damage annually, nearly equivalent to the combined damage of floods, hurricanes, tornadoes, and earthquakes (see table 19.1). The services of a professional geologist or soil scientist are required to identify this hazard, which is due to the presence of the expandable clay mineral montmorillonite. In such soils, special construction techniques, such as spread footings and floating slabs, must be used.

Certain soils, notably those containing gypsum, are mildly corrosive and can cause concrete foundations to deteriorate rapidly. Here, too, the hazard must be identified by a geologist or soil scientist. High water tables occur seasonally in a variety of settings, notably on floodplains, alluvial fans, and at the bases of slopes where seasonal springs occur. Flooded basements are an all too common problem experienced by homeowners in such areas.

### Oceanic Hazards

The coastal strip has troubles of its own. The two most destructive of these, tsunami and storm surges, are discussed under earthquakes and hurricanes, respectively.

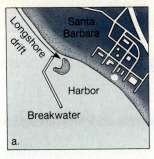

*Figure 19.26* Interference with the natural longshore drift at Santa Barbara, California, has resulted in unanticipated problems. The initial breakwater (*a*) allowed sand to drift into the harbor because it was not connected to the shore. After that connection was made (*b*), sand flowed around the breakwater, but accumulated near its downcurrent end to fill the mouth of the harbor. The sand must now be continually dredged and taken to beaches farther downcoast to replenish sand lost to erosion since the harbor was built.

Two other significant oceanic hazards, longshore drift and storm waves, have only become hazards as a result of human interference. This basic lesson in littoral hydraulics was learned at great expense by the city of Santa Barbara, California, which built a breakwater across the powerful, southward-flowing longshore current in 1925 to provide a sheltered harbor for deep-draft vessels (fig. 19.26). The breakwater was not tied to the shore, however, and great volumes of sand from the longshore drift poured through the gap, rapidly filling in the newly constructed harbor. The city responded by closing the gap, whereupon the sand, flowing at a rate of 592 m³/day, piled up against the north side of the breakwater, prograding the shoreline seaward until the flow curved around the outer end of the breakwater and poured sand into the harbor entrance. Thus deprived of its suspended sediment load, the longshore current then proceeded to erode beaches south of Santa Barbara, reducing their widths by as much as 75 m and causing over $2 million in damages within 12 years. The ultimate solution to this problem was to construct a dredging system to pump accumulating sand out of the harbor and return it, at a great annual expense, to the longshore current downcoast.

As equilibrium systems, beaches normally erode in winter and rebuild in summer for reasons explained in chapter 12. This natural process can be at odds with human interests, especially on wide beaches, where the process can extend inshore and offshore over many tens of meters. In places, seawalls have been built to stop beach erosion, but inasmuch as they concentrate wave attack, seawalls actually tend to accelerate erosion, which can affect adjacent segments of beach.

One approach to controlling beach erosion is the **groin,** a low, rock wall built out into the water from the beach and oriented perpendicularly to it (fig. 19.27). Groins collect sand from the longshore drift on their

*Figure 19.27* Groins on the seaward side of Fenwick barrier island in Maryland trap sand on their downcurrent sides. The longshore current flows away from the observer.

Photo by R. Dolan, U.S. Geological Survey

downcurrent sides. In robbing the current of sediment, however, they also increase its erosive power, and this normally results in increased erosion farther down the beach. Beachfront property owners in areas so affected usually respond to this problem by erecting groins of their own, thus shifting the problem still farther downcurrent. The result is a proliferation of groins along the coastline. Because of such problems, it is probably best to leave the beach to its drifting, cutting, and filling, and to restrict development to more inland regions.

### Atmospheric Hazards

Atmospheric phenomena are responsible for losses comparable in magnitude to those caused by earthquakes, mass wasting, and floods (see table 19.1). Because atmospheric phenomena are less localized than most other hazards, avoidance of high risk areas is a less viable option; therefore, safety is more reasonably assured by appropriate structural design and personal preparedness.

**Figure 19.28** A beach house in Kitty Hawk, North Carolina, hangs from a precarious perch after Hurricane Diane in 1955.

Photo by NOAA

**Figure 19.29** Most of the beach houses on this barrier island in the Gulf Shores area, Alabama, were destroyed during Hurricane Frederick in September 1979. Note that those houses which survived the storm surge are located nearer the lagoon (*foreground*) and are elevated on higher stilts.

Photo by U.S. Geological Survey

## Hurricanes

No natural phenomenon has any higher potential for wholesale and widespread destruction than the hurricane (fig. 19.28). These great tropical storms affect most of the eastern seaboard of North America, and because of their phenomenal energy and their erratic and unpredictable behavior, they have accounted for great losses of property and life (see table 19.1). The following are the most significant hazards associated with hurricanes:

**1. *Wind.*** The most powerful hurricanes have wind speeds in excess of 250 km/hr. Few human-made structures can withstand the forces generated by such intense winds.

**2. *Storm surge.*** The combination of high storm waves and an elevated sea surface due to extreme low pressure produces **storm surges** up to 12 m in height. In 1900, long before the advent of aerial and satellite surveillance, a hurricane struck Galveston, Texas, taking the city unaware, and 6000 people died in the accompanying storm surge. Seventy years later, in spite of highly sophisticated surveillance, a storm surge in Bangladesh claimed over 250,000 lives along the lowlands of the Ganges delta. In October 1985, Hurricane Juan pummelled the Gulf Coast of the United States, causing over a billion dollars worth of damage, but because of an effective advance warning network and an efficient evacuation plan, the human death toll was only seven. Three of these deaths occurred in the collapse of an off-shore oil rig that had not been completely evacuated.

**3. *Heavy rain.*** Rainfall rates in excess of 60 cm in 24 hours have been recorded in large hurricanes. Such torrential rains—the most intense known—have been responsible for severe flooding as hurricanes passed over land areas.

**Warning Signs.** The National Weather Service provides an early warning system for hurricanes on radio and television. In the absence of these media, the advent of a hurricane should be suspected whenever large swells approach the shore under a rapidly thickening and lowering deck of cirrostratus clouds. As the hurricane draws near, barometric pressure drops at an increasing rate, wind speed rises rapidly, and torrential rains begin.

**Precautionary Measures.** If possible, high-risk development of any coastal area less than 12 m above sea level should be avoided in regions where storm surges occur. All dwellings within such areas should be elevated on pilings with parking below although even this precaution is not foolproof (fig. 19.29). In heavily developed storm surge areas, populations should not exceed the capacities of local highway systems to evacuate the entire population efficiently within a 12 hr period. It is unwise to attempt to "sit out" a hurricane, but homes should be stocked with a two-week emergency supply of bottled water and *nonperishable* foodstuffs, a portable gas cookstove with fuel, two or more six-volt trouble lights with fresh batteries, and a small, portable toilet. Essential medicines and important valuables and papers should be placed in a strong, waterproof container secured with screws or a chain to a structural member that is unlikely to be blown or floated away. Large windows should be taped crosswise, corner to corner, with duct tape to minimize the risk of flying glass.

**Sources of Information.** A pamphlet titled *Storm Surge and Hurricane Safety with North Atlantic Tracking Chart* is available for a slight cost from the U.S.

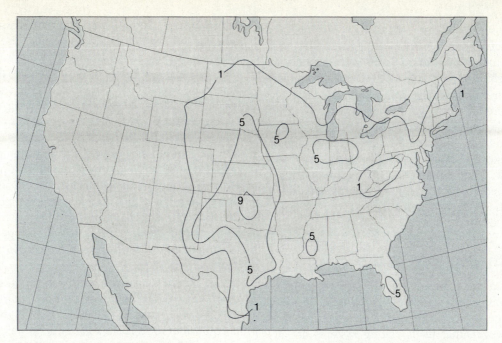

**Figure 19.30** Tornado frequency in number of tornadoes per year for the interval 1950–1980. In the south central United States, the essential elements for tornado occurrence—fast lapse rate, upper level inversion, moist air, and frontal storm activity—are often present.

Source: National Severe Storm Forecast Center.

Government Printing Office, Washington, D.C. This pamphlet provides information on preparedness and a chart for tracking the progress of hurricanes as they approach.

## Tornadoes

Although tornadoes are less destructive of property than hurricanes, they are on average, more destructive of life (see table 19.1). The principal hazards are the direct force of wind and the sudden pressure drop within the funnel, which can cause closed structures to explode from the higher pressure within. As noted in chapter 15, the appearance of pouchlike forms at the base of dark thunderclouds is the most reliable indicator of tornado weather.

**Precautionary Measures.** Because tornadoes can occur almost anywhere, the best preventive measure against them is to heed broadcast warnings and retreat to a designated public shelter or private storm cellar before a tornado strikes. If escape is impossible and there is no storm cellar, shelter can be taken in a basement, under a mattress if possible. Automobiles are no safer than houses. A motorist's best defense against tornadoes is to avoid driving when tornado warnings are in effect. If you sight a tornado while on the road, evasive driving in a direction at a right angle to the tornado's probable path can be appropriate, but if the funnel is too close for such tactics, the best strategy is to abandon the vehicle and find shelter in a ditch or similar depression.

**Sources of Information.** Up-to-the-minute information on tornado weather is provided on radio and television by the National Weather Service. Although tornadoes occur in all land areas subject to severe frontal thunderstorm activity, the United States is unusually blessed with them, averaging more than 700 per year. Figure 19.30 shows the frequency distribution of tornadoes in the United States. Note that the highest tornado frequencies correspond to the areas where all of the critical generative factors are present: high-latitude frontal storm activity, sources of warm, tropical, moist air, and sources of cold, dry, polar air.

## Lightning

Annual losses due to lightning in the United States are higher than those due to tornadoes (see table 19.1). The principal hazards from lightning are electrocution and fire. Traditionally, structures have been protected against lightning strikes by lightning rods, which conduct the current directly to the ground, but recently the value of this policy has been questioned because of the tendency of rods to draw lightning. Persons caught outdoors during thunderstorms should avoid high, exposed ridges and hilltops, trees, metal objects, and water bodies. If you should experience a buzzing sensation or static electric discharges on your body, move quickly to the lowest nearby spot and get down on all fours (do *not* lie down!). Automobiles provide some protection from lightning because of the insulation provided by their rubber tires.

## Hail

Annual damage to crops, machinery, automobiles, and structures by hail in the United States is comparable to damage by lightning (see table 19.1). Deaths from hailstorms are fairly common in some countries, notably India, but there are only two deaths on record due to hail in the United States. Hail is most problematic in the central United States but should be expected wherever large thunderstorms occur. During such storms, humans, pets, livestock, crops, ornamental plants, automobiles, boats, machinery, and other valuable items should be provided with whatever overhead protection is available.

## Routine Inspections and Service

Unlike nonliving mechanisms, our unique, living planet is self-maintaining and capable of operating indefinitely without the need of routine inspections or service. This assumes, however, that its operating systems are allowed to function according to their normal modes and that they are not significantly altered or modified. Earth is a fully integrated planet in which all operating systems function in coordination with one another. Modification of a given system inevitably produces changes in other systems, and when a system is forced to accept materials that do not fit its specifications, serious malfunction and possibly failure can result. Note that *evidence of such malfunctions and failures will not always be apparent until the system has been damaged beyond repair.* Note, also, that because of the interactive nature of planetary systems, the failure of any one system could, potentially, ruin the entire planet. With reasonable care, however, this problem should not arise. The following suggestions are offered in the interest of preventing such failure.

## Rules of Planetary Maintenance

**1. *Avoid excessive modifications.*** Work with the system instead of against it. The parts of any mechanism are located where they need to be in order to be able to perform their essential functions. The same is true of our planet. Insofar as possible, try to avoid moving mountains, rivers, and forests, etc. and exchanging them for valleys, lakes, and deserts. Try not to develop areas that lack adequate essential resources, as importing these can not only be prohibitively expensive, but it can also seriously disrupt the proper functioning of the areas from which such resources are taken. You will find that far less effort is needed to use the planet in unmodified form than to use it in any other way.

**2. *Avoid underestimating the importance of certain parts.*** Everything in the Earth system is there for a reason, even though in many cases the reason might not be immediately evident. There are no "extra" or "unnecessary" parts. If a certain part is eliminated, the entire system will change, and because the system is interactive, it can change in ways that are not always predictable. A good example is the wholesale cutting of hardwood forests and the consequently increased potential for greenhouse warming, flooding, and erosion.

**3. *Avoid overloading or underloading systems.*** Planetary systems are adjusted to accept a certain range of inputs and to produce a corresponding range of outputs. Because of the interactive nature of planetary systems, it is important to ensure that these inputs and outputs remain within their specified ranges, otherwise the flow of material among systems will become obstructed and inefficient. Global warming due to overloading of the atmosphere with carbon dioxide is a good example.

**4. *Avoid substitutions for natural inputs.*** Be careful not to feed systems materials, such as DDT, that they are not equipped to handle. DDT in the food chain is equivalent to sugar in a gas tank. Other examples include CFCs in the ozone layer, oil in the nearshore marine environment, and acid in the rain.

**5. *Avoid creating inconsistent systems.*** Although Earth is finely tuned and highly integrated, it has sufficient flexibility to allow the introduction of new systems as long as their input requirements and their outputs are consistent and compatible with existing systems (the consistency principle; see chapter 2). New systems that interfere with the functioning of existing systems set the stage for their own destruction, as they reduce the overall efficiency of the planet, on which they depend. Again, systems (such as the automobile) that overload the atmosphere with oxides of carbon sulfur and nitrogen are good examples. Another example is the centralized urban system that pulls needed resources from anywhere on the planet without regard to the potential impacts of such withdrawals.

**6. *Avoid creating unsustainable systems.*** New systems whose needs for input exceed the output capacity of supplying systems will fail when the output of those systems fails. Societies that become dependent on such unsustainable systems must also fail. Oil is a finite resource, and when it is finally exhausted, automobiles, airplanes, trains, trucks, motor ships, oil-fired power plants, and many other mechanisms that depend on oil will become useless unless they are converted to burn such substitute fuels as alcohol, a sustainable, but far less plentiful resource. When these mechanisms are no longer available (or *as* available), society will have to be drastically restructured to operate on a less energy-intensive base. Another example, the dependence of the green revolution on phosphate rock, is given in chapter 3.

**7. *Avoid spending Earth's "operating capital."*** Our planet is capable of yielding abundant resources, but avoid overtaxing those resources! Remember that to function properly, Earth needs a certain minimum for

## René Descartes (1596–1650) and Baruch Spinoza (1632–1677)

**F**ew people have had as powerful an influence on the attitudes that have driven modern technoindustrial society than René Descartes (*Day*-cahrt). At the heart of Descartes' philosophy is the concept of **dualism,** the idea that the Universe consists of two different principles, variously seen as mind and matter, sacred and profane, spiritual and worldly, etc., and that these principles are subject to different laws. That Descartes considered the mental/sacred/spiritual component to be dominant over the material/profane/worldly component is seen in his most famous statement, *Cogito, ergo sum* (I think, therefore I am). Thought, in Descartes' view, gives validity to matter. He considered animals to be incapable of thought and therefore to be mere *automata* (aw-*tom*-uh-ta), or mechanisms; hence, he judged them to be devoid of spirit, personality, or feeling. This difference, he felt, fully justifies human primacy, or dominance, over all other things, both living and nonliving.

Descartes adopted his way of thinking as a result of a deep mistrust of traditional knowledge, which he considered sterile and academic, certainly a valid judgement from our modern perspective! His introduction to that knowledge came through 10 years of traditional Jesuit schooling followed by the study of law at the University of Poitiers, from which he received a law degree in 1616. His father, who was a lawyer and a judge, had hoped that his son would also enter the legal profession, but Descartes disappointed and embittered him by becoming a wandering writer instead. Rejecting virtually all his schooling except mathematics, which he felt was sufficiently precise to be beyond suspicion, he developed a method for the pursuit of "true knowledge." In his *Discours de la Méthode* (1637), he presented his formula:

1. Accept as true only what the mind clearly sees as true.
2. Solve problems analytically (one part at a time), as opposed to synthetically (as a whole).
3. Start with what is simple and proceed to what is increasingly complex.
4. Review all steps carefully to make sure nothing has been overlooked.

This kind of rigorous, straight-line, one-step-at-a-time approach to the pursuit of knowledge actually proved to be of great value to early scientific investigations, and it is still used routinely in much of science today. Its virtue lies in that it focuses on one central phenomenon, which it isolates from the universe of related phenomena that surrounds it. The investigator can then study and describe the behavior of that one

*continued*

---

itself. A good example is the harvesting of hardwood forests, especially tropical rainforests. Care should be taken to preserve enough of these vital "planetary kidneys" to allow for the efficient recycling of carbon.

**8. *Avoid narrow perspectives.*** Planning for single-valued objectives is hazardous because of the interconnected nature of planetary systems. Try always to consider *all* possible ramifications of a particular use before instituting it. An example is the creation of **monoculture** (single-species) crops. The replacement of natural prairie ecosystems by cornfields, for example, has resulted in some of the most severe soil erosion in the United States, largely because of the large expanse of bare soil surrounding each corn plant. The establishment of monocultural stands of spruce and fir has resulted in devastating epidemics of spruce budworm and other pests. In particular, avoid short-term management. Decisions that seem to make sense in the short term can be ill-advised over a longer time frame because such decisions, including the invention of the automobile, often prove unsustainable.

**9. *Avoid growth-oriented activities.*** Inasmuch as Earth functions on the basis of steady states, expanding systems based on positive feedback (see chapter 2) are unsustainable in the long run. A good example is runaway human population. It would be far more prudent to exercise restraint now than to leave future generations to face wholesale famine, damaging pollution, and the draining of Earth's operating resources. Another example of an unsustainable activity based on positive feedback is the growth of capital, which depends on the increased production of resources. Here, too, there is a limit beyond which the activity becomes self-defeating.

As with any item of sophisticated equipment, our living planet should continue to provide us and our descendants with many years of satisfactory service if it is treated with the care and respect that it deserves.

phenomenon in isolation, without having to be concerned with the influences of the related phenomena. The obvious drawback of this approach is that it leads to gross oversimplifications of reality because nothing in Nature ever happens in isolation, free from external influences. It leads, in other words, to *idealized,* or *abstract* models of the real world, and not to the real world itself. The value of abstract models is that they are much easier to control and manipulate experimentally than the real phenomena they represent, and this seems to have been a quality that Descartes prized with an almost religious zeal. His abstract perfectionism is perhaps best exemplified in the Cartesian coordinate system, named after Descartes, who invented it. Portrayed on that system on page 515, Descartes' left eye has the Cartesian coordinates ($x = 1.2$, $y = 2.6$).

Descartes' childhood was disturbed and unhappy, and this undoubtedly contributed to his lifelong compulsion to control and maintain order in the world around him. It would be interesting to see what a modern psychiatrist would have to say about Descartes' preoccupation with control, but perhaps the following comment by Jesse Saunders, a modern student of Descartes, would serve as well: "Descartes' mistrust of his senses and

his need to redefine God, himself, and Nature suggest a certain degree of paranoia . . . Obviously, he was never in touch with the natural world . . . (He) adjusted Nature to fit his preconceived notions."

Oddly enough, one of the most brilliant students of Descartes' work arrived at a world view that is almost totally contrary to his. Baruch Spinoza, the Dutch-born son of Jewish refugees from religious persecution in Spain, blended Descartes' methodical approach to the study of Nature with the thinking of the ancient Greek philosopher Xenophanes (Ze-*noff*-uh-neez) and the sixteenth century Italian freethinker Giordano Bruno. Spinoza envisioned a reality in which all physical phenomena are interrelated parts of *one substance*. In his view, thought and behavior spring from the intrinsic natures of the various parts of that substance and from the combined nature of the whole. He found no logical basis for Descartes' dualistic distinction between the mental/sacred/ spiritual and the material/profane/ worldly components of existence. Rather, he saw the Universe as being both Creator and Creation in one. God is simply Nature and is therefore immediately accessible to the senses as well as to the mind. With this vision, Spinoza also resolved the troublesome problem of evil in what has no logical reason to be anything other than a

dynamically perfect Creation. What appears to be evil is simply local growing pains in the process of ongoing creation, and because everyone is a small part of God, we therefore all have the right, and indeed the responsibility, to weed "evil" out of the dynamic perfection as it unfolds.

In philosophic terms, Spinoza's system is known as **monism** ("one-ism" in contrast to dualism), and in religious terms it is called *pantheism* (Greek for "all-God"). Clearly, it is more compatible with the holistic, synthetic view of Nature that modern ecology affords us, whereas dualism is more compatible with the more traditional analytical view of Nature. Dualism holds that it is possible to describe an isolated system completely. Monism holds that the interactions among things make it impossible to define natural phenomena in isolation. The emerging theory of hierarchical systems is making the dualistic view increasingly difficult to defend, but it will probably be some time before monism gains equal time with dualism in scientific circles. At least the present intellectual climate is more tolerant of monist thinking than that of Rome in 1600 A.D., when Giordano Bruno was burned at the stake for refusing to recant his pantheistic views. No such evil fate befell the most notable pantheist of modern times: Albert Einstein.

## *Food for Thought*

One might well raise the question why the human species should need a service manual in order to "operate" Earth. We are, after all, products of Earth, and as such, we should be able to interact with it in a knowledgeable way, just as bears, fish, snakes, and eagles do. There is, however, an important distinction. Bears, fish, snakes, eagles, and in fact all other natural species except most of humankind have co-evolved with the environments within which they now

live and are therefore naturally equipped with instincts that allow them to interact successfully and efficiently within those environments. Our own species is the only one that has made a large-scale practice of living in environments that are foreign to its evolutionary biology. This habit has had far reaching consequences, among which are the development of language, intelligence, and the communication of abstract ideas, all of which can be viewed as substitutes for

instinct—as strategies for survival within environments for which we were not instinctively programmed. Perhaps the best way to appreciate these consequences is to take a brief look at two human societies that live in close proximity to one another, one still connected to the natural environment within which it evolved and the other quite disconnected.

In the 1950s, Colin Turnbull, a British anthropologist, lived with a hunting band of forest pygmies in

*continued*

northeastern Zaire and wrote about his experiences in a delightful book titled *The Forest People.* Because he allowed himself the "unprofessional luxury" of becoming emotionally involved with his hosts, Turnbull was able to gain a far more complete and accurate understanding of the lifestyle and attitudes of these elusive people than had ever been gained by previous students of pygmy ethnology. Interestingly, what he found was that the pygmies were, in turn, allowing themselves the luxury of being emotionally involved with their environment. To the Ituri rainforest, their home, these people look for provision, protection, guidance, reassurance, and friendship. They are constantly attuned to its changing moods, and to every subtlety of color, sound, scent, motion, humidity, temperature, and pressure. They know when and where the hunting will be good, and they go there then with nets and spears, and quickly and effortlessly build their camps of sapling-strutted, leaf-thatched dome huts to the accompaniment of much laughter and song.

Now and then, things go wrong in the pygmies' world, as, for example, when someone "dies." To a pygmy, death is a relative thing, expressed in terms of degree, as hot, feverish, ill, dead, completely dead, and dead forever. Only from the last of these is there no hope of recovery. Whatever the degree of death, however, it becomes the occasion for a raucous, rowdy ceremony called the *molimo,* in which the pygmies blare fearsome, realistic animal sounds through the heart of the forest with the aid of a great trumpet (traditionally wooden, but now often a length of iron drain pipe, as it "sounds just as good, and doesn't rot"). The function of all this ruckus is simply to wake up the forest and remind it of its duty to take proper care of its children. To us, it might seem unthinkable to yell at God for being asleep at the switch, but to the pygmies, who are pantheists (see "In the Spotlight"), there is no dualistic distinction between Creator and Creation. God, forest, and pygmies are all one and the same. Because they are a *dynamic* perfection, they are subject to local imperfections here and there, now and then, and these need to be corrected. As the pygmies see themselves as part of God, to fail to fix such glitches would be to shirk Creation's responsibility for its own self-maintenance.

This attitude is simply a reflection of the reverence and love the pygmies feel for the forest. Certainly the most poignant moment in Turnbull's book is when he happens upon his happy-go-lucky young assistant, Kenge, dancing by moonlight in the forest. When Turnbull asks him why he is dancing alone, Kenge laughs and replies that he isn't dancing alone (stupid!), he is dancing with the forest.

Here, then, is an example of a people who are intimately familiar with, comfortable within, and attuned to their surroundings, and consequently those surroundings hold few surprises for them and few mysteries. If you got eaten by a leopard, it's because you were stupid and weren't paying attention, and it served you right. The relationship between the pygmies and their forest home is not unlike that between a master auto mechanic and his/her serviceable old car. They know the forest system like the backs of their hands, and if something goes wrong, they can usually fix it themselves. That is undoubtedly the reason why Turnbull found the pygmies to be such a happy and self-confident group of people.

who live in clearings in the rainforest to which they were driven by other tribes who displaced them from their original homeland on the treeless plains. Because they did not evolve within the forest, as the pygmies did, these villagers are equipped with neither the instincts nor the understanding that would allow them to interact successfully within it. To them, the forest is a place of mystery, full of pitfalls, malificent monsters, and evil spirits. Their comfort zone is the clearings, which superficially resemble their former plains environment, yet in spite of their deep fear and mistrust of the forest, they are in many ways dependent on it, particularly for meat. Their situation is perhaps analogous to a maladroit (an inept person) and his/her Rolls-Canardley (rolls down one hill, canardley make it up the next). He/she hasn't the foggiest notion how the blinking thing works, could care less, and needs help fixing it whenever it blows up.

Similarly, the villagers rely on the pygmies to provide them with meat, for which they give them garden vegetables in return. In sharp contrast to the pygmies, the villagers are insecure, paranoid, and superstitious. They see themselves as powerless victims of the wrath of their ancestral spirits rather than as powerful and responsible parts of an all-pervasive God. They see their world as hostile and filled with evil spirits that need constant propitiation in order to ward off disaster. Their compulsive need to be in control of this unkind world leads them to treat the pygmies as servants, which the pygmies regard with great amusement, as they know full well that their "masters" would not dare pursue them into the forest.

Turning now to our own, technoindustrial society, it should be fairly obvious that our outlook is much closer to that of the villagers than it is to that of the pygmies. Most of us live in places where we would be hard put to survive if suddenly deprived of all props save the skin we were born with. If there is a difference, it would probably be that our level of insecurity and paranoia is lower because we have more props—and more reliable and sophisticated ones—to deal with environmental "hostilities" than do the villagers. Consequently, we are more arrogant and aggressive. Nonetheless, most of us certainly don't look to our natural surroundings for provision, protection, guidance, reassurance, and friendship. Rather, we espouse dualistic philosophies (see "In the Spotlight") that support our incessant struggle to alter Nature so that it will stop serving nonhuman species and

*continued*

serve us instead. To carry the car analogy one step further, we are not unlike the rock star who has the financial flexibility to dump last year's limousine if it doesn't work the way he or she wants it to, and replace it with a substitute that does. The obvious catch here, of course, is that there is no known viable substitute for Nature, at least not in the long term.

These three societies with their three different world views can be looked at as an evolutionary sequence: the happy, self-confident, pantheistic pygmies in tune with a beloved and benevolent natural environment; the paranoid, superstitious villagers defending their spare, artificial living environment against a hostile, ever-encroaching and detestable, but nonetheless indispensible wilderness; and the arrogant, aggressive, dualistic urbanites triumphantly brandishing bulldozers, chain saws, and pesticides against the ceaseless, but increasingly

futile attempts of "brute Nature" to resume its accustomed functions.

Perhaps being on the "upper" end of this sequence is more gratifying than being on the "lower" end, and perhaps it has been worth struggling through the disheartening middle ground of powerlessness and paranoia to reach the heights of Parnassus, from which we can look down on what the pygmies still look up to. Still, the underpinnings of those heights are made of that self-same Nature that comforts the pygmies and exasperates the villagers, and it is becoming increasingly evident from the lonely vantage of our growing Parnassian superstructure that those underpinnings have become less and less secure.

Because we look down on Nature, we avoid concerning ourselves with it. Often, we even deny that we are dependent on it and that we therefore need to ensure that its health and

welfare are adequately provided for. Those of us who adhere to the dualistic philosophy of René Descartes (see "In the Spotlight") view Nature as a mere mechanism placed here for the exclusive benefit of humankind, and we regard its functioning as a proper concern of its provider and not of ours. The archaeological record, however, is replete with the ruins of civilizations that paid more attention to their own world views than to stark reality, and as a result they, together with their world views, have vanished with the dust of their passing. If we are to hold on to Parnassus in the face of the present avalanche of environmental deterioration, we need desperately to avoid the mistakes of our proud antecedents and make our peace with reality. In short, it is becoming increasingly evident that we Parnassians need to start reading the service manual.

## Summary

Areas with a high **frequency** of hazardous events should not be developed. Areas with a low frequency and **intensity** of hazardous events can be developed with caution. Areas with a high intensity of hazardous events should be restricted to low-risk development. No area should be developed if to do so would increase the damage potential of an expectable hazardous event.

**Endogenetic hazards** originating beneath Earth's surface include earthquakes, volcanism, and natural radioactivity. Earthquake hazards include shaking, ground level change, **liquefaction, tsunami,** landslides, and human activities. Earthquake warning signs include ground tilting, microearthquakes, fall and rise in P wave velocity, changes in electrical and magnetic properties of rocks, *seismic gaps,* increased radon emission, and abnormal animal behavior. Precautionary measures include use of approved construction methods and materials and avoiding development within 30 m of an active fault, on unconsolidated materials, and beneath potentially unstable cliffs or slopes.

Volcanic hazards include tephra, ash flows, toxic gas, lava flows, lahars, and tsunami. Indicators of impending eruption include shallow microearthquakes and **harmonic tremor;** bulging of the cone; increased hydrochloric acid, hydrofluoric acid, and sulfur dioxide

emissions; increased heat flow; changes in magnetic, gravitational, and electric fields; and abandonment of the area by animals. Precautions include avoidance and evacuation. Radon seepage into well-sealed homes can be a health risk in areas of granite, shale, or phosphate bedrock.

**Exogenetic hazards,** originating at Earth's surface, include flooding, mass wasting, avalanches, ground subsidence and collapse, swelling and corrosive soils, high water tables, coastal erosion, hurricanes, tornadoes, lightning, and hail. Floods result from snowmelt or cloudbursts. Well-vegetated drainage basins normally experience water floods, in which sediment content is below 40%. Sparsely vegetated basins with deep regolith often experience mudflows, in which sediment content is above 40%. The **bankfull discharge** fills a stream channel to capacity at a **recurrence interval** of about 1.5 years. Higher flows with longer recurrence intervals inundate the floodplain. Floods of any magnitude can occur at any time. Land inundated by a **100-year flood** is ineligible for federal insurance and is unsuitable for all but low-risk uses.

Hazards due to large floods and mudflows include erosion, inundation, deposition, and human activities. Flood hazard indicators include sudden warming in winter or spring; intense rainfall on saturated soil; clear-cutting, forest fires, and paving; damming of a stream by a landslide; blockage of glacial drainage by ice; and the presence of floodplains and alluvial fans. Flood-prone areas are characterized by alluvial fans and flood plain features, alluvial soils, leaning and flood-damaged trees, and flood debris lodged in tree branches, etc. One can escape flash floods by climbing up a canyon wall.

**Avalanches** are falls of snow and ice. Landslides are falls of earth materials. **Potentially unstable slopes** can fail if existing conditions are altered. Avalanche hazard indicators include treeless, vertical swaths on steep slopes; a large bowl above timberline narrowing into a chute; broken tree debris with prostrate trees and shrubs; trees with no leaves or buds; rapid snow accumulation; fracture lines in snowfields; and mounds or blocks of snow. Indicators of mass wasting hazard include fresh talus, cracks and scarps, tilted trees, and hummocky ground. Indicators of potentially unstable slopes include displacement of trees, fences, etc.; cracks; and *soil ripples*. Artificial loading or undercutting of slopes and prolonged wetting add to the hazard potential. Avalanche and rockfall zones should not be developed. Sites with slide potential should be developed only if they can be adequately stabilized.

*Sinkhole collapse* can occur in soluble bedrock. **Ground subsidence** can occur in sediments affected by pumping of water or oil. *Swelling soil* warps and cracks foundations, necessitating special construction techniques. Soils containing gypsum are corrosive. Seasonal high water tables occur on floodplains, alluvial fans, and at the bases of slopes.

Hazards from longshore drift and storm waves result from human interference. Both seawalls and **groins** can increase erosion in adjacent areas. Beaches should not be subject to high risk development. Hurricane hazards include high winds, **storm surges,** and heavy rains. Developers of low-lying coastal areas with hurricane hazards should anticipate storm surges up to 12 m. Population in hurricane-prone areas should be limited by the capacity of existing escape routes. Houses should be adequately stocked with water and nonperishable food, and precautions should be taken against flying glass and loss of valuables.

Tornado hazards include high winds and sudden pressure drops. Pouchlike forms beneath thunderclouds indicate tornado weather. The best defense against tornadoes is to retreat to a designated public shelter or a storm cellar or basement. Avoid high, exposed ridges and hilltops, trees, metal objects, and water bodies during lightning weather. People and valuable items should be shielded against hail during large thunderstorms.

Natural Earth systems can be damaged beyond repair before evidence of malfunction or failure becomes apparent. Because of the interactive nature of Earth systems, the failure of one system could ruin the entire planet. To prevent this, avoid excessive modifications; avoid underestimating the importance of certain parts; avoid overloading or underloading systems; avoid substitutions for natural inputs; avoid creating inconsistent systems; avoid creating unsustainable systems; avoid spending Earth's "operating capital"; avoid narrow perspectives such as the creation of **monocultures;** and avoid growth-oriented activities.

The human species is unique in living in environments that are foreign to its evolutionary biology and in developing language, intelligence, and the communication of abstract ideas as survival strategies in environments for which it was not instinctively programmed. Disconnection from dependence on, and interaction with, the natural environment appears to have fostered superstition, insecurity, and paranoia in less technologically advanced societies, a complex which has evolved into **dualism,** arrogance, and anti-Nature prejudice in modern urban society. If untempered, that prejudice could well prove to be Nature's undoing, and ours as well. **Monism** holds promise for a more Earth-appropriate style of human management.

## Key Terms

| | |
|---|---|
| frequency | 100-year flood |
| intensity | avalanche |
| endogenetic hazard | potentially unstable |
| liquefaction | slope |
| tsunami | ground subsidence |
| harmonic tremor | groin |
| exogenetic hazard | storm surge |
| bankfull discharge | monoculture |
| recurrence interval | dualism |
| | monism |

## Questions for Review

1. In a contest judged on April 1, you win a choice of three dream houses: one situated astraddle the San Andreas fault in Palmdale, California; one on the flank of Kilauea volcano, Hawaii; and one within the limits of the 10-year flood on the Roaring Fork River in Aspen, Colorado. Which one would you choose, and why?
2. What two hazards are indicated by abnormal animal behavior?
3. What hazard is associated with granite bedrock?
4. At what times of the year should you be most on the alert for floods?
5. Give two reasons why mudflows are more common in the western United States than in the eastern United States.
6. In a second April 1 drawing, you are again the lucky winner, with a choice this time of three dream houses near Vail, Colorado, situated (a) on a gently sloping, fan-shaped gravel apron at the mouth of a mountain canyon, (b) at the base of a steep mountainside with an inspiring

**Table I.1**  Metric Conversions

## Length

| To | From | | | | | | | |
|----|------|----|----|----|----|----|----|----|
| | *km* | *m* | *cm* | *mm* | *µm* | *mi* | *ft* | *in* |
| *km* | 1. | 0.001. | — | — | — | 1.6093 | — | — |
| *m* | 1000. | 1. | 0.01. | 0.001. | $10^{-6}$ | 1609.34 | 0.3048 | 0.0254. |
| *cm* | 100,000. | 100. | 1. | 0.1. | 0.0001. | — | 30.48. | 2.54. |
| *mm* | — | 1000. | 10. | 1. | 0.001. | — | 304.8. | 25.4. |
| *µm* | — | $10^6$ | 10,000. | 1000. | 1. | — | — | 25,400. |
| *mi* | 0.6214 | 0.00062 | — | — | — | 1. | 0.0002 | — |
| *ft* | 3281 | 3.281 | 0.0328 | — | — | 5280. | 1. | 0.0833 |
| *in* | — | 39.37 | 0.3937 | 0.0394 | $3.9 \times 10^{-5}$ | 63,360 | 12. | 1. |

## Area

| To | From | | | | | | | | |
|----|------|----|----|----|----|----|----|----|----|
| | *km²* | *ha* | *m²* | *cm²* | *mm²* | *mi²* | *ac* | *ft²* | *in²* |
| *km²* | 1. | 0.01. | $10^{-6}$ | — | — | 2.5900 | 0.0040 | — | — |
| *ha* | 100. | 1. | 0.001. | — | — | 259.00 | 0.4047 | — | — |
| *m²* | $10^6$ | 10,000. | 1. | 0.0001. | $10^{-6}$ | — | 4046.9 | 0.0929 | — |
| *cm²* | — | — | 10,000. | 1. | 0.01. | — | — | 929.03 | 6.4516. |
| *mm²* | — | — | — | 100. | 1. | — | — | — | 645.16. |
| *mi²* | 0.3861 | — | — | — | — | 1. | — | — | — |
| *ac* | 247.105 | 2.4711 | — | — | — | 640. | 1. | — | — |
| *ft²* | | | 10.764 | 0.0011 | — | — | 43,560. | 1. | — |
| *in²* | — | — | 1550.0 | 0.1550 | 0.0016 | — | — | 144. | 1. |

## Volume

| To | From | | | | | | | |
|----|------|----|----|----|----|----|----|----|
| | *m³* | *l* | *cm³* | *ml* | *ft³* | *qt* | *in³* | *oz* |
| *m³* | 1. | 0.001. | $10^{-6}$ | $10^{-6}$ | 0.0283 | 0.0009 | — | — |
| *l* | 1000. | 1. | 0.001. | 0.001. | 28.317 | 0.9463 | 0.0164 | 0.0296 |
| *cm³* | $10^6$ | 1000. | 1. | 1. | — | 946.36 | 16.387 | 29.574 |
| *ml* | $10^6$ | 1000. | 1. | 1. | — | 946.36 | 16.387 | 29.574 |
| *ft³* | 35.315 | 0.0353 | — | — | 1. | 0.0334 | — | — |
| *qt* | 1056.7 | 1.0567 | 0.001 | 0.001 | 29.922 | 1. | 0.0173 | 0.03125. |
| *in³* | — | 61.024 | 0.0610 | 0.0610 | 1728. | 57.75 | 1. | 1.8047 |
| *oz* | — | 33.814 | 0.0338 | 0.0338 | 957.51 | 32. | 0.5541 | 1. |

## Astronomical Distances

| To | From | | | | |
|----|------|----|----|----|----|
| | *mi* | *km* | *AU* | *l.y.* | *ps* |
| *mi* | 1. | 0.6214 | $9.296 \times 10^7$ | $5.879 \times 10^{12}$ | $1.917 \times 10^{13}$ |
| *km* | 1.6093 | 1. | $1.4960 \times 10^8$ | $9.461 \times 10^{12}$ | $3.086 \times 10^{13}$ |
| *AU* | — | — | 1. | 63,280 | 206,293 |
| *l.y.* | — | — | $1.580 \times 10^{-5}$ | 1. | 3.2617 |
| *ps* | — | — | $4.850 \times 10^{-6}$ | 0.3067 | 1. |

## Mass

| To | From | | | |
|----|------|----|----|----|
| | *t* | *kg* | *g* | *mg* |
| *t* | 1. | 0.001. | $10^{-6}$ | $10^{-9}$ |
| *kg* | 1000. | 1. | 0.001. | $10^{-6}$ |
| *g* | $10^6$ | 1000. | 1. | 0.001. |
| *mg* | $10^9$ | $10^6$ | 1000. | 1. |

**Table I.1**  *continued*

## Time

| To | From | | | | | | | |
|----|------|------|------|------|------|------|------|------|
| | *y* | *y$_s$* | *mo* | *mo$_s$* | *d* | *hr* | *min* | *sec* |
| *y* | 1. | 1.00004 | — | — | — | — | — | — |
| *y$_s$* | 0.99996 | 1. | — | — | — | — | — | — |
| *mo* | 12.368 | 12.369 | 1. | 0.9252 | — | — | — | — |
| *mo$_s$* | 13.368 | 13.369 | 1.0808 | 1. | — | — | — | — |
| *d* | 365.242 | 365.256 | 29.5306 | 27.322 | 1. | — | — | — |
| *hr* | 8765.81 | 8766.15 | 708.734 | 655.72 | 24. | 1. | — | — |
| *min* | — | — | — | — | 1440. | 60. | 1. | — |
| *sec* | — | — | — | — | 86400. | 3600. | 60. | 1. |

## Pressure

| To | From | | | | |
|----|------|------|------|------|------|
| | *kb* | *bar* | *mb* | *atm* | *psi* |
| *kb* | 1. | 0.001. | $10^{-6}$ | 0.00101 | — |
| *bar* | 1000. | 1. | 0.001. | 1.01325 | 0.0689 |
| *mb* | $10^6$ | 1000. | 1. | 1013.25 | 68.948 |
| *atm* | 986.923 | 0.9869 | 0.001 | 1. | 0.0680 |
| *psi* | 14,504 | 14.504 | 0.0145 | 14.6960 | 1. |

## Temperature

| To | From | | |
|----|------|------|------|
| | °C | °K | °F |
| °C | 1. | K − 273.16 | (5/9)(°F − 32) |
| K | °C + 273.16 | 1. | (5/9)(°F + 459.69) |
| °F | (9/5) °C + 32 | (9/5) K − 459.69 | 1. |

Source: Data from *CRC Handbook of Chemistry & Physics,* 62d ed., 1982.

5b. Cleavage angle 65°; dark green; crystal faces striated; H = 6–7; G = 3.4:

### EPIDOTE

5c. Cleavage angle 56°; black, green, brown, or white; crystals elongate; H = 5–6; G = 2.9–3.6:

### AMPHIBOLE

1b. Microcrystalline or amorphous (no obvious crystal faces or cleavage):
2a. Luster metallic; streak colored:
3a. Color silvery gray to lead gray or black:
4a. Will not scratch a penny:
5a. Readily marks paper and soils fingers; feels greasy; streak black; H = 1–2; G = 2.2:

### GRAPHITE

5b. Does not readily mark paper; does not soil fingers; streak gray-black; G = 7.5:

### GALENA

4b. Will scratch a penny; streak black; strongly magnetic; G = 5.2–5.3:

### MAGNETITE

3b. Color golden yellow to brass yellow; streak greenish-black:
4a. Will not scratch glass or a knife; golden yellow; G = 4.2:

### CHALCOPYRITE

4b. Will scratch glass or a knife; brass yellow; G = 5.0:

### PYRITE

2b. Luster nonmetallic:
3a. Streak colored:
4a. Streak red-brown; color red to brown or black; H = 5.5–6.5; G = 5.3:

### HEMATITE

4b. Streak yellow-brown; color light to dark yellow-brown:
5a. Streak lighter than mineral; luster resinous; H = 3.5–4; G = 3.9–4.1:

### SPHALERITE

5b. Streak not lighter than mineral; luster earthy to brilliant; H = 5–5.5; G = 4.4:

### GOETHITE

3b. Streak colorless:
4a. Will not scratch a fingernail (if pure):
5a. Dark green; in metamorphic rocks:

### CHLORITE

5b. White, or variously stained other than dark green:

6a. Greasy feel; in metamorphic rocks; H = 1; G = 2.7–2.8:

### TALC

6b. Clay odor when breathed on; adheres to dry tongue; H = 2–2.5; G = 2.5–2.9:

### CLAY

6c. Reddish, pealike, round concretions present; in bauxites from lateritic soils; H = 1–3:

### GIBBSITE

6d. Luster oily to silky; H = 2; G = 2.3:

### GYPSUM

4b. Will scratch a fingernail, but will not scratch a penny:
5a. Color various shades of green; massive to fibrous; soapy feel; H = 3–5; G = 2.3–2.6:

### SERPENTINE

5b. Clay odor when breathed on; adheres to dry tongue; H = 2–2.5; G = 2.5–2.9:

### CLAY

5c. Reddish, pealike, round concretions present; in bauxites from lateritic soils; H = 1–3:

### GIBBSITE

4c. Will scratch a penny, but will not scratch glass or a knife:
5a. Color white to gray, or variously stained; bubbles in cold HCl:
6a. H = 3; G = 2.7:

### CALCITE

6b. H = 3.5–4; G = 3.0:

### ARAGONITE

5b. Color various shades of green; massive to fibrous; soapy feel; H = 3–5; G = 2.3–2.6:

### SERPENTINE

4d. Will scratch glass or a knife:
5a. Pistachio-green veins or disseminations; H = 6–7; G = 3.4–3.5:

### EPIDOTE

5b. White to gray or variously colored; luster dull, waxy; conchoidal fracture; H = 7; G = 2.7:

### CHERT

Source: Data from *CRC Handbook of Chemistry & Physics,* 62d ed., 1982.

# Key to the Common Rock Types

In identifying rock specimens, it is important to make certain that the features of the rock are observed on a fresh surface, free of saprolite. Weathering processes destroy or obscure many of the characteristic features of a given rock type. It is desirable also to determine the parent rock formation from which the specimen was derived, because the specimen might lack some broadly distributed characteristic features that could be readily identified in a larger outcrop. In this key, the name of each rock type is followed by a two-letter code in parentheses. The meanings of these codes are as follows:

| Igneous (I) | Sedimentary (S) | Metamorphic (M) |
|---|---|---|
| IP plutonic igneous | SB biological sediment | MI metamorphosed igneous |
| IV volcanic igneous | SC chemical sediment | MS metamorphosed sedimentary |
| IF fragmental (pyroclastic) igneous | SF fragmental (clastic) sediment | |

1a. Rocks consisting mainly of a glassy or an aphanitic (fine-grained) groundmass, with or without scattered crystals (phenocrysts or metacrysts):

2a. Groundmass glassy (crystals completely absent) or cryptocrystalline (crystals not distinguishable with a hand lens):

3a. Groundmass consisting of mineral glass:

4a. Brown to black; luster glassy; conchoidal fracture; sometimes flow banded; phenocrysts (if present) of K-feldspar or quartz; G = 2.3–2.4:

OBSIDIAN (IV)

3b. Groundmass cryptocrystalline (crystals not distinguishable with a hand lens):

4a. Groundmass jet black, consisting of organic material with or without plant fossils; luster dull to glassy; fracture irregular to conchoidal; G = 1.2–1.8:

COAL (SB)

4b. Groundmass light to dark green, rarely yellowish- to reddish-brown; luster greasy, waxy, or fibrous; fracture uneven:

SERPENTINITE (MI)

4c. Groundmass white, or variously lightly stained; milky luster; conchoidal fracture:

OPAL (SC)

2b. Groundmass microcrystalline (individual crystals distinguishable with a hand lens):

3a. Will not readily scratch a fingernail; colorless, white, or variously stained:

4a. Salty taste; water-soluble; H = 2.5:

SALT ROCK (SC)

4b. Slightly bitter taste; not readily water-soluble; H = 2:

GYPSUM ROCK (SC)

3b. Will readily scratch a fingernail; variously colored:

4a. Rock specimen not regularly laminated or foliated (use lens to be sure):

5a. Phenocrysts present:

6a. Phenocrysts olivine, pyroxene, or Ca-feldspar; amphibole absent; groundmass dark gray to black:

BASALT (IV)

6b. As in 6a, but phenocrysts altered to epidote, chlorite, or Na-feldspar:

SPILITE (MI)

6c. Phenocrysts amphibole, biotite, Ca,Na-feldspar, or pyroxene; olivine and quartz normally absent; groundmass gray, brown, or green:

ANDESITE (IV)

6d. Phenocrysts quartz, K-feldspar, or biotite; groundmass light gray to pink:
   7a. Groundmass hard and dense:

RHYOLITE (IV)

   7b. Groundmass chalky or earthy:

RHYOLITE TUFF (IF)

5b. Phenocrysts absent:
  6a. Bubbles in cold HCl:

LIMESTONE (SB)

  6b. Bubbles in cold HCl only when scratched or powdered:

DOLOMITE (SC)

  6c. Does not bubble in HCl even when scratched:
    7a. Hard; massive:
      8a. Black to brown; will scratch a penny but not a knife blade or glass; resembles limestone, and often contains fossils:

PHOSPHATE ROCK (CS)

      8b. Will scratch a knife blade or glass:
        9a. Waxy luster; conchoidal fracture:

CHERT ROCK (SB,SC)

        9b. Dull luster; irregular, rough fracture; igneous structures (vesicles, flow banding) can be present:
          10a. Groundmass dark:

TRAPROCK (basalt or andesite, IV)

          10b. Groundmass light:

FELSITE (andesite or rhyolite, IV)

7b. Soft; more or less crumbly:
  8a. Quartz silt visible with 10x lens:

SILTSTONE (SF)

  8b. Quartz silt not visible with 10x lens:
    9a. Gritty when nibbled:

MUDSTONE (SF)

    9b. Not gritty when nibbled:

CLAYSTONE (SF)

4b. Rock specimen laminated or foliated (use hand lens to be sure):
5a. Rock hard and dense:
  6a. Metacrysts or phenocrysts present:
    7a. Phenocrysts quartz, K-feldspar, or biotite; laminations formed by small glass shards which drape around phenocrysts:

ASH FLOW TUFF (IF)

  6b. Metacrysts and phenocrysts absent:
    7a. Rock splits readily along foliation planes; color gray to black, red, purple, brown, or green; may show relict beds:

SLATE (MS)

    7b. Rock does not split readily along foliation planes:
      8a. Color silvery; foliation planes wavy and shimmering:

PHYLLITE (MS)

      8b. Color various shades of green; sometimes has flattened, mineral-filled gas bubbles:

GREENSCHIST (MI)

      8c. Color blue:

BLUESCHIST (MS,MI)

5b. Rock relatively soft and friable:
  6a. Rock thinly laminated and fissile; often shows mudcracks, ripplemarks, and other sedimentary structures, sometimes contains fossils; color gray to black, brown, red, purple, or green:

SHALE (SF)

1b. Rocks consisting mainly of phaneritic (coarse-grained) aggregates of individual mineral grains or of rock fragments (clasts):

  2a. Rock consists of an aggregate of mineral grains:

    3a. One mineral species predominates (≥95%) (monomineralic rocks):

      4a. Mineral is olivine; granular:

**PERIDOTITE (IP)**

      4b. Mineral is Ca-feldspar; porphyritic:

**ANORTHOSITE (IP)**

      4c. Mineral is K-feldspar; granular or porphyritic:

**SYENITE (IP)**

      4d. Mineral is quartz:

        5a. Individual grains distinct; granular; rock fractures around grains:

**QUARTZ SANDSTONE (SF)**

        5b. Individual grains not distinct; recrystallized; rock fractures through grains:

**METAQUARTZITE (MS)**

      4e. Mineral is calcite or aragonite; crystalline:

        5a. Evidence of contortion or flowage lacking; fossils sometimes present; color usually light gray to black or brown but seldom white:

**LIMESTONE (SB)**

        5b. Evidence of contortion or flowage sometimes present; fossils lacking; color white or variously stained:

**MARBLE (MS)**

      4f. Mineral is dolomite; crystalline:

        5a. Evidence of contortion or flowage lacking:

**DOLOMITE ROCK (SC)**

        5b. Evidence of contortion or flowage sometimes present; color white or variously stained:

**DOLOMITE MARBLE (MS)**

      4g. Mineral is apatite (green to brown or black; hexagonal; H = 5; G = 3.2); crystalline:

**PHOSPHATE ROCK (SC)**

      4h. Mineral is gypsum; crystalline:

**GYPSUM ROCK (SC)**

      4i. Mineral is halite; crystalline:

**SALT ROCK (SC)**

    3b. Two or more mineral species are present (≥5%) (polymineralic rocks):

      4a. Minerals show no preferred orientation; phenocrysts sometimes present:

        5a. Rock contains >60% dark, mafic minerals:

          6a. Principal mafic mineral is olivine; pyroxene subordinate; no feldspar:

**PERIDOTITE (IP)**

          6b. Principal mafic mineral is pyroxene:

            7a. Pyroxene is jet black; amphibole subordinate; plagioclase feldspar a major constituent:

**GABBRO (IP)**

            7b. Pyroxene is bright green; red garnet is a major constituent:

**ECLOGITE (MI)**

          6c. Principal mafic mineral is amphibole; plagioclase feldspar is a major constituent:

**AMPHIBOLITE (MI,MS)**

        5b. Rock contains 30%–60% mafic minerals; principal mafic mineral is amphibole or biotite; Na,Ca-feldspar is predominant:

**DIORITE (IP)**

        5c. Rock contains <30% mafic minerals:

          6a. K-feldspar dominant; rock light-colored; Na-feldspar or Na,Ca-feldspar and quartz subordinate; muscovite, biotite, and amphibole minor:

            7a. Grains interlocking, not at all rounded:

              8a. Groundmass grains <1 cm diameter (phenocrysts can be much larger):

**GRANITE (IP)**

              8b. Groundmass grains mostly >1 cm diameter:

**PEGMATITE (IP)**

            7b. Grains not interlocking, some rounded:

**ARKOSE (SF)**

6b. Ca-feldspar dominant; rock dark-colored; often iridescent:

ANORTHOSITE
(IP)

4b. At least some mineral grains show preferred orientation; metacrysts sometimes present:

5a. Rock is dark with light bands of quartz and feldspar grains:

6a. Light bands are granite, clearly injected along foliation planes in schist:

MIGMATITE (MI, MS)

6b. Light bands show no evidence of injection:

GNEISS (MI,MS)

5b. Rock is not banded:

6a. Rock is composed dominantly of elongate minerals:

7a. Black amphibole (hornblende) is dominant; plagioclase feldspar subordinate:

AMPHIBOLITE
(MI,MS)

7b. Green amphibole (actinolite) is dominant:

GREENSCHIST
(MI)

7c. Blue amphibole (glaucophane) is dominant; elongate metacrysts of pale blue lawsonite often present:

BLUESCHIST
(MI,MS)

6b. Rock is composed dominantly of platy minerals:

7a. Muscovite is dominant; metacrysts of garnet or Al silicates are often present:

MUSCOVITE
SCHIST (MS)

7b. Biotite is dominant:

BIOTITE SCHIST
(MI)

7c. Chlorite is dominant:

CHLORITE
SCHIST (MI)

2b. Rock consists of an aggregate of rock fragments (clasts):

3a. Clasts are angular:

4a. Clasts of shale, slate, or phyllite are present (use hand lens):

5a. Clasts are mostly <1 cm diameter; graded bedding usually present:

GRAYWACKE
(SF)

5b. Some clasts are very large; groundmass sheared:

MÉLANGE
(MI,MS)

4b. Clasts are volcanic; groundmass of tuff:

VOLCANIC
BRECCIA (IF)

3b. Clasts are rounded:

4a. Rock breaks around clasts:

CONGLOMERATE
(SF)

4b. Rock breaks through clasts:

METACON
GLOMERATE
(MS)

Source: Data from *CRC Handbook of Chemistry & Physics*, 62d ed., 1982.

# Topographic and Geologic Maps

Indispensible tools in Earth science, topographic and geologic maps are produced and sold by the U.S. Geological Survey, by state geological surveys, and by some private mapping companies.

## Topographic Maps

The purpose of topographic maps is to portray the shapes of land surfaces conveniently on a two-dimensional surface. The way this is normally done is illustrated in figure IV.1, which shows a block diagram of a mountainous landscape through which a series of horizontal planes has been drawn at intervals of 100 m. The line of intersection of each of these planes with the land surface is called a *topographic contour line,* or simply a *contour.* Each contour represents a line of constant elevation above mean sea level.

Above the block diagram is a topographic contour map showing how these contours appear when viewed from above. Notice that the highest contours correspond to the three peaks, and that the lowest contours are located in the bottoms of the stream valleys. Several other important features of contour lines are evident in this figure:

1.  Contours close, or make circles, around peaks and hollows. Where they close around hollows, such as the volcanic crater (*rear corner*), they are hachured on the inside.
2.  Contours never cross each other, but two or more contours can blend into one another along a vertical cliff (*left front*).
3.  Closely spaced contours indicate steeper slopes (cliff and left side of volcano), whereas more widely spaced contours indicate gentler slopes (left corner and right side of volcano).

4.  Bends in contours that point toward higher elevations indicate stream valleys. Bends that point toward lower elevations indicate ridges (front and right side of volcano).

Figure IV.2*a* shows a portion of a U.S. Geological Survey topographic map of the vicinity of Marble, Colorado. Notice the steep horn of Elk Mountain with its three arétes (sharp ridges), indicated by closely spaced, sharply bent contours; the near-vertical cliff of Gallo Hill, indicated by even more closely spaced contours; the gently sloping alluvial fan of Slate and Carbonate Creeks, indicated by widely spaced contours; and the valleys of those two creeks and their tributaries, indicated by bends, or kinks, in the contours that point uphill. The mining town of Marble has been built on the alluvial fan, as a result of which it often experiences damaging floods and mudflows during runoff from rapid spring snowmelt and heavy summer cloudbursts.

Aside from contours (usually shown in brown), other noteworthy features of topographic maps include water bodies, shown in blue, vegetation, shown in green, and cultural (human-made) features, shown in black. Perennial streams, such as Carbonate Creek and the Crystal River (bordering the alluvial fan on its south side), are shown with solid blue lines. Intermittent streams, such as Slate Creek and some of the tributaries of Carbonate Creek, are shown with dot-dash lines. The marsh surrounding Beaver Lake is shown by a tufted line pattern. Snow avalanche chutes are shown as white strips cutting downslope through the green vegetation overprint.

The large squares, outlined in solid or dashed black lines and centrally numbered 14, 23, and 26 in figure IV.2*a* and *b*, are parts of the federal system of rectangular surveys, established by Congress with the Land

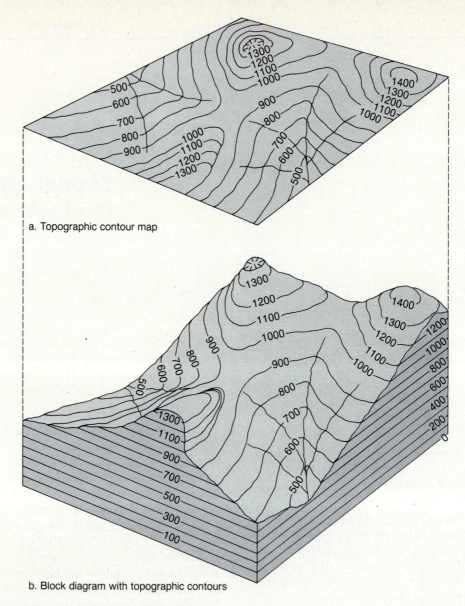

a. Topographic contour map

b. Block diagram with topographic contours

**Figure IV.1** The relationship between a topographic contour map (*a*) and the topography it portrays (*b*).

Ordinance of 1785. Under this system, Mississippi, Alabama, Florida, and all states west of the Ohio and Mississippi Rivers except Texas were divided into rectangular sections for purposes of development and administration (fig. IV.3*a*). The basic unit of this system is the *township,* a square 6 mi (9.66 km) on an edge (fig. IV.3*b*). These township units are numbered according to their positions east or west of a *principal meridian,* or local north-south survey line (heavy, solid lines in fig. IV.3*a*), and north or south of a local *base line,* or east-west survey line (heavy, dashed lines in fig. IV.3*a*). Thus, in figure IV.3*b,* the township labelled T2S, R3W is the second township south (T2S) of the base line in the third range west (R3W) of the principal meridian (a *range* is a north-south strip 6 mi wide). Each township is subdivided into 36 *sections,* each 1 mi (1.61 km) on

an edge (figure IV.3*b*). These sections are numbered in switchback fashion, beginning in the township's upper right corner and ending in the lower right. Each section can, in turn, be further subdivided in various ways, some of which are indicated for section 14.

Although it seemed a good idea at the time, the rectangular survey system has proved awkward in two respects. First, a rectangular grid does not fit well on a spherical surface; therefore, some townships have had to be eliminated from the ranks to compensate for the northward convergence of the principal meridians. Second, the grid ignores all topographic features, as a result of which property boundaries seldom coincide with such natural features as ridges, rivers, and shorelines. This often results in major complications in land use management.

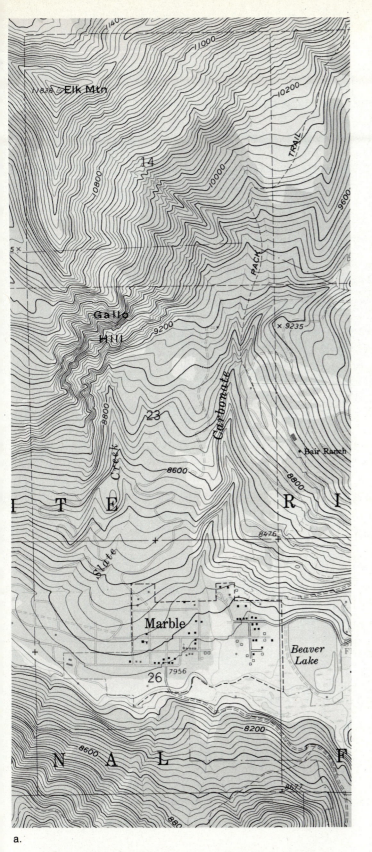

a.

b.

***Figure IV.2*** Topographic (*a*) and geologic (*b*) maps of
the vicinity of Marble, Colorado. Both are taken from 7.5
min quadrange maps published by the U.S. Geological
Survey.

Source: U.S. Geological Survey (GQ–512).

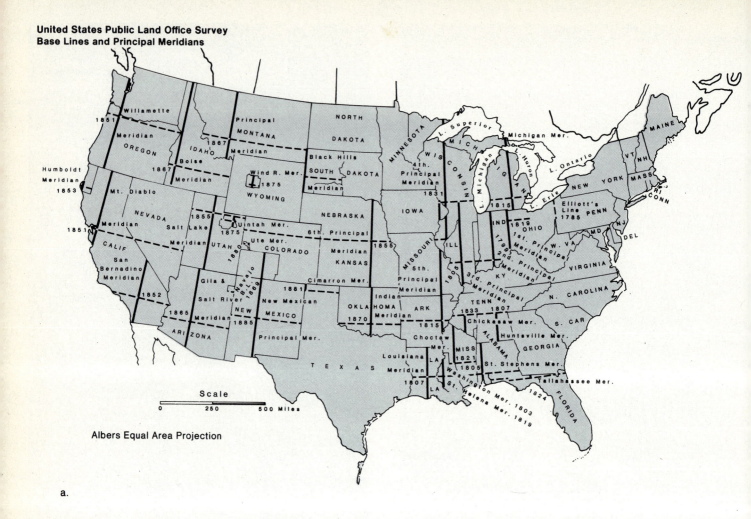

**United States Public Land Office Survey
Base Lines and Principal Meridians**

Scale

0     250     500 Miles

Albers Equal Area Projection

a.

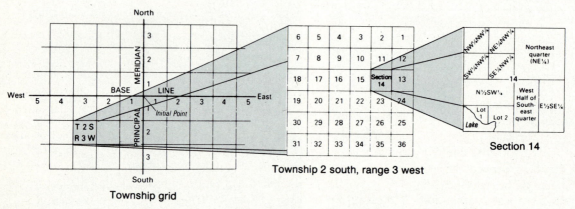

Township grid

Township 2 south, range 3 west

Section 14

b.

***Figure IV.3*** The federal system of rectangular surveys.
Bold vertical lines and dashed horizontal lines in *a* are
principal meridians and base lines, respectively. Boxes in
*b* illustrate the progressive division of a township, 6 mi
on each edge, from left to right.

Source: U.S. Geological Survey.

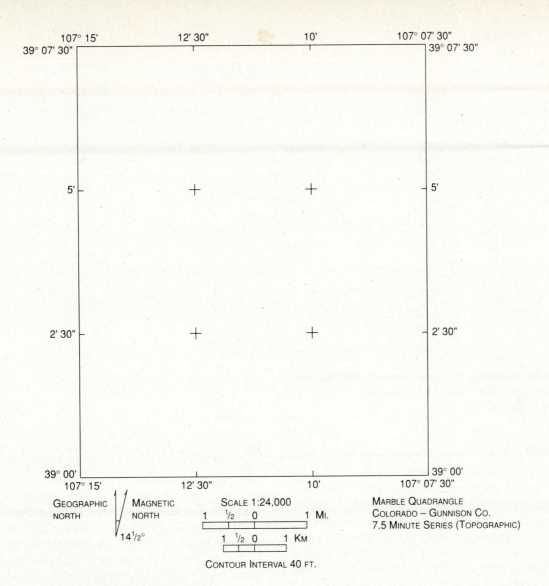

107° 15'　　　12' 30"　　　10'　　　107° 07' 30"
39° 07' 30"　　　　　　　　　　　　　　　39° 07' 30"

5'　　　　　　　　　　　　　　　　　　5'

2' 30"　　　　　　　　　　　　　　　2' 30"

39° 00'　　　　　　　　　　　　　　　39° 00'
107° 15'　　　12' 30"　　　10'　　　107° 07' 30"

GEOGRAPHIC　　MAGNETIC　　SCALE 1:24,000　　　MARBLE QUADRANGLE
NORTH　　　　　NORTH　　　1　½　0　　1 MI.　　COLORADO – GUNNISON CO.
　　　　　　　　　　　　　　　　　　　　　　7.5 MINUTE SERIES (TOPOGRAPHIC)
14½°　　　　　　　　　1　½　0　　1 KM

CONTOUR INTERVAL 40 FT.

*Figure IV.4*　**Essential features of a standard U.S. Geological Survey topographic map.**

In addition to the features already mentioned, every topographic map should be equipped with a scale; with arrows indicating true and magnetic north; with geographic coordinates; with the contour interval, or the difference in elevation between adjacent contours; and with a title that names the state and the county in which the mapped area is located and a prominent feature shown on the map (fig. IV.4). On the Marble map, the contour interval is 40 ft (12.19 m), and the scale is 1:24,000, indicating that one unit on the map corresponds to 24,000 units on the ground. The scale is also expressed by two bars that show the length represented by 1 km and 1 mi on the map. Geographic coordinates are given at each of the map's four corners and along its edges and are expressed in terms of degrees, minutes, and seconds north or south of the equator and west or east of the *prime meridian,* which passes through Greenwich, England. From the differences between the latitudes and the longitudes of opposite corners, you can see that the map covers an area that is exactly 7'30" (7 min, 30 sec) on an edge. Because of this, the map is referred to as a 7½ minute quadrangle. Most U.S. Geological Survey topographic maps are of this scale, but both larger scale maps, showing smaller areas, and smaller scale maps, showing larger areas, are also available.

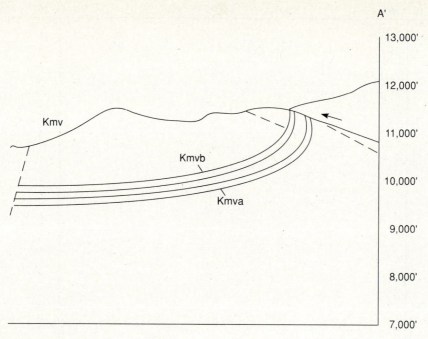

*Figure IV.5*  A geologic cross section along the diagonal line in the upper left corner of figure IV.2*b*.

Source: U.S. Geological Survey.

## Geologic Maps

Geologic maps produced by the U.S. Geological Survey are drawn on topographic base maps. Figure IV.2*b* shows the geologic map of the same area shown in figure IV.2*a*. Each mapped rock unit, usually a geologic formation, or member of a formation, is shown in a different color or pattern. In addition, the unit is given an identifying symbol. Kmu (*right of center*), for example, is the symbol for the upper member (u) of the Cretaceous (K) Mancos Shale (m; see fig. 4.22). Kmv (*upper left*) is the overlying Cretaceous Mesaverde Formation (see figs. 2.4 and 4.22). The symbol gp (*top*) stands for granodiorite porphyry, a plutonic rock type intermediate between granite and diorite. Ql (mottled, *center*) represents Quaternary landslide deposits from the Gallo Hill cliff (a further hazard for the town of Marble!). Qt (*top*) represents talus. Qm (open circles, *upper right*) represents a prominent lateral moraine (note sharply bent contours) from a valley glacier that flowed into the area from the northeast. Qf (stippled, *bottom*) represents deposits of the Slate Creek-Carbonate Creek alluvial fan. A key to these and other map units (not shown here) is included with the map, giving their age relations. A separate list of descriptions for each unit is also included.

Major geologic structures are usually evident from the map patterns of the rock units. Here, stratification in the Mesaverde Formation is indicated by two sandstone beds (Kmva and Kmvb) and by three overlying coal beds in black (thicknesses shown in feet). Several small extension faults (black lines) cutting Gallo Hill are shown, with their upthrown (U) and downthrown (D) sides indicated. A major compression fault has thrust the granodiorite porphyry southwestward over the Mesaverde Formation (barbed line, *top*). Numerous strike and dip symbols (see chapter 7) indicate the attitude of the strata in all portions of the map.

Most geologic maps are accompanied by geologic cross sections, which show the subsurface geologic structure. Figure IV.5 is a geologic cross section along the diagonal line east of the summit of Elk Mountain.

# Star Charts

Most of the heavens' brightest stars down to magnitude 4 are shown in figures V.1*a*-V.1*c*. The two polar charts (figs. V.1*a* and V.1*b*) include those portions of the celestial sphere that lie between 40° and 90° north and south declination. The equatorial-ecliptic chart (fig. V.1*c*) includes that portion of the celestial sphere between 40° north and 40° south declination. Right ascension is given on the horizontal axes of the equatorial-ecliptic chart and on the circumferences of the polar charts. The celestial equator horizontally bisects the equatorial-ecliptic chart, and the ecliptic circle makes an **S**-shaped curve about the celestial equator.

The positions and the magnitudes of the stars are indicated by a number from −1 to 4. All stars of magnitude 1 or greater are named on the charts. Some other stars of significance are also named (e.g., Polaris, Algol). Constellations are indicated by name and by tie-lines joining the brightest stars in each. The Milky Way is shown as an irregular, dark band crossing the charts.

If you live in the Northern Hemisphere, you should use the northern polar region chart (fig. V.1*a*) and the northern part of the equatorial-ecliptic chart (fig. V.1*c*). If you live in the Southern Hemisphere, you should use the southern polar region chart (fig. V.1*b*) and the southern half of the equatorial-ecliptic chart. Locate a prominent constellation in the sky (e.g., the Big or Little Dipper in the north or the Crux or Triangulum Australe in the south). Then, using a flashlight, locate the same constellation on the appropriate chart. Rotate the chart until the constellation is oriented the same way as it is in the heavens, and then the chart will be properly oriented for locating and identifying other features in the heavens.

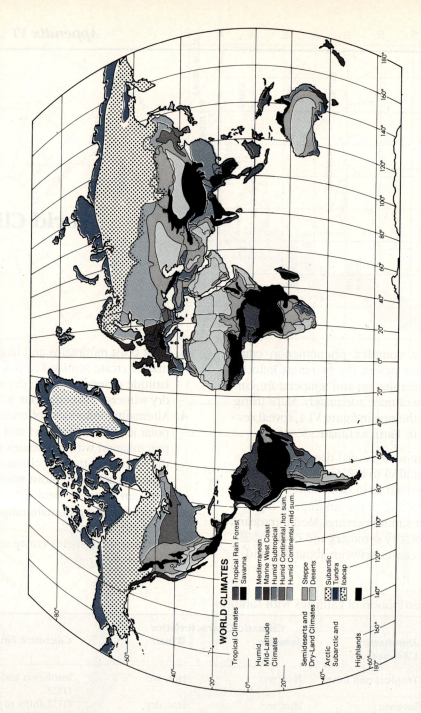

**WORLD CLIMATES**

Tropical Climates
- Tropical Rain Forest
- Savanna

Humid Mid-Latitude Climates
- Mediterranean
- Marine West Coast
- Humid Subtropical
- Humid Continental, hot sum.
- Humid Continental, mild sum.

Semideserts and Dry-Land Climates
- Steppe
- Deserts

Arctic Subarctic and
- Subarctic
- Tundra
- Icecap

- Highlands

*Figure VI.1*  **Climates of the world**.

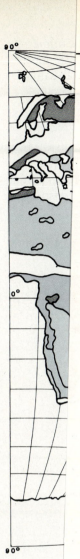

# World Soils

Two soil classification systems are in general use today. The newer Comprehensive Soil Classification System (CSCS), developed in the 1960s, classifies soils on the basis of observable physical and chemical characteristics, whereas the older Russian-American system, modified in 1938 by the United States Department of Agriculture (USDA), classifies soils on the basis of the climatic and vegetational environments in which they form. Although the CSCS system has certain practical advantages, especially with respect to engineering and commercial agriculture, it correlates poorly with natural environments; therefore, the USDA system is illustrated here.

A large variety of *great soil groups* has been recognized within the USDA system. The most important of these are illustrated in figure VII.1, which combines some of the groups into larger categories. An abbreviated USDA classification follows, with equivalent CSCS soil orders listed in parentheses after the great soil groups. Soil-forming processes (see fig. 8.15) are given in italics after the soil order names.

**Zonal soils.** Mature soils with well-developed soil horizons, developed in response to climate and vegetation.

  **Pedalfers.** Well-leached soils of humid regions in which aluminum and iron oxides accumulate in the B horizon.

    **Lateritic soils** (Oxisols) *Laterization.* Deep, highly leached, reddish to yellowish, mildly alkaline, infertile soils of tropical rain forests and savannas; rich in iron and aluminum oxides, poor in silica, humus, and nutrients. Horizons not sharply differentiated.

    **Red-Yellow Podzolic** (Ultisols) *Laterization and podzolization.* Moderately deep, well-leached, reddish to yellowish, neutral, relatively infertile soils of broadleaf evergreen forests in humid subtropical climates; rich in iron and aluminum oxides, richer than lateritic soils in silica, humus, and nutrients.

    **Gray-Brown Podzolic** (Alfisols, Ultisols) *Podzolization.* Moderately shallow, moderately leached, mildly acid, humus-rich, relatively fertile soils of deciduous forests in marine west coast and humid continental climates; gray-brown A horizon; clay-rich, yellow- to red-brown B horizon.

    **Podzolic** (Spodosols) *Podzolization.* Shallow, well-leached, strongly acid, infertile soils of coniferous forests in subarctic climates; litter- and humus-rich O horizon; ash-colored leached, silica-rich A horizon; yellowish-brown, clay-, iron-, and aluminum-rich B horizon.

    **Tundra** (Inceptisols) *Gleization.* Shallow, unleached, humus-rich, waterlogged soils of tundra climates; light to dark A horizon of little-decomposed humus; B horizon of bluish, sandy clay underlain by permafrost.

**Pedalfer/Pedocal Transition**

  **Prairie, Chernozem** (Mollisols) *Podzolization, calcification.* Thick, mildly leached, neutral to mildly alkaline, humus-rich, highly fertile soils of tall grass prairies in drier portions of humid continental climates; dark brown to black, humus-rich A horizon; brown B horizon with calcium carbonate (chernozem) or without it (prairie).

**Reddish Prairie, Reddish Chestnut, Reddish Brown, Terra Rossa** (Alfisols, Mollisols, Aridisols) *Calcification, podzolization.* Relatively thin, scarcely leached, neutral to mildly alkaline, humus-poor, reddish to brown, fertile soils of drought-deciduous and evergreen shrubland and prairie in steppe and savanna climates; calcium carbonate accumulations often present in B horizon.

**Pedocals.** Poorly-leached soils of semiarid or arid regions in which calcium carbonate accumulates in the B horizon.

**Chestnut, Brown.** (Mollisols, Aridisols) *Calcification.* Relatively thin, poorly leached, mildly alkaline, humus-poor, chestnut to brown, highly fertile soils of short grass prairie in steppe climates.

**Desert.** (Aridisols) *Calcification, salinization.* Thin, unleached, strongly alkaline, humus-poor, nutrient-rich, grayish, fertile to saline soils of desert climates; horizons obscure; heavy incrustations of calcium carbonate (caliche).

**Intrazonal and azonal soils.** Soils developed in response to local factors, such as high water tables, high evaporation rates, or unusual parent materials; and immature soils, often transported, lacking horizons.

**Mountain soils.** (Entisols) *Podzolization, gleization.* Thin, weakly developed soils of cold, steep, mountainous regions.

**Alluvium.** (Entisols) *Deposition.* Thin to thick, highly fertile soils of river floodplains. Soil horizons weakly developed.

*Figure*
Russia

Note: All glossary terms appearing in the following definitions are printed in *italic type* for easy cross-referencing.

# A

**aa**    a gas-poor form of *mafic lava* with a rough, clinkery surface.

**ablation**    the melting, evaporation, and wind erosion of *glacier* ice.

**abrasion**    the wearing away of rock due to grinding or rubbing by *clastic sediment* being transported by any of the fluid media: water, air, or ice.

**absolute geologic time scale**    a chronology of events recorded in the rocks of Earth's crust based on the application of *radiometric dating* to *geologic formations* whose positions on the *relative geologic time scale* are well-established.

**absolute humidity**    the amount of water vapor present in a given volume of air, usually measured in g/m³.

**absolute instability**    a condition in which the *environmental lapse rate* is greater than the *dry adiabatic rate*. Under these conditions, unsaturated air rises spontaneously.

**absolute magnitude**    the *apparent magnitude* of a star at a distance of 10 *parsecs*.

**absolute stability**    a condition in which the *environmental lapse rate* is less than the *moist adiabatic rate*. Under these conditions, air will not rise spontaneously.

**absolute zero**    the temperature at which all molecular motion stops; equivalent to 0 K or −273.16° C.

**absorption spectrum**    dark lines in a *light spectrum* resulting from the absorption of certain *wavelengths* of light by atoms located in the light path.

**abyssal hill**    a small *mafic*, volcanic *cone* rising less than 1000 m above the seafloor; Earth's most abundant landform.

**abyssal plain**    the flat, essentially featureless, sedimentary surface of the deep ocean floor lying between the *continental rise* and the *oceanic ridge* at depths of 3000–6000 m.

**accessory mineral**    a minor *mineral* constituent of a *rock,* usually much more variable in percentage than the rock's essential minerals.

**acid rain**    rainwater rendered acidic by absorption of nonmetallic oxides, such as $CO_2$, $SO_2$, or $NO_2$, or nonmetallic hydrides, such as $H_2S$ or HCl. Normal rain is slightly acid due to absorption of atmospheric $CO_2$.

**actualism**    the concept that Nature's laws are constant in time. Often expressed as "The present is the key to the past."

**adhesion**    a tendency of one substance to stick to another as a result of intermolecular attractions.

**advection fog**    fog caused by the *condensation* of water vapor from warm, moist air as it passes over a cold land or water surface, which cools the air below its *dew point temperature.*

**air mass**    a large body of air that acquires characteristics of temperature and *absolute humidity* from the underlying land or water surface and tends to retain these characteristics through subsequent migrations.

**albedo**    the ratio of the intensity of light energy reflected by a surface to the intensity of light that falls upon it.

**alluvial fan**    a fan-shaped body of coarse *clastic sediment* located at a point on a stream channel where the slope lessens abruptly or the channel broadens abruptly, or both.

**alluvial terrace**    a body of *alluvium* representing a portion of a former valley floor preserved after the *drainage system* has cut downward to a lower *base level.*

**alluvium**    *unlithified clastic sediment* deposited by streams.

**altitude**    in the *horizon system* of coordinates, the number of degrees at which a celestial body lies above an observer's horizon.

**altocumulus clouds** *middle clouds* in which the cloud deck has a repetitive, gray and white, roll-like or pillowlike structure.

**altostratus clouds** a uniformly textured, light gray deck of *middle clouds* through which the Sun or Moon is often dimly visible.

**anabatic wind** a local wind that blows upslope during the day as a result of *thermal convection* over radiatively heated highland surfaces.

**angle of repose** the maximum slope angle at which unconsolidated clastic material can stand: normally 33–37°.

**anion** an *atom* or group of atoms, to the outer shell(s) of which one or more *electrons* have been added, resulting in a negative electrical charge.

**annual snowline** the lower limit of unmelted snow and the upper limit of bare, exposed glacier ice at the end of the summer.

**Antarctic circle** the *latitude* of 66°32' south, poleward of which no *insolation* reaches Earth's surface at the *June solstice*.

**anticline** a fold in which older *strata* are enclosed within younger strata; an archlike fold if right side up.

**anticyclone** a mid-latitude or polar area of Earth's surface where air descends and *diverges* in a clockwise spiral (Northern Hemisphere) or a counterclockwise spiral (Southern Hemisphere).

**aphanitic** a rock texture in which the individual *mineral* grains, or *crystals,* are not visible to the naked eye.

**aphelion** the point on Earth's orbit at which it lies at its greatest distance from the Sun (152 million km). At present, aphelion falls on or about 4 July.

**apogee** the point on the Moon's orbit at which it is farthest from Earth (406,699 km, center-to-center).

**apparent magnitude** the brightness of a star measured by a scale on which a star of apparent magnitude $X$ is 2.512 times brighter than a star of apparent magnitude $X + 1$.

**aquiclude** a body of *impermeable rock* or unconsolidated *sediment* that resists the throughflow of water.

**aquifer** a body of *permeable rock* or unconsolidated *sediment* that permits a ready throughflow of water.

**Archeozoic Era** the first era of the *Cryptozoic Eon,* spanning the time interval from about 3800 million to about 2500 million years ago and characterized by one-celled *prokaryotic* organisms and an oxygen-free atmosphere.

**Arctic circle** the *latitude* of 66°32' north, poleward of which no *insolation* reaches Earth's surface at the *December solstice.*

**arête** a sharp, steep-walled, often jagged, rocky ridge between adjacent *cirques* or glaciated valleys.

**arid erosion cycle** a process of landscape evolution in arid regions, in which gently sloping *pediments* form at the expense of mountain fronts, which maintain a constant slope angle as they are eroded back.

**artesian system** an *aquifer* inclined parallel to the regional topographic slope, confined above by an *aquiclude,* and with a recharge area at its upper end.

**ash flow** an *avalanche* of gas-charged, frothing *magma* down the slopes of an *intermediate* or *felsic* volcano. Also known as a nuée ardente.

**asthenosphere** a zone of relatively low-strength *rock* in Earth's upper *mantle* between about 100 and 679 km depth.

**astronomical unit** the mean distance between Earth and Sun: 149.6 million km.

**atmospheric circulation** the network of pathways along which air moves both horizontally and vertically through the *troposphere.*

**atmospheric pressure** the weight of a column of Earth's atmosphere on an underlying surface (land, sea, or air) of specified area, usually expressed in *millibars.*

**atoll** a roughly circular coral *reef* developed from a reef that originally fringed a volcanic island. As the island sank, the reef grew upward in order to remain within the sunlit upper 80 m of the ocean.

**atom** the smallest characteristic unit of ordinary matter, consisting of an *atomic nucleus* containing one or more *protons* and zero or more *neutrons,* surrounded by a series of shells containing orbital *electrons* equal in number to the nuclear protons.

**atomic mass** the weighted average of the mass numbers of all *isotopes* of a given *element;* also called atomic weight.

**atomic nucleus** the central body of an *atom,* consisting of a single *proton,* in the case of hydrogen, or of a combination of protons and *neutrons,* as in all other *elements.*

**atomic number** the number of *protons* in the *nucleus* of an *atom.*

**aurora** a sheetlike or curtainlike glow in the night sky at high latitudes, occurring in the *ionosphere* at altitudes between 80 and 950 km and centered on the geomagnetic poles. Caused by the ionization of atmospheric gases by gusts of high-energy particles travelling on the *solar wind.*

**avalanche** a rapid, downslope flow of any loose, unconsolidated material, especially snow.

**axial plane** an imaginary surface including the *hinges* of all *strata* within a fold.

**azimuth** in the *horizon system* of coordinates, the number of degrees at which a celestial body lies to the east of the celestial meridian.

**Azoic Eon** the time interval before the appearance of life on Earth, between about 4600 million and 3800 million years ago.

## B

**backwash** the return flow of *swash* to the sea, directed perpendicularly to the shoreline.

**bajada** an extensive, gently sloping apron of *alluvium* formed by the coalescence of *alluvial fans* at the base of a mountain front in an arid region.

**banded iron formation** a deposit of alternating bands of chert (microcrystalline *silica*) and iron oxides formed in the *Cryptozoic Eon* (mainly) when oxygen wastes from photosynthetic microorganisms combined with soluble iron *ions* in seawater.

**bankfull discharge** that discharge that just fills a stream channel to the top of its banks; occurs on average about once every one or two years.

**barchan dune** a crescent-shaped sand dune the convex side of which faces the wind. Forms where sand supply is limited.

**barometer** a device for measuring *atmospheric pressure*.

**barred spiral galaxy** a *spiral galaxy* in which a single spiral arm springs from either end of a straight, central bar across the galactic nucleus.

**barrier island** an elongate, sandy island parallel to a shoreline; typically but not always, formed by the breaching of a *tidal inlet* through a *spit*.

**base level** the elevation of the larger water body into which the stream in question empties. Streams are unable to erode their beds below base level.

**basin** a roughly circular to elliptical structural depression in stratified rocks in which *strata dip* radially toward the center; also a topographic depression within which water and *sediment* can accumulate.

**batholith** a large *discordant pluton,* of which erosion has exposed a surface area in excess of 100 km².

**bathymetric map** a map of an *ocean basin* on which depths below sea level are indicated by depth contours, or isobaths.

**beach face** that portion of a beach that lies seaward of the *berm* crest and landward of the *surf zone*.

**bed load** that portion of a stream's *sediment* load that is moved along the channel by rolling, sliding, and bouncing rather than in suspension.

**berm** the high, landward edge of the *beach face*.

**big bang hypothesis** the concept that the Universe will continue expanding forever from the initial explosion of a point source.

**binary system** a star system in which two stars orbit one another.

**biochemical sedimentary rock** a *sedimentary rock,* such as limestone, chert, and siliceous ooze, that owes its origin to the activities of living organisms.

**black hole** a supermassive body in space whose gravitational field is so intense that neither matter nor light can escape if it passes within the event horizon of the black hole.

**block lava** viscous *intermediate* to *felsic* lava that breaks into large, angular blocks as it flows.

**body wave** a *seismic wave* that passes through Earth's body rather than travelling on the surface; includes *P* and *S waves*.

**Bowen reaction series** a theoretical diagram showing the sequences in which the *mafic minerals* and the *plagioclase feldspars* crystallize out of a cooling basaltic *magma* within which reactions between the liquid and solid phases are allowed to attain chemical *equilibrium*.

**braided stream** a stream whose *competence* or *capacity,* or both, have been so far exceeded by its sediment load that sediment accumulates as bars within the channel, forcing the flow to divide into multiple channels.

**brown clay** a deposit of fine *pelagic sediment* on the *abyssal plains* and *oceanic ridges;* largely derived from dust and volcanic ash blown from the continents.

## C

**calcification** the accumulation of calcium carbonate within a *soil profile* in semiarid to arid environments in which *leaching* is insufficient to remove calcium *ions* as fast as they accumulate from *weathering* and other processes.

**caldera** a large depression caused by volcanic explosion, by collapse of the crust into a partially evacuated *magma* chamber, by erosion, or (as is usually the case) by various combinations of the three.

**calorie** the quantity of heat required to raise the temperature of 1 cm³ of water by 1° C. Also called "gram calorie" to distinguish it from the kilogram calorie, or Calorie, which is 1000 times greater.

**capacity** the maximum concentration of *sediment* that a stream can transport (applies strictly only to *bed load* and not to *suspended load*).

**capillarity** the tendency of water to rise in a tube of small diameter due to the positive *adhesion* of water to glass.

**carbohydrate** a chemical *compound* consisting of carbon, hydrogen, and oxygen.

**carbonate buffer system** a natural regulation of the $CO_2$ content of seawater whereby increases in dissolved $CO_2$ convert calcium carbonate to calcium bicarbonate, whereas decreases in dissolved $CO_2$ stimulate the chemical precipitation and biological secretion of calcium carbonate.

**carbon cycle** the network of pathways along which carbon moves among its major reservoirs in the atmosphere, the oceans, carbonate *sediments* and *rocks,* and living organisms.

**catastrophism** a doctrine accounting for the geological features of Earth's surface by means of violent, worldwide, supernatural upheavals.

**cation** an *atom* or group of atoms from the outer shell(s) of which one or more *electrons* have been removed, resulting in a positive electrical charge.

**cation exchange** the substitution of one *cation* for another on the surface of a clay or soil *colloid* particle immersed in *groundwater.*

**cell** a biological *system* consisting of genetic material and mechanisms for food storage, energy production (in plants), energy transfer, protein synthesis, cell division, etc., enclosed within a cell membrane.

**cell nucleus** the central body of a biological *cell,* containing the cell's genetic material in the form of chromosomes and separated from the rest of the cell by a nuclear membrane.

**cementation** a *lithification* process in which *mineral matter* dissolved in seawater or *groundwater* is deposited within the void spaces among the mineral grains of a *sediment.*

**Cenozoic Era** the third era of the *Phanerozoic Eon*, spanning the time interval from about 66 million years ago to the present and characterized by the dominance of large mammals and deciduous trees in terrestrial environments.

**central eruption** ejection of *lava*, *pyroclastic* material, and gas from a vertical, pipelike opening or conduit, in Earth's crust.

**Cepheid variable** a class of *pulsating variable* star, named after ∂-Cephei, having periods ranging from one day to about one month.

**channel flow** water flow confined to a stream channel. Channel flow is more efficient than *sheet flow.*

**chemical sedimentary rock** a *sedimentary rock,* such as travertine, gypsum rock, salt rock, phosphate rock, and dolomite, produced by inorganic chemical precipitation from seawater.

**chemical weathering** any process that disintegrates rock by altering the chemical composition of its constituent minerals; includes *solution, hydrolysis, oxidation,* and *hydration.*

**chinook wind** a warm, dry wind, the moisture content of which has been depleted by precipitation as it passed over a mountain range.

**chlorofluorocarbon compounds (CFCs)** compounds of carbon with chlorine and fluorine; on photodissociation, they release monatomic chlorine, which destroys ozone.

**chromosphere** the lower layer of the Sun's atmosphere, about 6500 km thick, within which the Sun's *absorption spectrum* is formed. Overlies the *photosphere.*

**cinder cone** a small, steep, volcano consisting of *mafic* to *intermediate tephra* surrounding a *central eruption* vent.

**circle of illumination** the line on a planet or satellite that separates the sunlit side from the dark side. Also called the terminator.

**cirque** a scoop- or bowl-shaped depression in a mountain peak or ridge, carved by a *glacier* or permanent snowfield. Normally located high on a northeast-facing slope (southeast in the Southern Hemisphere).

**cirrocumulus clouds** *high clouds* in which the cloud deck has a fine, repetitive, white, roll-like or pillowlike structure. Also called mackerel sky.

**cirrostratus clouds** a uniformly textured, thin, whitish veil of *high clouds* composed of ice crystals through which the Sun is clearly visible and often surrounded by a single or double halo.

**cirrus clouds** wispy, feathery, high clouds, usually turned upward on their downwind ends. Also called mares' tails.

**clast** a loose fragment of *rock* of any size detached from a larger rock *formation;* also called a stone.

**clastic sedimentary rock** a *sedimentary rock* composed of *clasts:* includes conglomerate, breccia, quartz sandstone, arkose, graywacke, clastic limestone, and *mudstone (siltstone, claystone, and shale).*

**cleaveage plane** any plane within a *crystal* that includes a high concentration of **atoms.** Crystals split most readily along such planes because *covalent* and *ionic bonds* tend to form parallel to them rather than across them.

**climate** the characteristic, recurrent, annual patterns of temperature, precipitation, prevailing wind, and cloud cover in a given region.

**closed system** a *system* characterized by inputs and outputs of *energy* but not of *matter.*

**cloud base** in a given air mass, the altitude at and above which *condensation* and cloud formation can occur.

**clouds with vertical development** clouds, including *cumulus* and *cumulonimbus,* that form in *absolutely* or *conditionally unstable* air that is rising by *thermal convection.*

**cognitive selection** a process whereby individual humans, or groups of humans, who make unviable decisions regarding the use of their natural environments are eliminated from the population due to failure of the environments in question.

**cohesion** the degree to which the intermolecular bonds of a substance resist forces that tend to break the substance apart.

**cold front** in a *wave cyclone,* a fast-moving, relatively steep interface between cold, polar air to the west and warm, tropical air to the east. The cold front lies to the west of, and eventually overtakes, the *warm front.*

**collision-coalescence process** the growth of raindrops by the collision and coalescence of cloud droplets.

**colloid** a liquid in which particles 0.005–0.25 $\mu$m in diameter are suspended and in which electrostatic attractions among the particles produce a jellylike texture.

**color index** the *apparent magnitude* of a celestial object on a blue-sensitive meter minus its apparent magnitude on a yellow-sensitive meter.

**coma** a hazy, luminous shell of dust and gas surrounding the **nucleus** of a comet, formed as the *solar wind* vaporizes and ionizes frozen gas particles from the nucleus.

**comet** a "fuzzy snowball" of silicate dust, ices, and organic compounds. A halo of comets is thought to surround the Solar System, but some comets closely approach the Sun along highly eccentric orbits.

**compaction** the packing together of *clastic sedimentary* particles and the consequent reduction of interstitial pore spaces among them, due to the weight of overlying *sediment.*

**competence** the diameter of the largest *clast* that a given stream discharge can transport at a given location along a stream channel.

**composite volcano** a large volcanic *cone* composed of interstratified layers of *intermediate* lava and *tephra* with numerous *dikes* and *sills.*

**compound** an electrically neutral group of two or more *atoms* of the same or different *atomic number* that has characteristic physical properties.

**compression fault** a typically shallowly-dipping *fault* in which horizontal compression has caused the *hanging wall* to override the *footwall,* producing a shortening of the crust.

**compressive stress** an inward directed stress that tends to shorten an object to which it is applied.

**concordant pluton** an *intrusive* body of *igneous rock,* the borders of which are predominantly parallel with the *stratification* or *foliation* in the surrounding bedrock.

**condensation** a change from the vapor state to the liquid state, involving a release of *latent heat of evaporation.*

**condensation nucleus** a microscopic dust particle in the atmosphere with affinity for water molecules.

**conditional instability** a condition in which the *environmental lapse rate* lies between the *dry* and *moist adiabatic rates.* Under these conditions, air rises spontaneously if it is first forced to rise until it cools to its *dew point temperature.*

**conduction** the intermolecular transfer of heat energy, in which increased thermal motions are passed from one *molecule* to the adjacent ones.

**cone** a circular, conical deposit of *lava, tephra,* or both around a *central eruption.*

**cone of depression** a conical depression in the *water table* surrounding a well from which water is being pumped.

**confining pressure** confining *stress* that is uniform in all directions.

**conjunction** an alignment of two or more celestial objects at the same *azimuth* and *altitude.*

**consistency principle** the concept that the structure and function of any new subsystem that arises within a given *system* must be fully consistent with the structure and function of the larger system if it is to persist within it.

**constellation** a recognizable grouping of stars as seen from Earth.

**contact metamorphism** a localized, high-temperature, low-pressure *metamorphism* of bedrock caused by heat and hot-water *solutions* emanating from *intrusive igneous rocks.*

**continent** a large landmass of *felsic* composition that stands above sea level except for the marginal *continental shelves* and *continental slopes.*

**continental crust** a 30–80 km thick layer of *felsic rock* and overlying *clastic* and shallow marine *sediment* that lies above the Mohorovičić discontinuity and forms the *continents* and *continental shelves.*

**continental ice sheet** a massive blanket of glacier ice 2–4 km thick that flows outward from one or more centers of accumulation on a continental interior.

**continentality** the drying and cooling effect that a large landmass has on regional climate.

**continental rise** a thick, gently-sloping, *sedimentary* prism of mud and *turbidity flow* deposits located at the base of the *continental slope* and blending imperceptibly with the *abyssal plain.*

**continental shelf** the very gently-sloping, shallowly submerged margin of a *continent,* bounded on the landward side by the shoreline and on the seaward side by the *continental slope.*

**continental slope** a relatively steep slope on a continental margin separating the *continental shelf* above from either a *continental rise* or an *oceanic trench* below.

**continuous creation hypothesis** the concept that new *matter* is constantly being created as the Universe expands. A corollary is that the Universe had no beginning and will have no end.

**continuous light spectrum** the array of rainbow colors resulting from the *dispersion* of white light through a prism.

**continuous reaction series** that branch of *Bowen's reaction series* that illustrates the progressive compositional change in *plagioclase feldspar* from calcium-rich to sodium rich as it reacts with a cooling *magma* under conditions of chemical *equilibrium.*

**convection** see *thermal convection.*

**convection cell** see *convective circulation.*

**convective circulation** the flow of a convection current along a closed, cyclic path, such as a *Hadley cell* in the *troposphere.*

**convective zone** the outer shell of the Sun, from the *photosphere* down to a depth of about 150,000 km, within which the escape of *heat energy* from the Suns's core occurs mainly by *convective circulation.*

**convergence** the crowding together of wind or water currents, resulting in increased pressure that forces air or water to escape vertically upward or downward away from the convergence.

**convergent plate boundary** a line along which one *lithospheric plate* either butts against or sinks beneath the edge of another in the processes of continental collision and *subduction,* respectively.

**core** Earth's dense, metallic, central portion, consisting of a molten outer shell and a solid or semisolid nucleus.

**corer** a hollow, *tethered sampler* designed to plunge vertically into soft seafloors and retrieve samples of *sediment.*

**Coriolis effect** the tendency of air and water currents to veer to the right of their direction of travel in the Northern Hemisphere and to the left in the Southern Hemisphere. The effect increases with *latitude* and velocity.

**corona** the Sun's outer atmosphere, lying above the *chromosphere* and consisting of glowing, rarefied gas.

**cosmic radiation** very fast-moving *atomic nuclei* that bombard Earth from outer space.

**covalent bond** a mechanism for holding two or more *atoms* together by sharing pairs of outer shell *electrons.*

**crater** a roughly circular depression at the top of a *volcanic cone* representing the top of the underlying volcanic conduit; also a roughly circular depression caused by the impact of a large *meteorite.*

**creep** slow *deformation* of rock under constant *directed stress,* also the slow downslope movement of *soil* or *clastic* material.

**crevasse** an open crack within the upper, brittle zone of a *glacier.*

**cross-stratification**    layering within a *sediment* or *sedimentary rock* that lies at an angle to major bedding planes.

**Cryptozoic Eon**    the time interval from the end of the *Azoic Eon,* about 3800 million years ago, to the beginning of the *Phanerozoic Eon,* about 700 million years ago, characterized by one-celled organisms.

**crystal face**    a flat growth surface on the exterior of a crystal that lies parallel to a plane of aligned *atoms* within the *crystal structure.*

**crystal structure**    a repetitive, ordered arrangement of *atoms* within a crystal.

**cumulonimbus cloud**    an exceptionally tall, dense, thermally unstable, precipitation-producing *cloud of vertical development* with a flat base and either a cauliflowerlike or an anvil-shaped top.

**cumulus cloud**    a relatively short, thermally stable, rain-free *cloud of vertical development* with a flat base and a cauliflowerlike top.

**cybernetic potential**    a prede-termined, optimum quantity of gravitational or electromagnetic *potential energy* maintained within a *cybernetic system* by adjusting either *input* or *output* whenever *negative feedback* tells a sensory regulator that the potential is too high or too low.

**cybernetic system**    a biologically-created *system* that maintains one or more constant *cybernetic potentials* by varying either *input* or *output* by means of *negative feedback loops.*

**cyclone**    see *wave cyclone.*

# D

**debris avalanche**    the sudden, downslope flow of a loose mass of dry *regolith* under the influence of gravity.

**debris fall**    the free fall from a bank or cliff of a loose mass of dry *regolith* under the influence of gravity.

**debris slide**    the slow to rapid downslope sliding of a mass of loose *regolith* under the influence of gravity.

**December solstice**    in the *equatorial system* of coordinates, the *meridian* of 18 hours *right ascension,* located at the point where the Sun stands at its lowest point below the *celestial equator;* also the point on Earth's orbit at which the *Antarctic circle* receives 24 hours of sunlight a day.

**declination**    in the *equatorial system* of celestial coordinates, the position of an object north or south of the *celestial equator,* in degrees, minutes, and seconds. See also *geomagnetic declination.*

**deep-water wave**    a *water wave* whose base lies above the seafloor.

**deferent**    in the discredited hypothesis of *geocentrism,* an orbit around Earth followed by the center of the *epicycle* of a planet, the Sun, or the Moon.

**deflation basin**    a shallow depression in *regolith* or poorly *lithified* bedrock formed by wind action blowing away loose material.

**delta**    an accumulation of *clastic sediment* at the mouth of a stream where it empties into a body of standing water.

**dendritic drainage**    a treelike, branching *drainage pattern* assumed by a drainage *system* that has developed on homogeneous bedrock or *regolith.*

**density**    the *mass* of a given volume of matter. In the metric system, density is usually expressed as grams per cubic centimeter.

**deranged drainage**    an irregular pattern with many lakes assumed by a *drainage network* that has developed on recently glaciated bedrock or *till.*

**desertification**    conversion of land areas to wasteland through human agency.

**desert pavement**    a tightly interlocking surface layer of wind-polished *clasts,* formed in arid regions by the *deflation* of finer *clastic* material.

**dew**    water *condensed* on surfaces that have been radiatively cooled below the *dew point temperature* of the surrounding air during the night.

**dew point temperature**    the temperature to which an airmass must be cooled to raise its *relative humidity* to 100% and thus to bring about its *saturation* with water vapor.

**differential erosion**    the uneven lowering of the land surface as a result of differing susceptibilities of different *rock* types to *erosion.*

**dike**    a tabular (thin and flat) *discordant pluton.*

**dip**    the angle between an inclined surface, such as a bedding plane or *foliation* plane, and the horizontal, measured in the direction of steepest inclination, that is, perpendicular to the *strike.*

**directed stress**    a *stress* that is greater in one particular direction than in any other.

**discharge**    the volume of water flowing past a given point in a stream channel in a given time interval.

**discontinuous reaction series**    that branch of *Bowen's reaction series* that illustrates the sequential crystallization of the *mafic minerals* from olivine through pyroxene and amphibole to biotite as they react with a cooling *magma* under conditions of chemical *dynamic equilibrium.*

**discordant pluton**    a body of *intrusive igneous rock* that cuts across bedding planes or *foliation* in the surrounding rock.

**disjunct distribution**    the occurrence of a living or fossil species in two or more regions geographically isolated from one another.

**dispersion**    the progressive separation of *water waves* or light waves of different *wavelengths* as they radiate from a source due to the greater velocity of long waves than of short waves.

**diurnal tide**    a tide having only one crest and one trough per day.

**divergence**    the spreading apart of wind or water currents, resulting in reduced pressure that draws air or water vertically upward or downward toward the divergence.

**divergent plate boundary**    a plate boundary coincident with an active *median rift.*

**dome**    a roughly circular to elliptical structural upwarp in stratified rocks in which *strata dip* radially away from the center.

**Doppler effect**    a shortening of the observed *wavelength* of light or sound waves being emitted by an approaching object, or lengthening of the observed wavelength of light or sound waves being emitted by a receding object.

**drainage basin**    a scoop- or spoon-shaped region of Earth's surface containing a *drainage system* that removes precipitation from that region.

**drainage system**    an interconnected network of stream channels that gather water from a *drainage basin* and deliver it to a trunk channel that leaves the drainage basin at its lowest point.

**drift**    a general term for *clastic sediment* deposited either by *glacier* ice or by water flowing from melting glaciers.

**drumlin**    a low, elongated hill in *ground moraine,* the long axis of which is parallel to the flow direction of the *continental ice sheet* that produced the ground moraine.

**dry adiabatic cooling rate**    10° C/1000 m, the rate at which *unsaturated* air cools on rising in Earth's atmosphere.

**dualism**    the philosophical premise that the Universe consists of two separate and different principles, such as the sacred and profane; opposed to *monism.*

**dynamic equilibrium**    a condition in which the *equilibrium potential* of a *system* is so adjusted as to equalize *input* and *output.*

# E

**earthquake intensity**    the degree of destructiveness of an earthquake as measured by the extent of damage to human-made structures.

**earthquake magnitude**    the energy released by an earthquake as measured by the logarithm of the amplitude of the largest *seismic wave* produced by the earthquake on a *seismograph* located 100 km from the *epicenter.*

**ebb tide**    a minor, local lowering of sea level occurring once or twice daily.

**echo sounder**    a device for measuring ocean depth by determining the time required for a sound signal to return to the sounder after being reflected from the ocean floor.

**eclipsing binary**    a two-star system whose orbital plane is closely enough aligned with Earth that the stars periodically partially or totally eclipse one another, resulting in a regular fluctuation in the system's *apparent magnitude.*

**ejecta blanket**    a widespread deposit of *rock* debris surrounding the *impact crater* from which the debris was thrown by the impacting object.

**Ekman spiral**    a theoretical diagram illustrating the progressive deflection of wind-generated ocean currents with depth in the upper 100 m or so of the ocean due to the *Coriolis effect.* The spiral turns to the right in the Northern Hemisphere and to the left in the Southern Hemisphere.

**Ekman transport**    the net motion of the upper 100 m or so of the ocean at 90° to the right of the wind direction in the Northern Hemisphere and at 90° to the left of the wind direction in the Southern Hemisphere, as predicted by the *Ekman spiral.*

**elastic rebound**    the sudden return of the strained *rock* on either side of a *fault* to its original shape as the pressure bond between the opposing *fault blocks* is released during an earthquake.

**elastic strain**    recoverable deformation in *rock* resulting from *directed stress.*

**electromagnetic energy**    wavelike pulses of *energy* released from *matter* when electrically charged particles are set in motion.

**electromagnetic spectrum**    the full range of *wavelengths* assumed by *electromagnetic energy,* from *gamma rays* at the short end of the spectrum through X-rays light (ultraviolet, visible, and infrared) and microwaves to radio waves at the long end.

**electron**    a fundamental, subatomic particle having unit negative electrical charge and a *mass* equal to 1/1836 that of a *proton.*

**electron shell**    a region surrounding an *atomic nucleus* comprising a number of spherical or dumbbell-shaped "orbitals" within which *electrons* move.

**electrostatic force**    an attractive force that binds oppositely charged particles with an intensity that is directly proportional to the product of the intensities of the charges and inversely proportional to the square of the distance between the centers of the particles.

**element**    a substance composed entirely of *atoms* of the same *atomic number.*

**elliptical galaxy**    a circular to elliptical assemblage of millions of predominantly old, *population II* stars with relatively little interstellar gas and dust, and no spiral arms.

**emergent coastline**    a coastline, typically low and featureless, that is rising with respect to sea level due either to *tectonic* uplift or to a fall of sea level, as during an ice age.

**endogenetic hazard**    a geologic hazard, such as an earthquake or a volcanic eruption, that originates below Earth's surface.

**energy**    a field, or "presence," capable of changing the motion of *matter.*

**environmental lapse rate**    the rate of change of air temperature with altitude; can be positive (cooling upward), negative (warming upward, as in an *inversion*), or zero, depending on various weather phenomena, but averages about 6.5° C/1000 m.

**epicenter**    the point on Earth's surface that lies vertically above the *focus* of an earthquake.

**epicycle**    in the discredited hypothesis of *geocentrism,* a small, circular orbit followed by a planet, the Sun, or the Moon around a point in space that orbits Earth.

**equatorial system**    a system of coordinates in which celestial objects are located according to their *declination,* in degrees, north or south of the *celestial equator* and their *right ascension,* in hours, minutes, and seconds, east of the *meridian* of the *March equinox.*

**equilibrium potential** that quantity of gravitational or electromagnetic *potential energy* required to force *output* from a *system* at the same rate at which *input* is entering the system.

**equilibrium shoreline** a shoreline along which the overall rate of deposition has become equal to the overall rate of *erosion.*

**equilibrium system** a *system* in which *dynamic equilibrium* prevails.

**erosion** the removal of *weathered* or unweathered *rock* material or unconsolidated *sediment* from a position of rest by an erosional agent, such as river or ocean currents, waves, flowing ice, wind, or gravity.

**erosion cycle** the evolution of landscapes from youth, with *deranged drainage* and steep-walled valleys, through maturity, in which *floodplains* form and uplands are dissected by *headward erosion,* to old age, in which the landscape can be reduced to a *peneplain.*

**erratic** a large *clast* that has been moved by a *continental ice sheet* and randomly placed on the landscape when the ice sheet melted.

**esker** an elongated, sinuous ridge of *stratified drift* deposited by streams beneath or within a melting *glacier.*

**estuary** the mouth of a river that has been drowned by sea level rise and not subsequently filled with *delta* deposits.

**eukaryotic cell** a relatively large, advanced *cell* with linear chromosomes in a well-developed *cell nucleus* enclosed within a nuclear membrane and with various organelles that perform certain essential functions for the cell.

**evaporation** the escape of molecules from the surface of a liquid into a surrounding vapor.

**evaporation fog** a mist of fine cloud droplets formed when moist air containing water vapor *evaporated* from a warm water surface is cooled to its *dew point temperature* by overlying cold air or when warm rain is cooled by cold air.

**evaporite deposit** a *chemical sedimentary rock,* such as gypsum rock and salt rock, formed by the precipitation of *ionic minerals* out of evaporating seawater.

**exfoliation dome** a large, rounded outcrop of granite, or a similarly structureless rock type, from which thin, onionlike shells of rock gradually separate and fall away along *fractures* that develop parallel to the outcrop surface due to *unloading.*

**exogenetic hazard** a geologic hazard that originates at or above Earth's surface; includes flooding, *mass wasting,* ground subsidence and collapse, swelling and corrosive *soils,* high *water tables,* coastal *erosion,* tornadoes, lightning, and hail.

**exponential decay curve** a curve illustrating the decay, over successive *half-lives,* of one-half of the amount of material remaining in a sample of a radioactive *element* at the end of the preceding half-life.

**extension fault** a steeply-inclined *fault* in which horizontal extension has caused the *hanging wall* to pull apart from, and drop relative to, the *footwall,* producing a stretching of the crust.

**extrusive** said of an *aphanitic igneous rock* that formed on Earth's surface; the category comprises the *lavas* and *pyroclastic rocks.*

**eye** the central column of a *hurricane,* characterized by clear skies, calm air, and extremely low *atmospheric pressure.*

## F

**fall overturn** the sinking of the surface layer of a lake or ocean in fall as it becomes colder and denser than the underlying water.

**fault** a fracture in bedrock or *regolith* along which one side has moved with respect to the other.

**fault block** that portion of the bedrock adjacent to a *fault* that is affected by motion along the fault.

**fault plane** the surface separating adjacent *fault blocks.*

**felsic** said of a *rock* type that contains more than 66% silica.

**fissure eruption** eruption of *lava,* *pyroclastic* material, or gas from an open fissure in Earth's crust.

**floodplain** a flat-topped deposit of *alluvium* flanking a stream channel and produced by recurrent overbank flooding.

**flood tide** a minor, local rise of sea level occurring once or twice daily.

**focal length** the distance from the center of a lens to its *focus.*

**focus** the point at which light rays converge when *refracted* by a lens; also the point of origin of an earthquake.

**foliation** a layered structure within a *metamorphic rock* caused by the parallel arrangement of linear or platy *minerals,* or both.

**footwall** the *fault block* that lies beneath an inclined *fault* plane.

**fractional crystallization** any process that separates *crystals* from a cooling *magma,* thereby preventing *equilibrium* and inducing a change in the chemical composition of the magma.

**fracture** a break in the brittle bedrock near the surface of the Earth's crust.

**fracture zone** an apparent linear fracture in the seafloor—transverse to the axis of the *oceanic ridge*—whose midsection consists of a *transform fault* that offsets the *median rift,* and whose ends consist of sutures between sections of *lithospheric plate* that grew from adjacent offset rift segments.

**frequency** the average number of times that a particular event, such as a flood or the passage of a wave crest, occurs during a specified *time* interval.

**frontal wedging** the lifting of warm, tropical air by either a *cold front* or a *warm front.*

**frost** ice crystals deposited by *condensation* from moist air on a surface having a temperature below 0° C.

**frost wedging** the *mechanical weathering* of bedrock by the freezing of water in *fractures* and pore spaces.

## G

**gamma ray** a high-energy form of *electromagnetic* radiation released by *atomic nuclei* during *thermonuclear reactions* and having a *wavelength* between about $10^{-9}$ and $10^{-11}\mu m$.

**geocentrism** the hypothesis, held by Aristotle and many others, that Earth is the center of the Universe.

**geologic formation** a conveniently mappable, tabular or lens-shaped body of one or more *sedimentary rock* types usually deposited under environmental conditions distinct from those that prevailed during the deposition of underlying and overlying formations.

**geomagnetic anomaly** a positive or negative deviation from the mean strength of the *geomagnetic field* due to the various magnetic properties of underlying bedrock types or to the presence of bedrock with reversed geomagnetic polarity; also called magnetic anomaly.

**geomagnetic declination** the angular difference between true north and magnetic north.

**geomagnetic field** a powerful magnetic field generated within Earth's molten outer core.

**geomagnetic inclination** the angle between a dipping magnetic field (as indicated by a magnetic needle) and the horizontal.

**geomagnetic reversal** a reversal of the polarity of Earth's *geomagnetic field*.

**geostrophic flow** a wind or water current turned eastward by the *Coriolis effect* until the latter becomes equal to the *pressure gradient* force.

**geostrophic wind** see *geostrophic flow*.

**geothermal gradient** the increase in temperature with depth beneath Earth's surface; ranges upward from about 8° C/km within upper crustal *rocks* but falls to much lower values at greater depths.

**geyser** a hot *spring* that produces sudden, usually recurrent eruptions of superheated *groundwater* in an area with a high *geothermal gradient*.

**giant** a star swelled by core collapse until it has become too big, too bright, and too cool for its *spectral class*.

**glacial advance** a downslope advance of the *terminus* of a *valley glacier* or *continental ice sheet* in the direction of flow caused by an excess of accumulation over *ablation*.

**glacial outwash** *stratified clastic sediment* transported away from a *glacier terminus* by meltwater.

**glacial polish** a smooth *abrasion* surface developed on bedrock outcrops by the passage of dirt-bearing *glacier* ice.

**glacial recession** an upslope retreat of the position of the *terminus* of a *valley glacier* or *continental ice sheet* resulting from an excess of *ablation* over accumulation.

**glacial striations** scratches inscribed in the surface of *glacially polished* bedrock by small *clasts* embedded in the sole of a moving *glacier*.

**glacier** an accumulation of ice that flows under its own weight.

**glass** a body of *mineral matter* that lacks *crystal structure* because it has cooled too quickly from the liquid state for its *atoms* to assume ordered arrangements.

**gleization** the development in cold, poorly drained, polar environments of a "gley soil," consisting of a layer of peat overlying a layer of "gley" or reduced, blue-gray clay.

**globular cluster** a spherical cluster of about 1 million old, *population II* stars; many such clusters compose the halos that surround most galaxies.

**Gondwanaland** the southern portion of the supercontinent *Pangaea,* comprising South America, Africa, India, Australia, and Antarctica, and partially set off from the northern portion of Pangaea by the *Tethys Ocean*.

**graben** a downdropped *fault block* between two adjacent *extension faults*.

**graded bedding** a series of *strata* in which *clast* size grades upward from coarser to finer or, less commonly, the reverse.

**graded stream** a stream that has eroded its channel down to its *base level* and thus has reached *equilibrium* with respect to deposition and *erosion*.

**granule** one of millions of *convection cells* composing the Sun's *photosphere*.

**great planets** the Solar System's four large, outer planets, Jupiter, Saturn, Uranus, and Neptune, all of which have thick hydrogen and helium atmospheres.

**greenhouse effect** the absorption of outgoing infrared radiation from Earth's surface by atmospheric gases that are transparent to short wave *insolation* but opaque to long wave Earth radiation.

**Gregorian calendar** a calendar developed by Clavius in 1582, which adds one day every fourth year, called a leap year, to the 365 days of the year except for century years divisible by 400.

**groin** a low wall, usually of stone, built perpendicularly outward from a shoreline to prevent beach *erosion* by trapping *sediment* from *littoral drift*.

**ground moraine** an extensive sheet of basal *till* mainly deposited under high pressure beneath the sole of a *glacier* and often overlain by *ablation* till.

**ground subsidence** sinking of the ground surface due to *solution* of underlying bedrock, thawing of *permafrost,* withdrawal of *magma, tectonic stress,* mining, or *compaction* due to prolonged pumping of water or oil.

**groundwater** water in the *saturated zone* below the *water table*.

**guyot** a flat-topped *seamount*.

**gyre** a major, cyclical circulation, corresponding to an *ocean basin* in size and affecting the ocean water above the *permanent thermocline*. Gyres circulate clockwise in the Northern Hemisphere, counterclockwise in the Southern Hemisphere.

# H

**Hadley cell** one of two major *convective circulations* in Earth's *troposphere,* in which warm air rises over the equator, moves poleward, and descends at about 30° north and south latitude, giving rise to deserts and the *trade winds,* which blow toward the equator from the northeast and southeast.

**hail** pellets of ice with a concentrically layered structure formed within a *cumulonimbus cloud* when freezing raindrops are repeatedly lofted by strong updrafts.

**half-life** the *time* required for one-half of the *atoms* in a sample of a radioactive *isotope* to decay.

**hanging valley** a tributary glacial valley, the floor of which stands at a higher elevation than the floor of the trunk glacial valley at the confluence, or junction, of the two valleys.

**hanging wall** the *fault block* that lies above an inclined *fault plane*.

**hardness** the ability of a substance to resist scratching by another substance.

**harmonic tremor** a continuous, low-intensity seismic signal given off by rising *magma*.

**haze** a suspension in air of microscopic particles of dust, smoke, salt, organic particles, etc.; particles often serve as *condensation nuclei* (wet haze).

**headward erosion** the upslope extension of the channels of headwater tributary streams by accelerated *erosion* at the point where *sheet flow* becomes *channel flow*.

**heliocentrism** the hypothesis, held by Mikoiaj Kopernik and many others, that the Sun is the center of the Universe.

**Hertzsprung-Russell diagram** a graph, developed by Ejnar Hertzsprung and Henry Russell, on which stars are plotted according to *color index* (or surface temperature) and *absolute magnitude* (or luminosity).

**high clouds** clouds located in a shell of the *troposphere* that rises from between 3000 and 8000 m altitude in polar regions to between 6000 and 18,000 m altitude in the equatorial region.

**high tide** the *time* of the highest sea level between *flood* and *ebb* tide.

**hinge** the line of maximum flexure on a folded bedding plane.

**horizon system** a method whereby the position of any celestial object at a given time is located by its *altitude* in degrees above the observer's horizon and its *azimuth* in degrees to the east of the observer's celestial meridian.

**horn** a mountain peak that has been steepened by *erosion* within three or more glacial *cirques* developed on its slopes.

**horst** a raised *fault block* between two adjacent *extension faults*.

**hot spring** a *spring* whose water has been heated by a steep *geothermal gradient*.

**Hubble's law** the observation that the Universe is everywhere expanding at a uniform rate of 75 km/sec/mega*parsec*.

**humus** altered, decay-resistant organic material.

**hundred-year flood** a flood with a *recurrence interval* of 100 years.

**hurricane** a small, intense tropical *cyclone*, composed of *cumulonimbus clouds* and with wind speeds in excess of 120 km/hr, that travels west before veering poleward to redistribute tropical heat to the polar regions.

**hydration** the introduction of a fixed proportion of water *molecules* into the *crystal structure* of a *mineral*.

**hydraulic action** the *erosion* of material from a stream channel by the force of flowing water.

**hydrocarbon** a chemical *compound* consisting of carbon and hydrogen.

**hydrologic cycle** the network of pathways along which $H_2O$ moves among four major reservoirs: the ocean (as seawater), the atmosphere (as *water vapor*), the land surface (as streams, lakes and glaciers), and the land subsurface (as *groundwater*).

**hydrolysis** the chemical reaction of water with another *compound*.

**hygroscopic** showing an ability to *condense* water vapor.

# I

**ice crystal process** the growth of ice crystals by transfer of water *molecules* from water droplets, which have a higher *vapor pressure* than ice.

**ice storm** the freezing of raindrops as they fall through a cold air layer located at ground level, whereby solid surfaces, such as roads, tree branches, and buildings, become glazed with a thick coating of ice.

**igneous rock** a *rock* that solidified from a *magma*.

**impact crater** a *crater* caused by a *meteorite* impact.

**incised meander** a *meander* that has been eroded into the underlying bedrock due to *rejuvenation*.

**inner planets** the Solar System's four small, rocky, inner planets, Mercury, Venus, Earth, and Mars, all of which have relatively thin atmospheres composed of such relatively heavy gases as carbon dioxide, nitrogen, oxygen, and hydrogen oxide.

**input** *matter* or *energy* admitted into a *system*.

**insolation** solar energy striking Earth's surface.

**intensity** the degree to which a certain phenomenon is expressed.

**interlobate moraine** a *moraine* formed between two lobes of a *continental ice sheet*.

**intermediate** said of an *igneous rock* having a *silica* content between 52% and 66%.

**intermittent stream** a stream whose channel is dry for part of each year.

**intertropical convergence zone (ITCZ)** a near-equatorial belt of *convergence* where the *trade winds* of the Northern Hemisphere meet those of the Southern Hemisphere.

**intrusive** said of a body of *phaneritic igneous rock* emplaced below Earth's surface.

**ion** an *atom* that has gained or lost one or more *electrons* and so acquired a negative or positive electrical charge.

**ionic bond** a powerful, attractive *electrostatic force* between *atoms* that have become oppositely charged *ions* as a result of a transfer of one or more *electrons* from one to the other.

**ionosphere** a series of five layers within the upper *mesosphere* (above about 60 km) and *thermosphere* in which *insolation* ionizes atmospheric gases.

**irregular galaxy** an irregularly shaped assemblage of millions of predominantly young, *population I* stars with abundant interstellar gas and dust and no spiral arms.

**isobar** a contour line connecting all points of equal pressure on a map portraying pressure conditions (1) on a land or water surface, (2) at a specified altitude within the atmosphere, or (3) at a specified depth within the oceans or Earth's interior.

**isolated system**   a *system* that lacks *inputs* and *outputs* of *energy* and *matter*.

**isostasy**   the tendency of any point on Earth's surface to lie at a distance from Earth's center that is inversely proportional to the average *density* of the underlying *rock*.

**isostatic rebound**   the slow rise of an area of Earth's crust after the removal of a heavy load, such as a *continental ice sheet* or a large lake, as plastic rock flows into the underlying region of the *asthenosphere*.

**isotope**   a variety of an *atom* of a given *element* distinguished from other varieties of that element by the number of *neutrons* in its *nucleus*.

# J

**joint**   a *fracture* in bedrock or *regolith* along which there has been no relative motion of one side with respect to the other.

**June solstice**   in the *equatorial system* of coordinates, the *meridian* of 6 hours *right ascension,* located at the point where the Sun stands at its highest point above the *celestial equator;* also the point on Earth's orbit at which the *Arctic circle* receives 24 hours of sunlight a day.

# K

**kame terrace**   a flat-topped body of *stratified drift* deposited between melting glacier ice and a valley sidewall or *lateral moraine.*

**karst topography**   a landscape underlain by limestone bedrock and dominated by caverns and collapse structures, such as *sinkholes,* and a lack of surface drainage except in areas of high *water table.*

**katabatic wind**   a nighttime wind caused by the downslope flow of highland air cooled by loss of *radiant energy.*

**kettle**   a depression, usually in glacial *outwash,* formed by the melting of a partially or completely buried block of *glacier* ice. Often the site of a pond, swamp, or bog.

**kinetic energy**   a form of *energy* bound into *matter* by virtue of its motion and proportional to its *mass* and the square of its velocity.

# L

**laccolith**   a blisterlike *concordant pluton* that has arched overlying *strata.*

**lagoon**   a body of shallow, quiet water partially isolated from the open ocean by a *reef, spit,* or *barrier island.*

**lahar**   a volcanic *mudflow.*

**land breeze**   a light evening wind blowing from the land to the sea and constituting the base of a *thermal convection cell* that rises over the warmer sea surface and descends over the cooler land surface.

**latent heat of evaporation**   540 *calories* of *radiant energy* absorbed by 1 g of water on *evaporation* (at 100° C) and released by 1 g of water vapor on *condensation.*

**latent heat of fusion**   80 *calories* of *radiant energy* absorbed by 1 g of water on melting (at 0° C) and released by 1 g of water vapor on freezing.

**latent heat transfer**   the storing, within a liquid or a vapor, of *latent heat* energy that can later be released in a different location by a change of state back to a solid or to a liquid.

**lateral moraine**   a *moraine* ridge formed between *glacier* ice and a valley sidewall.

**laterization**   the formation of *laterite,* a rock-hard, brick-red, nodular deposit of iron and aluminum oxides within the thick, deep-red *B horizon* of tropical *soils,* due to *oxidation* and neutral leaching.

**latitude**   location on Earth's surface measured in degrees, minutes, and seconds north or south of the equator.

**Laurasia**   the northern portion of the supercontinent *Pangaea,* comprising North America and Asia and partially set off from the southern portion of Pangaea by the Tethys Ocean.

**lava**   molten or solidified *igneous rock* material that has erupted at Earth's surface and has lost most of its dissolved water and gases.

**law of constancy of interfacial angles**   the principle whereby equivalent *crystal faces* on different crystals of a given *mineral* always bear the same angular relationship to one another due to the arrangement of *atoms* in the *crystal structure* regardless of the sizes and shapes of the crystals.

**law of faunal succession**   a principle whereby biological species succeed one another without recurrence through geologic time; therefore, a given species assemblage is always diagnostic of the time interval during which it existed, because of which it can be used to date *rocks* that contain it.

**law of original horizontality**   *strata* are deposited with a horizontal attitude, perpendicular to the direction in which the force of gravity acts.

**law of original lateral continuity**   *strata* are laid down as continuous, uninterrupted "sheets" that thin gradually to a feather edge at the margins of the basin in which they are deposited.

**law of superposition**   in an undisturbed sequence of *strata,* the bottom layer is the oldest, and the top is the youngest.

**leap year**   every fourth year of the Gregorian calendar, including century years except those that are not evenly divisible by 400. In a leap year, the month of February has 29 days instead of 28.

**levee**   see *natural levee.*

**light horizon**   a boundary, approximately 13 billion *light-years* distant, beyond which objects would be receding faster than the speed of light due to *Hubble's law;* therefore, whatever light they might emit could never reach us.

**light-year**   the distance that *light,* travelling at a speed of 299,800 km/sec, covers in one *tropical year:* $9.46 \times 10^{12}$ km.

**limb**   the portion of a fold between adjacent *anticlinal* and *synclinal hinges.*

**liquefaction**   disruption of the microstructure of water-saturated, clay-rich materials in loose or poorly consolidated *sediments* by violent shaking of the ground during an earthquake, whereupon they assume the properties of a fluid.

**lithification**   the processes of *compaction, cementation,* and *recrystallization,* which act together to change *sediment* into *sedimentary rock.*

**lithospheric plate**   a rigid slab of *oceanic crust* and underlying upper *mantle* with or without a thick, uppermost layer of *continental crust.* Lithospheric plates are about 100 km thick, and each moves as an independent unit on the underlying, partially molten *low velocity zone.*

**littoral cell**   a shoreline *equilibrium system* that accepts *clastic sediment* as *input,* and stores it temporarily in such shoreline features as *beaches, spits, tombolos,* and *barrier islands* before expelling it as *output.*

**littoral drift**   the transport of *clastic sediment* along a shoreline by *swash* and *backwash,* which cause beach drift, and by *longshore currents,* which cause longshore drift.

**loam**   a *soil* containing a mixture of 23–52% sand-sized particles (1/16–2 mm), 28–50% silt-sized particles (1/256–1/16 mm), and 7–27% clay-sized particles (<1/256 mm).

**longitudinal dune**   a more or less smooth, straight sand ridge parallel with the prevailing wind direction and built by converging air flows between adjacent helical cells produced by high wind speeds aloft.

**longshore current**   an ocean current flowing adjacent to, and parallel with, a shoreline and resulting from the oblique approach of waves in the same general direction as the current.

**lopolith**   an enormous, *fractionally crystallized sill* intruded into a structural *basin.*

**low clouds**   clouds located at altitudes of less than 2000 m.

**low tide**   the time of the lowest sea level between *ebb* and *flood tide.*

**low velocity zone (LVZ)**   the partially molten, uppermost layer of the *asthenosphere* in Earth's upper *mantle,* within which the velocity of *seismic waves* is reduced. It is about 100 km thick and serves as a lubricating layer for the overlying *lithospheric plates.*

**lunar eclipse**   the passage of part or all of the Moon through Earth's *penumbra,* and sometimes also through Earth's *umbra,* during the *opposition* of Sun and Moon.

**luster**   the appearance of a *mineral* by light reflected from its surface.

# M

**mafic**   said of *rock* types that contain 45–52% *silica* or of *minerals* darkened by a high content of iron and magnesium.

**magma**   molten, subsurface *rock* containing dissolved water and gases.

**magmatic arc**   a chain of volcanic mountains or islands lying above and parallel to a *subduction zone,* and fed by *magma* rising from it; also called "volcanic arc."

**magmatic assimilation**   the incorporation into *magma* of *mineral matter* from partially or totally melted *clasts* of engulfed bedrock.

**magnetosphere**   the region of space influenced by the *geomagnetic field.* It extends to about 10 Earth radii on its Sunward side, where it is compressed by the solar wind, and to several hundred Earth radii on the side that faces away from the Sun.

**main sequence**   an alignment of stars on a *Hertzsprung-Russell diagram* indicating a general decrease in *absolute magnitude* (or a general increase in luminosity) with increasing *color index* (or surface temperature), i.e., hotter stars are brighter than cooler stars.

**manganese nodule**   a concretion, or concentrically layered deposit, of oxides of manganese, iron, and various other metals that forms on and slightly below the seafloor at the interface between reducing conditions below and oxidizing conditions above.

**mantle**   a thick shell of *ultramafic rock* in Earth's interior extending from the base of the *Mohorovičić discontinuity* at a depth of between 5 and 80 km to the outer boundary of the *core* at a depth of 2885 km.

**mantle plume**   a *convective circulation* of hot *mantle* material toward Earth's surface.

**March equinox**   in the *equatorial system* of coordinates, the *meridian* of 0 hours *right ascension,* located at the point where the Sun rises above the *celestial equator;* also the point on Earth's orbit at which the North Pole first receives *insolation.*

**mare**   (plural: **maria**) a large expanse of lunar flood basalt, typically within a *multiringed basin.*

**mass**   that property of *matter* that causes it to resist change of motion.

**mass wasting**   the detachment and downslope transfer of bedrock and *regolith* that occurs when gravitational force exceeds slope strength.

**matter**   anything that is capable of assuming definite physical form.

**meander**   a broad bend, or loop, in the *channel* of a *graded stream.*

**meander cutoff**   a *channel* segment cutting across the narrow neck of a large *meander;* shortcuts the stream's former course around the *meander* and locally steepens its gradient or slope.

**mean solar day**   the average time that it takes for Earth to make one complete rotation about its axis with respect to the Sun.

**mechanical metamorphism**   a localized, low-temperature, high-pressure *metamorphism* of bedrock caused by *faulting.*

**mechanical weathering**   any process that disintegrates *rock* without altering the chemical composition of its constituent *minerals.* Includes *frost wedging* and *root wedging,* growth of salt crystals, absorption of water, etc.

**medial moraine**   a *moraine* ridge formed by the coalescence of two *lateral moraines.*

**median rift** a valley or linear depression along the crestline of the *oceanic ridge*, characterized by *seafloor spreading, extension faulting*, and the *extrusion* of *mafic pillow lava*.

**meiosis** the division of one *cell* into four, each with half the number of chromosomes present in the original cell.

**mélange** a chaotically sheared and jumbled *breccia* within a *tectonic arc* including slices of ocean floor, *pelagic sediment*, and *clastic sediment* from the *magmatic arc*, all of which have been "offscraped" against the leading edge of the overriding *lithospheric plate*.

**meridian** any circle drawn on Earth's surface that passes through both the North Pole and the South Pole.

**mesopause** the upper surface of the *mesosphere*, located at an altitude of about 80 km.

**mesosphere** the layer of Earth's atmosphere lying between the *stratopause*, at about 48 km altitude and the *mesopause*, at about 80 km altitude, in which temperature falls upward and turbulence occurs.

**Mesozoic Era** the second era of the *Phanerozoic Eon*, spanning the time interval from about 245 million to about 66 million years ago; characterized by the dominance of large reptiles in both marine and terrestrial environments and of coniferous trees on land.

**metacryst** a large crystal that has grown within the *matrix* of a *metamorphic rock*; also called "porphyroblast."

**metallic bond** an "electron soup" that bathes all metallic *crystal structures* and holds them together.

**metamorphic rock** a body of *igneous, sedimentary*, or preexisting *metamorphic rock* that has undergone a change in mineralogy, chemistry, or both, under the influence of temperatures in excess of about 150° C, pressures in excess of about 2000 *atmospheres*, and chemical *solutions*.

**metamorphism** a change in the form and mineralogy of a *rock* under the influences of high temperature, pressure, and circulating *solutions*.

**meteor** a bright streak in the sky made by the passage through Earth's atmosphere of a meteoroid.

**meteorite** an extraterrestrial *clast* of *silicate rock* or nickel-iron alloy or both that has made contact with Earth's surface.

**middle clouds** clouds located in a shell of the *troposphere* whose base lies at 2000 m altitude and whose top rises from 4000 m altitude in polar regions to 8000 m altitude in the equatorial region.

**millibar** a pressure of 1000 dyn/cm².

**mineral** a naturally occurring, inorganic, solid substance with a definite range of chemical composition and a definite, ordered, internal structure.

**mitosis** simple *cell* division, beginning with division of the *cell nucleus*, and resulting in two new cells from a single parent.

**mixed tide** a tide having a higher and a lower crest and a higher and a lower trough in the daily cycle.

**Mohorovičić (M) discontinuity (Moho)** the boundary between Earth's crust and *mantle*. It varies in depth between 5 and 80 km and is marked by an abrupt downward increase in the velocity of *seismic waves*.

**moist adiabatic cooling rate** the rate at which *saturated* air cools on rising in Earth's atmosphere; increases from about 3° C/km in warm, moist air to about 10° C/km in cold, dry air.

**molecule** a group of two or more *atoms* joined together by *covalent bonds*.

**monism** the philosophical premise that the Universe consists of only one, indivisible principle and that everything is therefore sacred.

**monocline** an abrupt, local steepening of *dip* in otherwise horizontal or gently dipping *strata*.

**monoculture** a concentration of a single biological species to the virtual exclusion of all others within a given area.

**monsoon** in southern Asia, an alternation of (1) dry winters with a seaward flow of cool, dry air from central Asia and (2) rainy summers with a landward flow of warm, moist air from the Indian Ocean.

**moraine** an accumulation of *till* deposited by *glacier* ice.

**mountain belt** a linear region of Earth's crust, mountainous or otherwise, that has at some time undergone *orogeny* as revealed by the presence of folds, *faults*, and large *plutons* in its bedrock.

**mudcracks** polygonal networks of shallow, vertical shrinkage cracks in Sun-dried mud, often preserved by subsequent burial and *lithification*.

**mudflow** the rapid downslope flow of a mass of wet *regolith* having a water content of up to 60%.

**multiringed basin** a large *impact crater* in which the crust has been thrown into a series of concentric rims.

## N

**natural levee** a linear deposit of *clastic sediment* that raises the banks of a *graded stream* slightly higher than the adjacent *floodplain*. Natural levees are built during overbank flooding as the coarsest fraction of the stream's *suspended load* settles out immediately adjacent to the channel.

**natural selection** the concept that those individuals within a population who are best adapted to the prevailing environmental conditions are the ones who are most likely to survive and reproduce; because of this, their genetic traits tend to attain and maintain dominance until the environmental conditions change.

**neap tide** a tide of low amplitude that occurs during *quadrature* (quarter Moons).

**nebula** a cosmic cloud of dust and gas.

**negative feedback loop** in a *cybernetic system*, a continuous sampling of the level of the system's *cybernetic potential* used by a regulator to alter either *input* or *output* in such a way as to reverse any change in *cybenetic potential*.

**nekton** swimming organisms, such as most fish and cephalopods (squids, octopi, etc.).

**neutrino** a chargeless, apparently massless particle released during *thermonuclear reactions.*

**neutron** an electrically neutral particle found in *atomic nuclei* and having 1839 times the *mass* of the *electron.*

**neutron star** the rapidly spinning residue of a *supernova,* containing within a diameter of about 20 km about two solar masses of *matter,* in which *electrons* and *protons* have fused together to form a solid *mass* of *neutrons.*

**Newton's first and second laws of motion** unstressed objects do not accelerate in proportion to applied stress and in inverse proportion to their masses.

**nimbostratus clouds** *stratus clouds* from which rain or snow is falling.

**nova** an aging dwarf star that accumulates gas from a *giant* companion until *thermonuclear reactions* occur, about once every million years, blowing the accumulated gas clear of the star's surface and reducing its *apparent magnitude* by a factor of 7 to 10.

**nucleus** see *atomic nucleus, cell nucleus.*

# O

**occluded front** a complex front that develops when the *cold front* of a *wave cyclone* overtakes the *warm front* and raises the intervening wedge of tropical air aloft.

**ocean basin** a large, low-lying expanse of *mafic* rock and overlying *pelagic sediment* composing the deep seafloor and bounded by *oceanic trenches and continental slopes* or both.

**oceanic crust** a 5–15 km thick layer of *mafic rock* and overlying *pelagic sediment* that lies above the *Mohorovičić discontinuity* and forms the floors of the *ocean basins.*

**oceanic ridge** a continuous, 84,000 km long, submarine mountain chain traversing the ocean floors, along the axis of which *seafloor spreading* occurs. The ridge rises about 1–3 km above the ocean floor and is offset by hundreds of tranverse *fracture zones.*

**oceanic trench** a linear depression in the ocean floor caused by the *subduction* of a *lithospheric plate* occurs.

**open system** a *system* characterized by *inputs* and *outputs* of both *energy* and *matter.*

**opposition** an alignment of Earth with two celestial objects with Earth in the middle.

**ore mineral** a commercially valuable *mineral* consisting of *elements* that are not readily accommodated within the *crystal structures* of the rock-forming minerals; normally found in *veins* in bedrock.

**organic sedimentary rock** a *sedimentary rock,* such as coal, that consists mainly of organic remains.

**orogeny** mountain-building; includes folding, *faulting,* intrusion of large *plutons,* and *metamorphism.*

**output** *matter* or *energy* ejected by a *system.*

**outwash plain** an extensive deposit of *stratified drift* filling a valley bottom downstream from the *terminus* of a *glacier.*

**oxbow lake** a crescent-shaped lake remaining within a *meander* following the formation of a *meander cutoff.*

**oxidation** strictly, the removal of an *electron* from the outer *electron shell* of an *atom;* loosely, the chemical combination of an *element* or *compound* with oxygen, which is one of several causes of this effect.

**oxygen minimum zone** the region of the oceans between depths of 100 m and as much as 1 km, within which oxygen concentrations are low because mixing and *photosynthesis* are reduced but animal *respiration* continues unabated.

**ozone layer** a region of Earth's atmosphere, between about 15 and 55 km altitude, in which the action of solar ultraviolet light on oxygen produces triatomic oxygen, or ozone.

# P

**pahoehoe** a gas-rich, fluid variety of *mafic lava* with a surface that has been wrinkled into smooth-skinned, parallel rolls during flow.

**paleoclimate** an ancient climate, evidence of which is preserved in *sedimentary rocks.*

**paleolatitude** the *latitude* of a point on Earth's surface at a specified time in the geologic past as determined by *paleomagnetism.*

**paleomagnetism** study of the *magnetic inclination* of magnetic *minerals* within an oriented *rock* sample in order to determine the *paleolatitude* of the sample at the time of its formation.

**Paleozoic Era** the first era of the *Phanerozoic Eon,* spanning the time interval from about 700 million to about 245 million years ago and characterized by an oxygen-rich atmosphere and the rise of many-celled, *eukaryotic* organisms in both marine and terrestrial environments.

**Pangaea** the name given by Alfred Wegener to the supercontinent that comprised all of Earth's landmasses between the Pennsylvanian and Triassic Periods.

**parabolic dune** a crescent-shaped sand dune, the concave side of which faces the wind; develops around blowouts in areas of partly vegetated sand.

**parallax angle** the amount of apparent shift, in seconds of arc, of a nearby celestial object against the backdrop of more distant stars when that object is viewed from two different perspectives, usually opposite points on Earth's orbit.

**parallel** an imaginary line on Earth's surface connecting points of equal *latitude* north or south of the equator.

**parallel drainage** a *drainage pattern* developed on a uniformly sloping surface underlain by homogeneous bedrock or *regolith,* in which the channels of the *drainage system* are evenly spaced and aligned approximately parallel to the regional slope.

**parsec** a distance of 3.2 *light-years,* equal to the distance from Earth to a celestial object that has a *parallax angle* of 1 sec when Earth's orbit is used as a baseline.

**partial melting** the tendency of mixtures of the more *felsic minerals* in *rocks* to melt at lower temperatures than the remaining material; allows rock subjected to rising temperature to melt partially, generating *magma* of a more felsic composition than the rock itself.

**patterned ground** a geometric distribution of large *clasts* on the surface of ground subject to freeze-thaw cycles, especially if underlain by *permafrost.*

**pedalfer** a *soil* category characteristic of humid climates, in which percolation of neutral to acid water leaches iron and aluminum oxides from the *A horizon* and deposits them in the *B horizon.*

**pediment** a gently sloping erosional surface on bedrock bordering mountain ranges in arid regions, developed by the parallel retreat of mountain slopes.

**pedocal** a soil category characteristic of arid climates, in which calcium carbonate tends to accumulate in the *soil profile* due to the predominance of evaporation over percolation.

**pelagic sediment** *sediment* on *oceanic ridges* and *abyssal plains* that has settled out of suspension in the open ocean. Includes *brown clay,* siliceous and calcareous oozes, volcanic ash, and *manganese nodules.*

**peneplain** a flat to gently undulating land surface resulting from an extended period of stream *erosion* without *rejuvenation.*

**penumbra** an outward-widening cone of partial shadow extending from the dark side of a moon or planet; also the lighter, outer portion of a *sunspot.*

**perched water table** a local *water table* perched atop a local *aquiclude,* such as a shale lens, within an *aquifer.*

**perennial stream** a stream that flows year round without significant interruption.

**perigee** the point on the Moon's orbit at which it is closest to Earth (356,411 km, center-to-center).

**periglacial zone** a region immediately adjacent to a *glacier* margin, within which various phenomena characteristic of frozen and devegetated ground occur.

**perihelion** the point on Earth's orbit at which it lies closest to the Sun (147 million km). At present, perihelion falls on or about 3 January.

**permafrost** permanently frozen ground.

**permanent thermocline** a zone of rapid downward temperature decrease in the ocean between about 300 and 900 m depth.

**permeability** the ability to transmit fluids.

**phaneritic** a *rock* texture in which the individual *mineral* grains, or *crystals,* are visible to the naked eye.

**Phanerozoic Eon** the time interval from the end of the *Cryptozoic Eon,* about 700 million years ago, to the present, characterized by many-celled organisms.

**phases of the Moon** the cycle from new Moon at *conjunction* through first quarter Moon at first *quadrature,* full Moon at *opposition,* and third quarter Moon at second quadrature, and back to the new Moon.

**phenocryst** a large, early-formed crystal in a *porphyritic igneous rock* surrounded by a finer-*phaneritic* or *aphanitic* matrix.

**photon** a particle (or wave) of *radiant energy.*

**photosphere** the visible surface layer of the Sun, composed of *granules.*

**photosynthesis** the use of solar energy by green plants to reduce carbon dioxide to *carbohydrate* (sugar) by removing oxygen and adding hydrogen. The reverse of *respiration.*

**physical property** a characteristic way in which a *mineral* interacts with its environment. Includes *specific gravity,* color, *streak, luster,* magnetism, *hardness,* melting point, tenacity, *cleavage,* and *fracture.*

**pillow lava** pillowlike structures in *lava* that has been erupted under water. The individual pillows have water-chilled margins of finer grain than their interiors.

**plane of the ecliptic** the plane of Earth's orbit.

**planetary nebula** an expanding shell of gas expelled by a hot, dwarf star during a *nova* explosion.

**plankton** floating or weakly swimming organisms, especially nonrooted aquatic plants and the larval stages of many animals.

**plastic strain** nonrecoverable deformation in *rock* resulting from *directed stress.*

**plate tectonics** the growth of *lithospheric plates* by *seafloor spreading,* the sliding of such plates past one another by *transform faulting,* and the *subduction* and *partial melting* of the edge of one such plate beneath the edge of another.

**platform** a region of young, undeformed *sedimentary strata* overlying the deformed, eroded basement complex of an older *mountain belt.*

**playa lake** in an arid region, a shallow, ephemeral lake with no outlet that forms after heavy rains in the surrounding mountains and then dries up, leaving deposits of dissolved salts.

**plunging fold** a fold whose *hinge* is not horizontal.

**plutonic igneous rock** equivalent to a body of *intrusive igneous rock.*

**pluvial lake** a lake that forms in a *transglacial* environment and disappears when glaciation ceases.

**podzolization** the leaching of *cations,* iron and aluminum oxides, clay, and organic matter from the *A horizon* of a *soil* in a cool, humid region by percolating acid rainwater and the subsequent deposition of all these except the cations within the *B horizon.*

**polar front** the surface of interaction between cold, dry, polar air and warm, moist, tropical air, located between about 30° and 60° north and south *latitude.*

**polar jet stream** a high-velocity *geostrophic wind* that forms at the *tropopause* above the *polar front.*

**polar molecule** a *molecule* with a greater concentration of negative charge on one side than on the other due to an asymmetrical distribution of *electron* clouds.

**polar wandering curve** a line on a map showing the apparent locations of the north or south magnetic poles at various times in the geologic past.

**polymerization** the *covalent bonding* of *atoms* of the same *element* into long, chainlike *molecules* with atoms of different elements attached.

**population I**   stars of mixed age, characteristic of galactic spiral arms; includes young, hot, blue *giant* stars and abundant gas.

**population II**   older stars characteristic of the galactic hub and *globular clusters* surrounding the Milky Way; includes abundant red *giants,* but no blue giants and little gas.

**porosity**   the percentage of the volume of a body of *rock* or *regolith* that is occupied by pore space.

**porphyritic**   the *texture* of an *igneous rock* that contains *phenocrysts.*

**potential energy**   a form of *energy* bound into *matter* and emanating from it in all directions as gravitational or electromagnetic fields that diminish in strength as the square of the distance from the object.

**potentially unstable slope**   a slope that can undergo *mass wasting* if gravitational *stress* is increased or *cohesive* strength reduced.

**pothole**   a roughly circular depression in the bedrock bed of a stream channel caused by the *abrasive* action of boulders and sand swirling within it.

**preadaptation**   a feature of an organism that develops to serve one function and subsequently is found applicable, usually with modifications, to another function in a different environmental setting.

**precession of the equinoxes**   the clockwise shift of the *solstices* and *equinoxes* along Earth's orbit by about 50 sec of arc per year.

**pressure gradient**   the rate of pressure decrease experienced in moving directly away from a region of high *confining pressure.*

**prevailing westerlies**   mid-latitude winds that blow from the southwest to the northwest due to *cyclones* and *anticyclones* entrained within the *polar jet stream.*

**prokaryotic cell**   a small, primitive *cell* with one typically hoop-shaped chromosome and no *cell nucleus,* nuclear membrane, or organelles.

**Proterozoic Era**   the second era of the *Cryptozoic Eon,* spanning the time interval from about 2500 million to about 700 million years ago and characterized by one-celled *eukaryotic* organisms and an oxygen-bearing atmosphere.

**proton**   a particle found in *atomic nuclei* that has unit positive electrical charge and 1836 times the *mass* of the *electron.*

**proton-proton chain**   a *thermonuclear reaction* in stars which combines *protons* and *electrons* to form helium *nuclei, neutrinos,* and *gamma rays.*

**pulsar**   a dense, rapidly spinning *neutron star* that emits strong, directed radio beacons that sweep by Earth with high-*frequency* periodicities.

**pulsating Universe hypothesis**   the concept that the Universe undergoes endless cycles of expansion and contraction in which each *big bang* is both followed and preceeded by a "big crunch."

**pulsating variable**   a *variable star* whose *absolute magnitude* varies periodically as a result of variations in its rate of *thermonuclear reaction;* includes blue *giants,* RR-Lyrae variables, *Cepheid variables,* and longer-period variables.

**P wave**   the first, or primary, *seismic wave* to arrive at a *seismograph* station; a *body wave* that vibrates with a back and forth motion in the direction of propagation.

**pyroclastic**   said of an igneous *rock* composed of *tephra.*

# Q

**quadrature**   an arrangement of Earth and two other celestial objects such that the three objects form a right angle with Earth at the corner.

**quasar**   a distant, quasi-stellar object of phenomenal brightness thought to be a massive *black hole* in the center of a young galaxy, seen head-on.

# R

**radiation fog**   a mist of fine cloud droplets in ground-level air formed as a result of nighttime cooling of the underlying ground due to a loss of *radiant energy.*

**radiative zone**   the lower, thicker part of the Sun's mantle, in which a plasma of dismembered *atoms* absorbs *gamma radiation* from the outer core and reradiates it to the *convective zone* as *energy* of longer *wavelengths.*

**radio galaxy**   a galaxy that emits an abnormal quantity of radio energy.

**radiometric dating**   a method for determining the age of a *rock* by measuring the percentage of an original amount of a constituent radioactive *element* that has decayed since the rock was formed.

**radio telescope**   a movable, paraboloid, wire mesh surface for observing cosmic radio sources.

**rain shadow**   a dry region downwind of a mountain range affected by a *chinook wind.*

**ray**   a bright streak of ejected material extending radially away from an *impact crater.*

**recessional moraine**   a ridge of *moraine* deposited by the *terminus* of a receding *glacier* during a temporary stillstand.

**recharge**   the replenishment of water in an *aquifer* by rainfall or snowmelt.

**recrystallization**   a change in the shape of a crystal due to the addition of *atoms* to, or removal of atoms from, certain regions of its *crystal structure;* also, the growth of new minerals from materials released in the destruction of other minerals during *metamorphism.*

**recurrence interval**   the average *time* between successive occurrences of the same event.

**red bed**   a *clastic sedimentary* deposit that has been stained red by hematite (iron oxide).

**red shift**   an apparent increase in the *wavelength* of light being emitted by an object that is receding from the observer. See also *Doppler effect.*

**reef**    a shallowly-submerged, nearshore deposit of *biochemical sedimentary rock* consisting of intergrown calcium carbonate skeletons of marine animals, such as corals, sponges, and bryozoans, and of space fillings of *clastic* calcium carbonate debris and cement.

**reflecting telescope**    a light-collecting and image-magnifying device consisting of a parabolic mirror and an eyepiece.

**refracting telescope**    a light-collecting and image-magnifying device consisting of a doubly-convex lens and an eyepiece.

**refraction**    the bending of light as it changes velocity on crossing the boundary between two transparent media of different density (e.g., air and water).

**regional metamorphism**    widespread *metamorphism* deep within a *mountain belt* resulting from a combination of high temperature, high *confining pressure,* and high *directed stress.*

**regolith**    *unlithified mineral* material on Earth's surface, including *saprolite, sediment, tephra,* and *soil.*

**rejuvenation**    downcutting by a stream in response to uplift or a lowering of *base level.*

**relative geologic time scale**    a chronology of Earth's history based on the relative position in time of events recorded in the *rocks* of Earth's crust as determined chiefly by the *laws of superposition* and *faunal succession.*

**relative humidity**    the ratio of the *absolute humidity* of an air parcel at a given temperature to the absolute humidity it would have if it were *saturated* at that temperature.

**replacement deposit**    a chemical replacement of an organism or a *sedimentary rock.* Most fossils, much chert, and all dolomite rock and phosphate rock are examples.

**residence time**    the average *time* that a specified object, usually an *atom* or *molecule,* spends within a given *system,* such as the ocean or the atmosphere.

**residual bond**    a weak force, arising from the asymmetrical distribution of electric charge on *molecules* of certain *compounds,* such as hydrogen oxide, that binds molecules of the compound together.

**residual soil**    a *soil* developed on *regolith* that has formed in place from the *weathering* of underlying bedrock.

**respiration**    the *oxidation* of *carbohydrate* to carbon dioxide and water within the *cells* of a plant or animal. The reverse of *photosynthesis.*

**right ascension**    the angular distance of a celestial object to the east of the *March equinox,* measured in hours, minutes, and seconds.

**rip current**    a strong, narrow, seaward flow from the *surf zone* of up to 1 m/sec representing the escape of a *longshore current.*

**ripplemarks**    miniature, parallel waveforms in sand or mud preserved during *lithification* and produced by wind, waves, or currents moving at a right angle to the ripple crests.

**rock**    a continuous body of hard *mineral matter* forming a portion of Earth's crust. *Syn:* ledge.

**rock avalanche**    a sudden downslope transfer of loose *rock* debris under the influence of gravity.

**rock cycle**    the network of pathways along which *igneous, sedimentary,* and *metamorphic rocks* change into one another.

**rockfall**    the free fall of a large, detached mass of *rock* from a cliff.

**rock glacier**    a thick accumulation of *talus* that flows slowly downslope due to the freezing and thawing of interstitial ice.

**rockslide**    the rapid downslope sliding of a detached mass of *rock* on a sloping bedrock surface.

**Rossby wave**    a major, wavelike form in the *polar jet stream* that deflects its flow alternately poleward and equatorward.

## S

**salinity**    the concentration of salts dissolved in seawater, usually expressed as ‰ (per mil, or parts per thousand).

**salinization**    the deposition of soluble salts within a *soil profile* by *evaporation* of *groundwater* in arid regions.

**saltation**    a bouncing or jumping action of sand grains within a current of wind or water.

**sand dune**    a large mound of sand built by prevailing winds.

**Santa Ana wind**    a hot, dry wind that occurs in southern California when a high pressure cell lies over the western United States.

**saprolite**    *chemically weathered* bedrock that has not yet been *eroded* from the *weathering* site.

**saturated zone**    the zone below the *water table,* in which all open pore spaces are completely filled with water.

**saturation**    a condition in which no more of a given substance, or *solute,* can dissolve in another substance, or *solvent.* Examples include salt in water, water vapor in air, and carbon dioxide in water or *magma.* Saturation normally increases with temperature except for gases dissolved in liquids.

**saturation vapor pressure**    the *vapor pressure* over the surface of a liquid that is in *dynamic equilibrium* with its own vapor at a given temperature.

**scattering**    deflection of some of the *photons* of the solar *energy* beam through a wide range of angles by close encounters or collisions with atmospheric *atoms, molecules,* or particles.

**scour lag**    the tendency of a particle swept inshore to remain there because it requires a stronger current to erode it from the *beach face* or *shoreface* than that which deposited it.

**sea breeze**    a light, daytime wind blowing from the sea to the land and constituting the base of a *thermal convection* cell that rises over the Sun-warmed land surface and descends over the cooler sea surface.

**sea cave**    a cave eroded by wave action within zones of weakness at the base of a *wave-cut cliff* or *rock* promontory.

**seafloor spreading** the slow (1–10 cm/yr) spreading apart of two adjacent *plates* of *oceanic lithosphere* perpendicularly to the *oceanic ridge* as new material is added to their trailing margins by the upwelling of *magma* within the *median rift.*

**seamount** a submarine volcano that rises at least 1000 m above the seafloor.

**seamount chain** a line of *seamounts* resulting from the drifting of *oceanic lithosphere* over a *mantle plume,* as in the Hawaiian seamount chain.

**seasonal thermocline** a zone of rapidly declining temperature beneath a shallow, surface layer of warm, low-density, wind-mixed ocean water heated by warm rains in spring. Destroyed by *fall overturn.*

**sediment** *unlithified* material derived from the *weathering* and *erosion* of preexisting *rocks.* Includes both soluble and insoluble components.

**sedimentary rock** a body of *lithified sediment.*

**sedimentation** the deposition of *sediment;* more specifically, the combined processes of *erosion,* transportation, deposition, and *lithification.*

**seismic reflection profiling** determination of subsurface geologic structure by analysis of the travel times of sound pulses reflected from structural discontinuities.

**seismic refraction profiling** determination of subsurface geologic structure by analysis of travel times of sound pulses *refracted* through layers of different *density.*

**seismic wave** any of four different types of shock waves (*P, S,* L, and R) transmitted through solid *rock* when it is subjected to a sudden disturbance, such as an earthquake or explosion.

**seismogram** the evidence of the passage of *seismic waves* recorded by a *seismograph.*

**seismograph** a device for the recording of *seismic waves.*

**semidiurnal tide** a tide having two subequal crests and two subequal troughs in the daily cycle.

**sensible heat** heat that is *conducted* and *convected* from the ground into the atmosphere.

**September equinox** in the *equatorial system* of coordinates, the *meridian* of 12 hours *right ascension,* located at the point where the Sun sinks below the celestial equator; also one of the two points on Earth's orbit at which the South Pole first receives *insolation.*

**shadow zone** a region, on the opposite side of Earth from an earthquake *focus,* within which no *seismic waves* can be detected. The *P wave* shadow zone, between 103° and 143° from the *epicenter,* results from *refraction* (bending) of P waves by the *core;* the *S wave* shadow zone, between 103° and 180° from the epicenter, results from damping of S waves by the liquid outer core.

**shallow-water wave** a *shoaling wave* in which the water depth is less than 1/20 the deep-water *wavelength.*

**shear stress** horizontally *directed stress* that tends to slide one half of an object to which it is applied past the other half.

**sheet flow** flow of water over the land surface in a relatively uniform layer rather than concentrated into channels.

**shield** an extensive *peneplain* usually developed on *Cryptozoic rocks.*

**shield volcano** a broad, low *volcanic cone* around a *central eruption,* built by successive outpourings of fluid, *mafic lava.*

**shoaling wave** a wave whose *wave base* has touched the seafloor.

**sidereal day** the *time* required for Earth to make one rotation with respect to the fixed stars: 0.9973 *solar day.*

**sidereal month** the *time* required for the Moon to make one revolution with respect to the fixed stars: 27.32 *solar days.*

**sidereal year** the *time* required for Earth to make one revolution with respect to the fixed stars: 365.256 *solar days.*

**silica** silicon dioxide ($SiO_2$).

**silica tetrahedron** an *anion* consisting of a single silicon *atom* symmetrically surrounded by four oxygen atoms ($SiO_4^{-4}$); *polymerizes* readily and is the basic unit in the *crystal structures* of all *silicate minerals.*

**silicate mineral** a *mineral* that contains the *silica tetrahedron* in its *crystal structure,* either *polymerized* or as isolated tetrahedra.

**sill** a tabular (thin and flat) *concordant pluton.*

**sinkhole** a deep, circular depression formed by the collapse of the roof of an underlying cavern in limestone bedrock.

**sleet** rain that freezes as it falls through a ground-level *inversion,* forming ice pellets.

**slip face** the steeply-dipping, front face of an advancing sand dune, ripple, *delta,* or *spit,* down which sand *avalanches* as it is swept over the top by wind or currents.

**slump** a landslide during which the slope of the sliding mass is reduced.

**soil** a complex system of *saprolite* or *clastic* deposits, *humus,* air, water, and living organisms.

**soil horizon (O, A, B, C)** a layer within a *soil profile.* The O horizon contains organic material; the A horizon is a zone of leaching; the B horizon is a zone of accumulation; the C horizon is weathered bedrock or *sediment.*

**soil profile** the succession of *soil horizons (O, A, B, C)* characteristic of a mature *soil.*

**soil structure** the manner in which *soil* particles aggregate into small structural units called "peds."

**soil texture** the proportions of sand, silt, and clay within a *soil.*

**solar constant** the rate at which solar *radiant energy* is received at the top of Earth's atmosphere: 2.0 cal/cm²/min.

**solar eclipse** the passage of part or all of the Moon's *penumbra,* and sometimes also the Moon's *umbra,* across part of Earth's surface during the *conjunction* of Sun and Moon.

**solar flare** a solar explosion that erupts from the vicinity of a *sunspot,* sending out prodigious showers of *ionized* particles that result in *auroras* and magnetic storms on Earth.

**solar prominence**    a large loop of ionized hydrogen gas that arches up to 50,000 km above *sunspots,* following magnetic lines of force where they emerge from the Sun.

**solar wind**    a constant outpouring from the Sun of energized *matter,* consisting of helium *nuclei, electrons, protons* (hydrogen nuclei), and the nuclei of heavier *atoms.*

**solid solution**    the substitution of one *cation* for one another in all proportions within a given *crystal structure.*

**solifluction**    the slow flowage of water-saturated *regolith* in response to gravity, typically when overlying *permafrost.*

**solute**    a substance (solid, liquid, or gas) that has been dissolved in a *solvent* (solid, liquid, or gas).

**solution**    a *chemical weathering* process whereby soluble bedrock is dissolved along *fractures* by percolating water; also a *solvent* and its dissolved *solutes.*

**solvent**    a solid, liquid, or gas in which a *solute* (solid, liquid, or gas) can be, or has been, dissolved.

**sounding**    the remote probing of the atmosphere with instrumented rockets or balloons to determine environmental conditions at certain altitudes; also the measurement of depth in the ocean.

**specific gravity**    the weight of an object divided by the weight of an equal volume of water.

**specific heat**    the amount of heat required to raise the temperature of a substance by a certain number of degrees divided by the amount of heat required to raise the temperature of an equal mass of water by the same amount.

**spectral class**    one of a series of seven categories of star, O, B, A, F, G, K, and M, in which trends from O to M include lower temperature, blue to yellow to red color, and spectral lines of increasingly heavy elements and molecules.

**spectrograph**    a narrow slit and a prism attached to a telescope in order to *refract* starlight into its spectral colors.

**spheroidal weathering**    the tendency of more intense *weathering* at corners and edges to produce rounded boulders from originally angular blocks of bedrock.

**spin-orbit coupling**    synchronism of the periods, or multiples of the periods, of rotation and revolution of a planet or a moon.

**spiral galaxy**    a disk-shaped galaxy with spiral arms comprising mixed-age, *population I* stars with much interstellar gas and dust, a central, swollen hub of old, *population II* stars, and a halo of *globular clusters.*

**spit**    an offshore sandbar attached at one end to a headland. The spit platform is built by a *longshore current,* and on this platform, waves subsequently build a beach.

**spreading axis**    an imaginary line through Earth's center that defines the axis of rotation around which the motion of a given *lithospheric plate* must take place.

**spreading pole**    the points where a *spreading axis* emerges at Earth's surface.

**spring**    an outflow of *groundwater* where the *water table* intersects Earth's surface.

**spring tide**    a tide of high amplitude that occurs during *conjunction* (new Moon) and *opposition* (full Moon).

**stalactite**    a cone or spire of travertine, or calcium carbonate, hanging from the roof of a *cavern* and deposited by evaporating *groundwater.*

**stalagmite**    a cone or spire of travertine, or calcium carbonate, rising from the floor of a *cavern* and deposited by evaporating *groundwater.*

**standard atmosphere**    the average pressure of Earth's atmosphere at sea level: 1,013,000 dyn/cm², or 1013 mb.

**star dune**    a large many-ridged, stationary *sand dune* in a region of constantly shifting winds.

**steady state**    a condition, characteristic of *cybernetic systems,* in which *cybernetic potentials* and stored *matter* are held constant over *time.*

**stock**    a large *discordant pluton,* of which *erosion* has exposed a surface area of less than 100 km².

**storm surge**    a sudden, destructive rise of sea level at a shoreline during a *hurricane* resulting from reduced *atmospheric pressure,* high storm waves, high tides, and excess water driven onshore by the storm.

**stoss and lee topography**    bedrock outcrops that have been smoothed and rounded by glacial *abrasion* on their upflow sides and plucked, or quarried, on their leeward sides.

**strain**    deformation in *rock* resulting from the application of a *directed stress.*

**strata**    see *stratum.*

**stratification**    layering in a *sedimentary* or *pyroclastic rock.*

**stratified drift**    more or less sorted and *stratified* sand and gravel deposits that have been derived from *glaciers* but deposited by glacial meltwater rather than by glacier ice.

**stratocumulus clouds**    *low clouds* in which the cloud deck has a repetitive, gray, roll-like or pillowlike structure.

**stratopause**    the upper limit of the *stratosphere,* at about 48 km altitude, above which temperature decreases with altitude.

**stratosphere**    a stable region of Earth's atmosphere in which temperature increases upward due to heat produced in the creation and destruction of *ozone.* The base of the stratosphere varies from 8 km altitude at the poles to 18 km at the equator. Its top is at about 48 km altitude.

**stratum**    (plural, *strata*) a homogeneous layer in a *sedimentary* or *pyroclastic rock.*

**stratus clouds**    a uniformly textured, dark gray deck of *low clouds* through which the Sun or Moon is seldom visible.

**streak**    the color of the powder formed by rubbing a *mineral* on unglazed porcelain.

**stress**    a force acting on an area.

**strike**    the bearing of a horizontal line within an inclined stratification plane.

**strike-slip fault**    a *fault* with a vertical *fault plane,* along which *shear stress* has caused one *fault block* to slide horizontally either to the right (left-lateral fault) or to the left (right-lateral fault) with respect to the other fault block.

**stromatolite**   a laminated structure in limestone formed in shallow, nearshore marine water by the growth of cyanobacterial mats.

**strong force**   an extremely short-range, attractive force that binds *protons* together to form *atomic nuclei* during *thermonuclear reactions.*

**subduction**   the sinking of the edge of one *lithospheric plate* beneath the edge of another at a *convergent plate boundary.*

**subduction zone**   a plane, dipping gently to steeply beneath the edge of a *lithospheric plate,* along which a second plate descends into Earth's *mantle.*

**submarine canyon**   a deep, *dendritic* canyon, often corresponding in position to the mouth of a major river, that cuts across the boundary between the *continental shelf* and the *continental slope.*

**submarine fan**   a massive, submarine *alluvial fan* composed mainly of *turbidity flow* deposits.

**submergent coastline**   a coastline that has been lowered relative to sea level.

**subsolar point**   that point on Earth's surface at which the Sun's rays are striking the ground vertically.

**subtropical high pressure belts**   two regions of Earth's surface, centered approximately on 30° north and south *latitude,* where warm, dry air descends on the poleward sides of the *Hadley cells,* creating high pressure and desert conditions.

**subtropical jet stream**   a powerful westerly air flow aloft above the *subtropical high pressure belts* caused by the eastward turning of the *Hadley cell* circulation due to the *Coriolis effect.*

**sunspot**   a magnetic storm on the Sun, visible because its gas is cooler, and therefore darker, than the surrounding surface.

**supergiant**   an exceptionally large *giant* star.

**supernova**   the collaspe and catastrophic explosion of a massive star in which core temperature has risen high enough to produce iron and heavier *elements,* which absorb *energy* on formation and therefore further the collapse.

**surf zone**   the seaward portion of the *beach face,* shoreward of the breaker zone in which water travels on- and offshore as shallow-water waves, *rip currents,* and *backwash.*

**surface tension**   an apparent film on a water surface resulting from the *cohesion* of adjacent *molecules* on all sides except the upper side, creating a net attractive force downward that tends to shrink the water surface to a minimum area.

**surface wave**   a *seismic wave* that travels at Earth's surface rather than in its interior; includes L waves and R waves.

**suspended load**   that portion of a stream's *sediment* load that is transported in suspension in the turbulent stream water.

**S wave**   the second *seismic wave* to arrive at a *seismograph* station following an earthquake; a *body wave* that vibrates with a side-to-side motion across the direction of propagation.

**swash**   a sheet of water, rushing up the *beach face,* that represents the farthest shoreward advance of a *shallow-water wave.*

**syncline**   a fold in which younger *strata* are enclosed within older strata; a troughlike fold if right side up.

**synodic month**   the *time* between one *conjunction* of Moon and Sun and the next (or one new Moon and the next): 29.53 *solar days.*

**system**   any persistent, describable, and predictable arrangement of *matter, energy,* or both.

# T

**talus**   angular *rock* rubble deposited in steep, conical or apronlike slopes at the bases of cliffs during *rockfall* events.

**tarn**   a small lake filling a glacially scoured *rock* basin in the floor of a *cirque.*

**tectonic**   said of features of Earth's crust that have been produced by *directed stress.*

**tectonic arc**   a thick wedge of *pelagic* sediment mixed with slices of ocean floor and *clastic sediment* that accumulates in an *oceanic trench* adjacent to the edge of an overriding *lithospheric plate* during *subduction* of the underlying plate.

**temperature inversion**   a situation in which air temperature rises with altitude, the reverse of the normal condition in Earth's atmosphere.

**tensile stress**   an outward-*directed stress* that tends to stretch an object to which it is applied.

**tephra**   volcanic dust (<1/16 mm), ash (1/16–2 mm), lapilli (2–64 mm), and blocks and bombs (over 64 mm) consisting of glass fragments, *mineral* crystals; droplets or bombs of solidified *lava;* and fragments of shattered wallrock.

**terminal moraine**   a ridge of *moraine* deposited by the *terminus* of a *glacier* during a stillstand at its position of farthest advance.

**terminus**   the lower or outer margin of a *glacier* toward which ice flow occurs.

**tethered sampler**   a device, such as a Nansen bottle, secchi disk, or grab sampler, lowered by cable from a boat in order to sample seawater, marine life, or seafloor *sediment.*

**texture**   the size and shape of *mineral* grains in a *rock.*

**thermal conductivity**   the amount of heat, in *calories,* that can pass through a 1 cm thickness of a given material in 1 sec.

**thermal convection**   the tendency of warmer, less dense portions of a fluid, such as air or water, to rise relative to cooler, denser portions of the same fluid, which tend to sink.

**thermonuclear reaction**   the fusion of *atomic nuclei* under extremely high temperature to form nuclei of *elements* of higher *atomic number.*

**tidal inlet**   a storm-generated breach in a *spit* or *barrier island,* through which *tide* water flows.

**tide**   a giant *water wave* with a *wavelength* equal to half Earth's circumference; the tide is generated by the gravitational pulls of the Moon and Sun, and is driven by the relative motions of the Moon and Earth.

**till**   *unstratified* and unsorted *sediment* transported and deposited by *glacier* ice.

**time**   the measurement of *duration.*

**time horizon**    a boundary, approximately 13 billion *light-years* distant, beyond which the travel time to Earth for light being emitted by an object would exceed the age of the Universe.

**tombolo**    a necklike sandbar connecting the mainland with a nearby island.

**tornado**    a small, intense, destructive, funnel-shaped, *cyclonic* storm with a diameter of 100–600 m, wind speeds of ±200 km/hr, and an internal pressure up to 100 mb below the pressure of surrounding air.

**trade winds**    northeasterly (Northern Hemisphere) or southeasterly (Southern Hemisphere) winds representing the equatorward return flow of the *Hadley cells.*

**transform fault**    a vertical zone of relative motion between the margins of juxtaposed sections of two *lithospheric plates* spreading in opposite directions from offset segments of a *median rift,* which terminate both ends of the *fault* zone.

**transform fault boundary**    a line along which the edge of one *lithospheric plate* slides laterally along the edge of another in the process of *transform faulting.*

**transglacial zone**    a region far removed from *glacier* margins but subject to altered climatic and ecological effects, such as *pluvial lakes* and vegetational changes, attributable to the influence of glaciers.

**transported soil**    a *soil* developed on transported *regolith,* such as alluvium, landslide deposits, or volcanic ash.

**transverse dune**    a *sand dune* ridge with a scalloped crest oriented perpendicularly to the prevailing wind direction in an area of abundant sand supply.

**trellised drainage**    a drainage pattern in an area of dipping *stratified* rocks, in which the larger channels of the *drainage system* have carved valleys in the weaker *geologic formations,* whereas their tributaries tend to flow into them at right angles from the more resistant formations on either side.

**triple junction**    a meeting of three *lithospheric plates* and their common boundaries at a single point.

**tropical depression**    a tropical *cyclone* with wind speeds of 37–63 km/hr.

**tropical storm**    a tropical *cyclone* with wind speeds of 63–120 km/hr.

**tropical year**    the *time* between successive passages of the Sun through the *March equinox;* equal to 365.242 *mean solar days.*

**Tropic of Cancer**    the *parallel* of 23°27′ north *latitude,* on which the *subsolar point* is located on the *June solstice.*

**Tropic of Capricorn**    the *parallel* of 23°27′ south *latitude,* on which the *subsolar point* is located on the *December solstice.*

**tropopause**    the upper surface of the *troposphere,* located at an altitude of about 8 km in polar regions and at about 18 km over the equator.

**troposphere**    the shell of the atmosphere adjacent to Earth's surface, characterized by thermal instability due to a decrease in temperature with altitude.

**tsunami**    a destructive seismic sea wave with a *wavelength* of tens to hundreds of kilometers, produced by an earthquake, landslide, submarine eruption, etc.

**turbidity flow**    a submarine *mudflow* that moves downslope because its mixture of mud and water is denser than the surrounding seawater.

## U

**ultramafic**    said of an *igneous rock* type that contains less than 45% *silica.*

**umbra**    an outward-narrowing cone of full shadow extending from the dark side of a moon or planet; also the darker, inner portion of a *sunspot.*

**unconformity**    a surface of discontinuity between underlying and overlying sequences of *geologic formations* that records the deformation, *erosion,* or both, of the underlying rocks, or simply a long interval of nondeposition, prior to the deposition of the overlying *strata.*

**unloading**    the outward expansion of *rock* after confining material has been removed by *erosion.*

**unsaturated zone**    the zone between the *water table* and Earth's surface, in which pore spaces are filled with a mixture of air, *capillary* water, and percolating water.

**upslope fog**    a mist of fine cloud droplets in ground-level air formed as moist air flows from relatively warm lowlands to cooler highlands, expanding and cooling as it does so.

**urban island effect**    anomalously high cloudiness and precipitation downwind of an urban center due to a large concentration of particulate air pollutants that can act as *condensation nuclei.*

**usufruct**    the practice of caring for resources in such a way as to maintain or increase their health and productivity for future generations.

## V

**valley glacier**    a *glacier* that forms within a mountain *cirque* and flows down a stream valley.

**Van Allen radiation belts**    two bagel-shaped belts containing high concentrations of *protons* and *electrons* surrounding Earth's equator at distances of about 1.3–1.8 and 3.1–4.1 Earth radii.

**vapor pressure**    that portion of the total *atmospheric pressure* overlying a liquid that is due to the presence of *molecules* of the vaporized liquid.

**variable star**    a star that changes its *absolute magnitude.* The change can be nonrecurrent, as with *supernovae;* irregular, as with T-Tauri variables; or periodic, as with *novae* and *Cepheid variables.*

**varve**    one of many thin *strata* of silt or clay deposited on a glacial lake bottom, each of which represents an annual depositional cycle; the thicker, coarser, lighter, basal layer represents summer deposition, whereas the thinner, finer, darker, upper layer represents winter deposition.

**vein**    a thin, tabular *mineral* deposit within a *rock fracture.*

**volcanic dome**   a large, rounded mound of viscous *lava* overlying the vent of a *felsic* volcano.

**volcanic igneous rock**   *aphanitic lava* erupted on Earth's surface.

**volcanic outgassing**   the escape of gases from planetary interiors through volcanoes.

## W

**warm front**   in a *wave cyclone,* a slowly-moving, relatively gently inclined interface between warm, tropical air to the west and cold, polar air to the east. The warm front always lies to the east of, and is eventually overtaken by, a *cold front.*

**water mass**   a large body of ocean water that acquires characteristics of temperature and *salinity* from the region in which it originates and tends to retain these characteristics through subsequent migrations.

**waterspout**   a small, relatively slowly-rotating *tornado* over the ocean that does not last more than an hour.

**water table**   the upper surface of the *saturated zone.*

**water wave**   a vertical, oscillatory motion of a water surface that spreads radially outward from a point of disturbance and dies out with depth.

**wave base**   a depth of one-half the *wavelength* of a *water wave,* below which the oscillatory motion of water is negligible.

**wave-cut cliff**   a cliff formed along a shoreline as waves deepen a *wave-cut notch* at the base of a headland, inducing subsequent *rockfall* from above the notch.

**wave-cut notch**   the horizontal zone of most intense wave *erosion* at the base of a *wave-cut cliff.*

**wave-cut platform**   a shallowly submerged, gently sloping bedrock surface produced by the landward retreat of a *wave-cut notch.*

**wave cyclone**   a typically stormy, eastward-shifting, poleward-pointing wave form in the *polar front,* with a *cold front* on the west and a *warm front* on the east; created when divergence in the overlying *polar jet stream* draws air upward, inducing surface air to *converge* in a counterclockwise spiral (Northern Hemisphere) or a clockwise spiral (Southern Hemisphere).

**wavelength**   the distance between adjacent wave crests, measured perpendicularly to the crestlines.

**wave period**   the *time* between passages of two successive wave crests.

**wave refraction**   the bending of wave crests into approximate parallelism with the shoreline on approaching a headland or on approaching a straight shoreline at an angle because of reduction of *wavelength* and velocity as *wave base* touches the seafloor.

**weather**   the behavior of the *troposphere* at a given place and time.

**weathering**   the slow disintegration of *rock* in Earth's surface environment under the influence of various disruptive mechanical and chemical agents.

**west coast desert**   the strip of arid land along a mid-latitude west coast resulting from (1) subtropical high pressure, (2) seaward *Ekman transport* and cold water upwelling, resulting in an *inversion* that prevents *convective circulation,* and (3) a coastal, north-south mountain barrier that prevents westerly winds from dissipating the inversion.

**whirlwind**   a descending, spiralling wind that occurs on clear, hot, summer days when strong surface heating produces an *environmental lapse rate* high enough to create an upward increase in air *density.*

**white dwarf**   an old star about as massive as the Sun but about 1700 K hotter and with only about a thirtieth the Sun's diameter, a tenth its brightness, and 24,000 times its *density.*

**Wilson cycle**   the opening and closing of an *ocean basin* as a result of *plate tectonics.*

**wind**   a horizontal movement of air.

**wind shadow**   a region downwind of an obstacle in which turbulent eddies promote the deposition of windblown sand.

## Y

**yazoo stream**   a tributary that is forced to run parallel to the main channel of a *graded stream* for a considerable distance down its *floodplain* by *natural levees* that restrict its entrance into the main channel.

## Z

**zodiac**   the region of the celestial sphere that lies between 15° north and 15° south of the celestial equator.

**zone of ablation**   the region of a *glacier* below the *annual snowline* in which *ablation* exceeds accumulation.

**zone of accumulation**   the region of a *glacier* above the *annual snowline* in which accumulation exceeds *ablation.*

# Additional Readings

## Part One: Impressions of a Living Planet

Bertalanffy, Ludwig von. 1968. *General System Theory.* New York: George Braziller.

Clark, D. L. 1976. *Fossils, Paleontology, and Evolution.* 2d ed. Dubuque, IA: Wm. C. Brown Publishers.

Darwin, C. 1975. *The Origin of Species.* Edited by Philip Appleman. New York: W.W. Norton.

Dott, R. H., and Batten, R. L. 1988. *Evolution of the Earth.* 4th ed. New York: McGraw-Hill.

Lovelock, J. E. 1979. *Gaia: A New Look at Life on Earth.* Oxford: Oxford University Press.

Nass, G. 1970. *The Molecules of Life.* New York: McGraw-Hill.

Purves, W. K., and Orians, G. H. 1983. *Life: The Science of Biology.* Sunderland, MA: Sinauer Associates.

Siever, R. 1983. The dynamic Earth. *Scientific American,* Sept., 46–55.

Stanier, R. Y. et al. 1979. *Introduction to the Microbial World.* Englewood Cliffs, NJ: Prentice-Hall.

Strahler, A. N. 1971. *The Earth Sciences.* 2d ed. New York: Harper & Row.

Turekian, K. K. 1972. *Chemistry of the Earth.* New York: Holt, Rinehart, and Winston.

Volpe, E. P. 1981. *Understanding Evolution.* 4th ed. Dubuque, IA: Wm. C. Brown Publishers.

## Part Two: Earth's Solid Systems

Barker, D. S. 1983. *Igneous Rocks.* Englewood Cliffs, NJ: Prentice-Hall.

Billings, M. P. 1972. *Structural Geology.* 3d ed. Englewood Cliffs, NJ: Prentice-Hall.

Birkeland, P. W. 1974. *Pedology, Weathering, and Geomorphological Research.* New York: Oxford University Press.

Blackburn, W. H., and Dennen, W. H. 1988. *Principles of Mineralogy.* Dubuque, IA: Wm. C. Brown Publishers.

Blatt, H., Middleton, G., and Murray, R. 1980. *Origin of Sedimentary Rocks.* 2d ed. Englewood Cliffs, NJ: Prentice-Hall.

Bolt, B. A. 1982. *Inside the Earth.* San Francisco: W. H. Freeman.

Bowen, N. L. 1928. *The Evolution of the Igneous Rocks.* Princeton, NJ: Princeton University Press.

Brush, G. J., and Penfield, S. L. 1898. *Manual of Determinative Mineralogy.* 17th ed. New York: John Wiley.

Burchfiel, B. C. 1983. The continental crust. *Scientific American,* Sept., 130–42.

Chesterman, C. W. 1978. *The Audubon Society Field Guide to North American Rocks and Minerals.* New York: Alfred A. Knopf.

Compton, R. R. 1985. *Geology in the Field.* New York: John Wiley.

Condie, K. C. 1982. *Plate Tectonics and Crustal Evolution.* 2d ed. New York: Pergamon.

Dennis, J. G. 1987. *Structural Geology: An Introduction.* Dubuque, IA: Wm. C. Brown Publishers.

Dewey, J. F. 1972. Plate tectonics. *Scientific American,* May, 56–68.

Doerr, A. H. 1990. *Fundamentals of Physical Geography.* Dubuque, IA: Wm. C. Brown Publishers.

Ehlers, E. G., and Blatt, H. 1982. *Petrology: Igneous, Sedimentary, and Metamorphic.* San Francisco: W. H. Freeman.

Flint, R. F. 1971. *Glacial and Quaternary Geology.* New York: John Wiley.

Francheteau, J. 1983. The oceanic crust. *Scientific American,* Sept., 114–29.

Freeze, R. A., and Cherry, J. A. 1979. *Groundwater.* Englewood Cliffs, NJ: Prentice-Hall.

Friedman, G. M., and Sanders, J. E. 1978. *Principles of Sedimentology.* New York: John Wiley.

Glen, W. 1975. *Continental Drift and Plate Tectonics.* Columbus, OH: Merrill.

Gordon, R. B. 1972. *Physics of the Earth.* New York: Holt, Rinehart, and Winston.

Gregory, K. J., and Walling, D. E. 1973. *Drainage Basin Form and Process: A Geomorphological Approach.* New York: John Wiley.

Hunt, C. B. 1974. *Natural Regions of the United States and Canada.* San Francisco: W. H. Freeman

King, P. B. 1977. *The Evolution of North America.* Rev. ed. Princeton, NJ: Princeton University Press.

Klein, C., and Hurlbut, C. S., Jr. 1985. *Manual of Mineralogy.* 20th ed. New York: John Wiley.

McKenzie, D. P. 1983. The Earth's mantle. *Scientific American,* Sept., 66–78.

Plummer, C. C., and McGeary, D. 1988. *Physical Geology.* 4th ed. Dubuque, IA: Wm. C. Brown Publishers.

Ritter, D. F. 1986. *Process Geomorphology.* 2d ed. Dubuque, IA: Wm. C. Brown Publishers.

Sorrell, C. A. 1973. *Rocks and Minerals.* New York: Golden Press.

Volk, T. 1989. Rise of angiosperms as a factor in long-term climatic cooling. *Geology* 70: 107–110.

Wilson, J. T., ed. 1976. "Continents Adrift and Continents Aground." In *Readings from Scientific American.* San Francisco: W. H. Freeman.

## Part Three: Earth's Oceanic Systems

Anikouchine, W. A., and Sternberg, R. W. 1981. *The World Ocean.* Englewood Cliffs, NJ: Prentice-Hall.

Davis, R. A., Jr. 1987. *Oceanography: An Introduction to the Marine Environment.* Dubuque, IA: Wm. C. Brown Publishers.

Duxbury, A. C., and Duxbury, A. B. 1989. *An Introduction to the World's Oceans.* 2d ed. Dubuque, IA: Wm. C. Brown Publishers.

Holland, H. D. 1972. The Geologic History of Sea Water: An Attempt to Solve the Problem. *Geochimica et Cosmochimica Acta* 36: 637–651.

Turekian, K. K. 1968. *Oceans.* Englewood Cliffs, NJ: Prentice-Hall.

Valentine, J. W. 1973. *Evolutionary Paleoecology of the Marine Biosphere.* Englewood Cliffs, NJ: Prentice-Hall.

## Part Four: Earth's Atmospheric Systems

Ahrens, C. D. 1982. *Meteorology Today.* St. Paul, MN: West.

Battan, L. J. 1961. *The Nature of Violent Storms.* Garden City, NJ: Anchor Books, Doubleday.

Byers, H. R. 1974. *General Meteorology.* New York: McGraw-Hill.

Davis, R. J., and Grant, L. O., eds. 1978. *Weather Modification, Technology, and Law.* Boulder, CO: Westview Press.

Fairbridge, R. W. 1967. *Encyclopedia of Atmospheric Sciences and Astrogeology.* New York: Reinhold.

Ludlam, F. H. 1980. *Clouds and Storms: The Behavior and Effects of Water in the Atmosphere.* State College, PA: The Pennsylvania State University Press.

Namias, J. 1983. The History of Polar Front and Air Mass Concepts in the United States—An Eyewitness Account. *Bulletin of the American Meteorological Society,* 64.

Palmen, E., and Newton, C. W. 1969. *Atmospheric Circulation Systems.* New York: Academic Press.

Reiter, E. R. 1967. *Jet Streams: How Do They Affect Our Weather?* Garden City, NY: Anchor Books, Doubleday & Co.

Schaefer, V. J., and Day, J. A. 1981. *A Field Guide to the Atmosphere.* Boston: Houghton Mifflin.

Sellers, W. D. 1965. *Physical Climatology.* Chicago: University of Chicago Press.

Simpson, R. H., and Riehl, H. 1981. *The Hurricane and Its Impact.* Baton Rouge, LA: Louisiana State University Press.

## Part Five: Beyond Earth

Chartrand, M. R. 1982. *Skyguide: A Field Guide for Amateur Astronomers.* New York: Golden Press.

Hoyle, F. 1975. *Astronomy and Cosmology.* San Francisco: W. H. Freeman.

Jastrow, R., and Thompson, M. H. 1984. *Astronomy: Fundamentals and Frontiers,* 4th ed. New York: John Wiley.

Meadows, A. J. 1967. *Stellar Evolution.* Oxford: Pergamon.

Menzel, D. H., and Pasachoff, J. M. 1983. *A Field Guide to the Stars and Planets.* 2d ed. Peterson Field Guide Series, vol. 15. Boston: Houghton Mifflin.

Murray, B., ed. 1983. "The Planets." In *Readings from Scientific American.* San Francisco: W. H. Freeman.

Murray, B., Malin, M. C., and Greeley, R. 1981. *Earthlike Planets: Surfaces of Mercury, Venus, Earth, Moon, and Mars.* San Francisco: W. H. Freeman.

Pasachoff, J. M. 1983. *Astronomy: From the Earth to the Universe.* 2d ed. Philadelphia: Saunders.

## Part Six: Earth and Humanity

Cohen, M. J. 1988. *How Nature Works: Regenerating Kinship with Planet Earth.* Walpole, NH: Stillpoint.

Costa, J. E., and Baker, V. R. 1981. *Surficial Geology: Building with the Earth.* New York: John Wiley.

Dennen, W. H., and Moore, B. R. 1986. *Geology and Engineering.* Dubuque, IA: Wm. C. Brown Publishers.

Ehrlich, P. R., Ehrlich, A. H., and Holdren, J. P. 1977. *Ecoscience: Population, Resources, Environment.* San Francisco: W. H. Freeman.

Getis, A., Getis, J., and Fellmann, J. 1988. *Introduction to Geography.* 2d ed. Dubuque, IA: Wm. C. Brown Publishers.

Harris, M. 1975. *Culture, People, Nature: An Introduction to General Anthropology.* 2d ed. New York: Thomas Y. Crowell.

Lundgren, L. 1986. *Environmental Geology.* Englewood Cliffs, NJ: Prentice-Hall.

Montgomery, C. W. 1989. *Environmental Geology.* 2d ed. Dubuque, IA: Wm. C. Brown Publishers.

Strahler, A. N., and Strahler, A. H. 1974. *Introduction to Environmental Science.* Santa Barbara, CA: Hamilton.

Thomas, W. L., Jr., ed. 1956. *Man's Role in Changing the Face of the Earth.* Chicago: University of Chicago Press.

Turnbull, C. 1961. *The Forest People.* New York: Simon and Schuster.

# Credits

## Illustrations

**Fig. 1.17:** From Arthur Getis, Judith Getis, and Jerome Fellmann, *Introduction to Geography,* 2d ed. Copyright © 1988 Wm. C. Brown Publishers, Dubuque, Iowa. All Rights Reserved. Reprinted by permission. **fig. 1.19:** From Arthur Getis, Judith Getis, and Jerome Fellmann, *Introduction to Geography,* 2d ed. Copyright © 1988 Wm. C. Brown Publishers, Dubuque, Iowa. All Rights Reserved. Reprinted by permission.

**Fig. 2.3:** From Carla W. Montgomery, *Physical Geology,* 2d ed. Copyright © 1990 Wm. C. Brown Publishers, Dubuque, Iowa. All Rights Reserved. Reprinted by permission. **fig. 2.12:** From Richard A. Pimentel, *Natural History.* Copyright © Van Nostrand Reinhold Company, New York, NY. **fig. 2.15:** From *Biological Science: An Ecological Approved BSCS Green Version,* 4th ed. Copyright © Riverside Publishing Co., Chicago, IL. **fig. 2.16:** From Morris S. Peterson, J. Keith Rigby, and Lehi F. Hintze, *Historical Geology of North America.* Copyright © 1973 Wm. C. Brown Publishers, Dubuque, Iowa. All Rights Reserved. Reprinted by permission. **fig. 2.17:** From L. W. Mintz, *Historical Geology,* 3d ed. Copyright © Charles E. Merrill, Columbus, OH. **fig. 2.18:** From L. W. Mintz, *Historical Geology,* 3d ed. Copyright © Charles E. Merrill, Columbus, OH. **fig. 2.19:** From *Biological Science: An Ecological Approved BSCS Green Version,* 4th ed. Copyright © Riverside Publishing Co.,

Chicago, IL. **fig. 2.20:** From David L. Clark, *Fossils, Paleontology, and Evolution,* 2d ed. Copyright © David L. Clark. Reprinted by permission of the author.

**Fig. 3.9:** From Carla W. Montgomery, *Physical Geology,* 2d ed. Copyright © 1990 Wm. C. Brown Publishers, Dubuque, Iowa. All Rights Reserved. Reprinted by permission. **fig. 3.10:** From William H. Blackburn and William H. Dennen, *Principles of Minerology.* Copyright © 1988 Wm. C. Brown Publishers, Dubuque, Iowa. All Rights Reserved. Reprinted by permission. **fig. 3.11:** From William H. Blackburn and William H. Dennen, *Principles of Minerology.* Copyright © 1988 Wm. C. Brown Publishers, Dubuque, Iowa. All Rights Reserved. Reprinted by permission.

**Fig. 5.1:** From Alfred Wegener, *The Origin of Continents and Oceans,* 4th ed. Copyright © Methuen Publishing Group, London, England. **fig. 5.2:** From R. S. Dietz and J. C. Holden, *Journal of Geophysical Research,* Vol. 75:4939–56, 1970, copyright by the American Geophysical Union. **fig. 5.3:** From L. W. Mintz, *Historical Geology,* 3d ed. Copyright © Macmillan Publishing Company, New York, NY. **fig. 5.6:** From R. J. Foster, *Physical Geology,* 3d ed. Copyright © Charles E. Merrill, Columbus, OH. **fig. 5.7a:** *Principles of Physical Geology,* 1965, by A. Holmes, Van Nostrand Reinhold (UK) Ltd., London. **fig. 5.7b:** From Charles C. Plummer and David McGeary, *Physical Geology,* 4th ed. Copyright © 1988 Wm. C. Brown Publishers, Dubuque, Iowa. All Rights Reserved. Reprinted by permission. **fig. 5.9:** From R. L. Larson

and W. C. Pittman, III, *Geological Society of American Bulletin,* 1972, originally published by The Geological Society of America. **fig. 5.10:** From Norris S. Petersen, J. Keith Rigby, and Lehi F. Hintze, *Historical Geology of North America,* 2d ed. Copyright © 1980 Wm. C. Brown Publishers, Dubuque, Iowa. All Rights Reserved. Reprinted by permission. **fig. 5.16:** From A. A. Meyerhoff, ''Arthur Holmes: Originator of Spreading Ocean Floor Hypothesis'' in *Journal of Geophysical Research,* 73:6563–6565, 1968, copyright by the American Geophysical Union. **fig. 5.17:** From Alyn C. Duxbury and Alison B. Duxbury, *An Introduction to the World's Oceans,* 2d ed. Copyright © 1989 Wm. C. Brown Publishers, Dubuque, Iowa. All Rights Reserved. Reprinted by permission. **fig. 5.20:** From Charles C. Plummer and David McGeary, *Physical Geology,* 4th ed. Copyright © 1988 Wm. C. Brown Publishers, Dubuque, Iowa. All Rights Reserved. Reprinted by permission. **fig. 5.23:** From Charles C. Plummer and David McGeary, *Physical Geology,* 4th ed. Copyright © 1988 Wm. C. Brown Publishers, Dubuque, Iowa. All Rights Reserved. Reprinted by permission. **fig. 5.24:** From Charles C. Plummer and David McGeary, *Physical Geology,* 4th ed. Copyright © 1988 Wm. C. Brown Publishers, Dubuque, Iowa. All Rights Reserved. Reprinted by permission. **fig. 5.28:** From Charles C. Plummer and David McGeary, *Physical Geology,* 4th ed. Copyright © 1988 Wm. C. Brown Publishers, Dubuque, Iowa. All Rights Reserved. Reprinted by permission. **fig. 5.29:** From Dewey and

Intrazonal soils, 544
Intrusions, 176
  igneous, 178–80
  salt-water, 244
Intrusive rocks, 178
Inundation, 506
Inversion, 399
  temperature, 343
  trade wind, 411
Io, 10, 446–47, 459, 460, 461
Ion, 58–62, 98
Ionic bond, 61
Ionic reservoirs, 302–3
Ionization, 346
Ionosphere, 346
Iron, 6, 106
Iron meteorites, 458
Irregular galaxies, 488, 490
Irrigation, 214
Island, urban, 393
Island arcs, 9
Islands, barrier. *See* Barrier islands
Isobaric surfaces, 365–67
Isobars, 319, 365–67
Isolated silicates, 63, 69
Isolated systems, 47–52
Isostacy, 7, 66
Isostatic rebound, 269
Isothermal temperature, 322
Isotherms, 355
Isotopes, 59, 442
ITCZ. *See* Intertropical convergence zone
    (ITCZ)

## J

Jadeite, 77, 188
Jansky, Karl, 429
Janssen, Hans, 426
Janssen, Zacharias, 426
Japanese islands, 11
Japan Trench, 290
*Jeanette*, 318
Jeans, James Hopwood, 465
Jefferson, Thomas, 219
Jeffreys, Harold, 111, 465
Jeffries, John, 339
Jenny, Hans, 206
Jet streams, 368, 370, 372–73, 405
*Job*, 4
Johnson, Samuel, 156
John Wesley Powell Survey, 238
JOIDES. *See* Joint Oceanographic Institutions
    for Deep Earth Sampling (JOIDES)
*JOIDES Resolution*, 122–23, 283
Jointed internal skeletons, 42, 43
Joint Oceanographic Institutions for Deep
    Earth Sampling (JOIDES), 122–23
Joints, 143
Julian calendar, 437
Junctions, triple. *See* Triple junctions
June solstice, 351, 352
Jupiter, 5, 10, 19, 22, 114, 342, 446–47,
    459–61
Jurassic Period, 31, 119, 249, 383
Juvenile stage, 405, 406

## K

Kame, 266–67
Kame terraces, 266–67
Kansan glaciation, 268
Kansu, 499
Kant, Immanual, 465
Karroo basalts, 172
Karst, and cavern, 245–46
Karst topography, 245
Katabatic wind, 376–77
Kelper, Johann, 424, 425
Kelvin, William Thomson, 29, 124, 469
Kenai Range, 495
Kenge, 517
Kepler, Johannes, 474, 487
Kermadec trench, 123
Kettles, 265, 266
Key to the common minerals, 524–26
Key to the common rock types, 527–30
K-feldspar, 65, 69, 71, 72, 88, 92, 106, 525
Kilauea volcano, 170, 174, 175, 176, 196,
    503
Kilroyite, 83
Kimberlite pipes, 171
Kinetic energy, 47
Kirchhoff, Gustav, 478
Kohala, 175
Kolinite clay, 69, 74
Komatite, 86, 90, 91
Kopernik, Mikoiaj, 423
Krakatao, 504
Krypton, 60
K-spar, 68
Kuiper, Gerard Peter, 465
Kurile Trench, 290
Kyanite, 77, 102, 525
Kyushu, 11

## L

Laccolites, 238
Laccoliths, 178–79, 238
Lacustrine environment, 180
Lag concentrate, 381
Lagoon, 331, 512
Lagrange, Joseph, 426
Lahar, 176, 177, 503
Lake Champlain, 11
Lake environments, 180
Lake Morat, 255
Lake Nyos, 503
Lakes
  oxbow, 234
  playa, 239
  pluvial, 272
Lambeck, K., 438
Lamont-Doherty Geological Observatory,
    127
Land, water, and gravity, 223–33
Land breeze, 376
Landscapes
  glaciated, 267–71
  middle youth, mature, and old age, 235
  and streams, 235–39
Landslide. *See* Mass wasting
Lao Tzu, 222, 420, 496
Lapilli, 171

Laplace, Pierre-Simon, 465, 469
Laramide orogeny, 163
Latent heat of evaporation, 294
Latent heat of fusion, 294
Latent heat transfer, 350
Lateral continuity, 27, 29
Lateral moraines, 264, 265
Lateral planation, 235
Lateral till, 262–63
Laterite, 211
Lateritic soils, 541
Laterization, 211, 541
Latitude, 115, 539
Latosol, 211
Laurasia, 111, 112
Laurentia, 111
Lava, 169–70, 176, 178, 191, 243, 503
Lava dome, 176
Lava flows, 176, 503
Lava tubes, 170
Lavoisier, Antoine, 16
Law of faunal succession, 31–32
Law of original horizontality, 27, 29
Law of original lateral continuity, 27, 29
Law of superposition, 27, 29
Law of universal gravitation, 425
Lawrence Scientific School, 255
League of Nations, 318
Leap year, 437
Leaves, 44
Leavitt, Henrietta S., 475
Leclerc, Georges Louis, 29
Ledge, 83
Lee. *See* Stoss and lee topography
Left-lateral faults, 144
Lehmann, Inge, 150
Leibnitz, Gottfried Wilhelm von, 426
Length, metric conversion, 522
Leptons, 491
Lesser Mellanic Cloud, 475
Levees, natural, 233
Lichens, 13
Life
  and life cycles of organisms, 40
  quality of, 496–520
  time, systems, 26–53
Light, 57
  and continuous light spectrum, 477, 479
  visible, 349, 350
Light gases, 19
Light horizon, 480
Lightning, 404, 497, 513–14
Light-year, 474
Lignite, 101
Limbs, 154
Limestone, 22, 94, 96, 100, 106, 107, 184,
    188, 243, 246, 327, 528, 529
Limestone caverns, 246
Linear chromosomes, 38
Linear minerals, 101
Line spectra, 477–80
Lippershey, Hans, 426
Liquefaction, 499
Liquid, 16, 292–96
Lithic sandstones, 96
Lithification, 15, 84
Lithium, 58, 59